Natural Disasters: Challenges and Management

Natural Disasters: Challenges and Management

Edited by Alfred Scott

SYRAWOOD
PUBLISHING HOUSE
New York

Published by Syrawood Publishing House,
750 Third Avenue, 9th Floor,
New York, NY 10017, USA
www.syrawoodpublishinghouse.com

Natural Disasters: Challenges and Management
Edited by Alfred Scott

International Standard Book Number: 978-1-68286-465-4 (Hardback)

Cataloging-in-publication Data

Natural disasters : challenges and management / edited by Alfred Scott.
p. cm.
Includes bibliographical references and index.
ISBN 978-1-68286-465-4
1. Natural disasters. 2. Emergency management. I. Scott, Alfred.
GB5014 .N38 2017
363.34--dc23

Printed in China.

TABLE OF CONTENTS

PREFACE

This book provides comprehensive insights into the field of natural disaster, its challenges and management. The subject area deals with study of events that are caused naturally resulting in damage to the natural life and the human life. The text aims to highlights the different types of natural disasters like floods, earthquakes, volcanic eruption, etc. This book is a compilation of different challenges faced and management strategies that are widely used across the globe. The objective of this book is to give a general view of the different phases of natural disasters. The book will help new researchers by foregrounding their knowledge in this branch.

Significant researches are present in this book. Intensive efforts have been employed by authors to make this book an outstanding discourse. This book contains the enlightening chapters which have been written on the basis of significant researches done by the experts.

Finally, I would also like to thank all the members involved in this book for being a team and meeting all the deadlines for the submission of their respective works. I would also like to thank my friends and family for being supportive in my efforts.

Editor

1

PM_{10} measurements in urban settlements after lava fountain episodes at Mt. Etna, Italy: pilot test to assess volcanic ash hazard to human health

D. Andronico[1] **and P. Del Carlo**[2]

[1]Istituto Nazionale di Geofisica e Vulcanologia, Osservatorio Etneo, Piazza Roma 2, 95125 Catania, Italy
[2]Istituto Nazionale di Geofisica e Vulcanologia, Sezione di Pisa, Via Della Faggiola 32, 56126 Pisa, Italy

Correspondence to: D. Andronico (daniele.andronico@ingv.it)

Abstract. We have carried out a preliminary study on the potential risks caused by the sub-10 μm fraction of volcanic ash (particulate matter, PM_{10}) after the basaltic explosive eruptions from Mt. Etna volcano (Italy), which have dramatically increased in frequency over the last 20 years. We present results deriving from the study of the ash concentration in the air following the lava fountain episode from the New Southeast Crater of Etna on 15 November 2011, which caused tephra fallout over the eastern slope of the volcano. Short-duration tests of PM_{10} measurements were carried out at three different sites using a TSI® DustTrak™ aerosol monitor a few hours after the end of the eruption, and readouts of the air quality were repeated at the same sites a month later without volcanic activity. Furthermore, ash samples were characterized by grain size, componentry and morphological and petrochemical analyses.

By comparing PM_{10} levels measured a few hours after the 15 November lava fountain and on 15 December, we found that relatively low amounts (500–1500 g m^{-2}) of tephra fallout cause high levels of PM_{10} in the air. This is because the coarse particles, particularly basaltic ash, are readily broken up by traffic and hence remobilized into the air. We believe the impact from ash fallout in the Etnean territory should receive greater attention, especially regarding potential health problems. Simple but effective actions can be implemented to reduce eventual risks, first and foremost the prompt removal of the ash deposits from the urbanized areas.

1 Introduction

The effects of volcanic ash on people's health have been widely acknowledged and documented in the literature (e.g. Damby et al., 2013; Hincks et al., 2006; Horwell et al., 2003, 2007, 2010, 2013; Le Blond et al., 2010; Searl et al., 2002; Wilson et al., 2011). The finer particles, especially particulate matter with dimensions ≤ 10 μm or PM_{10} in aerodynamic equivalent diameter (i.e. the diameter of a spherical particle with density 1000 kg m^{-3} and the same settling velocity as the airborne particle is considered), remain suspended in the air and may be inhaled, causing different degrees of injury depending on the particle size (Horwell and Baxter, 2006). Coarse particles of 4–10 μm aerodynamic diameter are also called "thoracic" because they may reach the bronchioles of the lungs via the nose and throat, while fine ($< PM_4$ and $PM_{2.5}$) particles are called "respirable" since they can more readily penetrate into the lungs, thus involving the alveolar region where the gas exchanges take place (QUARG, 1996). The limit between thoracic and respirable particles at the sub-4 μm ash fraction also reflects the greater hazard from the fine particles during long-term exposure (Expert Panel on Air Quality Standards, 1995, 2001). In the first Daughter Directive of the EU Air Quality Framework Directive (EC, 1999), the European legislation establishes the limit values of PM_{10} concentrations not to be exceeded over 24 h and yearly for the protection of human health, i.e. 0.050 mg m^{-3} as the mean measured value over 24 h which must not be exceeded more than 35 times per year and 0.040 mg m^{-3} as the maximum annual mean concentration. It is worth noting that the EU directive 99/30 (EU, 2008) had fixed 1 January 2010 as the starting date from which more severe limits would have to be respected, i.e. a limit value of 0.050 mg m^{-3} should not be exceeded more than 7 times per year, while 0.020 mg m^{-3} was the mean limit value for the calendar year.

Among basaltic volcanoes, Mt. Etna (Italy; Fig. 1) has proved to be a great producer of ash in recent decades. Since 1989, Etna has undergone recurrent explosive activity consisting of more than 200 paroxysmal episodes, and two main long-lasting explosive eruptions in 2001 (Taddeucci et al., 2002; Scollo et al., 2007) and 2002–2003 (Andronico et al., 2009a and reference therein) that produced prolonged ash emissions. Paroxysmal events are characterized by explosive activity ranging in style from Strombolian to lava fountains, often as sequences of episodes within periods of weeks to months at the summit craters (Branca and Del Carlo, 2005). Recently, these paroxysmal sequences have been considered single "episodic" eruptions by Andronico and Corsaro (2011). Most of these paroxysms are typically associated with the production of tephra (from ash to lapilli to bombs) injected into the atmosphere and dispersed all around the volcano and further, where they may cause light to relatively heavy and serious fallout. In urbanized areas, in fact, paroxysms characterized by a low mass eruption rate (i.e. values of $\sim 2 \times 10^2$ kg s^{-1}; 16 November 2006; Andronico et al., 2009b) have produced deposits prevalently made of ash, while paroxysms with high mass eruption rate (i.e. $4.5 \pm 3.6 \times 10^5$ kg s^{-1}; 23 November 2013; Andronico et al., 2015) have produced lapilli fallout deposits, differing also from a hazard point of view.

Studies on the health effects of particulate matter in the Etnean area have been conducted by Fano et al. (2005), who compared the levels of PM_{10} in the cities of Catania and Acireale (Fig. 1) during the intense fallout from the 2002–2003 eruption with PM_{10} values measured in the same season but during years without volcanic activity. These authors concluded that although a significant increase of PM_{10} in the air was recorded for several days that greatly exceeded the threshold fixed by law at that time (0.050 mg m^{-3} over 24 h), there was no evident direct cause of respiratory diseases from the ash fallout. They did note a temporary increase in cardiovascular diseases, particularly among the elderly, which however they assumed to be related uniquely to the stress caused by the eruption. Such effects on the cardiovascular system have also been documented in non-volcanic areas (WHO, 2006; Delfino et al., 2005). However, a retrospective study by Lombardo et al. (2013) reports, contrary to the results by Fano et al. (2005), on an increase in hospital admissions for cardiovascular morbidity along with a decrease in the rate of admission for respiratory diseases, finding a significantly higher frequency of visits to the emergency department for acute respiratory and cardiovascular diseases, and ocular disturbances during the ash exposure time period in the three main hospitals of Catania in 2002, as compared to the same period of the previous year. Furthermore, Barsotti et al. (2010) simulated the concentration of 10 μm volcanic particles (both in the air and on the ground) in several towns and key infrastructure around the volcano, showing that the remobilization of ash may considerably increase the concentration of the study class of particles in the air.

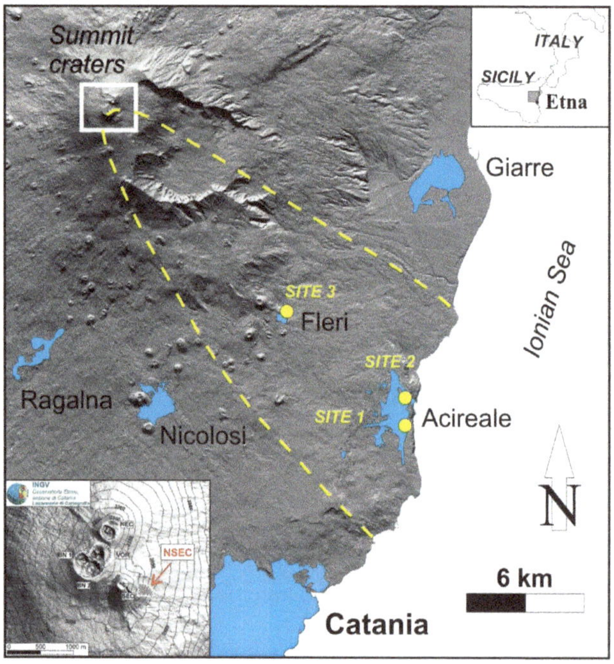

Figure 1. Digital elevation map of Mt. Etna showing the sites (full yellow circles) chosen for PM_{10} measurements and the collected samples (SITE 1, SITE 2 and SITE 3). The white rectangle indicates the summit crater area. The dashed yellow line shows the extent of tephra fallout from the 15 November 2011 lava fountain. Blue areas represent the main cities in this volcano sector, especially those reported in the text. Right-top inset: location of Sicily. Left-bottom inset: map of the summit craters (white rectangle) on September 2011 (courtesy of INGV Cartography Laboratory). NEC denotes Northeast Crater, VOR denotes Voragine crater, BN-1 and BN-2 represent Bocca Nuova pit craters, SEC represents Southeast Crater and NSEC represents the New Southeast Crater.

In this work, we quantitatively evaluate the effects of ash fallout on the air quality during explosive activity from Etna, namely the 2011–2012 and 2013 episodic eruptions (25 and 21 paroxysmal episodes, respectively; Andronico et al., 2014a) from the New Southeast Crater (Fig. 1). The problem for human health is potentially large because approximately 900 000 people live on the slopes of Etna, Catania being the largest city with 300 000 inhabitants. The area around Etna comprises another 12 towns, each of which has a population of more than 20 000.

To this end, we carried out short-duration PM_{10} surveys of particulate matter in the air at three sites with different environmental conditions: (i) within a few hours after ash fallout following the 15 November 2011 lava fountain, and (ii) in similar weather conditions but without any evident air pollution by volcanic ash. Sedimentological, morphological and petrochemical analyses were performed on ash samples collected at the measurement sites, in order to determine the grain size, componentry, shape and composition of particles. Furthermore, we discuss the potential risk caused by tephra fallout in the Etnean urban areas, suggesting simple measures to prevent and mitigate such effects.

2 The 15 November 2011 lava fountain episode

On 11 January 2011, the Southeast Crater (SEC; Fig. 1) resumed its eruptive activity that gave rise to a lava fountain episode in the night between 12 and 13 January 2011, the first of a sequence of 25 events over the next months. The intense activity gradually built up a new cone on the east flank of SEC, named the New Southeast Crater (hereafter NSEC; Fig. 1; Andronico et al., 2014a, b; Behncke et al., 2014).

The 15 November 2011 episode studied here was the 18th and occurred in partially cloudy conditions around the volcano which prevented full visibility of the eruptive phenomena, in particular of the eruption column. The eruptive activity resumed in the morning around 05:45 GMT (local time: GMT + 1). A thermal anomaly was observed by video recordings from a network of cameras managed by Istituto Nazionale di Geofisica e Vulcanologia, Sezione di Catania-Osservatorio Etneo (INGV-OE). The anomaly, located on the eruptive fissure cutting the new cone, indicated the downward advancement of a lava flow that had been evident after 06:05 GMT, while discontinuous, low explosions became visible from the upper vent within the new cone after 08:20 GMT (*resumption phase* of Alparone et al., 2003; Fig. 2a). The Strombolian explosions slowly increased in frequency and intensity, also extending along the eruptive fissure, until about 11:15 GMT, when they began to produce almost continuous magma jets up to at least 300 m high (Fig. 2b), forming (since 11:30 GMT) an eruption column up to several kilometres above the volcano (*paroxysmal phase*; Fig. 2c). This activity lasted with similar intensity and features for more than 1 h; after 12:20 GMT, the lava fountains dramatically dropped in intensity and in a few minutes, the activity passed to lower-intensity Strombolian explosions (*conclusive phase*; Fig. 2d); also the eruption column, no longer fed, became exhausted in a few minutes. All the effusive and explosive phenomena ceased completely after 13:00 GMT.

The eruption column was driven eastward by winds blowing above the volcano, causing tephra fallout on the southeast flank of Etna and further towards the coastline and over the Ionian Sea (Fig. 1).

3 Methods

3.1 DustTrak™ measurements

PM_{10} measurements were carried out by a TSI® DustTrak™ (hereafter DT) aerosol monitor (Model 8520), a portable laser photometer providing real-time digital readout of airborne particulates (Fig. 3a). In volcanology, the DT was extensively used in the Caribbean island of Montserrat to evaluate the impact of the ash erupted by the Soufrière Hills Volcano between 1996 and 1999 and particularly the exposure of islanders to respiratory hazards (Moore et al., 2002; Searl et al., 2002). These authors produced guidelines for air quality surveys in volcanic areas characterized by ash fallout in urban settlements. Moore et al. (2002), in particular, provided two strategies for investigating PM_{10} levels in Montserrat: a network of static test sites by which to carry out continuous monitoring and short-duration tests, ranging between 15 min and 1 h. In both cases they set the DT for 1 min average recording.

In this work, we measured PM_{10} concentration in the air at three different sites a few hours after the 15 November 2011 lava fountain. The sites were selected on the eastern flank of the volcano in some inhabited areas affected by the ash fallout (Fig. 1). Short-duration tests were done during the mapping and sampling of the tephra fallout performed soon after the eruption to prevent the removal of the original amount of tephra on the ground.

Site 1 is on the SS114 road linking Acireale with Catania at 21.3 km from the eruptive vent, site 2 is about 1.5 km north on the same road and 20.4 km from NSEC, while site 3 is on the SP148 within the Fleri village, 12.8 km from NSEC. These sites were selected on the basis of their different features in terms of road traffic. Site 1 is on a road with "average" but continuous, fairly smooth-flowing (both cars and lorries) traffic. Measurements were carried out at the roadside at about 1 m of height. Site 2 is at a main crossroads where cars must stop at the traffic lights. There is more traffic here than at site 1, and the DT operated at 80 cm of height just below a traffic light pole. Site 3, finally, is located on a road with regular traffic flow crossed by a secondary, quiet to busy road with a stop sign. The DT operated at a height of 40 cm. The traffic at this site is roughly intermediate in relation to site 1 and site 2. All the measurements were carried out at the same 1 m horizontal distance from the carriageway.

We measured PM_{10} over 10 to 16 min periods; the DT was set with a time constant of 10 s, i.e. the display showed readings averaged over the previous 10 s, while the frequency/averaging period was fixed to 1 min, i.e. readings

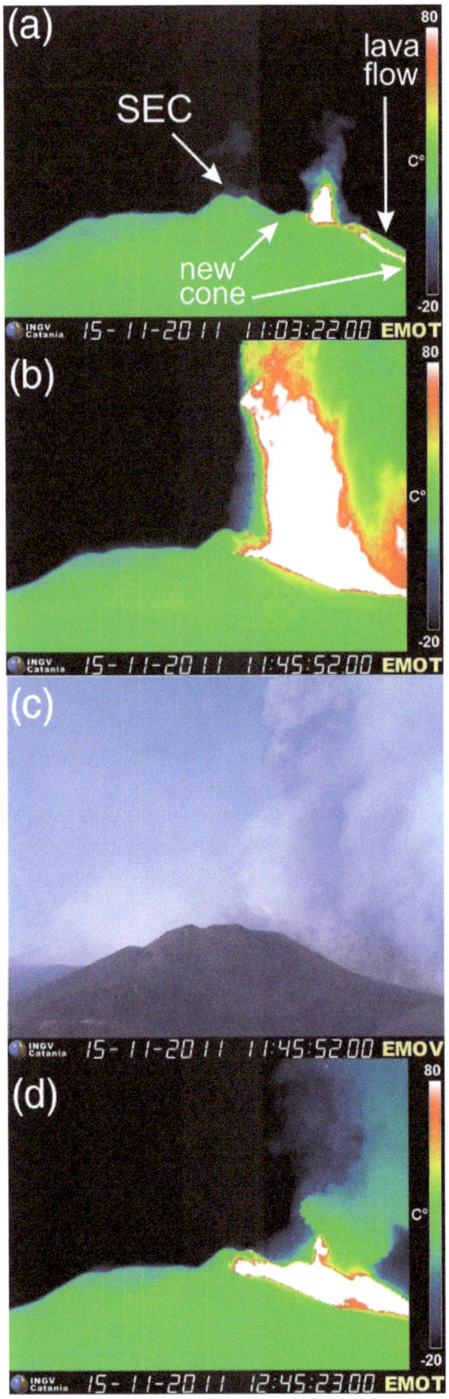

Figure 2. Images extracted from the INGV-OE video camera recordings of the La Montagnola site – (**a, b, d**) by the thermal camera; (**c**) by the true-colour camera, showing the main phases of the 15 November 2011 lava fountain at the new cone on the east flank of the SEC. Panel (**a**) shows the beginning of the explosive and effusive activity in the upper vent and eruptive fissure of the new cone, respectively; (**b**) lava fountains from the eruptive fissure; (**c**) the eruption column formed above; (**d**) the last explosions before the cessation of all eruptive phenomena.

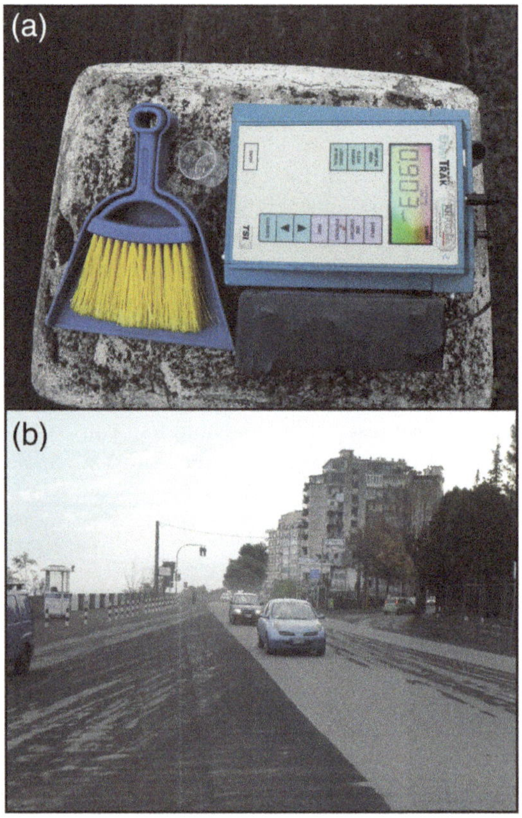

Figure 3. PM_{10} measurement carried out on 15 November 2011 in Acireale at site 2 (SS114 road): (**a**) DustTrak™ in operation; the display shows the averaged reading over the previous 10 s; (**b**) transit of cars at the crossing.

were recorded at 1 min intervals. We observed that just a few hours after the tephra fallout, ash was swept laterally by the cars, causing remobilization phenomena and accumulation at both sides of the carriageway (Fig. 3b).

We repeated measurements a month later, on 15 December 2011, at the same time and with the same instrument set up and similar weather conditions.

The methodology used does not entirely follow the exposure guidelines for PM_{10} specifically for volcanic environments. Our focus was primarily on measuring the increase in PM_{10} concentrations during and immediately after tephra fallout. In the future, we plan to undertake more detailed surveys with more attention to a precise procedure. The instrument should be set up at a standard level in all the sites to facilitate comparisons, and the DT set at an appropriate height for human exposure, e.g. 1.5 m, which is considered the "breathing zone" for adults. The duration of measurements should be in 24 h periods for comparison with EU standards. In the future, it might also be helpful to measure the PM_{10} at different heights at the same site to quantify a vertical gradient, thus evaluating if there is any variation in PM_{10} concentration with height.

3.2 Sample analyses

A tephra sample (NSEC1, NSEC2 and NSEC3) for each site was collected on a measured surface to evaluate the mass per square metre. NSEC2 and NSEC3 samples were selected for textural investigation and morphological observation, since NSEC1 was collected very close to NSEC2. At the Laboratory of Sedimentology, INGV-OE, we carried out grain-size analysis by CAMSIZER (Lo Castro and Andronico, 2008) at $1/2-\varphi$ intervals (φ being the $-\log_2 d$, where d is the particle diameter in millimetres), estimating the percentages of different size classes in volume %, and ash componentry characterization under a binocular microscope (Nikon SMZ1500). At Istituto Nazionale di Geofisica e Vulcanologia, Sezione di Pisa Laboratory, we performed morphological analyses of clasts and chemical analyses of glass and minerals using scanning electron microscopy and energy dispersive spectroscopy (SEM-EDS), Zeiss EVO MA 10, equipped with an Oxford ISIS microanalytical system.

4 Results

The mass per square metre of the samples is significantly different from sites 1 and 2 with respect to site 3. In the most distal and lateral sample (NSEC1), the weight amounted to $\sim 490\,\mathrm{g\,m^{-2}}$, a value fairly similar to the close sample NSEC2 ($\sim 440\,\mathrm{g\,m^{-2}}$), while NSEC3 was $\sim 1500\,\mathrm{g\,m^{-2}}$.

4.1 DustTrak™ measurements

On 15 November, we measured PM_{10} particulate matter between 2 h (site 1) and 5 h (site 3) after the end of the lava fountain. All the loggings show a few peaks in the pattern (Fig. 4a–c, blue lines), indicating the significant influence of passing vehicles as observed during the readings.

At site 1, the least affected by car traffic, we recorded an average of $0.084\,\mathrm{mg\,m^{-3}}$, with a maximum value of $0.125\,\mathrm{mg\,m^{-3}}$. At site 2, the average value is considerably higher ($0.642\,\mathrm{mg\,m^{-3}}$) and the same is the case for the maximum value ($1.285\,\mathrm{mg\,m^{-3}}$). Remarkably, the minimum value recorded is $0.337\,\mathrm{mg\,m^{-3}}$, far higher than the maximum value recorded at site 1. Finally, at site 3, all the statistic values are very high: average at $0.935\,\mathrm{mg\,m^{-3}}$, minimum at $0.379\,\mathrm{mg\,m^{-3}}$ and maximum at $1.755\,\mathrm{mg\,m^{-3}}$.

Measurements carried out at the same sites a month later gave surprisingly low values of PM_{10} (Fig. 4a–c, red lines). Site 1 confirmed the lowest values of particulate matter: average of $0.033\,\mathrm{mg\,m^{-3}}$, minimum of $0.023\,\mathrm{mg\,m^{-3}}$ and maximum of $0.045\,\mathrm{mg\,m^{-3}}$. These values show a very limited variability between these three statistic values. At site 2 the differential among the statistics is increased, though not significantly: average $0.060\,\mathrm{mg\,m^{-3}}$, minimum $0.026\,\mathrm{mg\,m^{-3}}$ and maximum $0.149\,\mathrm{mg\,m^{-3}}$. Site 3, finally, was again the site with the highest PM_{10} values, with an average of $0.081\,\mathrm{mg\,m^{-3}}$, a minimum of $0.061\,\mathrm{mg\,m^{-3}}$ and a maximum of $0.192\,\mathrm{mg\,m^{-3}}$.

4.2 Grain-size, textural and petrochemical data

The grain-size distribution of NSEC2 and NSEC3 samples shows both a Gaussian pattern peaked on 1 and 2 mm, respectively (Fig. 4d). The most proximal sample (NSEC3) is obviously coarser, with only 17 % of the collected tephra <1 mm, with respect to the distal sample (NSEC2) which amounts to a good deal more than 60 % of particles below the same size of 1 mm. It should be noted that very fine ash particles (less than 0.125 mm) are practically absent in the proximal sample (just 0.1 % of clasts in the range 0.09–0.125 mm), while we measured a higher, even if poorly significant value of 0.9 % between 0.045 and 0.125 mm in sample NSEC2.

NSEC2 and NSEC3 samples are both mainly made up of basaltic sideromelane clasts (~ 90 %) consisting of light brown, fresh, vesicular clasts with irregular to subrounded shape and smoothed surface (Figs. 5 and 6). The rest of the sample is formed of black, poorly vesicular, blocky clasts (tachylite), rare lithic particles (reworked lavas and scoriae) and loose crystals (plagioclase).

Scanning electron microscope (SEM) analyses of ash indicate that clasts are from highly to moderately vesicular, with elongated or blocky shapes (Fig. 6a–c). Clasts show smoothed surfaces and delicate edges. The finer fraction of the deposit (Fig. 6d) is formed mainly by fragments of larger clasts with similar features. Furthermore, shape parameters obtained by CAMSIZER analyses for NSEC2 and NSEC3 samples give aspect ratios of 0.69 and 0.70, and sphericities of 0.71 and 0.78, respectively, indicating on average that particles are moderately elongated and subrounded. These morphological characteristics are typical of most of the lava fountain products.

Bulk rock composition of scoriae is K-trachybasaltic like the other recent volcanic rocks of Etna (Viccaro et al., 2015). Ash samples are poorly porphyritic with phenocrysts of plagioclase and clinopyroxene, and microphenocrysts of olivine and oxides (titaniferous-magnetite). The groundmass is glassy with a large abundance of microlites of plagioclase, clinopyroxene, olivine and oxides (Fig. 7). Composition of glass is more evolved and alkali-rich as we can observe in the total alkali silica diagram (Fig. 8 and Table 1), where the 15 November 2011 samples are plotted between the basaltic trachyandesite and the phonotephrite fields.

5 Discussion

5.1 The 15 November 2011 PM_{10} test

Measurements carried out shortly after the 15 November 2011 Etna lava fountain have shown that the values of PM_{10} are higher than those measured a month after at the same sites. This survey has documented that tephra fallouts

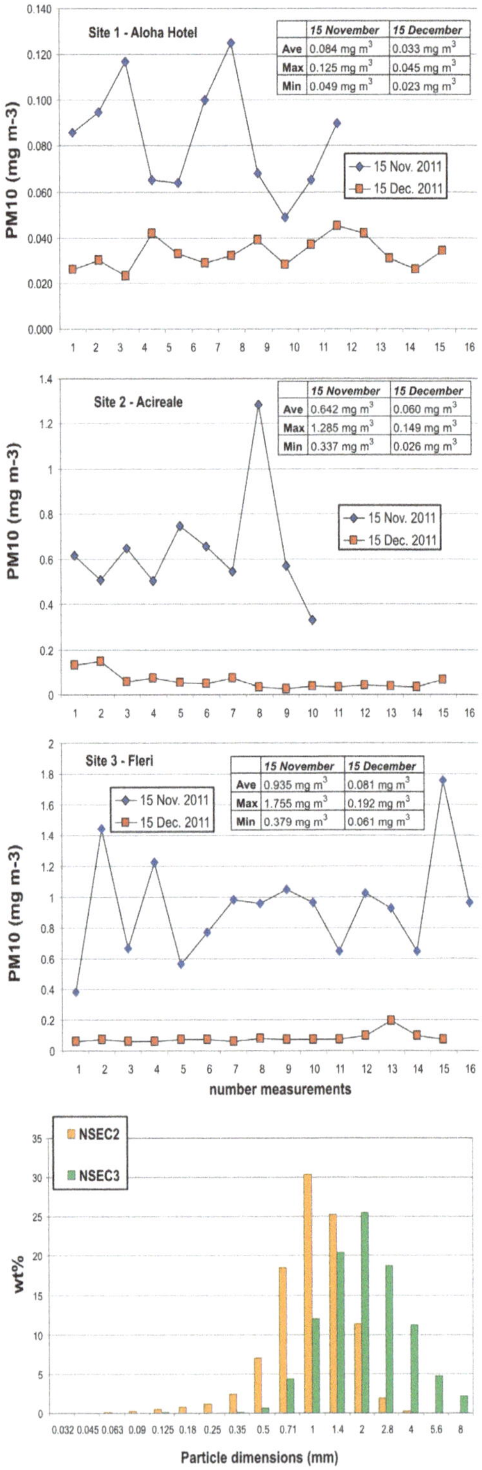

Figure 4. Panels (**a, b, c**) show PM_{10} measurements in three studied sites: (**a**) site 1, Aloha Hotel; (**b**) site 2, Acireale; (**c**) site 3, Fleri. The time interval of the number measurements is 1 min. Blue lines and diamonds represent values measured in the afternoon of 15 November 2011, a few hours after the tephra fallout; red lines and squares represent values measured on 15 December 2011, in absence of a volcanic event. At the side of each plot the main PM_{10} statistics data are reported. Ave is the average value; Max is the maximum value; Min is the minimum value. (**d**) Grain-size histograms concerning samples NSEC2 and NSEC3.

Figure 5. Binocular microscope images showing different morphological types of ash particles from NSEC2 (**a, b**) and NSEC3 (**c**) samples.

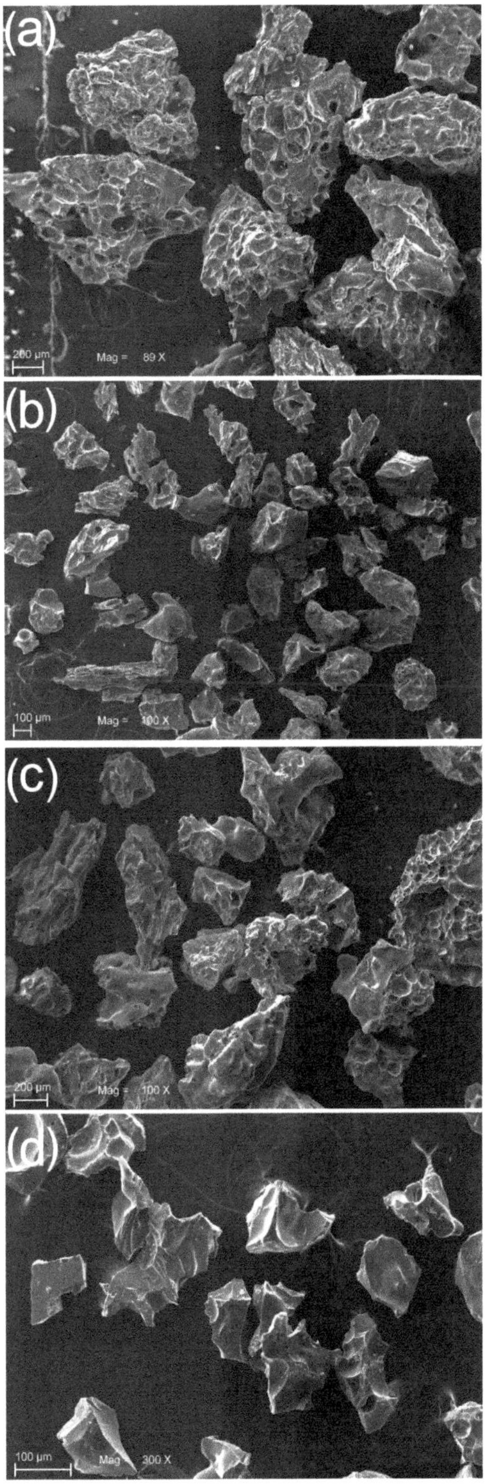

Figure 6. Scanning electron microscope secondary electrons (SEM-SE) images representative of the textures identified in studied tephra, showing moderately vesicular, glassy scoriaceous fragments with fluidal texture (sideromelane) and scarce poorly vesicular, blocky clasts (tachylite). **(a, b)** NSEC3 sample at Fleri; **(c, d)** NSEC2 sample at Acireale.

Table 1. SEM-EDS analyses of major elements in glass of 15 November 2011 ash fragments (weight % average value); no. denotes the number of analyses; SD denotes standard deviation value.

	No. 10	SD
SiO_2	49.29	0.72
TiO_2	2.19	0.22
Al_2O_3	16.11	0.40
FeO	10.13	0.37
MgO	3.68	0.22
MnO	0.26	0.10
CaO	7.56	0.30
Na_2O	4.59	0.15
K_2O	3.32	0.15
P_2O_5	0.82	0.20
Total	98.06	

on the ground may pollute the air quality, with PM_{10} levels (at least during the short-duration tests) far exceeding the recommended limit for a 24 h exposure. In particular, PM_{10} values are higher though within the same order of magnitude in site 1, the location least affected by vehicular traffic. Conversely, site 2 and 3 are both characterized by average, maximum and minimum values of PM_{10} 6–13 times higher a few hours after the 15 November lava fountain. Interestingly, sites 1 and 2, though affected by similar amounts of tephra, show substantial differences in the production of finer particles and this may be related to the varying intensity of vehicular traffic, higher at site 2 with respect to site 1. Further, Fig. 4 highlights that the 1 min, see-saw averaged line is heavily dependent on the vehicular flow, which by disturbing the tephra deposit is able to raise a clearly visible, irritating cloud of fine particles including a high percentage of PM_{10}.

Given that the local authorities in charge of the territory should take these data into consideration, we think they should reflect especially on the following: (1) the average PM_{10} value recorded during our tests, which remains well over the 24 h limit (0.050 mg m^{-3}) fixed by law at all three sites, i.e. only twice at site 1 but 13 and 19 times at sites 2 and 3, respectively, and (2) the minimum values, which, though not so high as at site 1 (0.49 mg m^{-3}), reached very high readings at sites 2 and 3 (0.337 and 0.379 mg m^{-3}, respectively), i.e. $\sim$7 times the limit fixed to safeguard human health over periods of 24 h. Sites usually characterized by low PM_{10} values proved very sensitive after the fallout from lava fountains, because relatively low quantities of tephra deposit per square metre (from $\sim$500 to $>$ 1500 g m^{-2}) were able to produce significant concentrations of particulate matter in the air.

The grain-size distributions of collected tephra show that 99.8 and 94.8 % of the volume for samples NSEC3 and NSEC2 (respectively) is coarser than 1 mm, and all the samples are composed of particles coarser than 10 μm. Hence,

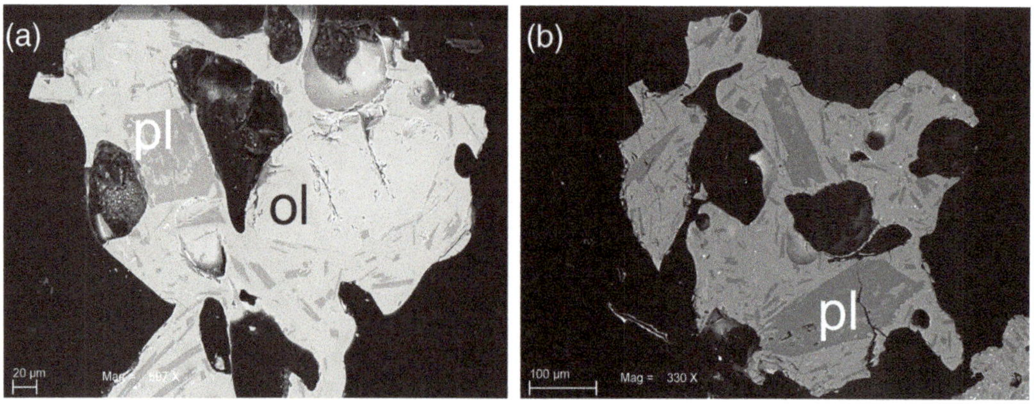

Figure 7. Scanning electron microscope secondary electrons (SEM-SE) images representative of textures identified in the 15 November 2015 ash; pl denote plagioclase; ol denotes olivine.

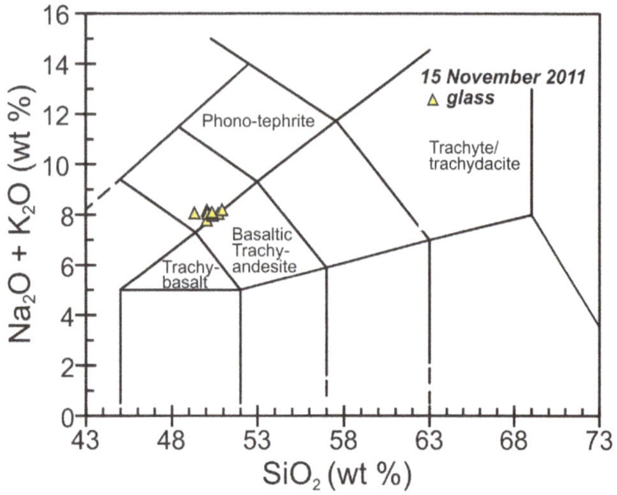

Figure 8. Total alkali–silica classification diagram of Le Bas et al. (1986).

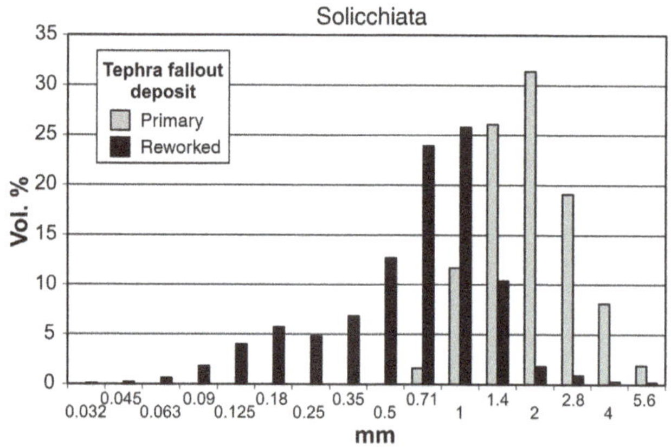

Figure 9. Comparison between the grain-size distribution of two samples collected at the same site in Solicchiata village (14 km northeast of the vent): primary (grey) and reworked (black) tephra fallout deposit after grinding by passing cars.

we infer that the air contamination by PM_{10} is not due to the direct ash fallout but to the break-up of the coarser clasts after mechanical pressure from vehicular transit. The breakage could be enhanced to the fragility of the sideromelane particles, which represent ca. 90 % of the samples as morphological and component analyses indicate.

5.2 Quantitative estimation of the tephra grinding by car passage

The passage of the cars on the roads covered by tephra fallout causes the grinding of the original tephra on the ground itself and its accumulation along the edge of the carriageway (Fig. 3b). This simple as well as reiterated observation suggested we evaluate the phenomenon quantitatively to understand if the original grain size of the tephra on the ground could be significantly modified.

We tested this observation following the lava fountain at SEC on 23–24 November 2007 (Andronico et al., 2007). The paroxysmal episode produced tephra fallout more than 80 km NE away from Etna, reaching the Calabria region. At 14 km from the vent along the dispersal axis (Solicchiata village), we collected two distinct samples, one unaltered ($\sim 1120\,\mathrm{g\,m^{-2}}$) and another one from the edge of the carriageway, clearly related to the passage of the cars which had ground down and accumulated a macroscopically finer deposit.

The comparison between the grain-size distribution of the two samples (Fig. 9) shows that the mode of the "ground" sample is at least one-phi shifted toward finer classes with respect to the original deposit (from 1.4–2 to 0.71–1 mm). What is more striking, however, is that while the grain-size distribution of the original sample indicates that only 1.6 % was lower than 1 mm, the "ground" sample was formed by more than 60 % of particles with dimensions < 1 mm.

Further, to better quantify the finer particles, the fraction < 2 mm (i.e. 96.8 % of the total sample) was analysed using the CILAS laser diffraction analyser (http://www.cilas.com/granulometrie.html) at the University of Geneva; the cumulative grain-size distribution curve shows that it contains 2.8 % of particles with a dimension of < 8 μm. In other terms, for each 100 g m^{-2}, the grinding produced ~2.5 g m^{-2} of PM_{10} particles in the ground deposit. This is indeed a high percentage of potentially dangerous particles for human exposure directly produced by passing cars.

5.3 Tephra fallout in urban settlements

In the past, high PM_{10} values have been measured by a network of air monitoring stations located in the city of Catania, which have exceeded the EU set threshold even up to some tens of times greater after the long-lasting fallout (2002–2003 eruption; Rete di monitoraggio ambientale, 2002) or single paroxysmal events (e.g. July 2011). These data, easily accessible at the web page maintained by the Environmental Pollution Office of the municipality of Catania, confirm our warning on the potential effects of tephra fallout in the Etna region.

These results are also supported by measurements taken on 6 October 2011 in the town of Ragalna (Fig. 1), 8 days after the tephra fallout on 28 September 2011 during the lava fountain episode. Here, the tephra cover (96 % of which was made up of 0.25–1 mm sized particles) had not been completely swept from the streets, reaching significant thicknesses (up to several centimetres) at specific points. Also in this case, we measured high PM_{10} peaks up to 0.300 mg m^{-3} correlated with the transit of cars over the ash mounds. In other terms, as occurred during our measurements following the 15 November 2011 lava fountain, the grinding by car passage was able (a week after the end of the tephra fallout) to cause the remobilization of volcanic particulate matter and its resuspension in the air, the effects of which have been quantified by the high PM_{10} concentrations measured by DT.

Recently, the villages of Zafferana Etnea and Piedimonte Etneo, located 10 km eastward and 17 km north-eastward of the NSEC, were covered by unusually thick tephra fallout deposits (2–3 cm depth) produced by high-energy lava fountain events. These cumulated > 7000 g m^{-2} at Zafferana on 16 March 2013, and ~6000 g m^{-2} at Piedimonte on 23 November 2013, respectively (Lo Castro et al., 2013; Andronico et al., 2015), causing a few days of temporary closure of traffic. Notably, both during the removal of the accumulated tephra and the remobilization by car traffic, a high resuspension of ash affected the roads, causing the formation of a PM_{10} cloud, thus increasing the ash exposure and inevitable inhalation for several days (Fig. 10).

This is also demonstrated by Thorsteinsson et al. (2012), who reported that ash resuspension near the ground was able to generate levels of particulate matter similar to those caused by direct tephra fallout during the 2010 Eyjafjallajökull eruption in Iceland. The wind erosion due to strong storm episodes remobilized the ash deposited during this eruption, generating resuspension events with higher concentration of PM_{10} also several tens of kilometres away from the deposit (Leadbetter et al., 2012; Arnalds et al., 2013). Furthermore, the sweeping of tephra from the road several days after the fallout had also increased the resuspension effects (Leadbetter et al., 2012).

Figure 10. The remobilization of ash in the air after the 16 March 2013 tephra fallout at Zafferana Etnea.

Studies by Hincks et al. (2006) and Barsotti et al. (2010) also linked the resuspension to the amount of tephra cover and showed the direct correlation between the thickness of tephra deposit to the higher concentrations of PM_{10} and 10 μm volcanic particles, respectively.

Finally, it should be remembered that physicochemical analyses carried out on ash samples from the 2002–2003 Etna eruption also showed that fine ash particles may generate hydroxyl radicals, an additional source of respiratory toxicity (Horwell et al., 2007), thus increasing the potential health hazard.

6 Conclusions

Tephra fallouts are a source of particulate matter in the air and their high frequency in the east sector of Etna may maintain high levels of PM_{10} for several days in the urban settlements around the volcano. Our study, in particular, suggests that the presence of fine-grained tephra on the ground may be hazardous in terms of PM_{10} concentrations in the air when not quickly removed, and matches well with a medical study on acute health effects due to volcanic ash exposure during the 2002 Etna eruption.

Although our measurements represent preliminary short-duration tests to evaluate eventual exposure to volcanic ash at Etna, they may nonetheless represent a starting point for future research. Data showed that after the 15 November 2011 lava fountain, the PM_{10} concentrations exceeded the EU limit of 0.050 mg m^{-3} (fixed over 24 h) 91, 90 and 94 % of the time in the three investigated sites. Our tests urge the planning of a systematic monitoring of the PM_{10} concentrations, especially after tephra fallout from paroxysmal activity. Measuring and assessing the real levels of exposure needs to be done to understand eventual respiratory (and cardiovascular) hazards to the Etnean population.

It is noteworthy that all the sites selected for our PM_{10} measurements (sites 1, 2, 3 and Ragalna) are not located in highly urbanized areas in terms of people passing (whether walking, cycling, motorbiking or driving), so that the risks and thus the effects on the human health could be significantly higher in the town centres. Here, the continuous vehicular traffic could lead, on the one hand, to the formation of a quasi-permanent PM_{10} cloud within the first 2–3 m of height above the ground, and on the other hand, to a continuous and marked exceedance of the PM_{10} limits set to protect human health.

We conclude by hoping that the impact from ash fallout in the Etnean territory will not be underestimated but given due attention to avoid potential health risks. Although difficult, if not impossible, to prevent exposure to high concentrations of PM_{10} entirely, this goal might be achieved by carrying out simple but effective actions and precautions; first and foremost, the swift and effective removal of the ash deposit in the urbanized areas and public roadways within a few hours from the tephra deposition on the ground, limiting the resuspension of high amounts of particulate matter into the air as well the grinding of tephra by passing cars.

Acknowledgements. We thank S. Scollo for her collaboration during the PM_{10} measurements carried out on 15 November 2011, A. Di Roberto for assistance in laboratory procedures and F. Alfano for carrying out a grain-size analysis with the CILAS laser diffraction analyser at the University of Geneva. E. Privitera, Director of INGV Sezione di Catania OE, is kindly acknowledged for supporting our research. The DustTrak™ instrument was funded by the Italian "Dipartimento della Protezione Civile" in the frame of the 2004–2006 Agreement with the Istituto Nazionale di Geofisica e Vulcanologia – INGV (V3-Etna Project, A. Neri supervisor). S. Conway is acknowledged for revising the English language of the text.

We also thank E. Liu and three anonymous reviewers for constructive revisions and the editor G. Macedonio for his helpful handling of the paper.

References

Alparone, S., Andronico, D., Lodato, L., and Sgroi, T.: Relationship between tremor and volcanic activity during the Southeast Crater eruption on Mount Etna in early 2000, J. Geophys. Res., 108, B52241, doi:10.1029/2002JB001866, 2003.

Andronico, D. and Corsaro, R. A.: Lava fountains during the episodic eruption of South–East Crater (Mt. Etna), 2000: insights into magma-gas dynamics within the shallow volcano plumbing system, Bull. Volcanol., 73, 1165–1178, doi:10.1007/s00445-011-0467-y, 2011.

Andronico, D., and Cristaldi, A.: Il parossismo del 23–24 novembre 2007 al Cratere di SE: caratteristiche del deposito di caduta, Internal report no. UFVG2007/073, available at: http://www.ct.ingv.it (available on demand), 2007.

Andronico, D., Scollo, S., Cristaldi, A., and Ferrari, F.: Monitoring ash emission episodes at Mt. Etna: The 16 November 2006 case study, J. Volcanol. Geoth. Res., 180, 123–134, doi:10.1016/j.jvolgeores.2008.10.019, 2009a.

Andronico, D., Cristaldi, A., Del Carlo, P., and Taddeucci, J.: Shifting styles of basaltic explosive activity during the 2002–03 eruption of Mt Etna, Italy, J. Volcanol. Geoth. Res., 180, 110–122, doi:10.1016/j.jvolgeores.2008.07.026, 2009b.

Andronico, D., Scollo, S., Cristaldi, A., and Lo Castro, M. D.: Representivity of incompletely sampled fall deposits in estimating eruption source parameters: a test using the 12–13 January 2011 lava fountain deposit from Mt. Etna volcano, Italy, Bull. Volcanol., 76, 861, doi:10.1007/s00445-014-0861-3, 2014a.

Andronico, D., Scollo, S., Lo Castro, M. D., Cristaldi, A., Lodato, L., and Taddeucci, J.: Eruption dynamics and tephra dispersal from the 24 November 2006 paroxysm at South-East Crater, Mt Etna, Italy, J. Volcanol. Geoth. Res., 274, 78–91, doi:10.1016/j.jvolgeores.2014.01.009, 2014b.

Andronico, D., Scollo, S., and Cristaldi, A.: Unexpected hazards from tephra fallouts at Mt Etna: The 23 November 2013 lava fountain, J. Volcanol. Geoth. Res., 304, 118-125, doi:10.1016/j.jvolgeores.2015.08.007, 2015.

Arnalds, O., Thorarinsdottir, E. F., Thorsson, J., Waldhauserova, P. D., and Agustsdottir, A. M.: An extreme wind erosion event of the fresh Eyjafjallajokull 2010 volcanic ash, Scient. Rep., 3, 1–7, doi:10.1038/srep01257, 2013.

Barsotti, S., Andronico, D., Neri, A., Del Carlo, P., Baxter, P. J., Aspinall, W. P., and Hincks, T.: Quantitative assessment of volcanic ash hazards for health and infrastructure at Mt. Etna (Italy) by numerical simulation, J. Volcanol. Geoth. Res., 192, 85–96, doi:10.1016/j.jvolgeores.2010.02.011, 2010.

Behncke, B., Branca, S., Corsaro, R. A., De Beni, E., Miraglia, L., and Proietti, P.: The 2011–2012 summit activity of Mount Etna: birth, growth and products of the new SE crater, J. Volcanol. Geoth. Res., 270, 10–21, 2014.

Branca, S. and Del Carlo, P.: Types of eruptions of Etna volcano AD 1670–2003: implications for short-term eruptive behaviour, Bull. Volcanol., 67, 732–742, 2005.

Damby, D. E., Horwell, C. J., Baxter, P. J., Delmelle, P., Donaldson, K., Dunster, C., Fubini, B., Murphy, F., Nattrass, C., Sweeney, S., Tetley, T., and Tomatis, M.: The respiratory health hazard of tephra from the 2010 Centennial eruption of Merapi with implications for occupational mining of deposits, J. Volcanol. Geoth. Res., 261, 376–387, doi:10.1016/j.jvolgeores.2012.09.001, 2013.

Delfino, R. J., Sioutas, C., and Malik, S.: Potential Role of Ultrafine Particles in Associations between Airborne Particle Mass and Cardiovascular Health, Environ, Health Perspect., 113, 934–946, doi:10.1289/ehp.7938, 2005.

EC: Council Directive 1999/30/EC, relating to limit values for sulphur dioxide, nitrogen dioxide and oxides of nitrogen, particulate matter and lead in ambient air, Office J EU, L 163, 29 June 1999, 41–60, http://eur-lex.europa.eu/oj/direct-access.html (last access: December 2015), 1999.

EU: Directive 2008/50/EC of the European Parliament and of the Council of 21 May 2008 on ambient air quality and cleaner air for Europe, Office J EU, L 152, 11 June 2008, 1–44, http://eur-lex.europa.eu/oj/direct-access.html (last access: December 2015), 2008.

Expert Panel on Air Quality Standards, Particles: Department of the Environment, Her Majesty's Stationery Office, London, 1995.

Expert Panel on Air Quality Standards, Airborne particles: what is the appropriate measurement on which to base a standard? A discussion document, Department for Environment, Food & Rural Affairs, London, http://webarchive.nationalarchives.gov.uk/20060715141954/http:/www.defra.gov.uk/environment/airquality/aqs/air_measure/index.htm (last access: 21 December 2015), 2001.

Fano, V., Cernigliaro, A., Scondotto, S., Cuccia, M., Forestiere, F., Nicolosi, A., Oliveti, C., Scillieri, R., Di Stefano, P., and Peducci, C. A.: Health effects of environmental contamination due to volcanic ash of Mount Etna in autumn, Epidemiol. Prev., 29, 180–187, 2005.

Hincks, T. K., Aspinall, W. P., Baxter, P. J., Searl, A., Sparks, R. S. J., and Woo, G.: Long term exposure to respirable volcanic ash on Montserrat: a time series simulation, Bull. Volcanol., 68, 266–284, doi:10.1007/s00445-005-0006-9, 2006.

Horwell, C. J. and Baxter, P. J.: The respiratory health hazards of volcanic ash: a review for volcanic risk mitigation, Bull. Volcanol., 69, 1–24, doi:10.1007/s00445-006-0052-y, 2006.

Horwell, C. J., Sparks, R. S. J., Brewer, T. S., Llewellin, E. W., and Williamson, B. J.: The characterization of respirable volcanic ash from the Soufrière Hills Volcano, Montserrat, with implications for health hazard, Bull. Volcanol., 65, 346–362, 2003.

Horwell, C. J., Fenoglio, I., and Fubini, B.: Iron-induced hydroxyl radical generation from basaltic volcanic ash, Earth Planet. Sc. Lett., 261, 662–669, 2007.

Horwell, C. J., Stannett, G. W., Andronico, D., Bertagnini, A., Fenoglio, I., Fubini, B., Le Blond, J. S., and Williamson, B. W.: A mineralogical health hazard assessment of Mt. Vesuvius volcanic ash, J. Volcanol. Geoth. Res., 191, 222–232, 2010.

Horwell, C. J., Baxter, P. J., Hillman, S. E., Calkins, J. A., Damby, D. E., Delmelle, P., Donaldson, K., Dunster, C., Fubini, B., Hoskuldsson, A., Kelly, F. J., Larsen, G., Le Blond, J. S., Livi, K. J. T., Mendis, B., Murphy, F., Nattrass, C., Sweeney, S., Tetley, T. D., Thordarson, T., and Tomatis, M.: Physicochemical and toxicological profiling of ash from the 2010 and 2011 eruptions of Eyjafjallajökull and Grímsvötn volcanoes, Iceland using a rapid respiratory hazard assessment protocol, Environ. Res., 127, 63–73, 2013.

Leadbetter, S. J., Hort, M. C., Löwis, S., Weber, K., and Witham, C. S.: Modeling the resuspension of ash deposited during the eruption of Eyjafjallajökull in spring 2010, J. Geophys. Res., 117, D00U10, doi:10.1029/2011JD016802, 2012.

Le Bas, M. J., Le Maitre, R. W., Streckeisen, A., and Zanettin, B. A.: Chemical classification of volcanic rocks based on the total alkali-silica diagram, J. Petrol., 27,745–750, doi:10.1093/petrology/27.3.745, 1986.

Le Blond, J. S., Horwell, C. J., Baxter, P. J., Michnowicz, S. A. K., Tomatis, M., Fubini, B., Delmelle, P., Dunster, C., and Patia, H.: Mineralogical analyses and in vitro screening tests for the rapid evaluation of the health hazard of volcanic ash at Rabaul volcano, Papua New Guinea, Bull. Volcanol., 72, 1077–1092, doi:10.1007/s00445-010-0382-7, 2010.

Lo Castro, M. D. and Andronico, D.: Operazioni di base per la misura della distribuzione granulometrica di particelle vulcaniche tramite il CAMSIZER, Rapporti Tecnici INGV, 79, 1–35, 2008.

Lo Castro, M. D., Coltelli, M., and Scollo, S.: Il parossismo del 16 marzo 2013 al Nuovo Cratere di SE: caratteristiche del deposito di caduta, Rapporto UFVG del 26 Marzo 2013, Rapporti, Vulcanologia, available at: http://www.ct.ingv.it/it/rapporti/vulcanologia.html?view=docman&start=15 (last access: 21 December 2015), 2013.

Lombardo, D., Ciancio, N., Campisi, R., Di Maria, A., Bivona, L., Poletti, V., Mistretta, A., Biggeri, A., and Di Maria, G.: A retrospective study on acute health effects due to volcanic ash exposure during the eruption of Mount Etna (Sicily) in 2002, Multidiscip. Respir. Med., 7, 51, doi:10.1186/2049-6958-8-51, 2013.

Moore, K. R., Duffell, H., Nicholl, A., and Searl, A.: Monitoring of airborne particulate matter during the eruption of Soufriere Hills Volcano, Montserrat, in: The Eruption of Soufriere Hills Volcano, Montserrat, From 1995 to 1999, edited by: Druitt, T. H. and Kokelaar, B. P., Geological Society, London, Memoir, 21, 557–566, 2002.

QUARG – Quality of Urban Air Review Group: Airborne particulate matter in the United Kingdom, Third Report of the Quality of Urban Air Review Group, Dept. Environ., University of Birmingham, Institute of Public and Environmental Health, School of Chemistry, Birmingham, 1996.

Rete di monitoraggio ambientale: Rapporto annuale 2002 sulla qualità dell'aria (D. M. n. 163 del 21/4/1999, all.2), available at: (last access: 21 December 2015), 2002.

Scollo, S., Del Carlo, P., and Coltelli, M.: Tephra fallout of 2001 Etna flank eruption: analysis of the deposit and plume dispersion, J. Volcanol. Geoth. Res., 160, 147–164, 2007.

Searl, A., Nicholl, A., and Baxter, P. J.: Assessment of the exposure of islanders to ash from the Soufriere Hills volcano, Montserrat, West Indies, Occup. Environ. Med., 59, 523–531, 2002.

Taddeucci J., Pompilio, M., and Scarlato, P.: Monitoring the explosive activity of the July–August 2001 eruption of Mt. Etna (Italy) by ash characterization, Geophys. Res. Lett., 29, 1029–1032, 2002.

Thorsteinsson, T., Jóhannsson, T., Stohl, A., and Kristiansen, N. I.: High levels of particulate matter in Iceland due to direct ash emissions by the Eyjafjallajökull eruption and resuspension of deposited ash, J. Geophys. Res., 117, B00C05, doi:10.1029/2011JB008756, 2012.

Viccaro, M., Calcagno, R., Garozzo, I., Giuffrida, M., and Nicotra, E.: Continuous magma recharge at Mt. Etna during the 2011–2013 period controls the style of volcanic activity and compositions of erupted lavas, Miner. Petrol., 109, 67–83, doi:10.1007/s00710-014-0352-4, 2015.

WHO: Air quality guidelines for particulate matter, ozone, nitrogen dioxide and sulfur dioxide, Global update 2005, Summary of risk assessment, World Health Organization, Geneva, 1–22, 2006.

Wilson, T. M., Cole, J. W., Stewart, C., Cronin, S. J., and Johnston, D. M.: Ash storms: impacts of wind-remobilised volcanic ash on rural communities and agriculture following the 1991 Hudson eruption, southern Patagonia, Chile, Bull. Volcanol., 73, 223–239, doi:10.1007/s00445-010-0396-1, 2011.

2

Seeking key meteorological parameters to better understand Hector

S. Gentile[1] **and R. Ferretti**[1,2]

[1]Department of Physical and Chemical Sciences/CETEMPS, University of L'Aquila, Via Vetoio, 67010 Coppito, L'Aquila, Italy

[2]Danish Meteorological Institute, Lyngbyvej 100, 2100 Copenhagen, Denmark

Correspondence to: S. Gentile (sabrina.gentile@aquila.infn.it)

Abstract. Twelve Hector events, a storm which develops in northern Australia, are analyzed with the aim of identifying the main meteorological parameters involved in the storm's convective development. Based on Crook's ideal study (Crook, 2001), wind speed and direction, wind shear, water vapor, convective available potential energy and type of convection are the parameters used for this analysis. Both the European Centre for Medium-Range Weather Forecasts (ECMWF) analysis and high-resolution simulations from the Fifth-Generation Mesoscale Model (MM5) are used. The MM5 simulations are used to connect the mean vertical velocity to the total condensate at the maximum stage and to study the dynamics of the storms. The ECMWF analyses are used to evaluate the initial conditions and the environmental fields contributing to Hector's development. The analysis suggests that the strength of convection, defined in terms of vertical velocity, largely contributes to the vertical distribution of hydrometeors. The role of total condensate and mean lifting versus low-level moisture, convective available potential energy, surface wind and direction is analyzed for shear and no-shear conditions to evaluate the differences between type A and B for real events. Results confirm the tendency suggested by Crook's analysis. However, Crook's hypothesis of low-level moisture as the only parameter that differentiates between type A and B can only be applied if the events develop in the same meteorological conditions. Crook's tests also helped to assess how the meteorological parameters contribute to Hector's development in terms of percentage.

1 Introduction

Hector is a vigorous convective system that develops on the Tiwi Islands, two islands included in the "Maritime Continent" (Ramage, 1968), an area extending across the Indonesian archipelago, north Australia and New Guinea. This is one of the primary regions of global latent heat release contributing to the forcing of planetary-scale circulations (e.g., Hadley and Walker cells). The Tiwi Islands, located in the northern tropical part of Australia, produce regular tropical convection during the pre-monsoon and monsoon "break" seasons (from November to March) in response to the latent heat released during the diurnal cycle (Keenan and Carbone, 1992).

This storm has been analyzed during observing campaigns like ITEX (Island Thunderstorm Experiment, 1988), MCTEX (Maritime Continent Thunderstorm Experiment, 1995), SCOUT-O3 (Stratospheric-Climate links with emphasis On the Upper Troposphere and lower stratosphere, 2005) and TWP-ICE (Tropical Warm Pool – International Cloud Experiment, 2006) whose aims were to better understand the triggering mechanisms and the meteorological parameters favorable to the convective development. Particularly the MCTEX campaign collected many environmental factors that are known or believed to influence the initiation, organization, propagation and intensity of deep convection. In addition this data set allowed two distinct forcing regimes leading to Hector to be defined (Carbone et al., 2000):

1. type A: resulting from the confluence and convergence of the sea breeze fronts;

2. type B: rising from the interaction between sea breeze and gust front, convectively generated by cold pools.

Type A forcing may be viewed as nature's backup mechanism when the meteorological conditions do not allow type B development.

Following these campaigns, some ideal and real numerical studies have been performed to understand the forcing and the triggering mechanisms of the Hector storm (e.g., Golding, 1993; Crook, 2001; Saito et al., 2001; Ferretti and Gentile, 2009; Gentile et al., 2014).

Golding (1993) used the UK Met Office's mesoscale model at 3 km resolution initialized by a sounding to examine two cases from ITEX. The results suggest that the model qualitatively reproduces the diurnal evolution of Hector, showing a clear relationship between the storm development and the island topography. Saito et al. (2001) used the Japanese Meteorological Institute's mesoscale model at 1 km resolution to simulate a case from MCTEX. The study highlights a good agreement between the simulations and observations and focuses on the five stages of the convective life cycle.

Chemel et al. (2009) simulated the 30 November 2005 Hector event using two models, the Advanced Research Weather Research and Forecasting (ARW) model, and the Met Office Unified Model, with a resolution of 1 km. Both models reproduce the development of Hector fairly well, even though the two simulated surface heat fluxes are very different. This would mean that the intensity of the storm is not only controlled by this factor. The aim of the paper is to investigate the role of deep convection in the vertical transport of tropospheric air into the lower stratosphere. Chemel et al. (2009) conducted a further simulation with ARW in large eddy simulation (LES) mode, refining the grid spacing to 250 m, and concluded that the characteristics of the Hector storm are basically similar in time and space to those obtained in the 1 km resolution. Therefore a 1 km resolution is fine enough to simulate the timing, the structure and strength of deep convection when compared with the field campaign observations (Chemel et al., 2009).

In the study by Ferretti and Gentile (2009) two Hector events (one observed during SCOUT-O3 and one during TWP-ICE) have been investigated, analyzing the dynamics and thermodynamics. Using the Fifth-Generation Mesoscale Model (MM5) at 1 km resolution over the Tiwi Islands, several numerical experiments have been performed with the aim of understanding the forcing and triggering conditions for the development of Hector. The study demonstrates the key role of the sea breeze, water vapor content and soil moisture content in the growth of Hector. Moreover, Gentile et al. (2014) carried out a study for highlighting both the triggering factors and microphysical structure of a Hector event. The event was analyzed using MM5 model simulations, ground-based radar and TRMM (Tropical Rainfall Measuring Mission) satellite data, with the aim of understanding the mechanisms leading to the convective development. The analysis of the horizontal and vertical structure at high temporal and spatial resolution produced by MM5 allows the mechanisms for triggering Hector to be established: sea breeze, a gust front from previous convection and the channeling effect by topography.

Zhu et al. (2013) simulated four cases of the Hector storm by running the ARW model with a maximum horizontal resolution of 1 km, incorporating and not incorporating the observations collected during the ACTIVE campaign. Only one (30 November 2005) of the four cases was well simulated by the run without the inclusion of observations. Three events (16 November, 6 and 10 February 2006) can be simulated only if the model was run incorporating observations. The major deficiency deduced by Zhu et al. (2013) in the simulations of Hector is the smaller size and the weaker intensity in comparison with the observations.

Simulations were performed for a Hector event observed on 30 November 2005 by Dauhut et al. (2014) using the Meso-NH (mesoscale non-hydrostatic) model, performed with a grid spacing of 1600, 800, 400, 200 and 100 m. The updraft generally decreases with reduced resolution due to the reduced entrainment into the base of the updrafts. Indeed, the strong updrafts in the boundary layer obtained by the three finest simulations reinforce the updrafts in the upper troposphere.

Crook (2001) performed an ideal study using both a linear and nonlinear flow model for assessing the most important parameters of the Hector convective system. The low-level moisture is found to be an important parameter for differentiating between type A and B. High values of low-level moisture correspond to earlier convection, then the associated evaporational cooling produces cold pools that retard the further inland progress of the sea breezes. Hence, Hector type B develops because of the convergence of the sea breeze and the gust front related to previous convection. Hector type A, which is associated with low values of low-level moisture, develops when the generation of precipitating cold pools is delayed so that the sea breeze fronts have time to converge. Moreover, Crook (2001) performed sensitivity tests to surface heating, wind speed and direction. The results show a strong link between convective available potential energy (CAPE), wind speed and direction and total condensate (sum of all hydrometeors) of Hector cells. The relationship between couples of meteorological parameters was investigated using diagrams that allowed it to be assessed that the convective strength, in terms of vertical velocity, increases as the wind speed decreases and as the wind direction turns toward the major axis of the Tiwi Islands (Crook, 2001).

In this study, 12 Hector events (from November 1995 to November 2008) are used to investigate the transferability of the conclusions of Crook's study to real events. With this aim, the relationship with the same meteorological parameters used by Crook is investigated for each real Hector event, each one characterized by its own boundary and initial conditions. The case studies are simulated using the MM5 as described in Ferretti and Gentile (2009) and Gentile et al. (2014), and the results are investigated to establish the contribution of water vapor, surface wind speed and direction to the convective strength. These previous works (Ferretti and Gentile, 2009; Gentile et al., 2014) allow the model's ability to reproduce the dynamics and correctly detect the triggering factors leading to the development of Hector to be assessed, by performing a detailed comparison with observations. However, a temporal and spatial shift is found for MM5. This is also a common issue found for the WRF (Weather Research and Forecasting) model by Chemel et al. (2009) and Zhu et al. (2013). The main focus of this study, as already pointed out, is to investigate the role of a few key meteorological parameters for Hector's development by using Crook's diagrams, which are independent of time. Therefore, a possible temporal or spatial shift in the MM5 simulations of Hector does not affect the results.

The study is organized as follows. A meteorological analysis of the events is presented in Sect. 2 as a function of wind speed, wind direction and shear, CAPE and water vapor and convection modes A or B, with a brief description of the model configuration. The comparison, in terms of cloud total condensate and vertical velocity profiles among the 12 events, is shown in Sect. 3. The fourth paragraph describes Crook's test and outlines the main features in terms of percentage involved in Hector's development. Conclusions are drawn in Sect. 5.

2 Meteorological characteristics of the Hector events

The convective strength of the tropical storm Hector is evaluated using the meteorological variables suggested by Crook (2001). In this study, 12 real events (eight single-cell and two double-cell) are selected. Some of them (20 November 1995; 23 November 1995; double-cell 27 November 1995; 1 and 4 December 1995) were observed during MCTEX, a campaign held in late 1995 on the Tiwi Islands with the goal of monitoring the convective life cycle of the mesoscale convective system (Keenan and Carbone, 1992; Carbone et al., 2000). Four of the remaining events (the double-cell 30 November 2005, 6 February 2006 and 29 November 2007) have already been analyzed by Ferretti and Gentile (2009) and Gentile et al. (2014). The events are named with the acronyms as follows:

1. 20 November 1995: N20;
2. 23 November 1995: N23;
3. 27 November 1995: N27 (double-cell);
4. 1 December 1995: D1;
5. 4 December 1995: D4;
6. 30 November 2005: N30 (double-cell);
7. 6 February 2006: F6;
8. 29 November 2007: N29;
9. 11 November 2007: N11;
10. 17 November 2008: N17.

To simulate the events, the same configuration is used as in Ferretti and Gentile (2009) and Gentile et al. (2014). The previous works (Ferretti and Gentile, 2009; Gentile et al., 2014) allow the model's ability to reproduce the dynamics and correctly detect the triggering factors leading to the development of Hector to be assessed, by performing a detailed comparison with radar and satellite observations. The mesoscale model, MM5V3, is a non-hydrostatic, fully compressible, primitive equation model with a terrain following vertical coordinates (Dudhia et al., 2004). Four nested domains and 58 vertical levels are used. The mother domain has a 27 km grid, covering the tropical part of Australia. The finest domain has a horizontal grid of 1 km and it is centered over the Tiwi Islands. The following parametrizations are used: the Gayno–Seaman scheme for the planetary boundary layer; the MM5 cloud radiation scheme for radiative transfer processes; the Kain–Fritsch cumulus convection parametrization for domains 1, 2 and 3 (though there is no cumulus convective parametrization for the finest domain); the Reisner 2 parametrization as a microphysical scheme. To improve the meteorological analysis on the mesoscale grid, direct surface and radiosonde observations have been incorporated using the objective analysis based on the Cressman scheme (Faccani et al., 2003). The simulations are initialized using ECMWF (European Centre for Medium-Range Weather Forecasts) analysis at 0.25° and they last 24 h for all the events. The boundary conditions are upgraded every 6 h.

In the following subsections, the ECMWF analysis is used to analyze the main dynamical aspects of the Hector events. The analysis is performed, evaluating the role of the following parameters in the development of the Hector storm: the wind speed and direction at three different levels (lower (LL, 950 hPa), medium (ML, 700 hPa) and upper levels (UL, 300 hPa)), the CAPE and the water vapor content extracted at 950 hPa (mixing ratio). The values of these parameters for the Hector events are summarized in Table 1. To better understand the dynamical conditions for the storm development, two parameters are added: the shear occurrence and the typology of the events (definition based on Carbone et al., 2000). All these quantities are derived from the ECMWF analysis at 00:00 UTC (09:30 LST). For the second cells (marked in the table as 2), the meteorological parameters are extracted 6 h later, that is, at 06:00 UTC (15:30 LST). Indeed, the ECMWF analysis is only provided for base times of 00:00, 06:00, 12:00 or 18:00, so it is not possible to obtain meteorological information for a time interval of less than 6 h (the first cell of N27 reaches maximum development at 13:10 LST and the second one, 3 h later, at 16:10 LST; for

N30, the maximum of the first cell is reached at 14:30 LST and the second one at 15:50 LST).

2.1 Wind speed

The wind speed at the surface controls the magnitude of convective instability over the Tiwi Islands; indeed, as the wind speed decreases, the low-level air mass spends more time over the heated and moistened island, increasing its instability (Crook, 2001).

Hector D1 and N29 present weak wind at the surface with a maximum of 2.5 $m s^{-1}$ (Table 1) from south for D1 and from east for N29 (Fig. 1a, e). N17 and N20 are also characterized by very weak wind at the lower level with a maximum speed of 2 $m s^{-1}$, from the southeast and the north, respectively, but in addition, weak wind is also found at 700 hPa (Fig. 2c, d, e, f). These conditions are favorable for increasing the instability which allows for the vertical growth of the tropical thunderstorm. On the contrary, N11 and F6 show a very strong southeasterly wind up to 16–18 $m s^{-1}$ and a moderate easterly wind of 8 $m s^{-1}$, respectively, at the middle level (Figs. 2b, 3b), allowing more stable conditions and an unfavorable environment to be supposed for the vertical growth. D4 and N23 have very similar wind structure; both events have a weak westerly flow (less than 3 $m s^{-1}$) at the lower levels (Figs. 1c, 3c) and a sharp change of wind direction at the middle level, with a speed of approximately 5 $m s^{-1}$ (Figs. 1d, 3d). Finally, the initial conditions of N27 (double-cell event) show a very weak surface wind, characterized by a speed of 1.5–2.5 $m s^{-1}$ (Fig. 4e, Table 1), produced by an area of high pressure centered on the Tiwi Islands (Fig. 4f). In contrast, the double-cell event (N30) shows a moderate wind speed (approximately 5 $m s^{-1}$) at the three levels, changing direction at higher altitudes (Table 1, Fig. 4a, b). For these two events, the environmental conditions prior to the organization and the development of the second convective cell are more unstable and disorganized than the single-cell events, as suggested by the fast low-level wind (Table 1, Fig. 4c, g, d, h) produced by the gust front of the previous cell.

2.2 Wind direction and shear

The wind direction is another meteorological parameter affecting the development of Hector. Assuming that the Tiwi Islands have an ellipse shape, if the air mass blows along the major axis (east–west), the low-level convergence, produced by the sea breeze and the surface wind, is maximized because of the longer time spent by the air mass over the heated and moistened surface of Tiwi Islands. In contrast, if the air mass blows along the minor axis (north–south), the low-level convergence is reduced, both by the shorter time spent by the air mass over the heated surface, and by the overlapping of the surface wind in the same direction of the sea breeze, producing a much weaker convection (Crook, 2001). In addition, the vertical wind shear (change of direction) enhances the instability, allowing for the vertical growth of the cell (Crook, 2001).

The D4, F6 and N23 events show a similar flow structure characterized by a strong and sharp vertical wind shear (Table 1); the westerly surface wind (Figs. 1c, 3a, c) turns 180°, becoming easterly at middle level (Fig. 1d, 3b, d). Hectors N11 and N20 show shears from a different direction: a surface flow from the southwest and from the north (Fig. 2a, e for N11 and N20) and a strong southeasterly (N11, Fig. 2b) and moderate south wind (N20, Fig. 2f) at the middle level, respectively. Figures 1a, e and 2c show the lack of a change in the wind direction between the low and the middle level, which confirms the absence of the shear (Fig. 1b, f and 2d) for D1, N29 and N17. Finally, for the two double-cell events, no wind shear is detected for N27 (Fig. 4e, f): the easterly wind is constant up to middle level. A strong vertical wind shear is found for N30 (Fig. 4a, b): the lower level wind turns from a northwesterly to an easterly direction at 700 hPa. This structure lasted until the onset of the second cell (Fig. 4c, d). For what concerns the environment in which the second convective cell develops, it is more heterogeneous: the onset of a weak wind shear helps to develop the N27 second cell (Table 1, Fig. 4g, h) and the stable presence of a strong vertical wind shear contributes to the growth of the N30 second cell (Table 1, Fig. 4c, d).

2.3 CAPE and water vapor

The CAPE is the vertical integral of positive buoyancy of an air parcel and it is an indicator of atmospheric instability. Results from the MCTEX campaign showed that the variability of CAPE is mainly due to the variability of the low-level moisture. Therefore, the two parameters are directly proportional (Crook, 2001).

Four single-cell cases are characterized by high values of CAPE (greater than 1200 $J kg^{-1}$ and lower than 2500 $J kg^{-1}$): D1, D4, N23 and N29 (Table 1). All these events show a remarkable convective activity, with several cells developing before and/or after the main Hector cell. The double-cell events present wet conditions with a high value of CAPE (2000 $J kg^{-1}$) and mixing ratio of water vapor (18–19 $g kg^{-1}$, Table 1) for N30, and a dry environment with low CAPE (450 $J kg^{-1}$) and water vapor mixing ratio (16–18 $g kg^{-1}$, Table 1) for N27. However, the second convective cell develops in a more unstable environment for both events; CAPE remains close to 2000 $J kg^{-1}$ for N30 and increased up to 1650 $J kg^{-1}$ for N27 (Table 1).

The following events were characterized by low values (between 150 and 650 $J kg^{-1}$) of CAPE (Table 1): N11, N17, N20 and F6, inferring a weak convective activity.

Table 1. Meteorological characteristics of the Hector events extracted from ECMWF analysis at 00:00 UTC (09:30 LST) for all the single and first cells of the double-structure events (marked in the table as 1) and at 06:00 UTC (15:30 LST) for the second cells (marked in the table as 2). In the heading LL stands for low level (950 hPa), ML for medium level (700 hPa) and UL for upper level (300 hPa). Wind shear is denoted by "yes" if there is a change of wind direction between LL and ML.

Events	Wind speed/direction ($m\,s^{-1}$ – deg)			Shear	CAPE ($J\,kg^{-1}$)	q ($g\,kg^{-1}$)	Type
	LL	ML	UL				
D1	1.5–2.5 – S	8–7 – SE	0.5–1.5 – E	yes	1350	17–20	B
D4	2–2.5 – W	5.5–6 – E	6.5–7.5 – W	yes	1170	18–19.5	B
F6	2.5–3.5 – SW	7–8 – E	5–5.5 – E	yes	175	16.5–17.5	B
N11	3–4 – SSW	16–18 – SE	7–9 – NW	yes	650	17.5–21	B
N17	1–2 – SE	1.5–2 – E	8–9 – W	no	450	16–18	B
N20	<1 – N	3–4.5 – S	12–16 – WSW	yes	650	16.5–17.5	B
N23	2.5–3.5 – W	4–5 – ESE	10–12 – NW	yes	2500	18.5–19.5	A
N27-1	1.5–2.5 – ENE	6–7 – E	2.5–4 – W	no	450	16–18	A
N27-2	2.5–3.5 – NNE	4.5–5.5 – ENE	3–5 – W	yes	1650	18–20	B
N29	1–2.5 – E	15–16 – ESE	7–8.5 NE	no	2000	18–19.5	A
N30-1	3.5–4 – NW	4–5 – ESE	3–5 – SW	yes	2000	18–19	A
N30-2	3–5 – W	3–4 – ENE	2–5 – SW	yes	1800	16–18.5	B

2.4 Convection type A or B

Convection of type A or B is a simple way to differentiate the dynamical development of the storm. Type A convection is generated by the convergence of two sea breeze fronts (Carbone et al., 2000), whereas type B is generated by the convergence of a single sea breeze front and a cold pool produced by previous convection (Crook, 2001; Gentile et al., 2014).

For both double-cell events (N27, N30), the first convective cell develops from the convergence of the two sea breeze fronts (type A) and the second one from the interaction of the gust front of the first decaying cell (type B) with the north and the south sea breeze front, for N27 and N30, respectively (an exhaustive description of the meteorological characteristics of N30 is given by Ferretti and Gentile, 2009). The dynamics of the storm is very similar for both of the events. The first precipitating cells develop in the northeastern part of Melville Island at approximately 12:00 LST. In the following hours, the convective system reaches a first maximum of reflectivity of 55–60 dBz, associated with a strong convective cell (Figs. 5b, 14b of Ferretti and Gentile, 2009) that reaches a height of 16 km at 13:10 LST for N27 (Fig. 5a) and 14 km at 14:30 LST for N30 (Fig. 14b of Ferretti and Gentile, 2009). The maximum development of the second cells is reached at 16:10 LST with a height of 16–17 km for N27 (Fig. 5c, d) and at 15:50 LST with a height of 16 km for N30 (Fig. 14c, d of Ferretti and Gentile, 2009).

For D1, the maximum development of the storm is reached at 15:10 LST with a height of 16 km (Fig. 5e, f) after the organization and aggregation of several convective cells. The precipitation starts at 13:00 LST in the northern part of the islands, and a first deep cell develops at 13:50 LST; this last one contributes to the growth of Hector (type B).

Two different Hector developments are found for D4 and N23, although the rain starts with the front of the south sea breeze for both events (at 13:00 LST for the first and 10:30 LST for the second event). This leads to a first convective tower reaching 10–11 km at 13:10 LST that finally reaches maximum development (type B) at 16:50 LST with a height of 17 km for D4 (Fig. 5g, h); whereas for N23, the convective line moves quickly to the north west of the Tiwi Islands and interacts with the north breeze front, triggering the development of the Hector cell (maximum of 15 km at 12:50, Fig. 6a, b). This is why N23 can be classified as type A (Table 1).

N29 (a detailed description of this event can be found in Gentile et al., 2014) is also type B; the development of this convective event is characterized by non-precipitating and well-organized cells during the first stage that ends as weak precipitation starts. Hence, the cells merge into a convergence line, that, interacting with the south sea breeze front and strengthened by a channeling effect, produces an intense growth of the convection. This phase corresponds to the mature stage, which is characterized by a cloud top height reaching 18 km (Fig. 13a, b of Gentile et al., 2014).

The N11 dynamical evolution is characterized by intense convective activity leading to type B development: a first cell appears in the northern part of the islands at 15:50 LST that triggers the vigorous Hector cell. A maximum height of 19–20 km is reached at 17:10 LST in the middle area of the Tiwi Islands (Fig. 6c, d). Similarly for N17 (type B), the gust front, related to a first precipitating cell which developed at 15:50 LST in the eastern part of the Tiwi Islands, interacted with the south sea breeze front, leading to a maximum height of 16–17 km at 16:30 LST (Fig. 6e, f). N20 is also classified as type B (Table 1); its first stage is characterized by

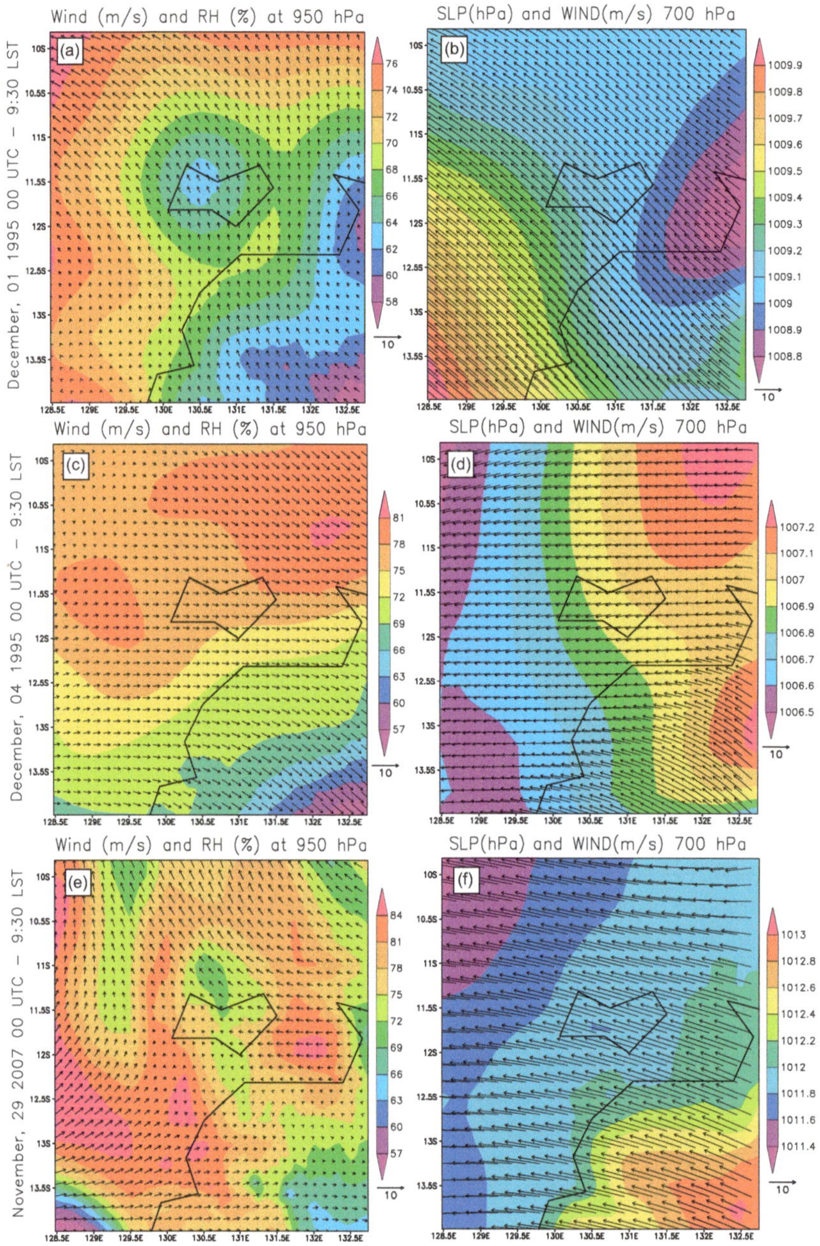

Figure 1. ECMWF analysis over the Tiwi Islands at 00:00 UTC for D1, D4 and N29. Panels **(a, c, e)** report surface wind and relative humidity at 950 hPa; panels **(b, d, f)** report sea level pressure in filled contours and wind flow vectors at 700 hPa.

aligned non-precipitating convective cells. These cells lead to an initial double structure that merges into a unique Hector cell at 17:10 LST. The maximum development shows a height of 17 km (Fig. 6g, h). Finally, F6 (a detailed description of the meteorological characteristics is given in Ferretti and Gentile, 2009) is characterized by two precipitating cells: the first one developing at 12:30 LST in the eastern part of Tiwi Islands, then decaying at 14:30 LST in the central area, and the second deep cell reaching maximum reflectivity of 45–50 dBz at 15:30 LST with a maximum cloud top of 16 km (Fig. 14e, f of Ferretti and Gentile, 2009). The interaction between the gust front of the decaying first cell with the south breeze front is the triggering mechanism for this Hector event (type B).

3 Cloud total condensate and vertical velocity profiles for the events

In order to better understand the mechanisms leading to different convective structures for these Hector events, the vertical structures of the storms are analyzed in terms of total condensate (Crook, 2001). With this aim, MM5 simulations

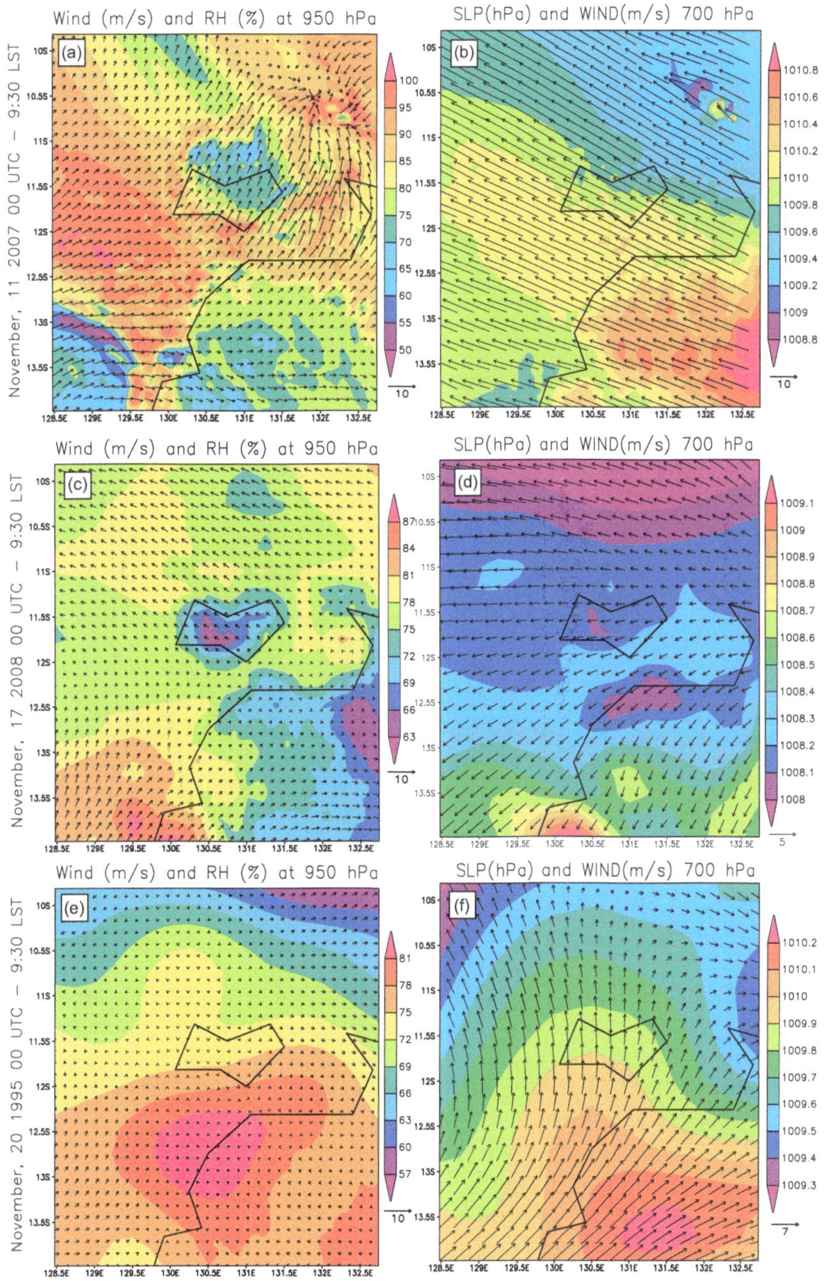

Figure 2. ECMWF analysis over the Tiwi Islands at 00:00 UTC for N11, N17 and N20. Panels **(a, c, e)** report surface wind and relative humidity at 950 hPa; panels **(b, d, f)** report sea level pressure in filled contours and wind flow vectors at 700 hPa.

are used to extract the mean vertical profile of the cloud total condensate (sum of all hydrometeors) and the vertical velocity for each event at the maximum stage (Fig. 7). The maximum stage is selected based on the time of the storm maximum height. The profiles are spatially averaged for each layer within the volume encapsulating Hector at a specific time. The maximum of the mean vertical velocity profile is approximately 0.9 $m s^{-1}$ for N11 (Fig. 7d), and the largest vertical velocity is 35 $m s^{-1}$ for the same event. D1, D4, N29, N17 and the first and the second cell, for N27 and N30, respectively, have a maximum updraft value that exceeds 20 $m s^{-1}$. If it is spatially averaged, it does not exceed 0.6 $m s^{-1}$. The values of the maximum vertical velocity for N30 are very close to those obtained with the LES simulations by Dauhut et al. (2014). The maximum updraft obtained by MM5 is approximately 22 $m s^{-1}$, sustained for a height up to 8–16 km (not shown) and the structure is very close to the one simulated by Meso-NH, using horizontal resolutions of 400, 200 and 100 m (Fig. 3a in Dauhut et al., 2014).

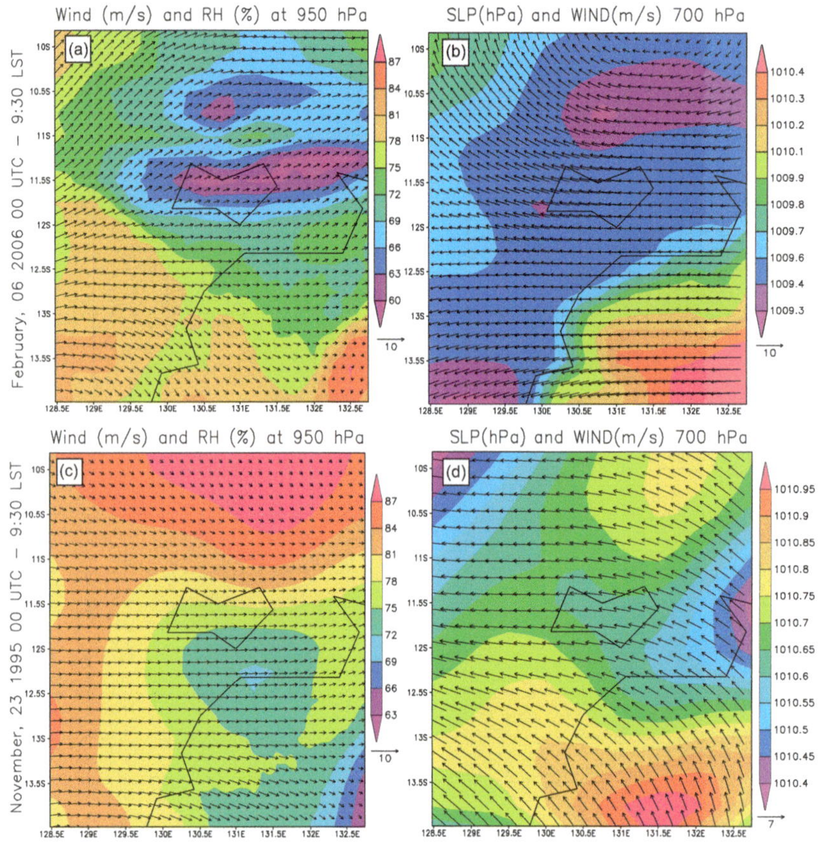

Figure 3. ECMWF analysis over the Tiwi Islands at 00:00 UTC for F6 and N23. Panels **(a, c)** report surface wind and relative humidity at 950 hPa; panels **(b, d)** report sea level pressure in filled contours and wind flow vectors at 700 hPa.

The absolute maximum of the total condensate matter is approximately 1.5 $g\,m^{-3}$ for the first cell of N27 (Fig. 7e), but the largest vertical velocity is not reached by this event.

The vertical velocity profiles (Fig. 7b, d, f) present common features for most of the events. At the lower level, a weak downdraft related to the precipitating hydrometeors prevails, and at the upper level, a very strong updraft can be detected, associated with the latent heat release due to the condensation process. In addition, a downdraft peak in the vertical velocity profile is found at the same level of a relative maximum in the total condensate profile for N11 at 2 km (Fig. 7c, d), the second cell of N30 at 3 km (Fig. 7e, f) and the first cell of N27 at 3–4 km (Fig. 7e, f). These features would suggest that the downdraft is related to the sinking due to melting or evaporation cooling, and the total condensate maximum corresponding to the production of rain or melted graupel. Moreover, the total condensate maxima are at higher altitude than the updraft peaks as for N29, N20 and N11 (Fig. 7b, d); these events have a maximum updraft at approximately 12 km and show a relatively large amount of total condensate up to 14–16 km (Fig. 7a, c). However, if the maximum vertical velocity is positioned at lower levels, the most part of the hydrometeors' distribution is at lower levels too. A clear example is the first cell of the N27 (Fig. 7e, f): the maximum updraft is located at approximately 4 km and the largest part of the hydrometeors is below 10–12 km.

The maximum total condensate and its vertical distribution may be related to the strength of convection, which is generally stronger if generated by the convergence of downdraft of previous cells and the sea breeze front (type B) than the one generated by the convergence of the two sea breeze fronts (type A) (Crook, 2001). Therefore, larger vertical velocities are expected for type B events; indeed, the largest vertical velocity is found for N11, which is type B. The previous analysis suggests that the strength of convection largely contributes to the vertical distribution of the total condensate. Therefore, the structure of these Hector events agrees with the hypothesis of Crook (2001) and allows it to be established that the strength of the event is proportional to the total condensate. However, the large variability of the total condensate vertical distribution among type A and B events suggests that other parameters play an important role beside the strength of convection.

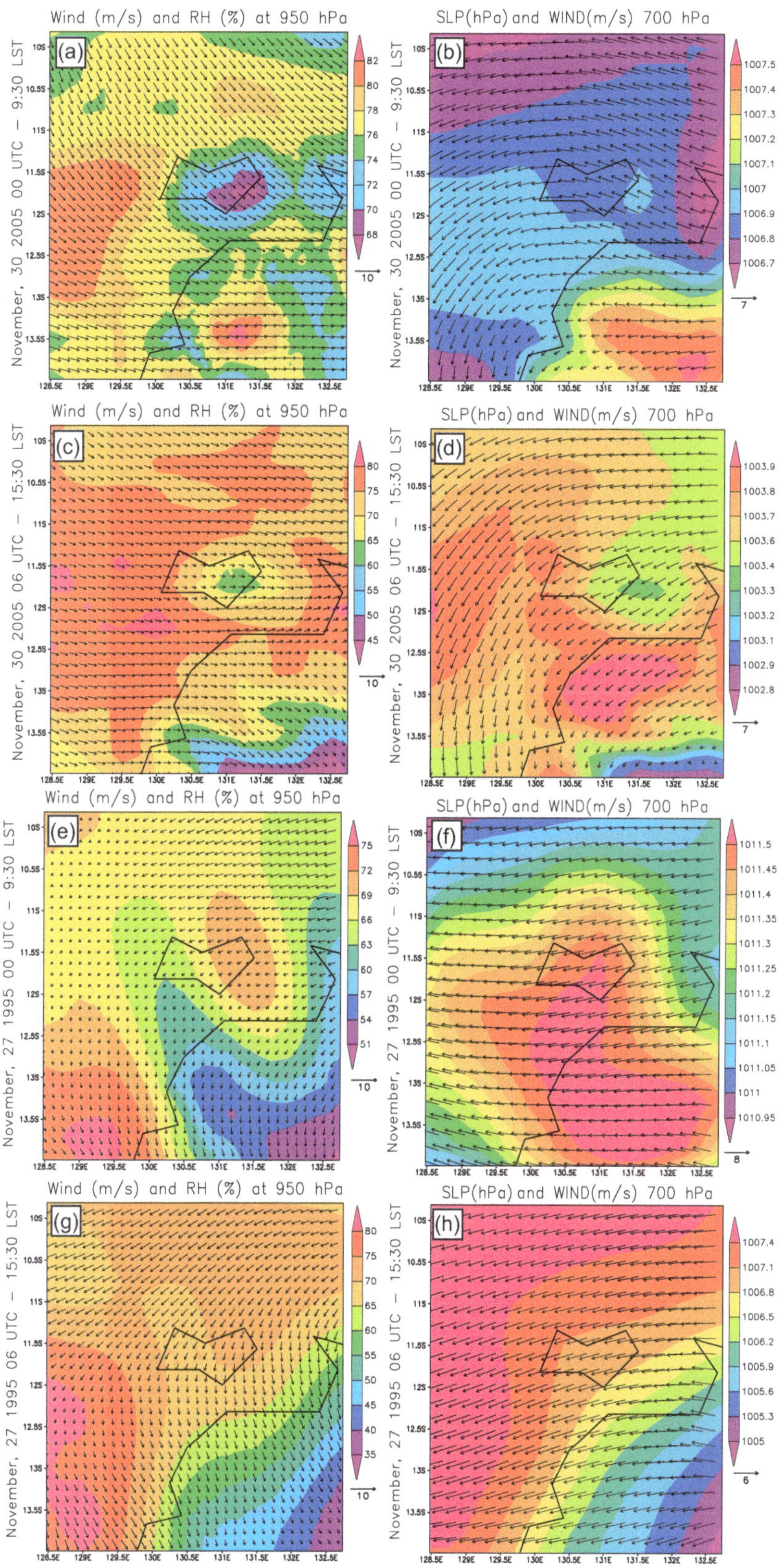

Figure 4. ECMWF analysis over the Tiwi Islands for N30 and N27 at 00:00 UTC for the first cell and at 06:00 UTC for the second one. Panels **(a, c, e, g)** report surface wind and relative humidity at 950 hPa; panels **(b, d, f, h)** report sea level pressure in filled contours and wind flow vectors at 700 hPa.

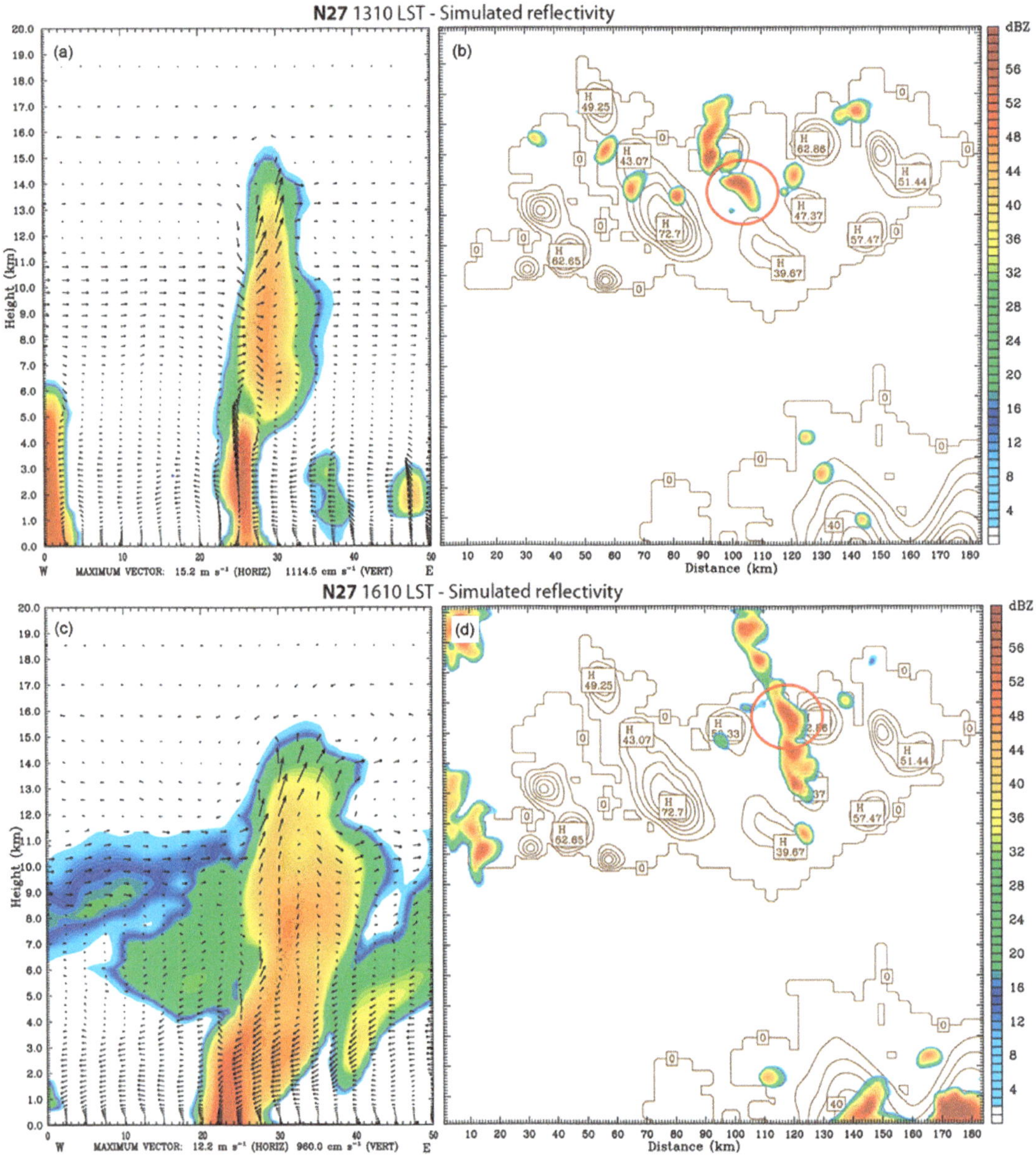

Figure 5.

4 Crook's test to detect triggering factors

In this section the analysis of the real events is carried out, using Crook's diagrams as a benchmark.

Based on ideal studies of Hector, Crook (2001) suggested that the amount of total condensate is strongly related to the low-level moisture, in terms of CAPE, as well as to the surface wind velocity and direction. Therefore, a model-aided analysis of the total condensate and of a few meteorological parameters (wind speed and direction, and CAPE), as used by Crook (2001), may help to highlight the most important factors for previous Hector events. With this aim, the same analysis performed by Crook (2001) is applied to the real Hector events analyzed in this study, but some differences are obviously present. The possibility of changing the meteorological parameters, for example, keeping one field constant as is done by Crook (2001), is not applied because its disruptive effect on Hector has already been verified. Indeed, for real events, the variation of a parameter causes a lack in the development of Hector. For example, in the work of Ferretti and Gentile (2009), the halving and the increasing of the initial water vapor content disabled the development of Hector.

The following MM5 meteorological parameters are used for the analysis: low-level moisture, in terms of CAPE, surface wind speed and direction. Following Crook (2001), the variables are analyzed at the model start time, which is sev-

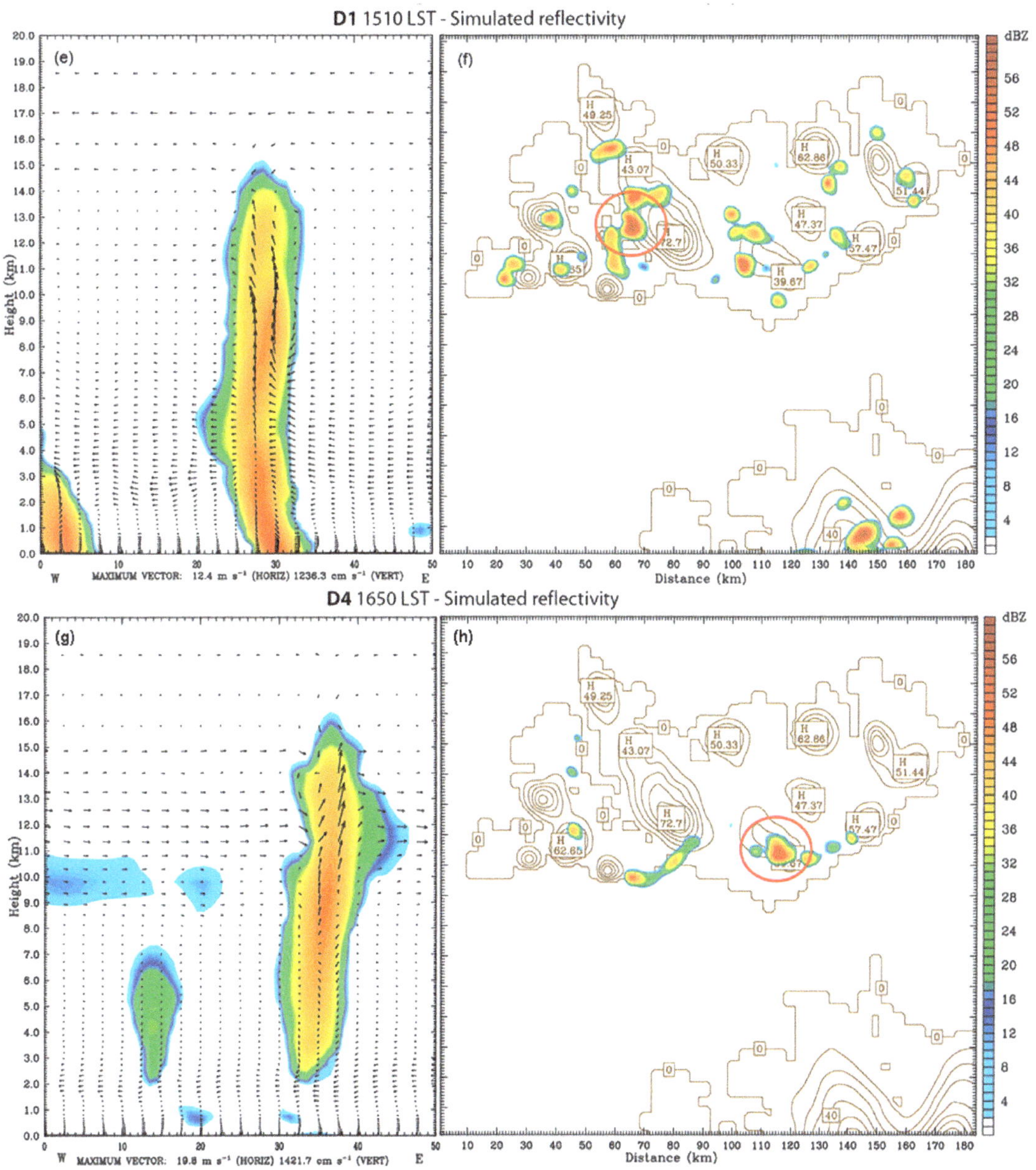

Figure 5. Sections at the maximum development of the Hector events N27 (double-cell), D1 and D4. **(a, c, e)** Simulated vertical radar reflectivity (dBZ; filled color) and vertical wind; the section is taken longitudinally along the red circle reported in the right panel. **(b, d, f)** Horizontal radar reflectivity (dBZ; filled color) and topography (brown).

eral hours before the development of Hector. For the single events and for the first cell of double events, the analysis is performed at 08:30 LST, whereas for the second cell of double events, the ending time of the first cell is taken as a reference: 15:10 LST for N30 and 15:30 LST for N27.

All the meteorological parameters are analyzed versus total condensate and mean lifting (vertical velocity at 500 m) as in Crook (2001). The vertical velocity is extracted 3 h before the maximum development of Hector and is averaged all over the island surface. The total condensate, conversely, is averaged within the volume encapsulating Hector at the maximum stage. CAPE, surface wind speed and direction are also averaged all over the surface of the Tiwi Islands. With the aim of understanding the convective response to the flow direction either along the major (90°) or minor axes (0°) of the islands, the wind direction is projected in the first quarter of the wind rose.

Crook (2001) analyzed the vertical velocity wind with respect to the surface wind velocity and direction (Fig. 7b, c in Crook, 2001), related to shear and no-shear conditions. Hence, Fig. 8 shows the results by Crook and the Hector events, indicated by gray lines, and 12 symbols, represent-

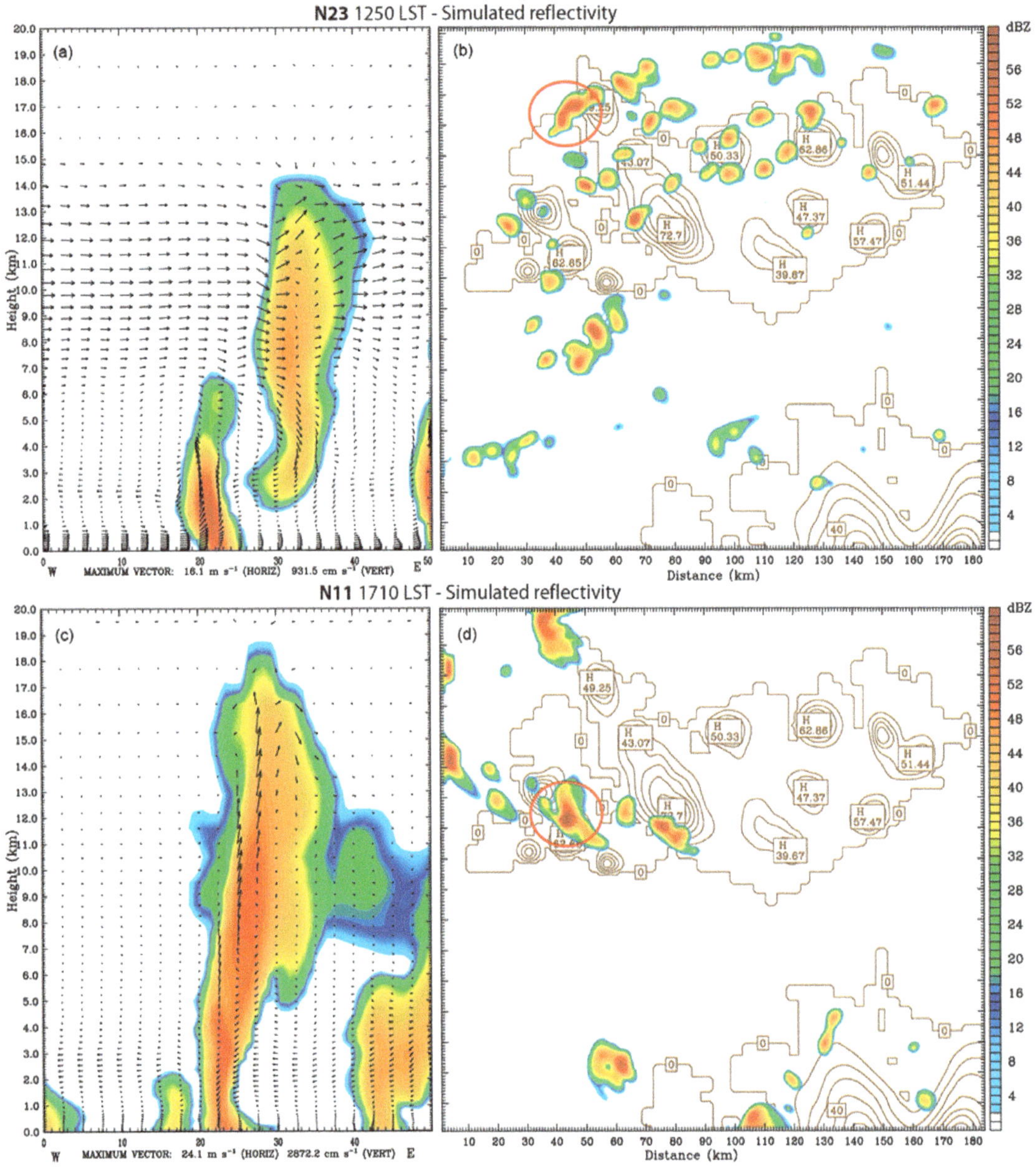

Figure 6.

ing the Hector events. A few events closely follow the shear (F6, N20 and the second cell of N27 and N30) and no-shear (D1, N29) lines (Fig. 8a) or they belong to the right zone (i.e., no shear zone for the first cell of N27), confirming for the shear events the correlation between the decrease of the vertical velocity and the surface wind increase. A few cases do no show any specific signal, probably because the real atmosphere is more complex than the ideal one, and more than one parameter contributes to the vertical lifting, as for example the topography or the convergence line. Therefore, the ideal response can be used to sort the events: the closer the position of the event in the diagram is to the "ideal" one, the more the meteorological parameter contributes to the convective strength.

The second cells of both the double structure events show a stronger surface wind speed than the corresponding first cell; this is due to the gust front associated with the downdraft of the previous convective cell. This is why the double-cell events are not aligned with the others.

A similar analysis is performed for the vertical velocity as a function of the surface wind direction. The results clearly show (Fig. 8b) that most of the events are located in the right position except for four events: D1, N17, N29 and N23, suggesting difficulties in the real atmosphere to completely separate the two regimes. However, it roughly confirms the

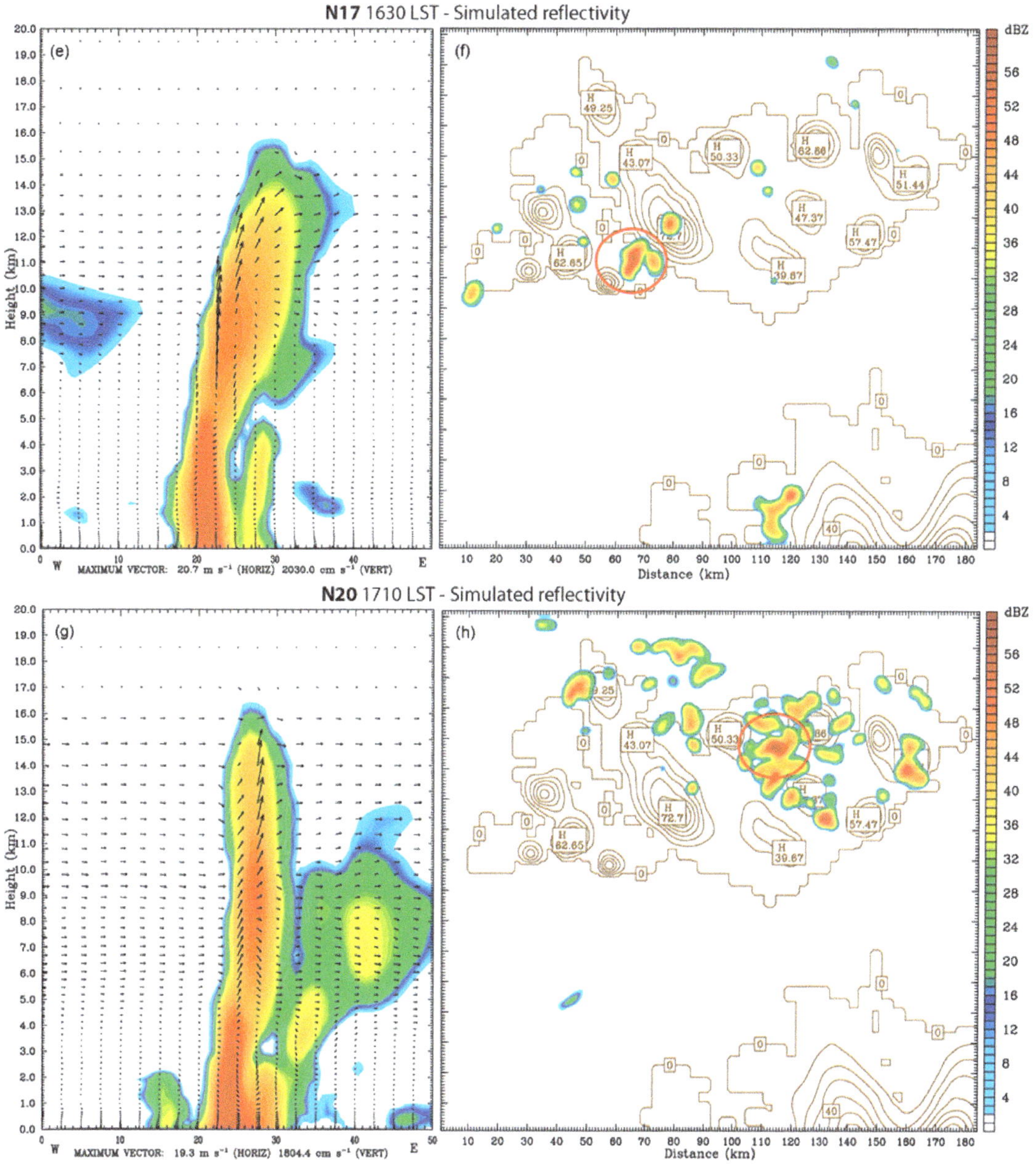

Figure 6. Sections at the maximum development of the Hector events N23, N11, N17 and N20. **(a, c, e)** Simulated vertical radar reflectivity (dBZ; filled color) and vertical wind, the section is taken longitudinally along the red circle reported in the right panel. **(b, d, f)** Horizontal radar reflectivity (dBZ; filled color) and topography (brown).

increase of the vertical velocity when the flow is eastward (90°). This supports the hypothesis of greatest lifting when the flow is along the major axis of the island as assessed by Carbone et al. (2000) and Crook (2001). It is useful to highlight that the wind direction related to the second cells of the double events does not show a clear signal, because at this time, the sea breeze regime is either well developed or destroyed, and "leftovers" from the first cell affect the environment. Therefore, their positions in the graph have an uncertainty larger than the one for the single-cell events.

The analysis of the total condensate versus wind speed (Fig. 9a) shows that most of the Hector events are aligned along a line with a slope close to that of Crook's study, but an intercept that is smaller (gray dotted line in Fig. 9a), suggesting a sort of bias between Crook's ideal and the real atmosphere. This disagreement can be explained by speculating that the real events need weaker surface wind than the ideal ones to produce the same total condensate. Based on this hypothesis, a new reference line can be assumed, then only two events appear outside of the distribution, the second events of the double Hector N27 and N30 (Fig. 9a, white square

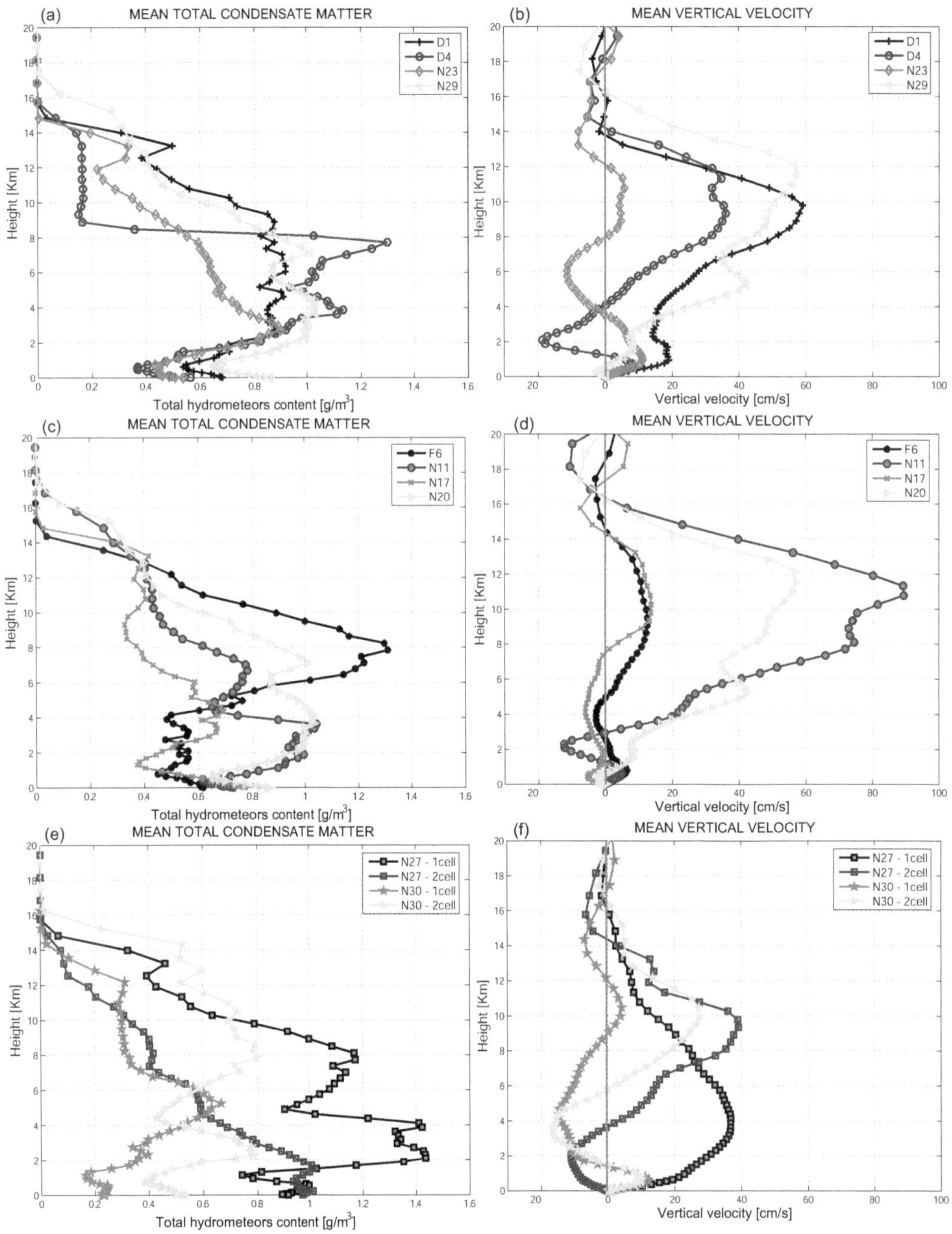

Figure 7. MM5 mean vertical profiles of cloud total condensate (sum of all hydrometeors) and vertical velocity at the maximum stage. Profiles are averaged for each layer within the volume encapsulating Hector.

and white star, respectively). Both events have larger wind speed than the expected one on equal terms of normalized total condensate. The absolute maximum of total condensate is reached by the first cell of N27, whereas the second cell shows approximately a 65 % of the first cell total condensate. On the contrary, for N30 the second cell is stronger than the first one, in terms of cloud total condensate. This is partly due to the leftovers of the previous cell because of the very short time interval (1 h and 20 min) that occurred between the two maxima; whereas for N27, the second cell develops 3 h later than the first one, making the two cells more independent than the previous event. Hence, the hypothesis of the increase of the total condensate as the surface wind speed decreases (Fig. 9a) is still confirmed, but below $4\,\mathrm{m\,s^{-1}}$ for all single-cell events and the first cell of double ones. Based on the results of the dry linear and non-linear models, Crook (2001) assessed that the relationship between convective strength and low-level convergence (i.e., surface

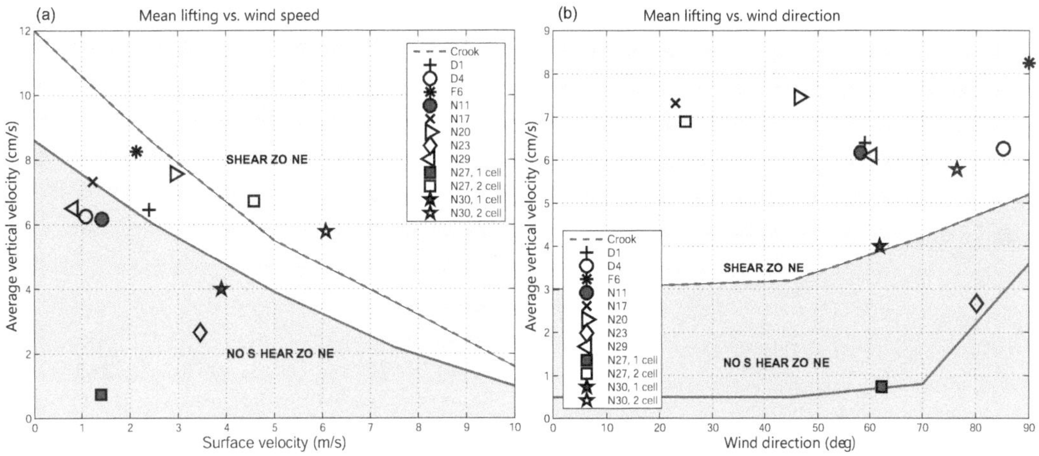

Figure 8. Crook's test: panels **(a)** and **(b)** show vertical velocity at 500 m extracted 3 h before the maximum development versus surface wind speed **(a)** and surface wind direction **(b)**, both extracted at the start time. The figure reports two different regimes with a sheared (light gray, dashed line) and unsheared flow (dark gray, solid line) as studied by Crook.

wind speed) is not strictly monotonic because the convective strength did not continue to increase as the flow decreases below 4 $\mathrm{m\,s^{-1}}$. The 12 Hector events also reproduce a monotonic relationship also below 4 $\mathrm{m\,s^{-1}}$ (Fig. 9a). This discrepancy between the simulations of the real events and the Crook experiments is not surprising and it may be due to the differences between the model's assumptions and the use of an idealized sounding in the Crook study.

All Hector events, except the second cell of N27 and N17, confirm the increase of the total condensate if the low-level flow is along the major axes as shown by the plot of the total condensate versus the wind direction (Fig. 9b), but a spread along the Crook "theoretical" line is found for the real events. As for the mean lifting versus wind direction (Fig. 8b), the major difference between Crook ideal behavior and real behavior is found for the surface wind, of which direction is close to the minor axes. However, Crook's hypothesis of maximizing the low-level convergence if the flow is aligned along the major axis of the Tiwi Islands is confirmed.

Finally, similarly to what was done by Crook (2001), the total condensate versus the low-level moisture, expressed in terms of CAPE, is analyzed. The Hector events do not show a clear signal, but they show a slight increase of the total condensate as the CAPE increase is found, except for the first cell of N27. This is completely outside of Crook's line (Fig. 9c, gray square). Moreover, on the contrary to what was found by Crook (Fig. 13a in Crook, 2001), no maximum is found for the total condensate versus CAPE because of the lack of Hector's values around the theoretical maximum; hence it is not possible to assess its occurrence.

4.1 Main features of the Hector events

Based on the previous analysis and on the brief summary of the main characteristics of these Hector events given in Table 1, some important highlights can be inferred using the surface wind speed, surface wind direction and CAPE. Each event seems to be driven by particular meteorological conditions, whose contributions to the convective strength have been estimated and summarized in Table 2. This computation is performed by evaluating the distance between the real point and its corresponding ideal one. The lines reported in Figs. 8 and 9 show the ideal conditions for the development of Hector. As the real points get closer to the ideal lines, the more the meteorological conditions are suitable for the development of Hector. Once the percentages from surface wind speed, surface wind direction and CAPE are obtained, the total contribution from the three parameters is normalized to 100. Therefore, the percentage of influence for each parameter is calculated and the following conclusions can be drawn.

1. D1 and N29 events, both type B, produce a large amount of total condensate (respectively 75 and 82 % of the maximum), suggesting a strong convective strength. A similar contribution ($\approx$ 33 %) to the development of Hector is found for the three meteorological parameters, with a prevalence of favorable surface flow direction for the first event and slow surface wind velocity for the second one (Table 2).

2. D4, F6, N17, N23 and the first cell of N27 have a total condensate ranging between 55 and 65 % of the maximum (except for the first cell of N27 that is the maximum). The convective development is due mainly (percentage of its influence from 47 to 62 %) to just one meteorological parameter: surface flow direction for D4, F6 and N24, and slow wind speed for N17 and N27.

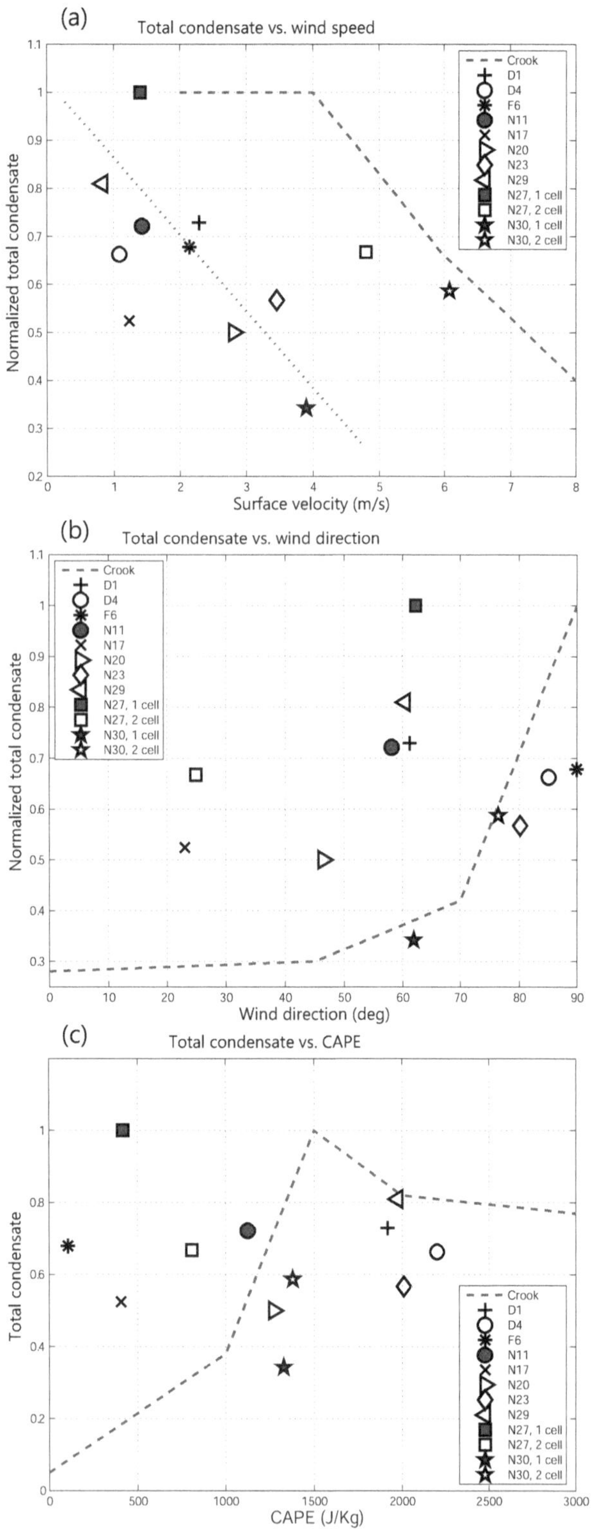

Figure 9. Crook's test: normalized total condensate extracted at the maximum stage and averaged into the Hector volume versus surface wind speed **(a)**, surface wind direction **(b)** and CAPE **(c)**, extracted at the start time. The dashed gray line reports the Crook ideal trend, and the light gray dotted line in panel **(b)** reports the derived real trend.

Table 2. Percentage of influence for the meteorological parameters on the convective development of the Hector events.

Events	Surface wind speed	Surface wind direction	CAPE
	(%)		
D1	33	35	32
D4	26	47.5	26.5
F6	39	57	4
N11	27	34	39
N17	62	18	20
N20	32	25.5	42.5
N23	13	50	37
N27-1	47	38.5	14.5
N27-2	42	20	38
N29	37	31	32
N30-1	17.5	36.5	46
N30-2	22	38	40

Another important contribution for three of these events is also found: the surface wind speed for F6, the low-level moisture for N23 and the wind direction for the first cell of N27.

3. The development of N11 and both cells of N30 is sustained by the "right" CAPE value (Crook, 2001) and by the wind direction. Indeed, the flow blows along the major axis of the Tiwi Islands, maximizing the low-level convergence. The N30 (type B) produces a larger amount of total condensate (58 % for the second cell vs 35 % of the first one) than type A, suggesting a stronger convective strength; the most important parameter to justify it is the low-level wind shear. Indeed, Crook's analysis allows the change in the regime from no-weak shear for the first cell to strong shear for this second cell to be highlighted; whereas the other meteorological parameters are similar for both. The N30 (type A) produces the smallest amount of total condensate, suggesting a weak convective strength; several meteorological parameters justify this. The large surface velocity, the wind direction and a weak shear do not sustain Hector, whereas CAPE is the only parameter which acts positively.

4. N20 and the second cell of N27 are characterized by a total condensate around 50–60 % of the maximum content but the events present a strong mean lifting. The convective strength for both events mostly depends on two parameters with different weights. The main contribution (approximately 42 %, see Table 2) comes from the slow wind speed for N27 and CAPE for N20. Moreover, slow wind speed and CAPE contribute 35 % to N20 and N27 development, respectively.

In summary, the previous analysis highlights the role of the meteorological parameters in defining the Hector convective strength, but it does not allow a specific parameter to be highlighted, as Crook assessed for the low-level moisture, to establish the Hector typology. Moreover, the relationship between both the total condensate and mean lifting and several meteorological parameters for the Hector events confirms what was found by Crook (2001) within this sensitivity analysis.

5 Conclusions

In this study, 12 Hector events are analyzed using MM5 model simulations, with the aim of highlighting the main meteorological parameters and their role in triggering convection. A brief meteorological analysis of the events is performed using CAPE, water vapor, wind speed and direction and typology of convection. Moreover, a comparison in terms of mean total condensate vertical profiles and mean vertical velocity at the maximum development is carried out. The applicability of Crook's hypothesis to real cases is explored, verifying the linear relationship between both the convective strength and the total condensate versus the low-level moisture, expressed in terms of CAPE, surface wind speed and direction. Crook's tests allow the following to be concluded.

1. The strength of convection, in terms of mean lifting and total condensate, increases if the wind direction tends to be parallel to the major axes of the Tiwi Islands and if the wind speed surface decreases.
2. Crook's assumption on the low-level moisture as the parameter that differentiates between type A and B modes of convection is not confirmed.
3. The previous hypothesis is verified for the two N30 Hector cells, where the second cell (type B) has a larger low-level moisture and convective strength than the first cell (type A). This would suggest the applicability of a type A or B classification based on the low-level moisture for events developing in the same meteorological conditions only. That means it cannot be generalized to all real cases.
4. The meteorological parameters contributing to the Hector development are only one with an influence coefficient up to 47 % for five events, and two with influence coefficients from 32 % to 46 % for five events. All the parameters contribute by a percentage of 31–37 % for only two cases.

Thanks to their simple orography and shape, the Tiwi Islands can be used as a laboratory to study the triggering factors contributing to convection. Hence, in this context, this study will allow for a better understanding of different meteorological parameters occurring at the onset of convection, even in complex orography regions.

Acknowledgements. NCAR is acknowledged for the MM5 model. ECMWF is acknowledged for data analysis.

References

Carbone, R., Wilson, J., Keenan, T., and Hacker, J.: Tropical island convection in the absence of significant topography. Part I: Life cycle of diurnally forced convection, Mon. Weather Rev., 128, 3459–3480, 2000.

Chemel, C., Russo, M., Pyle, J., Sokhi, R., and Schiller, C.: Quantifying the Imprint of a Severe Hector Thunderstorm during ACTIVE/SCOUT-O3 onto the Water Content in the Upper Troposphere/lower Stratospere, Mon. Weather Rev., 137, 2493–2514, doi:10.1175/2008MWR2666.1, 2009.

Crook, N.: Understanding Hector: the dynamics of Island thunderstorm, Mon. Weather Rev., 129, 1550–1563, 2001.

Dauhut, T., Chaboureau, J., Escobar, J., and Mascart, P.: Large-eddy simulations of Hector the convector making the stratosphere wetter., Atmos. Sci. Lett., 16, 135–140, doi:10.1002/asl2.534, 2014.

Dudhia, J., Dave, G., Manning, K., Wang, W., and Bruyere, C.: MM5 Modeling System Version 3, PSU/NCAR Mesoscale Modelling System Tutorial Class Notes and User's Guide, Mesoscale and Microscale Meteorology Division of National Center for Atmospheric Research, Boulder, CO, USA, 2004.

Faccani, C., Ferretti, R., and Visconti, G.: High-resolution weather forecasting over complex orography: sensitivity to the assimilation of conventional data, Mon. Weather Rev., 131, 136–154, 2003.

Ferretti, R. and Gentile, S.: A study of the triggering mechanisms for deep convection in the Tropics using a Mesoscale Model: Hector events during SCOUT-O3 and TWP-ICE campaigns, Atmos. Res., 93, 247–269, 2009.

Gentile, S., Ferretti, R., and Marzano, F.: Investigating Hector convective development and microphysical structure by high resolution model simulations, ground-based radar data and TRMM satellite data, J. Atmos. Sci., 71, 1353–1370, 2014.

Golding, B.: A numerical investigation of tropical island thunderstorms, Mon. Weather Rev., 121, 1417–1433, 1993.

Keenan, T. and Carbone, R.: A preliminary morphology of precipitation system in tropical northern Australia, Q. J. Roy. Meteor. Soc., 118, 283–326, 1992.

Ramage, C.: Role of a tropical "Maritime Continent" in the atmospheric circulation, Mon. Weather Rev., 96, 365–370, 1968.

Saito, K., Keenan, T., Holland, G., and Puri, K.: Numerical simulation of the diurnal evolution of tropical island convection over the maritime continent, Mon. Weather Rev., 129, 378–400, 2001.

Zhu, M., Connolly, P., Vaughan, G., Choularton, T., and May, P.: Numerical simulation of tropical island thunderstorms (Hectors) during the ACTIVE campaign, Meteorol. Appl., 20, 357–370, doi:10.1002/met.1295, 2013.

3

Mobile augmented reality in support of building damage and safety assessment

W. Kim, N. Kerle, and M. Gerke

Faculty of Geo-Information Science and Earth Observation, University of Twente, Enschede, the Netherlands

Correspondence to: W. Kim (jinsilto@gmail.com)

Abstract. Rapid and accurate assessment of the state of buildings in the aftermath of a disaster event is critical for an effective and timely response. For rapid damage assessment of buildings, the utility of remote sensing (RS) technology has been widely researched, with focus on a range of platforms and sensors. However, RS-based approaches still have limitations to assess structural integrity and the specific damage status of individual buildings. Structural integrity refers to the ability of a building to hold the entire structure. Consequently, ground-based assessment conducted by structural engineers and first responders is still required. This paper demonstrates the concept of mobile augmented reality (mAR) to improve performance of building damage and safety assessment in situ. Mobile AR provides a means to superimpose various types of reference or pre-disaster information (virtual data) on actual post-disaster building data (real buildings). To adopt mobile AR, this study defines a conceptual framework based on the level of complexity (LOC). The framework consists of four LOCs, and for each of these, the data types, required processing steps, AR implementation and use for damage assessment are described. Based on this conceptualization we demonstrate prototypes of mAR for both indoor and outdoor purposes. Finally, we conduct a user evaluation of the prototypes to validate the mAR approach for building damage and safety assessment.

1 Introduction

Natural disasters bring about economic damage and victims, and despite signs of a decline in the number of events in recent years, between about 300 and 500 events are recorded worldwide annually, with total damage frequently exceeding USD 100 billion (Kerle and Alkema, 2012). After the occurrence of a disaster event, in particular in urban areas, rapid and accurate building damage and safety evaluation is critical. The former only provides information on the physical level of impairment of a structure, while the latter assesses to what extent buildings can be safely entered, which also takes the state of adjacent buildings into account. Rapid building assessment plays a major role for initiating effective emergency response actions (Brunner et al., 2010), and also constitutes the basis for post-disaster needs assessment (PDNA). Building safety inspection is related to building usage after disaster events; hence time delays before buildings can be declared safe again also translate into economic losses (Dai et al., 2011). Thus, a comprehensive and rapid assessment of building stock following disaster events is a prerequisite for effective rehabilitation and reconstruction.

Remote sensing (RS) technology has been widely used for rapid building assessment, with many sensor/platform combinations having been tested and with principal focus on seismic damage (e.g. Dell'Acqua and Gamba, 2012). Although RS technology has improved the speed and accuracy of damage mapping at a macro level, e.g. at a city block scale, challenges persist at more detailed, per-building levels. For instance, while RS-based approaches can detect the extreme damage categories (none or complete) quite accurately, detection of intermediate damage states continues to be marked by great uncertainty (Kerle, 2010; Fernandez Galarreta et al., 2015).

Given the problems in assessing damage and safety and structural integrity of individual buildings with RS data (Kamat and El-Tawil, 2007), ground-based appraisals remain necessary. This is usually conducted by reconnaissance teams comprising certified inspectors or structural engineers, and using assessment guidelines such as the ATC-20 field manual (ATC, 2005) or the European Macroseismic Scale (EMS-98). For example, in the aftermath of the 2010 Haiti earthquake, some 300 engineers mapped approximately 400 000 buildings using an adapted version of ATC-20, focusing on determining the residual vertical capacity of each structure, which even with airborne multi-perspective oblique RS images is challenging to determine (Gerke and Kerle, 2011; Kerle and Hoffman, 2013; Dell'Acqua and Gamba, 2012). ATC-20 classifies the building state into safe, unsafe and limited use, while EMS-98 uses damage scores ranging from 1 (no damage) to 5 (heavily damaged). However, ground surveys are inefficient in terms of cost and time of evaluation. In addition, the ground surveys rely on human skills and knowledge, so that human-induced errors can affect the quality of data during the mapping or subsequent transcription and analysis process (Kerle, 2010).

Ground-based surveys are mainly hindered by (i) limited site access, (ii) problems with orientation in the field with landmarks and street signs also having been affected and (iii) the lack of efficient tools to provide pre-disaster reference data to facilitate a pre- and post-disaster data comparison, as is done in RS-based change detection. Therefore, this study proposes an approach based on mobile augmented reality (mAR) that can efficiently deliver pre-disaster reference data of various types and levels of complexity, provide location and navigation information, but also support additional situational awareness, such as by warning of the presence of nearby hazardous facilities or substances. Augmented reality (AR) superimposes computer-generated graphics or contents on real-world imagery, for example as captured by a camera of a mobile device. The main purpose of AR is to combine real and virtual world information and provide real-time interaction to users (Azuma, 1997). While a conventional data overlay (e.g. as done in geographic information systems (GISs)) replaces reality with virtual data, AR supplements reality (Azuma, 1997). The roots of augmented reality go back to the 1990s (Valentini et al., 2010), and the first prototype of mAR was developed by Feiner et al. (1997) to provide tour information for buildings based on 3-D graphical information.

The main objective of this study is to demonstrate an mAR concept that can increase the efficiency and safety of ground-based building damage and safety assessment. We first identify the main obstacles of ground-based building assessment that are directly connected with the user requirements. Based on this analysis a conceptual framework is defined, which defines specific methods and processes to apply mAR for building damage and safety assessment. In order to validate the efficiency of mAR, a prototype is designed and implemented, showing how information ranging from simple labels to complex 3-D building models can be provided to a user. Based on an online survey that demonstrates the prototype, we analyse to what extent current limitations of ground-based damage and safety assessment can be assisted with mAR, qualitatively and quantitatively.

2 Overview of previous work

Building damage and safety assessment was the focus of a number of earlier studies. Much research has focused on remote-sensing-based damage inventorization (for a recent overview see Dell'Acqua and Gamba, 2012), while work focusing on assessing building safety is comparatively scarce. Altan et al. (2001) used ground-based photogrammetry to determine critical damage and deformation that may compromise structural integrity, though in an approach that required entering a potentially unstable structure. Schweier and Markus (2006) analysed 3-D CAD models derived from laser scanning data to identify geometric changes between pre- and post-disaster buildings, including volume or height reduction, as well as orientation changes, which were linked to stability. The main challenge of their method is to obtain pre- and post-disaster 3-D data at sufficient accuracy to support change quantification. Curtis and Mills (2012) developed a video-based field reconnaissance system that allows tornado damage to be mapped, though no pre-event reference information is provided. Adams et al. (2009) introduced VIEWS (Visualizing Impacts of Earthquake With Satellites). This is a video-based scene assessment system that not only uses satellite imagery, but supports it with video data and photos that are taken in the field after a disaster. The captured video data and photos are georeferenced, hence the user can combine and compare the reference satellite data with video footage and photos for a comprehensive damage assessment. Although VIEWS provides an interface that can readily visualize all data on one screen, it requires post-processing of the video data and photos. Moreover, the actual damage assessment task is not meant to be done in the field.

Only few studies have investigated the utility of AR for post-disaster situations. Leebmann (2004) designed software and hardware of an AR system to overlay disaster-related information onto the real-world view through a digital video camera. Leebmann's research demonstrated that AR is accurate enough to register virtual information onto real objects in the context of disaster response. At that time focus was mostly on how to improve matching accuracy between real and virtual objects, and how to integrate hardware components such as video camera, GPS and laptop. Tsai et al. (2012) proposed mAR to support evacuation of nuclear power plants following events such as earthquakes. The system displays direction and distance to shelters on a live camera image taken with a smartphone, using Google Maps and the electronic compass function. Boddhu et al. (2013) introduced context-aware event detection with mAR for first responders. This system gathers and analyses data that are distributed by social networks such as Twitter and Facebook. Because SNs can deliver important information, such as the user's location, in real time, they can play a role as a sharing tool during or after a disaster. In gathering and analysing these spatiotemporal data, the system tried to support first responders in managing the disaster situation (Boddhu et al., 2013). To display analysed data for first responder in the field, the system used AR that superimposes virtual data regarding historic or live events occurring around a user's location. Wani et al. (2013) proposed a workflow for collaborative AR in a fire emergency situation. In the workflow a wearable computer was adopted for the collaboration between experts in an office and the operator in the field. For instance, a medical doctor in the office can, through the AR device, interactively indicate parts of a patient's body which are in need of an initial treatment.

While the above systems simply focus on the visualization of exiting information, Kamat and El-Tawil (2007) tried to generate new information from AR to support seismic building damage assessment. Their approach focused on comparing baseline information from pre-disaster 3-D building data with the actual post-event structure to identify differences, detecting deformation through the interstory drift ratio (IDR). For the IDR measurements, the pixel offset between the image of the real building and the 3-D model baseline was counted, using both sensor-based and vision (marker)-based AR. The former uses the sensors included in mobile devices, such as GPS, compass and gyro sensors, to obtain the user's location and orientation, while vision-based AR analyses images taken by the camera identify unique features of real objects, using computer-vision technology (Rabbi and Ullah, 2013). While Kamat and El-Tawil (2007) measured displacement between the pre-disaster 3-D model and a real building, Dong et al. (2013) determined IDR without a pre-disaster 3-D model, but instead also using vision-based AR. They used edge detection and corner detection methods (e.g. a line segment detector) to extract building baselines from reference images of an affected building. Subsequently, the generated baseline model was superimposed on the real (post-event) object to calculate the IDR. Although these approaches provide efficient methods for damage assessment, they have traditionally required an array of separate pieces of equipment and sensors (e.g. laptop, camera, GPS). In addition, their approaches only provide one indicator, IDR, which is too narrow to improve the overall assessment procedure of building damage and safety in the field. For instance, standard evaluation forms used in rapid damage assessment ask assessors to provide comprehensive information, such as building address, type and height, which existing systems cannot provide. Additionally, response to large disasters is marked nowadays by a large variety of stakeholders, lack of cooperation and coordination and limited situational awareness (Kerle, 2013). mAR has the potential to improve this situation by not only allowing orientation or providing a certain type of reference information. Instead, it can provide pre-event reference data on a number of scales of complexity (from simple labels of where buildings used to be or the function of a given building, to complex geometry fitted over building remains), in addition to data on environmental parameters that directly affect the safety of the responder, or information on other response teams in the area or on prior assessment work. Given the trend in miniaturization in mobile devices, some of the mAR features discussed in this work will likely eventually be available for AR platforms such as Google Glass, which has great potential to provide critical, customized information to disaster responders, with the additional benefit of hands-free operation.

3 Conceptualization of mAR for building damage and safety assessment

First we define the overall concept of mAR for building damage and safety assessment. It shows how mAR can support ground-based building assessment, and which methods and data sets are required by different users, who we also review and categorize. The stakeholders we envision for this system comprise structural or geotechnical engineers, architects, search and rescue teams, or professional groups who carry out building damage and safety assessment in the field (see for example Ghosh et al., 2011). Based on the defined concept, prototypes of indoor (marker-based) AR and outdoor (sensor-based) AR are developed. Subsequently the concepts and prototypes are evaluated through a user evaluation.

3.1 Overall concept of mAR for building damage and safety assessment

To apply mAR for building damage and safety assessment in the field, we conceptualize the relationship between AR and the different levels of assessment as shown in Table 1. AR superimposes various reference data sets onto real buildings, which include simple location information (e.g. building coordinates), attributes (e.g. building information), linked attributes (e.g. street-level imagery) and geometry (e.g. 3-D building frame).

To conceptualize the utility of mAR in a post-disaster situation for different complexity levels, we use a term, level of complexity (LOC), which ranges from LOC1 (low complexity) to LOC4 (high complexity; Table 1). The complexity levels relate to increasing number of data dimensions, levels of accuracy, etc. LOC1 simply indicates the existence of a building. This is useful in case of a completely collapsed building, or for seismic or tornado damage, where vast debris fields challenge the identification of individual buildings that existed before. After recognizing the existence of a building, LOC2 provides information on the type of structure and basic attributes (e.g. building classification, materials, height and story). Subsequently, LOC3 provides pre-event reference imagery (e.g. photos of the building, or Google Street View imagery, GSV), while LOC4 includes pre-event 3-D information. This may include imagery of the building exterior,

Table 1. Level of complexity (LOC) in augmented reality.

	LOC1	LOC2	LOC3	LOC4
Purpose	Visualization of existing information			
Definition	Existence of building	Building attributes	Linked building attributes	Building geometry
Concept	Was it there before?	What type of building existed before?	What did it look like before?	What was the geometry before?
Data sets	Georeferenced building database	Building information	Photos Video clips Street-level imagery	3-D building model, possibly model of interior elements
Example	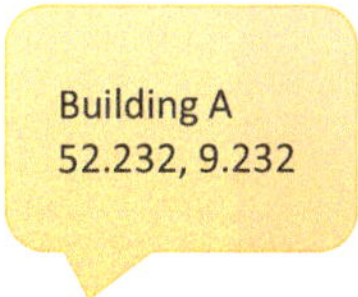			

but also data on building frames and internal structures. Each LOC level is defined in detail in terms of procedure, flow, accuracy and uncertainty in the next section.

3.1.1 Simple location-based (LOC1) and attribute-based (LOC2) visualization

The purpose of LOC1 is to provide simple but essential information on the (prior) existence of a structure. Meanwhile, LOC2 information helps to reduce the building assessment duration, since the most time-consuming task is building type classification (Flesch, 2007). Both concepts, including the required data processing steps, are illustrated in Fig. 1.

Procedure and flow

Building information can be extracted from several sources, such as cadastral databases maintained by local governments, private building management companies and emergency response organizations. Following the 2010 Haiti earthquake, organizations such as the Remote Sensing Laboratories at the University of Zurich and Swisstopo provided extensive building data (Corbane et al., 2011). A common challenge is that those data sets come in different formats, so that data need to be preprocessed and converted in order to be processed through AR. In our LOC1 concept, the data type is mainly text format that is very irregular with respect to data providers. Therefore, it is important to refine raw data that can be interpreted by AR. In terms of data clutter on the screen, AR needs to adjust the display range (radius), i.e. AR only displays information that is within a given radius from the user location. Lastly, LOC1 and LOC2 are interactively connected, as shown in Fig. 1. Initially, only the building name is displayed on a target building to let the user recognize the building existence (LOC1). If the user needs additional building attributes, LOC2 information is provided on a touch basis.

Accuracy

As mentioned earlier, text format data are usually irregular, necessitating data refinement, including georeferencing. Both processes may cause data changes or loss. However, text information is a form of point data, and as long as it is displayed within the building boundary, it is still interpretable. Hence, the georeferencing only needs to fall anywhere within the building footprint.

Uncertainty

Uncertainty refers to uncontrollable and unpredictable parameters that can affect a given procedure. Irregularity of data sources and formats can affect data availability and usability and thus all subsequent LOC processes. Information sharing itself can be restricted because of political and privacy issues. GPS signal availability may be compromised in densely built-up areas, undermining the ability to associate database information with a given structure in the field, also affecting all LOCs.

3.1.2 Linked attribute-based (LOC3) and geometry-based (LOC4) visualization

The concept of LOC3 is to provide multimedia data, such as photos, video clips and sounds, that are linked (related) to a specific building. The data sets needed for LOC3 can be obtained through internet services such as YouTube, Flickr or GSV.

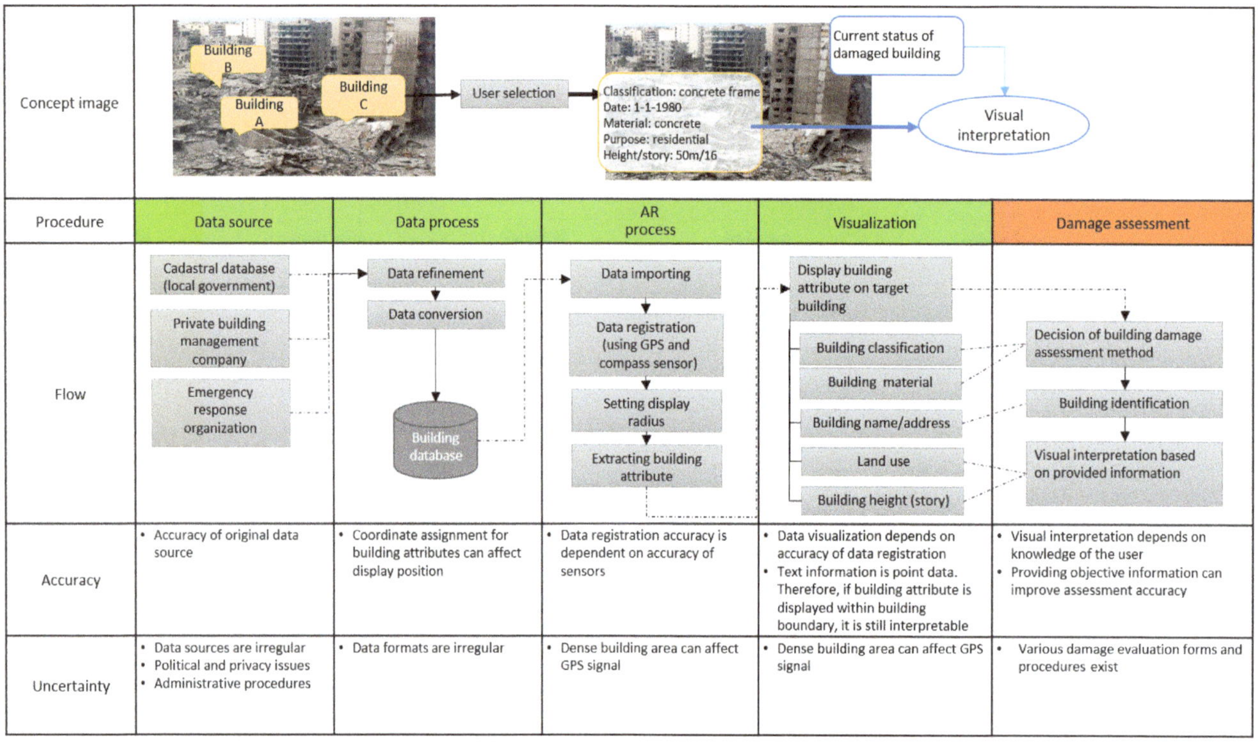

Figure 1. The concept of simple location-based (LOC1) and attribute-based visualization (LOC2), data processing workflows, and aspects that determine accuracy and uncertainty of the superimposed information.

The concept of LOC4 is to provide building geometry that includes 3-D information of the structure's exterior and possibly the interior. The 3-D building exterior shows the building's original shape and frame structure, while 3-D building interior displays features within the building, such as stairs, elevators, rooms and so on, facilitating better damage interpretation, or evacuation planning. The detailed data sources and processing steps are shown in Fig. 2.

Procedure and flow

LOC3 mainly depends on external APIs to extract data from service providers. These APIs require query parameters such as coordinates and keywords, meaning the results will vary depending on a specific search string used. After locating and extracting a data set, those data need to be converted to a data format understood by the AR platform. For instance, Metaio API used in this study adopts MPEG4 and 3G2 format to display video clips while others use AVI format. Lastly, AR imports the converted data. Meanwhile, GSV employs its own registration process – each panorama is georeferenced; hence GSV data can be expected to be directly integrable. Therefore, GSV can be visualized in a separate window without an AR process. With LOC3, a user can call up pictures of buildings, as well as of the surrounding area.

LOC4 uses 3-D data sets generated by or extracted from various sources that can be grouped into primary and secondary data sets. The former contain 3-D buildings, used directly without any data generation processes, for instance, 3-D buildings extracted from Google Earth or from a building management company. Conversely, secondary data sets require a process to generate 3-D building information, for example from 2-D building footprints, or lidar data. The secondary data sets are typically processed on a PC, and subsequently converted to the 3-D data format required by the AR platform. In order to visualize these 3-D buildings through AR, the following parameters should be defined: coordinates, scale, rotation of building, exchange data format and level of detail.

Accuracy

Accuracy of data sources for LOC3 depends on the query condition to extract meaningful data from APIs (e.g. keyword). Additionally, the locational accuracy of geotagged photos and video clips is not guaranteed. Although GSV does not rely on the registration process of AR, it also uses its own registration process, meaning that data accuracy can be affected by accuracy of the mAR device sensors. Multimedia files are basically point data that can be meaningfully associated with a building, as long as they are displayed within the building boundary.

The accuracy of the original 3-D data and the data conversion process can affect both data source and data process

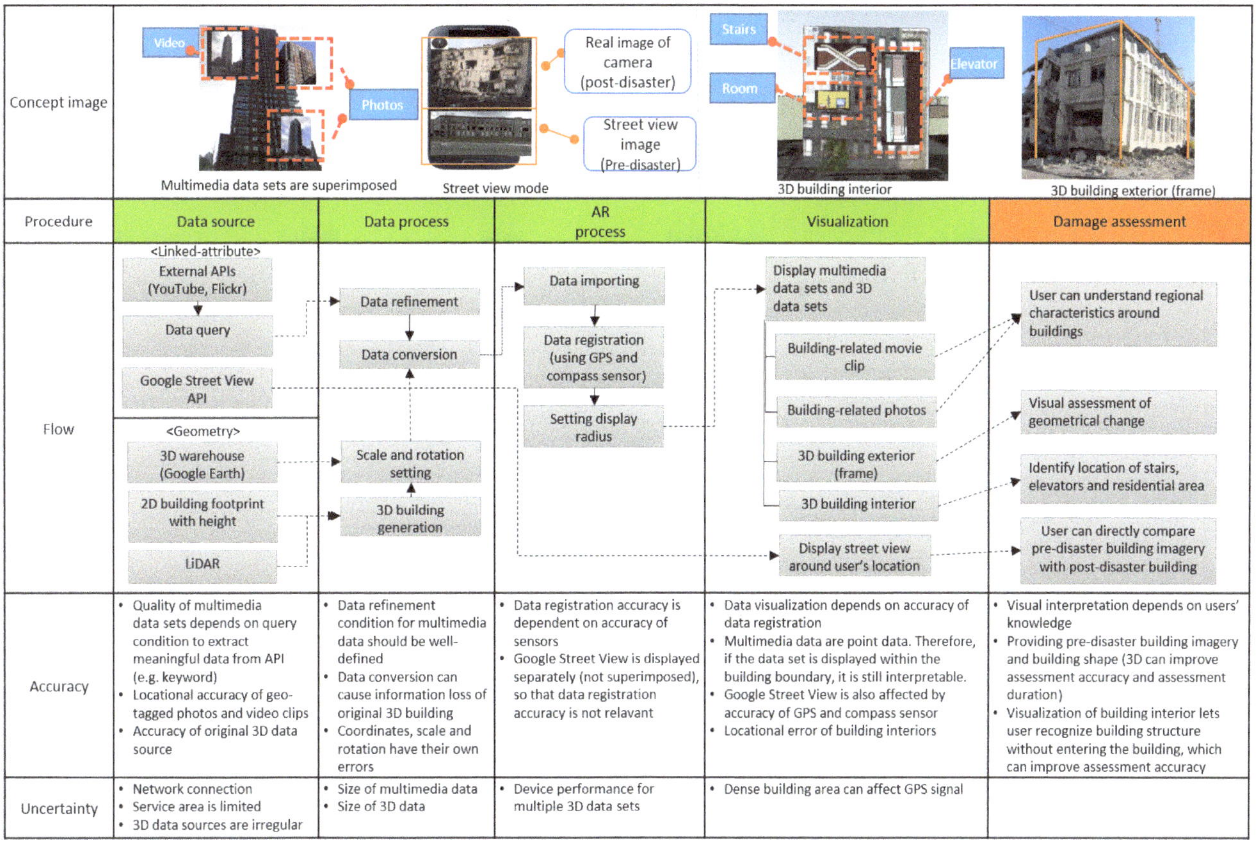

Figure 2. The concept of linked attribute-based (LOC3) and geometry-based visualization (LOC4).

accuracy for LOC4. The accuracy of the primary data set is a direct function of the accuracy of the original data generation, while the preprocessing or conversion steps listed above determine the accuracy of the secondary data sets. For the AR process of building exterior and interior, as long as a user can identify which 3-D building belongs to which real building, the accuracy of data registration is acceptable (see for example the concept image for LOC1 in Fig. 1). Data meant to facilitate a more detailed damage analysis (e.g. inclination; Dai et al., 2011), such as building frames, need to be more accurately registered.

Uncertainty

In the concept of LOC3, the main uncertainty stems from the network connection and service coverage, since the data sets are extracted by external APIs through a live internet connection. Another uncertainty results from data size, since storage on a mobile device is limited. Although in a disaster situation network connection might be limited, several alternatives for internet access are already in practical use. For instance, in the case of the Great East Japan Earthquake, 2011, portable satellite internet Earth stations and satellite phones were installed to use satellite-based internet connection (Sakurai et al., 2014).

In the concept of LOC4, file size of the 3-D data can affect the performance of both data process and AR process. In addition, the number of 3-D buildings that can be displayed simultaneously can be limited depending on the performance of the mobile device. The higher the 3-D data precision and accuracy, the more performance and storage of the mobile device is required.

3.2 Prototype implementation

For the user evaluation of the LOCs developed, we designed and implemented an mAR prototype. It was developed based on the following APIs: Android API, AR API (http://www.metaio.com/products/sdk/) and Google Maps Android API (https://developers.google.com/maps/documentation/android/). Android API is a framework to develop applications, involving core modules of the operating system. In this research, Android API 4.3 (code name: Jellybean) was used. The AR API is a core part of mAR, and includes registration and visualization (superimposition) functions. In this research an AR API called Metaio was used, which supports common 3-D data formats (e.g. obj format) without any conversion, and also supports video clip display. In order to display base maps in separate windows, we adopted Google Maps Android API V2.

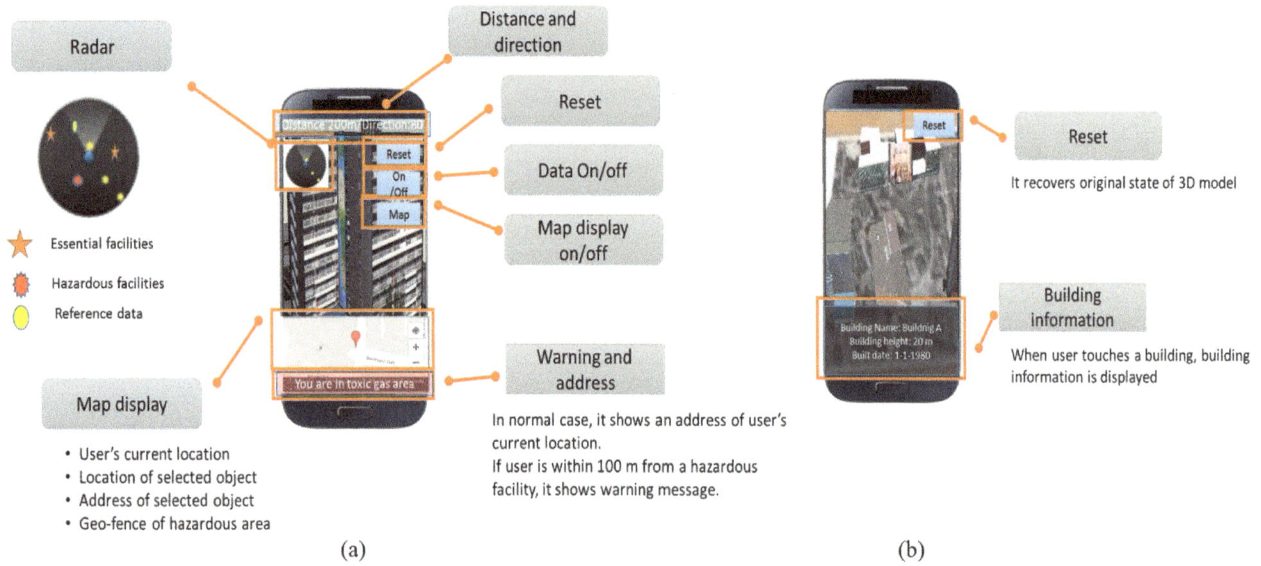

Figure 3. User interface of the **(a)** outdoor mAR and **(a)** indoor mAR.

The prototype comprised two systems, an indoor and an outdoor mAR. Figure 3 shows the user interface of each system. The indoor mAR uses markers such as image and fiducial marker to recognize the plane of AR visualization. With predefined markers, the required rotation, scale and translation of the 3-D object can be defined. The indoor AR is proposed for stakeholders who work inside, e.g. of an emergency headquarters or organization. The prototype of indoor mAR simply shows 3-D buildings and a video clip on a satellite image that covers the city of Enschede, the Netherlands. It also shows specific building information when the user touches a virtual building.

The outdoor mAR uses the GPS to obtain the user's current location (coordinates), and the gyro sensor to determine the orientation and movements of the mobile device. The outdoor mAR visualizes the data for LOC1 to LOC4 with geographic coordinates within the city centre of Enschede, the Netherlands. The outdoor mAR has the following functions: a radar and a map display, warning for hazard zones, display of current address and 3-D model controls (e.g. rotation, zoom-in, zoom-out by touch control).

The 3-D models of building exterior, interior and frame were created with SketchUp (www.sketchup.com) for demonstration purposes. To visualize a 3-D building exterior that has an accurate scale, some buildings were extracted from the Google Earth warehouse. For visualization of the multimedia data set, the prototype used video clips that were downloaded from YouTube.

A demonstration of the specific functions and operations of indoor and outdoor mAR can be accessed through the following links and QR codes (Fig. 4): indoor mAR (http://youtu.be/fL1yaPnBoI4) and outdoor mAR (http://youtu.be/0tCzXaGMx4Y).

(a) (b)

Figure 4. QR codes for video clips of AR demonstration. **(a)** Indoor mAR, **(b)** outdoor mAR.

4 Result of user evaluation

Based on the prototypes, an online survey was conducted to evaluate how much mAR can improve building damage and safety assessment in a disaster situation. The online questionnaire form was distributed to several associations of structural engineer (e.g. National Council of Structural Engineers Associations in the United States) and online communities (such as Facebook, LinkedIn) of structural engineers and first responders. The questionnaire consisted of six parts: (1) demonstration of indoor and outdoor mAR for each concept (online video clips were used), (2) questions about experience of building damage (safety) assessment, (3) efficiency of mobile AR, (4) locational awareness, (5) usability and functionality and (6) essential information. To evaluate the efficiency of mobile AR, after playing video clips of AR demonstration, we asked them how much the mobile AR can reduce assessment time of building damage, and how much AR can improve assessment accuracy with five scales between 10 % and more than 50 %. To evaluate locational (and situational) accuracy, we showed them two illustrations. The first one displayed a map of the current location with its address. The other one uses the so-called radar function to show a user's current location and relative distance and direction of hazardous facilities (e.g. a chemical plant) and essential facilities (e.g. hospital, power station). Then we asked how much it can improve the locational awareness if they can identify the current street name or address in a disaster area using such a system. In addition, we referred to the improvement of situational awareness, if they know the location of a target facility in a disaster situation. For the usability evaluation of AR based on video clips that we showed in the first stage, we explained to them how the user can efficiently view the 3-D model or map using the touch screen of a mobile device. We then asked them how much this kind of usability could improve operational performance.

In total 34 responses were collected from users with experience in building damage assessment. Of those, 24 responses (70 %) came from structural engineers with more than 5 years experience in building damage and safety assessment. Most of the responders were structural engineers (91 %), with the remaining responses (9 %) coming from post-graduate students and building managers. Eight (24 %) responses indicated prior experience with mAR. Table 2 shows the results of the online survey.

To understand the level of importance for each type of reference data, the survey showed simple illustrations of each concept to the users, covering building information, street-level imagery, 3-D building exterior, 3-D building interior, 3-D building frame and automatic change detection, increasing in order of complexity. The users were asked to rate them from 1 (not necessary) to 5 (very important). Table 3 shows the result of the survey about what is considered essential information.

The results show that users consider the 3-D building frame (4.2), followed by building information (4.1), to be the most important reference data types. They also indicate that there is a strong need for street-level imagery (4.1). On the other hand, multimedia data (3.1) are considered to be less useful. However, the score of these data still shows they are necessary data for building damage assessment.

5 Discussion

A comprehensive and rapid assessment of building stock should be conducted quickly after disaster events. In this paper, we developed an mAR concept for post-disaster field reconnaissance, conceptualizing and testing how various reference data, at increasing levels of complexity, can be delivered to field staff to increase efficiency and safety of ground-based building damage and safety assessment. For this, we defined a conceptual framework in terms of LOCs. Subsequently, based on the framework, prototypes were developed and evaluated through the online survey taken by experienced users.

The results of the survey showed that mAR can improve the assessment accuracy (objectiveness) and time. Although the results indicate the potential of mAR to facilitate field-level damage assessment, the following factors need to be considered carefully. First, there is no standard range for accuracy and speed of the building assessment. The procedure tends to be ad hoc, and strongly shaped by specific circumstances of the event (location of the disaster, accessibility, extent of the damage, presence of skilled responders and required infrastructure, etc.). This means that no benchmark in terms of time needed or accuracy achieved exists against which to measure any improvement afforded by the concept defined in this study objectively. Secondly, indoor AR did not get the positive feedback we expected (only 3.07 out of 5), likely because the work of responders typically takes place outside, where site assessment challenges are greatest. However, indoor AR might still be a useful solution for someone managing the overall disaster situation and response.

The survey about locational (situational) awareness showed interesting results that indicate that the user wants geographic information that identifies the location of both the user and the building. It implies that mAR gains in value when combined with a map system (GIS). The utility of map information is linked to the need for locational awareness that decreases especially in a severely damaged area. Furthermore, disaster responders are typically dispatched to unfamiliar places at short notice, resulting in limited locational and situational awareness. To deliver various maps to mobile devices, service-oriented architecture (SOA) and Open Geospatial Consortium (OGC) web services can be a good approach. Both SOA and OGC web services can distribute or share data between different systems working on different OS and hardware environments.

In addition, users rated the usability of mAR based on touch gestures (e.g. zoom-in/out or rotation of 3-D building with two fingers) highly that let users manipulate contents on the screen interactively. The results showed that the interaction can improve the operational performance in a disaster situation. In addition to touch handling, current mobile devices are increasingly adopting voice recognition functions, allowing yet further means to manipulate the data and functions. Although the performance of voice recognition in current mobile devices is still being improved, this is strongly driven by developments of next-generation consumer devices such as Google Glass. Thus, voice recognition is also expected to increase in usability in a disaster situation.

In terms of essential information, the results showed that the most important data type is building frame, followed by building attribute and street-level imagery. In particular, the strong need for large data sets such as the latter can pose a challenge in real disaster situations, where suitable mo-

Table 2. Results of the survey (statistical results): mean (M), median (Md), standard deviation (SD).

$N=34$	Questionnaire	Scale	M	SD	Md
Efficiency of mobile AR (Scale 1 to 5, 1: not necessary 5: very important)	How much do you think indoor AR can improve familiarity with the disaster area?	1–5	3.1	1.1	3
	How much do you think mobile AR can reduce the assessment time compared to traditional methods of building damage assessment?	%	24.5	15.4	20
	How much do you think mobile AR can improve assessment accuracy (objectiveness) compared to traditional methods of building damage assessment?	%	23.9	13.7	20
Locational awareness	If you can identify the current street name of your position in a disaster area, do you think it can improve your locational awareness?	%	35.7	13.8	35
	If you know the location of a target building in the disaster situation, do you think it can improve your locational awareness?	%	32.7	13.1	40
Usability and functionality of mobile AR	How much do you think touch handling of mobile device can improve operation performance in the field?	%	34.4	13.2	35
	Do you think mobile AR can be a practical tool for building damage assessment in a real disaster situation?	1–5	3.6	1.7	4

Table 3. Results of the survey on the value of different LOC concepts and data types: mean (M), median (Md), standard deviation (SD).

$N=34$	LOC	Data type	M	SD	Md
Essential information (level of importance) (Scale 1 to 5 1: not necessary 5: very important)	LOC1/LOC2	Building information (building classification, material, number of floors etc.)	4.1	1.2	5
	LOC3	Street-level imagery	4.1	1.2	4.5
	LOC3	Building-related multimedia data (video clips, photos)	3.1	1.3	3
	LOC4	3-D building frame	4.2	1.1	5
	LOC4	3-D building exterior	3.7	1.1	4
	LOC4	3-D building interior	3.7	1.3	4

bile internet access is quickly compromised. GSV recently introduced a new function that shows multi-temporal imagery. With this function, a user can directly compare the current building status with street-level imagery taken in dif-

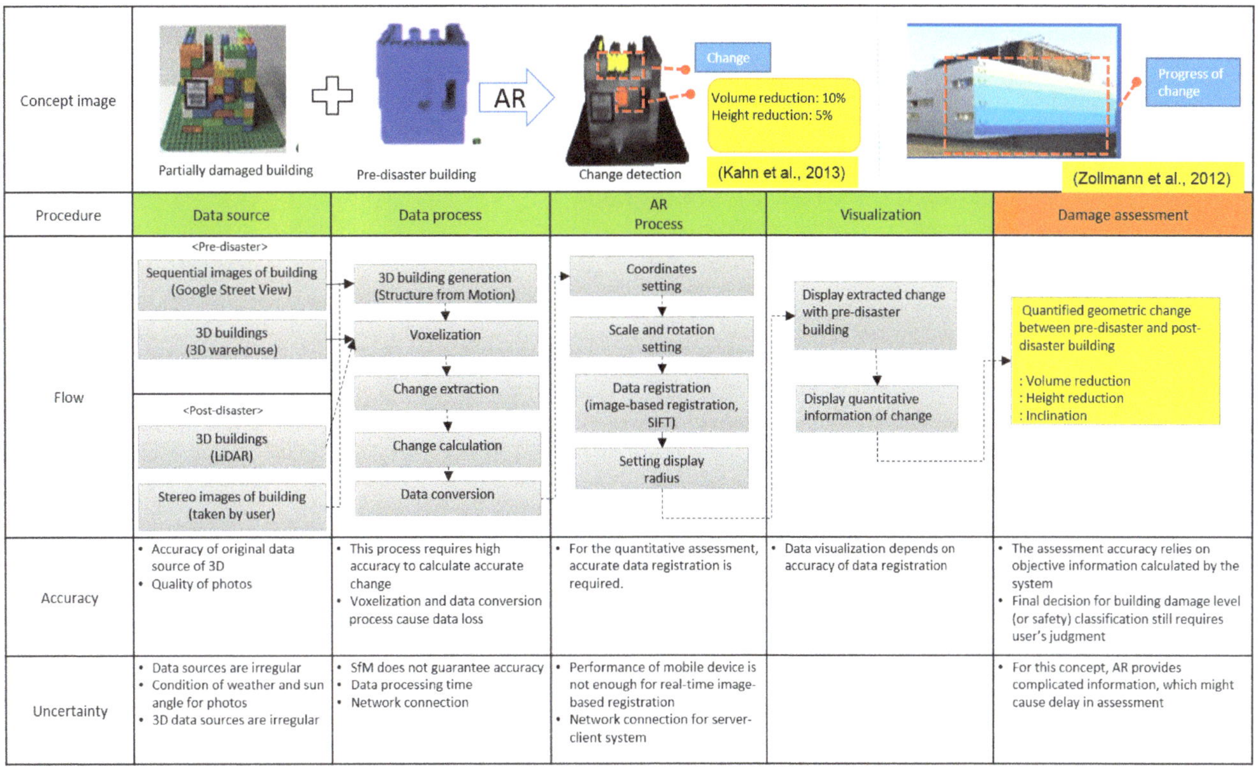

Figure 5. The concept of change-based (LOC5) and continuous change-based visualization (LOC6).

ferent time periods. Based on the results of the user evaluation and prototype test, we defined additional LOC concepts. Because the current LOC1 to LOC4 focus on the visualization of existing information, they strongly depend on the existence and ready availability of relevant data sets. Thus we conceptualized LOC5 and LOC6, which focus on the generation of new information. The concept of LOC5 is to provide the result of change detection between pre- and post-disaster data, while LOC6 traces continuous changes between different post-event time periods. The latter includes geometric change, such as volume or height reduction. The other concepts (LOC1 to LOC4) allow users to interpret change visually with references of pre-disaster data, while LOC5 and LOC6 can detect change automatically and superimpose them on the imagery of the real building. The concepts and specific processes for LOC5 and LOC6 are shown in Fig. 5.

With sequential images, mAR can extract 3-D geometry using structure from motion (SfM) methods that are widely used in several application fields (Quan and Wu, 2013; Verstockt et al., in press). Using SfM in combination with dense image matching, 3-D models of post-disaster buildings can be reconstructed. Since mobile AR uses the camera of the mobile device, a user can directly take photos of the target building in the disaster area. If a network connection is available, photos can be uploaded directly to a server to generate 3-D building models (for example, using freely available smartphone apps such as Autodesk's 123D Catch; www.123dapp.com/catch). Instead of using sequential images, tools such as "Spike" (http://www.ikegps.com/spike/) showed the possibility of scanning 3-D objects directly to generate 3-D data from an active scanner mounted onto the mobile device. Spike combines imagery with a laser range finder and position sensors to generate real-time 3-D data of building objects. Google have also been testing similar functions in its "Project Tango" (http://www.google.com/atap/projecttango/#project) to extract 3-D geometry directly from a smartphone. Tango can give accurate information (e.g. building shape, height, inclination) of real buildings without relying on GPS or gyro sensors as it scans in real time. In LOC5 and LOC6, users can straightforwardly determine the extent of building damage, and which parts of the building have changed during a given period based on automatically calculated information.

Since LOC5 and LOC6 only emerged as possibilities from the user evaluation of our prototypes, in this paper they were only conceptualized but not yet implemented and tested. However, with progress in 3-D reconstruction technologies based on smart devices, it is thus reasonable to expect that mAR in various forms will soon become a common information generation and communication method, including in post-disaster settings.

6 Conclusions

AR is being increasingly adopted in various fields. Especially the rapidly growing use of mobile devices means that mAR can be adopted at a lower cost but with higher usability than traditional AR. With the development of not only new technologies such as wearable devices and 3-D mobile scanners, but also improvements in the mobile devices themselves, it is expected that mAR will play a major role in comprehensive and rapid building assessment.

In this paper we developed a number of concepts of increasing complexity to provide stakeholders operating in a disaster area with pertinent reference information to facilitate their orientation, the assessment of damage and safety, and to improve their own safety in a potentially hazardous environment. The potential of mAR in a disaster situation was confirmed by experienced disaster responders. Nevertheless, a number of limitations persist. First, our utility assessment was based on an online survey, meaning that the concept, not the actual practical usability, was assessed. Secondly, because mAR delivers and processes 3-D data, it is important to control performance given available resources. The experts emphasized the need for fast and responsive mAR. Current consumer devices, which can readily be used as mAR platforms, struggle when confronted with the variety and size of the data streams and processing steps required for more advanced LOC performance (in particular photogrammetric processing). Lastly, obtaining and sharing of data acquired in a disaster situation raises privacy issues that need to be considered carefully.

In this study, AR-based change detection (LOC5/LOC6) was only conceptualized. A comprehensive implementation will require components in vision recognition, photogrammetry, 3-D data reconstruction and multi-data integration, vision-based AR, etc. However, rapid developments in mobile 3-D reconstruction (e.g. Spike or Google's Project Tango) will provide ready solutions for some of the above, and also reduce the need for server-based processing that requires large bandwidth. In addition, we also need to expand our mAR concepts in terms of not only delivering reference data, but also allowing further augmentation with information directly fed into a mobile device in the disaster area. Even though satellite imagery obtained following a disaster (e.g. through the Disaster Charter) is currently not made available to response organizations, in principle, such a data stream can also be meaningfully incorporated into an mAR framework.

References

Adams, B., Ghosh, S., Wabnitz, C., and Alder, J.: Post-tsunami urban damage assessment in Thailand, using optical satellite imagery and the VIEWSTM field reconnaissance system, in: The 1755 Lisbon Earthquake: Revisited, Geot Geol Earthquake, Springer Netherlands, 523–539, 2009.

Altan, O., Toz, G., Kulur, S., Seker, D., Volz, S., Fritsch, D., and Sester, M.: Photogrammetry and geographic information systems for quick assessment, documentation and analysis of earthquakes, ISPRS J Photogramm, 55, 359–372, 2001.

ATC: ATC-20-1 Field Manual: Postearthquake safety evaluation of buildings, Applied Technology Council, 2005.

Azuma, R.: A survey of augmented reality, Presence-Teleop Virt, 6, 355–385, 1997.

Boddhu, S. K., Dave, R. P., McCartney, M., West, J. A., and Williams, R. L.: Context-aware event detection smartphone application for first responders, Proc. SPIE 8742, Ground/Air Multisensor Interoperability, Integration, and Networking for Persistent ISR IV, 874213-874213-9; doi:10.1117/12.2016352, 2013.

Brunner, D., Lemoine, G., and Bruzzone, L.: Earthquake damage assessment of buildings using VHR optical and SAR Imagery, IEEE T. Geosci. Remote, 48, 2403–2420, 2010.

Corbane, C., Saito, K., Dell'Oro, L., Gill, S., Piard, B., Huyck, C., Kemper, T., Lemoine, G., Spence, R., and Krishnan, R.: A comprehensive analysis of building damage in the January 12, 2010 Mw7 Haiti earthquake using high-resolution satellite and aerial imagery, Photogramm. Eng. Rem. S., 77, 997–1009, 2011.

Curtis, A. and Mills, J. W.: Spatial video data collection in a post-disaster landscape: The Tuscaloosa Tornado of April 27th 2011, Appl. Geogr., 32, 393–400, 2012.

Dai, F., Dong, S., Kamat, V., and Lu, M.: Photogrammetry assisted measurement of interstory drift for rapid post-disaster building damage reconnaissance, J. Nondestruct. Eval., 30, 201–212, 2011.

Dell'Acqua, F. and Gamba, P.: Remote sensing and earthquake damage assessment: Experiences, limits, and perspectives, Proc. IEEE, 100, 2876–2890, 2012.

Dong, S., Feng, C., and Kamat, V. R.: Sensitivity analysis of augmented reality-assisted building damage reconnaissance using virtual prototyping, Automat. Constr., 33, 24–36, 2013.

Feiner, S., MacIntyre, B., Höllerer, T., and Webster, A.: A touring machine: Prototyping 3-D mobile augmented reality systems for exploring the urban environment, Pers. Technol., 1, 208–217, 1997.

Fernandez Galarreta, J., Kerle, N., and Gerke, M.: UAV-based urban structural damage assessment using object-based image analysis and semantic reasoning, Nat. Hazards Earth Syst. Sci., 15, 1087–1101, doi:10.5194/nhess-15-1087-2015, 2015.

Flesch, R.: European Manual for in-situ Assessment of Important Existing Structures, LESSLOS Report, ISBN 9788861980068, 2007.

Gerke, M. and Kerle, N.: Automatic structural seismic damage assessment with airborne oblique pictometry, Photogramm. Eng. Rem. S., 77, 885–898, 2011.

Ghosh, S., Huyck, C. K., Greene, M., Gill, S. P., Bevington, J., Svekla, W., DesRoches, R., and Eguchi, R. T.: Crowdsourcing for rapid damage assessment: The Global Earth Observation Catastrophe Assessment Network (GEO-CAN), Earthq Spectra, 27, S179–S198, 2011.

Kahn, S., Bockholt, U., Kuijper, A., and Fellner, D. W.: Towards precise real-time 3-D difference detection for industrial applications, Computers in Industry, 64, 1115–1128, 2013.

Kamat, V. R. and El-Tawil, S.: Evaluation of augmented reality for rapid assessment of earthquake-induced building damage, J. Comput. Civil. Eng., 21, 303–310, 2007.

Kerle, N.: Satellite-based damage mapping following the 2006 Indonesia earthquake – How accurate was it?, Int. J. Appl. Earth Obs., 12, 466–476, 2010.

Kerle, N.: Remote sensing based post-disaster damage mapping with collaborative methods, in: Intelligent Systems for Crisis Management, Springer, 121–133, 2013.

Kerle, N. and Hoffman, R. R.: Collaborative damage mapping for emergency response: the role of cognitive systems engineering, Nat. Hazards Earth Syst. Sci, 13, 97–113, 2013.

Leebmann, J.: An augmented reality system for earthquake disaster response, XXth ISPRS Congress, The International Archives of the Photogrammetry, Remote Sensing and Spatial Information Sciences, Istanbul, 2004,

Quan, H. and Wu, M.: A real-time SFM method in augmented reality, in: Proceedings of the 2012 International Conference on Information Technology and Software Engineering, edited by: Lu, W., Cai, G., Liu, W., and Xing, W., Lecture Notes in Electrical Engineering, Springer Berlin Heidelberg, 841–848, 2013.

Rabbi, I. and Ullah, S.: A survey on augmented reality challenges and tracking, Acta Graphica, 24, 29–46, 2013.

Sakurai, M., Watson, R. T., Abraham, C., and Kokuryo, J.: Sustaining life during the early stages of disaster relief with a frugal information system: learning from the great east Japan earthquake, Commun. Magaz., IEEE, 52, 176–185, 2014.

Schweier, C. and Markus, M.: Classification of collapsed buildings for fast damage and loss Assessment, Bull. Earthquake Eng., 4, 177–192, 2006.

Tsai, M., Lee, Y., Lu, C., Chen, M., Chou, T., and Yau, N.: Integrating geographical information and augmented reality techniques for mobile escape guidelines on nuclear accident sites, J. Environ. Radioactiv., 109, 36–44, 2012.

Valentini, P. P., Gattamelata, D., and Pezzuti, E.: Virtual engineering in Augmented Reality, in: Computer Animation, Nova Science Publisher Inc., New York, 2010.

Verstockt, S., Gerke, M., and Kerle, N.: Geo-localization of crowd-sourced images for 3-D modeling of city points of interest, IEEE Geosci. Remote Sens., 12, 1670–1674, 2015.

Wani, A. R., Shabir, S., and Naaz, R.: Augmented reality for fire and emergency services, Int. Conf. on Recent Trends in Communication and Computer Networks, Byderabad India, 2013,

Zollmann, S., Kalkofen, D., Hoppe, C., Kluckner, S., Bischof, H., and Reitmayr, G.: Interactive 4D overview and detail visualization in Augmented Reality, IEEE International Symposium on Mixed and Augmented Reality, WOS:000319516900019, 167–176, 2012.

4

The 9 September 2010 torrential rain and flash flood in the Dragone catchment, Atrani, Amalfi Coast (southern Italy)

C. Violante[1], G. Braca[2], E. Esposito[1], and G. Tranfaglia[2]

[1]Istituto per l'Ambiente Marino Costiero (IAMC), Consiglio Nazionale delle Ricerche (CNR), Naples, Italy
[2]Istituto Superiore per la Protezione e la Ricerca Ambientale (ISPRA), Rome, Italy

Correspondence to: C. Violante (crescenzo.violante@cnr.it)

Abstract. In this paper we use a multi-hazard approach to analyse the 9 September 2010 flash flood in the Dragone basin, a 9 km^2 catchment located along the Amalfi rocky coastal range, southern Italy. In this area, alluvial fan flooding has been the most frequent and destructive geologic hazard since Roman times. Sudden torrents of water (flash floods) are caused by high-intensity and very localized cloudbursts of short duration, inducing slope erosion and sediment delivery from slope to stream. The elevated bed load transport produces fast-moving hyperconcentrated flows with significant catastrophic implications for communities living at the stream mouth.

The 9 September 2010 rainstorm event lasted 1 h with an intensity rainfall peak of nearly 120 $mm\,h^{-1}$. High topographic relief of the Amalfi coastal range and positive anomalies of the coastal waters conditioned the character of the convective system. Based on geological data and post-event field evidence and surveys, as well as homemade videos and eyewitness accounts, it is reported that the flash flood mobilized some 25 000 m^3 of materials with a total (water and sediment) peak flow of 80 $m^3\,s^{-1}$. The estimated peak discharge of only clear water was about 65 $m^3\,s^{-1}$. This leads to a sediment bulking factor of 1.2 that corresponds to a flow with velocities similar to those of water during a flood.

1 Introduction

The Amalfi Coast consists of a steep mountain front (up to 1444 m a.s.l.) that rises abruptly from the Tyrrhenian Sea (Fig. 1). It is a rocky coast mostly formed by a pile of Mesozoic carbonate rocks, covered by Tertiary to Quaternary siliciclastic and pyroclastic units which have been tectonically uplifted since the lower Pleistocene. Bedrock rivers and channels deeply dissect the carbonate bedrock, forming a complex fluvial system characterized by small catchments that are very high relative to the base sea level. These rivers show a distinct seasonality and torrential behaviour, with main delivery areas into the adjacent marine shelf (Fig. 2; Esposito et al., 2004; Budillon et al., 2005; Violante, 2009; Violante et al., 2009).

During the last millennia this area has been repeatedly mantled by the pyroclastic products of the Somma-Vesuvio that create favourable conditions for volcaniclastic debris to generate rainfall-triggered mass flows and flash floods. The Plinian eruption, that destroyed the Roman cities of Pompeii, Stabiae and Herculaneum in AD 79, deposited up to 2 m of erosion-prone volcaniclastic material (Sigurdsson et al., 1985) on the steep coastal slopes, causing conditions of increased geomorphic instability.

Geologic evidences for rapid slope erosion following the Pompeii pyroclastic fall include alluvial reworked volcaniclastic sequences (locally called Durece) occurring as residual outcrops along narrow stream valleys (Cinque and Robustelli, 2009) and coastal fan deltas fed by small alluvial fans at the mouth of the main streams (Sacchi et al., 2009; Violante et al., 2009). The latter are composed of wedge-shaped coarse-grained alluvial deposits that thicken towards the sea and represent the subaqueous counterpart of small fans at the river mouths (Fig. 2, inset).

Pyroclastic air-fall tephra derived from the late Quaternary activity of Somma-Vesuvio still represent unstable sedimentary covers on top of the steep carbonate slopes of the Amalfi Coast. These deposits create conditions of elevated

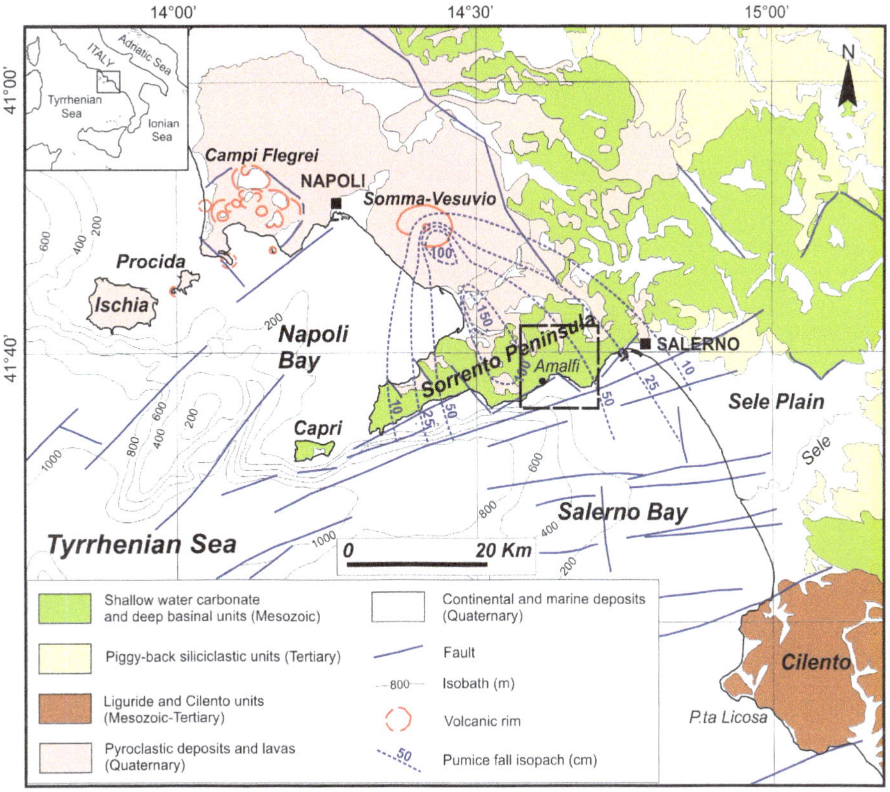

Figure 1. Geological sketch map and location of the study area (dashed box).

slope instability in conjunction with rainstorm events that frequently hit the Amalfi Coast. The slides are mostly shallow and very wide, extending all the way to the mountain ridge and crest, and largely belong to the categories of soil slip and debris/earth flow phenomena. In addition to rapid sediment transport along the valley flanks, landslide debris flowing from slope to streams produces fast-moving large debris torrents (flash floods) with significant catastrophic implications for local communities mostly living on alluvial fans at stream mouths. Here flood-prone streams have been artificially forced to flow underneath roads and squares to exploit the whole delta surfaces for urban development (Fig. 2).

Human interference plays an important role in the generation of flood hazard (e.g. Smith and Ward, 1998; Plate, 2002). In particular, the mismanagement and lack of maintenance of channel sections and the deforestation of water catchment areas can significantly increase the risk of flash floods. In the study area, extensive terracing through deforestation and use of rocks to build retaining walls has generated a cultural landscape strictly depending on the correct management of mountain slopes and channel beds. Interference by human actions has been documented for past floods in the Sarno-Quindici area (Calcaterra et al., 2003), located a few kilometres north of Atrani, which was hit by a destructive flooding event in 1998.

In this paper we analyse the 9 September 2010 rainfall event that hit the Amalfi Coast and its effects on the Dragone catchment and the village of Atrani. Direct field observations that include geological investigations and damage to property and infrastructures have been combined with meteorological and hydraulic/hydrological analyses. Reconstruction and recurrence of past events based on different historical sources and marine geophysical and geological data of the Dragone submerged delta have also been taken into account.

2 The Dragone catchment fan delta system

The Dragone catchment drains an area of $9.3\,km^2$ along the steep coastal slopes of the Amalfi Coast. The basin develops in a north–south direction and is strongly asymmetric, with the eastern flank composed of a short and abrupt slope corresponding to a fault scarp and the western flank formed by four main sub-basins: Scalandrone, Nocelle, Senite, and S. Caterina. A fifth sub-basin, Frezzi, develops at the head of the Dragone stream (Fig. 3a). The drainage area rises up to 1420 m a.s.l. and cuts into Mesozoic limestone discontinuously mantled by Quaternary volcaniclastic and alluvial deposits. A low drainage frequency ($5\,km^{-2}$) and a subdendritic pattern characterize the hydrographic network with the main stream, the Dragone, discharging directly into the Tyrrhenian Sea. Slope analysis based on a 5×5 m cell size

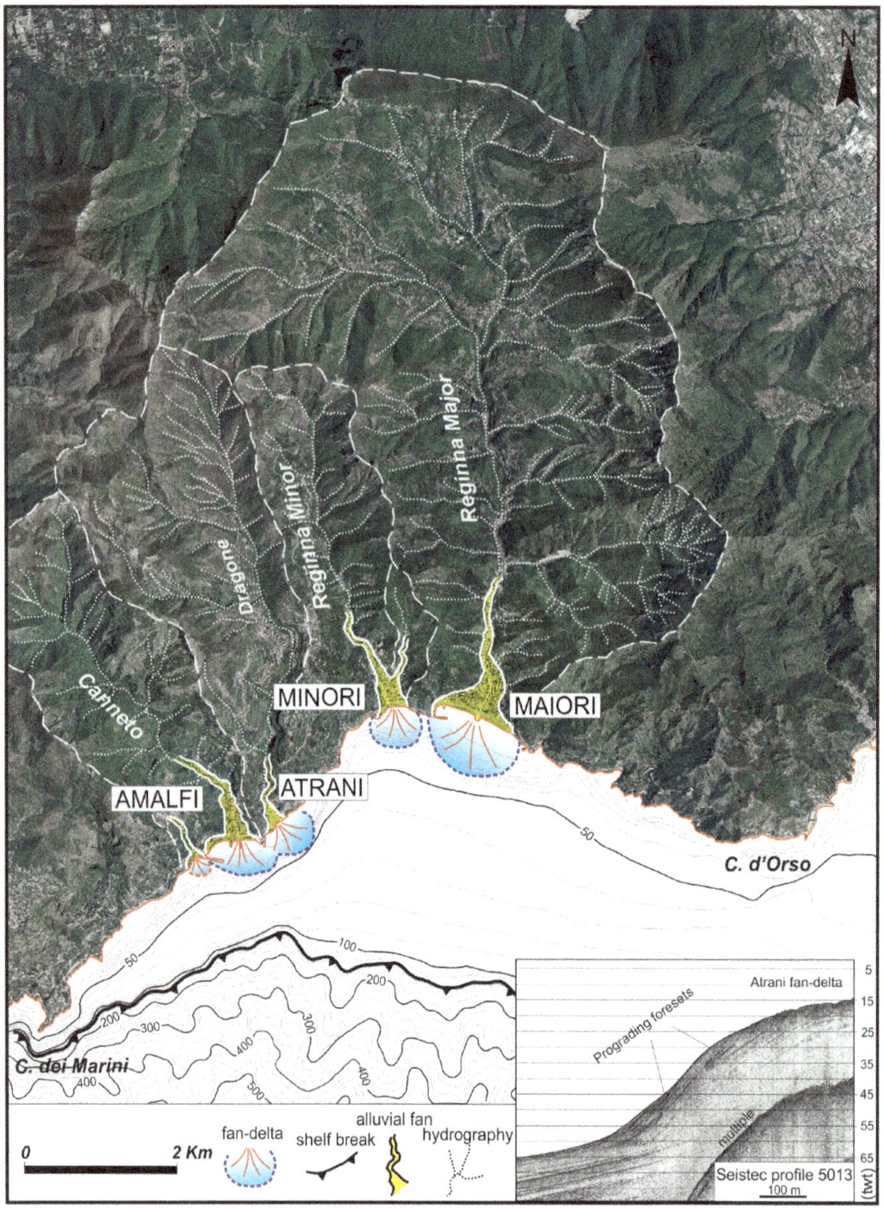

Figure 2. The Amalfi rocky coast system characterized by steep and high watersheds, urbanized alluvial fans and fan deltas at the mouth of main streams, reduced continental shelf engraved by canyons, and abrupt shelf break (fault-controlled). The fan deltas are composed of prograding clinoforms resulting from flood activity as revealed by the high-resolution seismic profiles (inset map in the lower right corner). Modified after Violante (2009) and Sacchi et al. (2009).

DTM (digital elevation model) indicates that the topographic gradient of the catchment area is arranged into two main slope classes, ranging from 15 to 35° and from 35 to 50° (golden/yellow and brown colours, respectively, in Fig. 3b). The mean slope is 30°.

The Dragone stream is 6.8 km in length with the terminal section covered by a roadway crossing the village of Atrani. The covered section of the water course has an input section of 3 × 9 m ($h \times w$) that reduces to 1.80 × 5.50 m at the closing section, and has a total length of about 300 m. Runoff waters are regulated by concrete levees and check dams that extend over two-thirds of the hydrographic network.

At sea a submerged fan delta is present at the closing section of the Dragone catchment (Sacchi et al., 2009; Violante et al., 2009). The delta body is 0.2 km^2 wide and reaches a maximum thickness of 25 m. It displays a generally conical morphology with a delta front slope of 30° and foreset inclination ranging from 15 to 30°. This structure is composed of alluvial sequences that coincide with significant changes in the river activity during streamflow phenomena.

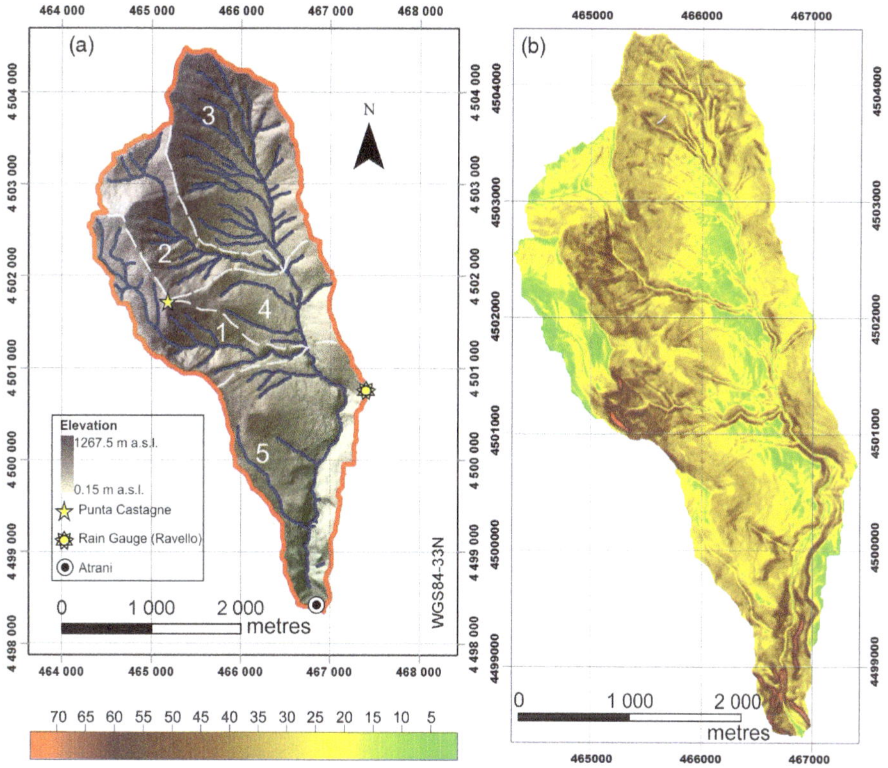

Figure 3. The Dragone catchment. **(a)** Basin elevation (DEM) and sub-basins of the Dragone stream. 1. Scalandrone, 2. Nocelle, 3. Frezzi, 4. Senite, 5. S. Caterina. **(b)** Slope map. The scale for the slope map is in the left lower corner.

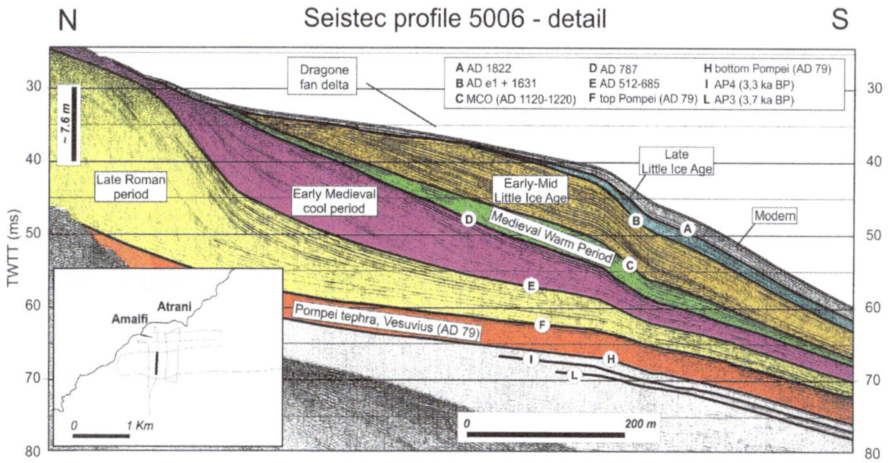

Figure 4. The Dragone–Canneto fan delta body found off the village of Atrani. Detail of a very high-resolution seismic profile showing flood-controlled seismic-stratigraphic units and their inferred association with major climatic changes of the last 2000 years. Letters A to L represent age-dated stratigraphic horizons. See inset map for location (modified after Sacchi et al., 2009).

Increases in fluvial sedimentary discharge are recorded as successive phases of delta growths with which temporary shoreline progradations are associated.

The detailed study of the internal stratigraphic architecture of the Dragone fan delta indicates various depositional phases following the main AD 79 alluvial crisis, possibly modulated by the interplay between the availability of loose pyroclastic covers and the varying erosional rates due to the climatic oscillations occurring in the last millennia (Fig. 4; Sacchi et al., 2009; Violante et al., 2009). The major change detectable in the Amalfi fan deltas occurred in the Early Medieval Cool Period (ca. AD 500–AD 800), that developed immediately after the Roman Warm Period. Further changes in the stratal patterns of the delta foresets, indicative of high

Figure 5. Geo-environmental effects induced by the 9 September 2010 rainstorm in the Dragone catchment. **(a)** Linear erosion engraving a tributary channel up to 2 m. **(b)** Soil slip (arrows) at P. Castagne. **(c)** Partial breaking up of the main road (Via dei Dogi) in the village of Atrani. Red arrows indicate maximum height of the flow, reaching ca. 1 m above the street pavement. Location is shown in Fig. 13. **(d)** Site of temporary damming in correspondence of a man-made structure built in the stream bed. **(e)** Terminal fan at the mouth of the Dragone stream. See Fig. 3 for location.

streamflow activity, may be correlated with the Medieval Warm Period (ca. AD 900–AD 1100) and the Little Ice Age (ca. AD 1400–AD 1850).

3 Rainstorm-induced geological effects

A field survey was undertaken soon after the 9 September 2010 rainstorm in the Dragone catchment and village of Atrani. The observed rainstorm-induced geological effects include shallow landslides and sediment removal along channels, a temporary dam in the mid-lower section of the Dragone stream, the partial breaking up of the main road crossing the village of Atrani, and the deposition of a coarse terminal fan at the mouth of the Dragone stream (Fig. 5).

The field data indicate that slope erosion triggered by the 9 September 2010 rainstorm was predominantly linear. Sediment removal by linear erosive processes significantly engraved tributaries and the main stream up to a depth of 2 m (Fig. 5a). The displaced materials were mostly composed of pyroclastic deposits and solid wastes occurring along the channel beds, often above the hydraulic check dams. A large number of tree trunks of different sizes, used to build bridges across ditches and tributaries, were also included in the transported material.

Minor soil slips have been observed uphill on the high slope at Punta delle Castagne, a rock crest connecting the S. Eustacchio, S. Caterina, and Scalandrone watersheds (Fig. 5b) and along the western flank of the Frezzi watershed. Here displacement of the channel sides occurred both at channel heads and, locally, at the middle–upper reaches of small tributaries.

Apart from overspill deposits composed of white pumices observed at break slopes just below the watershed areas and along the middle reach of the Dragone stream, no significant aggradation was observed in the catchment area. Removal of materials from the stream bed was produced by a fast-moving debris torrent with high erosive capacity, transporting the mobilized materials all the way down to the coast. Here the Dragone stream is artificially forced to flow underneath the main road crossing the village of Atrani, resulting in siphoning and consequent breaking up of the above roadway (Fig. 5c). Once it reached the coastline, the hyperconcentrated flow engraved the Atrani beach and dumped the transported materials into the sea in the form of a coarse alluvial fan (Fig. 5e). This induced a shore progradation of about 30 m. The alluvial deposits entering into the sea were mostly composed of white-grey pumices including tree trunks, wastes, and rock boulders of different sizes reaching up to 1 m in diameter. Some cars parked along the main road were transported up to the beach area and beyond (Fig. 5e).

Despite the carbonate nature of the study area, there was no evidence of karst features that could have influenced the flooding mechanism. In fact the drainage of waters in the Amalfi Coast is mainly by relatively deep underground paths towards offshore springs located in correspondence of the main neotectonic lineaments (Celico et al., 2005). This may, at least in part, prevent direct interactions between groundwater and surface water during rainfall events as typically occurs in karst settings (Jourde et al., 2007; Gutierrez et al., 2014).

Discharge and depth of flow downstream was probably increased by abrupt draining of a temporary dam reported by eyewitnesses in the lower section of the Dragone stream. Such damming was favoured by a narrow flow section and enhanced by a man-made structure built in the stream bed (Fig. 5d). The failure of temporary debris dams and the draining of ephemeral lakes have been described for different flood events that have occurred in the study area (Passerini, 1925; Penta et al., 1954). These phenomena can produce exceptional temporary discharges and highly destructive peak flows, reaching depths as high as 8–10 m (Larsen et al., 2001; Perez, 2001; Esposito et al., 2004; Violante, 2009).

4 Synoptic description and physical features of the meteorological event

The first days of September 2010 in the Mediterranean area were characterized by a drastic seasonal transition. The first depression from the Atlantic was favoured by the formation of a high-latitude anticyclone over Scandinavia. Such an anticyclone forced the oceanic perturbed airstream towards the medium European latitudes and the Italian peninsula. The unsettled air of Atlantic origin headed later for the low Tyrrhenian basin where very humid airstreams coming from south/south-west were forced by an area of low pressure centred in the northern Tyrrhenian Sea. This unstable scenario was fed by the Mediterranean Sea, which created conditions for a secondary area of low pressure to develop in the lower layers of the atmosphere.

The front was followed on 9 September 2010 by a far more organised instability in the southern Tyrrhenian basin with well-defined storm cells localized along the Campania, Sicily, and Calabria coastal areas. The high humidity associated with such a positive vorticity led to a mesoscale convective system (Fujita, 1986) over the southern Tyrrhenian Sea and southern Italy, as captured by a Meteosat image in the visible channel at 17:00 UTC (time hereafter expressed in UTC) on 9 September in Europe (Fig. 6a). The warm-humid flow associated with this system hit the Campania region, inducing a strong maritime thunderstorm between 10:00 and 22:00, with maximum intensity on the Lattari Mountains, as shown in the map of ground lightings (Fig. 6b).

The character of the convective system was conditioned by the Picentini and Lattari mountains that forced the air masses to follow the orography of the coastal range (see Fig. 1) and the positive anomalies of the coastal waters. Data from the Italian Oceanographic Network (ISPRA; Fig. 7) indicate that air temperature was lower than sea-surface temperature during the rainstorm, maintaining the thermodynamic conditions that allowed the convective system to sustain the storm cell.

Physical analysis of the rainstorm was based on the real-time data recorded by the rain gauges provided by the Centre for weather forecast and monitoring of the Campania regional authority (CEMPID, 2010; Table 1). On the coast the rainfall lasted 4 h, starting at 14:10 east of Salerno (Fig. 8). At the Ravello rain gauge, located along the watershed of the Dragone basin, the rain started at 15:50 and persisted for about 1 h (cumulative rainfall 80.8 mm) with a maximum rainfall intensity (19.4 mm over 10 min; Table 1) from 15:50

Table 1. Rainfall intensity (mm) registered at time intervals of 10 min and 1 h on September 9, 2010, by rain gauges in Salerno and the Sorrento peninsula. The maximum values are reported in bold. Data are from the Centre for weather forecast and monitoring of the Campania regional authority (CEMPID, 2010).

Rain gauge	Altitude (m)	Latitude N	Longitude E	Distance from Ravello (km)	Rainfall intensity (mm) 10′	1 h
Pontecagnano	36	40°38′36.9″	14°52′02.2″	21.8	15.0	53.6
Salerno Meteo	16	40°38′37.7″	14°50′11.5″	20.3	16.0	47.0
Ravello	390	40°39′24.3″	14°36′52.5″	0.0	19.4	80.8
Agerola Meteo	848	40°38′48.6″	14°32′26.2″	6.0	**26.2**	80.8
Agerola	623	40°38′21.3″	14°32′44.8″	5.7	21.6	66.8
Moiano	485	40°39′12.6″	14°27′50.0″	13.5	21.0	78.0
Pimonte	437	40°40′27.8″	14°30′17.4″	10.2	23.2	**92.2**
Maiori	10	40°39′05.7″	14°38′24.6″	3.0	12.0	43.0
Gragnano	195	40°41′15.1″	14°31′38.1″	8.6	14.8	70.0
Lettere	312	40°42′15.9"	14°31′58.3″	9.8	16.2	45.6
Corbara	424	40°43′32.8"	14°36′07.5″	9.1	18.4	45.0
Tramonti	422	40°42′13.9″	14°38′49.3″	6.9	12.4	42.6
Amalfi	114	40°37′23.7″	14°34′49.8″	1.4	11.0	28.6
Cetara	140	40°39′04.0″	14°42′12.5″	8.6	9.8	27.2

Table 2. Main morphometric parameters used in the hydrologic model.

Sub-basin	Name	A (km^2)	L (m)	S (%)	H_{mean} (m a.s.l.)	H_{min} (m a.s.l.)
Basin_1	Scalandrone	1.39	3600	56.7	873	389
Basin_2	Nocelle	1.46	2450	60.5	922	479
Basin_3	Frezzi	2.66	2600	54.3	803	479
Basin_4	Senite	1.12	1000	46.7	570	389
Basin_5	Sant'Eustacchio	2.71	3200	60.3	411	0
Whole basin	Dragone	9.33	6800	56.5	692	0

Table 3. Estimated lag-time values and the curve number (CN) parameter used in the hydrologic model.

Sub-basin	CN (–)	t_c (min)	t_{lag} (min)
Basin_1	66	44	26
Basin_2	66	32	19
Basin_3	66	34	20
Basin_4	66	17	10
Basin_5	66	38	23
Whole basin	66	72	43

to 16:00. At the Agerola weather station, located about 5 km from the Dragone basin, the same event lasted about 1 h with an intensity rainfall peak of 157.2 mm h^{-1} from 16:00 to 16:10, showing a significant similarity with the cumulative rainfall of the Ravello gauge (see Fig. 8). This suggests that the Atrani event is consistent with a storm cell characterized by a very limited areal extent, as confirmed by the Meteosat visible image that shows a single cell with a very flat elliptical shape elongated in a NE–SW direction (inset in Fig. 6a).

The size of the storm cell that produced the Atrani event was calculated (1) by considering the 50 mm isohyet corresponding to the downdraught in the time span ranging from 16:00 to 16.50 (Fig. 9); and (2) on the basis of the number of pixels forming the storm cell visible in the Meteosat images at 17:00. In the first case the area was about 80 km^2, while in the latter it ranged from 56 to 75 km^2. Therefore, the Atrani storm cell can be ascribed to a mesoscale convective system (MCS) with a horizontal scale of about 20–30 km and duration of less than 1 h (lower bound of the MCS mesoscale β; Orlansky, 1975; Thunis and Borstein, 1996). These events typically occur in the Mediterranean Sea between April and September, with a higher frequency mostly concentrated during September (Lionello et al., 2006), and include multiple storm cells in different evolutionary stages (Morel and Senesi, 2002) that are strongly influenced by local orography.

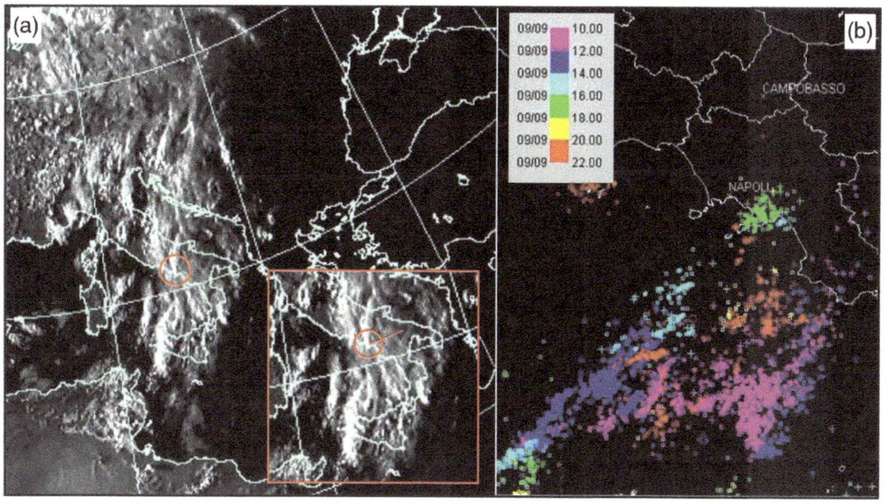

Figure 6. **(a)** Meteosat image captured in the visible channel on 9 September 2010 at 17:00 UTC (modified) showing a mesoscale convective system (MCS) over the southern Tyrrhenian Sea and southern Italy. Inset: detail showing the Atrani storm cell. **(b)** Thunderstorm activity on 9 September 2010. Intense electric activity is recorded between 14:00 and 16:00 (sky blue) near the city of Salerno and between 16:00 and 18:00 (green) on the Amalfi Coast.

5 Hydrological model

In order to evaluate the hydrological response of the Dragone basin at the stream mouth, a semi-distributed rainfall-runoff model has been used. For this purpose, each of the five sub-basins of the Dragone stream (see Fig. 3a and Table 2) were analysed by Soil Conservation Service (SCS) dimensionless unit hydrograph rainfall-runoff transformation, and SCS curve number (CN) loss method (USDA SCS, 1986a) implemented in HEC-HMS (USACE HEC, 2010).

The SCS dimensionless unit hydrograph procedure is one of the most well-known methods for deriving synthetic unit hydrographs, especially for small basins. The rainfall-runoff model is based on the lag time parameter, that is the time interval between the centroid of the effective rainfall hyetograph and the peak of discharge (Table 3; USDA SCS, 1986b; Singh, 1989).

The effective precipitation depth, according to the SCS-CN procedure, is

$$P_e = \begin{cases} \dfrac{(P - I_a)^2}{(P - I_a + S)} & \text{if} \quad P > I_a \\ 0 & \text{if} \quad P \leq I_a \end{cases}. \tag{1}$$

P is the total precipitation depth; P_e is the depth of excess precipitation or direct runoff; I_a is the "initial abstraction" or the amount of rain before the runoff starts, which infiltrates or is intercepted by vegetation; and S is the potential maximum soil moisture retention during the runoff.

The value of S for a given soil is related to the curve number (CN) which is a function of the hydrologic soil-cover complexes as (if S is expressed in mm)

$$\text{CN} = \frac{25\,400}{S + 254}. \tag{2}$$

In the case of the Dragone basin, the adopted CN value is 66, corresponding to forest-like cover (woods–grass combination), considering hydrological soil group B (moderate infiltration) with antecedent moisture condition II (average) and $I_a = 0.2\,S$.

The lag time t_{lag} has been estimated from the concentration time t_c calculated by the SCS formula (Chow et al., 1988):

$$t_{lag} = 0.6 \times t_c = 0.6 \times 0.00227 \times L^{0.8} \left(\frac{1000}{\text{CN}} - 9\right)^{0.7} i^{-0.5} \text{ [hour]}. \tag{3}$$

L is the hydraulic watershed length expressed in metres (m) and i the mean basin slope as a percentage (%). Figure 10 reports the results of the hydrological model where the estimated peak discharge of the clear water is about 65 $m^3\,s^{-1}$ occurring around 17:00.

6 Hydraulic model and sediment transfer

Due to the lack of stream gauge data, we analysed the hydraulic response of the Dragone catchment to the 9 September 2010 rainstorm event and the sediment transfer during the flood on the basis of (a) morphology of the Dragone basin, (b) pluviometric data, (c) field evidences, (d) home-made videos and photos, and (e) eyewitness accounts.

According to local eyewitness accounts, the peak flow of the Atrani flood event occurred between 16:50 and 17:10, which is about 40 min later than the hyetograph centroid, at about 16:25, while the flood duration ranged from 40 to 60 min (Fig. 11). In the Atrani urban area the covering of the Dragone stream caused the flood wave to be split

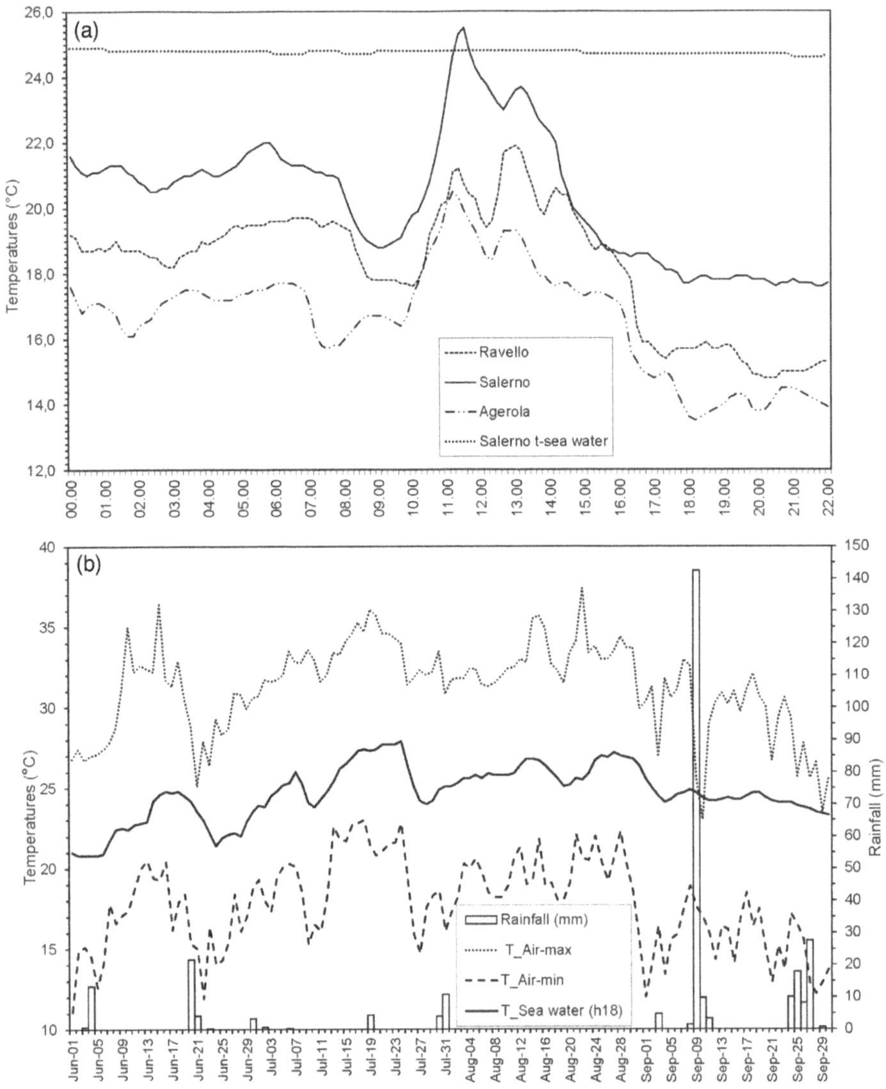

Figure 7. **(a)** Rainfall intensity and air (min and max) and sea water temperatures measured at Salerno from 1 June to 30 September 2010; **(b)** air and sea water temperatures at 10 min intervals on 9 September 2010. Data are from the Centre for weather forecast and monitoring of the Campania regional authority (CEMPID, 2010).

into two different currents: a main flow underground, below the road level (Via Dei Dogi), along a closed section varying from 3(h) m × 9(w) m at the input section to 1.80(h) m × 5.50(w) m at the closing section; and a second flow at the surface, along the road, constrained by almost continuous man-made structures and buildings 5.5 m away from each other (see Fig. 5c). The maximum depth of the upper flow was about 1 m, i.e. 3.5 m on average above the stream bed.

The highest flow velocity for both flows was estimated from amateur videos by tracking particles transported by the flood through the analysis of video frames with an approach similar to that used in the particle image velocimetry technique (Fig. 12; Raffel et al., 2007). Peak discharge was obtained by multiplying estimated flow velocity and the flow section.

The estimated flow velocity along Via Dei Dogi is about 3–4 $m\,s^{-1}$ and, consequently, the peak discharge is approximately 20 $m^3\,s^{-1}$. For the underground flow the estimated peak velocity is about 6–7 $m\,s^{-1}$, while the peak discharge is on the order of 60 $m^3\,s^{-1}$, taking into account that the flow fills roughly 80 % of the closed section. Then the total estimated peak discharge is 80 $m^3\,s^{-1}$ (water + sediment; see Fig. 11), which is 15 $m^3\,s^{-1}$ more than the estimated clear water peak discharge (65 $m^3\,s^{-1}$) due to sediment load (Q_S). Similar values for total peak discharge (98.4 $m^3\,s^{-1}$) have been reported by Ciervo et al. (2014), who used a dedicated numerical code for the hydraulic modelling of the 9 September 2010 Atrani flash flood.

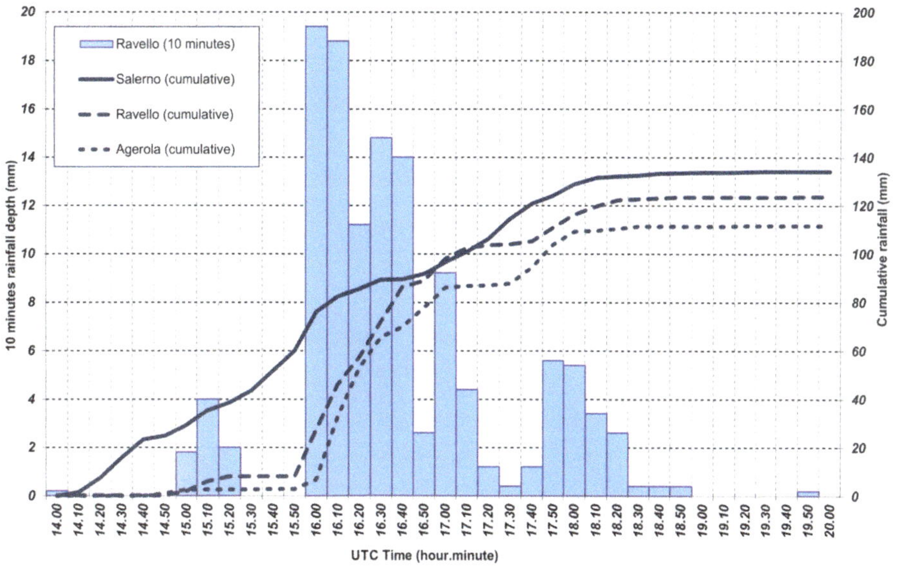

Figure 8. Cumulative rainfall at 10 min intervals on 9 September 2010, from 14:00 to 20:00 UTC, as recorded by the rain gauges of Salerno, Ravello, and Agerola. Rainfall intensity at Ravello rain gauge is also reported. Data are from the Centre for weather forecast and monitoring of the Campania regional authority (CEMPID, 2010).

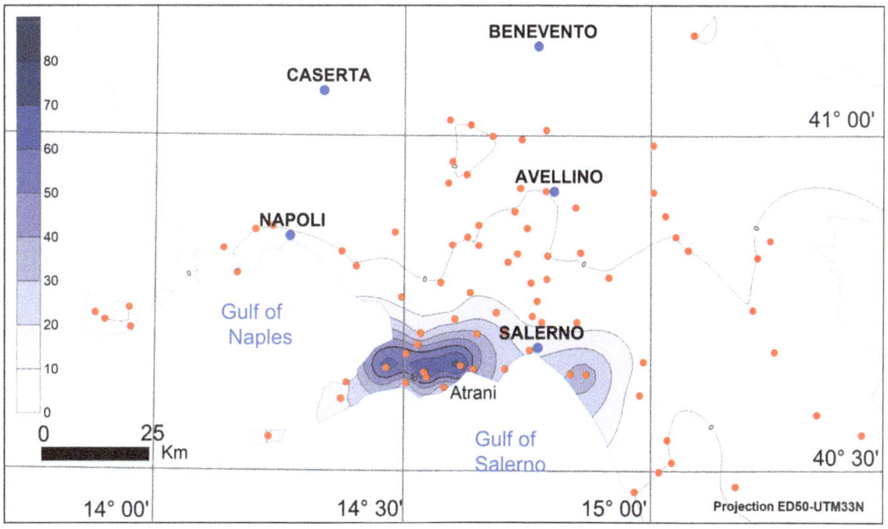

Figure 9. Isohyetal map of the cumulated rainfall from 16:00 to 16:50 UTC. Red dots denote rain gauges.

Assuming that active sediment removal occurred in a time span of 40 ÷ 60 min, the calculated sediment volume mobilized during the event is

$$V_s = Q_s \times t = 15 \times \frac{(40 \div 60)}{2} \times 60$$
$$= 18\,000 \div 27\,000\,\mathrm{m}^3. \qquad (4)$$

This value is in good agreement with the volume estimation of the sediments deposited in the form of an alluvial fan delta at the Dragone mouth and on the street and square that cover the stream path (Fig. 13). Comparison of bathymetric data collected soon after the flood event with an older data set owned by IAMC-Napoli enabled the measurement of the sediment volume that flowed into the sea, at about 14 000 m^3 (Fig. 13). On land, sediment thickness along Via Dei Dogi and Umberto I square reached depths averaging 0.5 m, while beach aggradation was about 1 m, with a volume of sediment accumulation of about 7000 m^3. Therefore, the estimated volume of the sediment transported to the terminal section of the Dragone stream is about 21 000 m^3. Taking into account the additional volumes (estimated at about 20 % of the measured volume) removed by the sea currents, or related to dispersal of finer sediments at sea, a total volume of about 25 000 m^3 can be obtained.

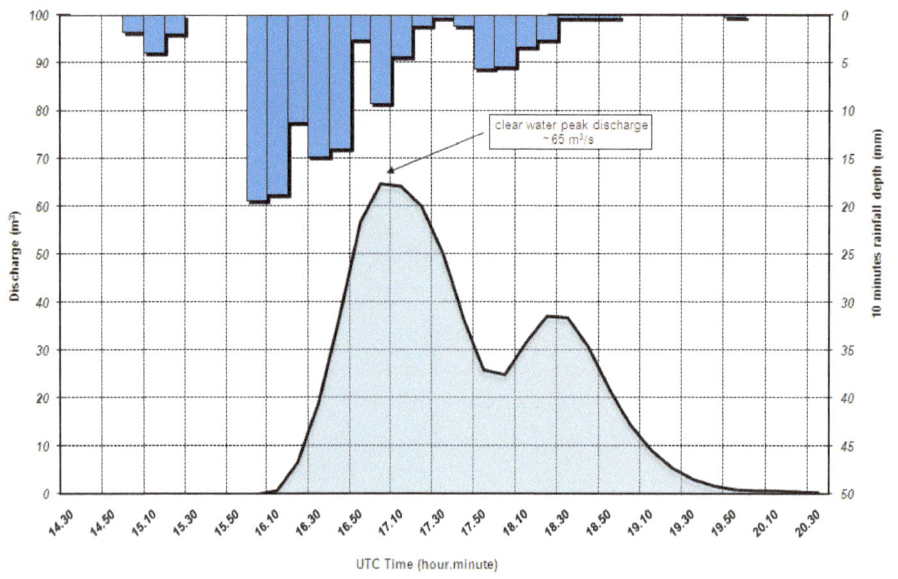

Figure 10. Clear water hydrograph resulting from the hydrological model.

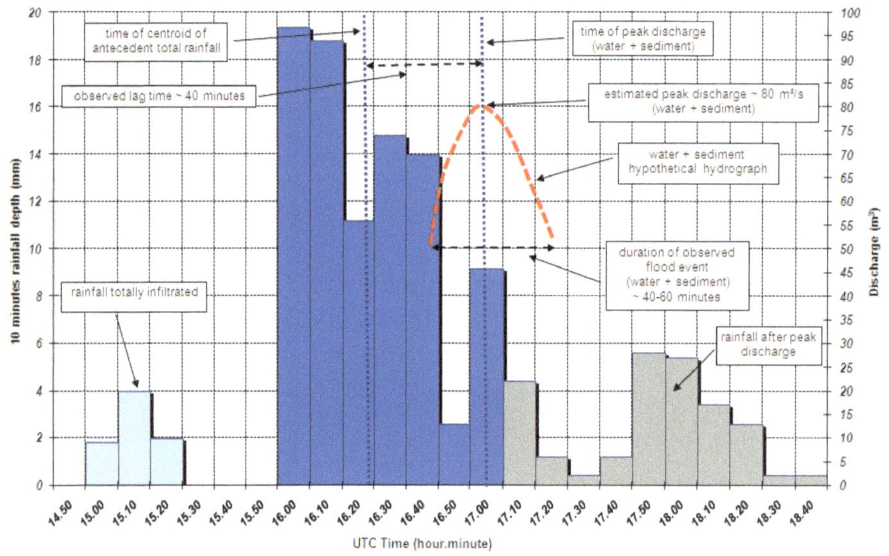

Figure 11. Hyetograph and estimated flood peak discharge.

Sediment volume can be expressed as sediment bulking factor (BF; Gusman et al., 2009) that defines the ratio between the peak flood discharge Q_B and clear water discharge (Q_W):

$$\mathrm{BF} = \frac{Q_B}{Q_W} = \frac{Q_W + Q_s}{Q_W}. \quad (5)$$

In terms of sediment load, bulking factor can be also expressed as

$$\mathrm{BF} = \frac{1}{1 - \frac{C_V}{100}}. \quad (6)$$

C_V is the volume concentration of sediment.

In the case of the 9 September 2010 flood event the estimated BF to the flood peak is 1.2 (5) that leads to a sediment concentration on the order of 20 % in volume (6). This value is close to the lower limit of hyperconcentrated flow (Costa, 1988; Jakob and Hungr, 2005) and it corresponds to a flow in which peak discharge is comparable to that of a clear water flood, and velocities are similar to those of water during a flood (Hungr et al., 2001; Pierson, 2005). This is also confirmed by amateur videos that show a very turbulent flow.

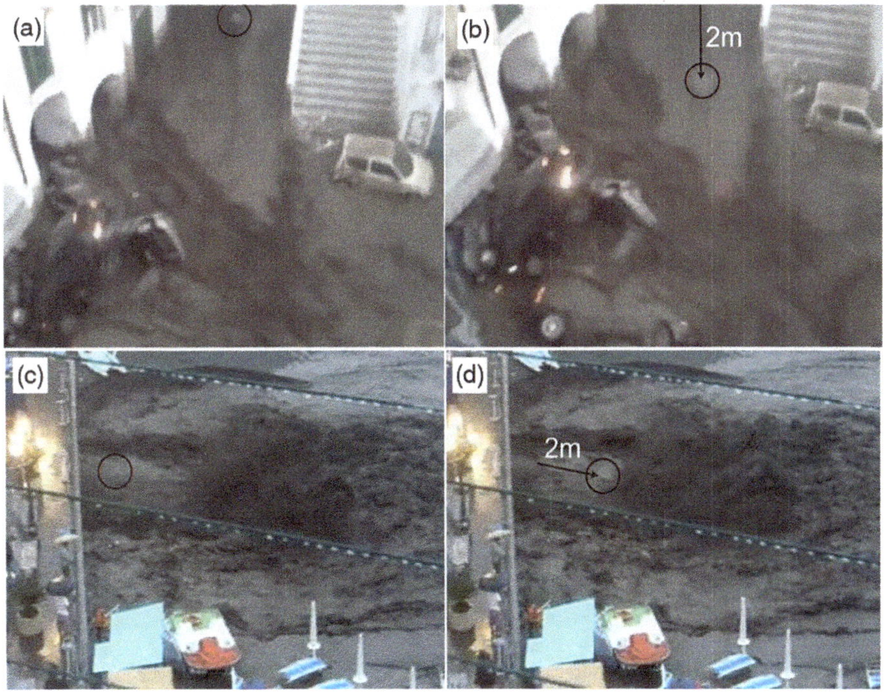

Figure 12. Estimation of flow velocity from amateur videos by tracking selected particles (black circles) transported by the floods. Panels **(a)** and **(b)** show the upper flow along Via Dei Dogi. Panels **(c)** and **(d)** show underground flow at the Dragone stream mouth. Location in Fig. 13. See text for discussion.

7 Historical documentation of past floods

Historical records are an important source of information on floods and fundamental to establish a reliable flood return frequency. The potential of historical data is based on their value as repositories of the magnitude and characteristics of flood events by indicating the level of damage, the number of victims, as well as the type of the induced geological effects (e.g. Calcaterra and Parise, 2001; Tropeano and Turconi, 2004; Marchi and Tecca, 2006). This is particularly true for the Campania region, an area with a long history and much documentation, where historical research is crucial for hydrogeological risk assessment in the perspective of proper land use planning, damage mitigation, and reduction of human fatalities.

In the study area the systematic analysis of more than 3000 documents, including documentary and bibliographic sources, as well as newspapers at the state archives in Naples and Salerno (Esposito et al., 2003, 2004; Porfido et al., 2009), complemented by data attained from scientific papers and national and international projects (Guzzetti et al., 1994; Guzzetti and Tonelli, 2004), allowed the time–space distribution and the characteristics of flood events that have occurred since 1540 to be reconstructed. The quality and completeness of the various sources were evaluated and carefully analysed in their historical context, to obtain the best information rather than the best data set quality (Calcaterra and Parise, 2001; Barriendos et al., 2003). Eighteen events, excluding the one discussed in the present paper, were identified and characterized on the basis of (a) the distribution of the flooded areas, (b) distribution of damaged localities, (c) duration and timing of the event, and (d) number of casualties. The historical information shows that most of the events (13 events) took place at the transition between summer and autumn, four events were in winter, while only one was recorded during the spring season (Table 4).

The intensity and the duration of the rainfall, the level and distribution of damage of man-made structures, the number of victims, and the induced geological effects have been considered to distinguish major flash floods from minor flash floods (Table 4; Casas Planes et al., 2003; Llasat et al., 2005; Barnolas and Llasat, 2007). In particular, shoreline progradation has been considered as a key element for major flash floods as it results from significant bed load transport and hyperconcentrated streamflows. Based upon the above assumptions, the following 6 flood events out of the total of 18 have been classified as major flash floods.

- October 1540: not much information is available for this event, but it is reported as "the great Atrani flood". This indication, along with the occurrence of severe damage and extensive landslides and inundation, allows it to be classified as a flash flood.

- August 1588: this was a flood that produced severe damage to properties, extensive inundation, landslides and shoreline progradation, as reported by the Cronaca

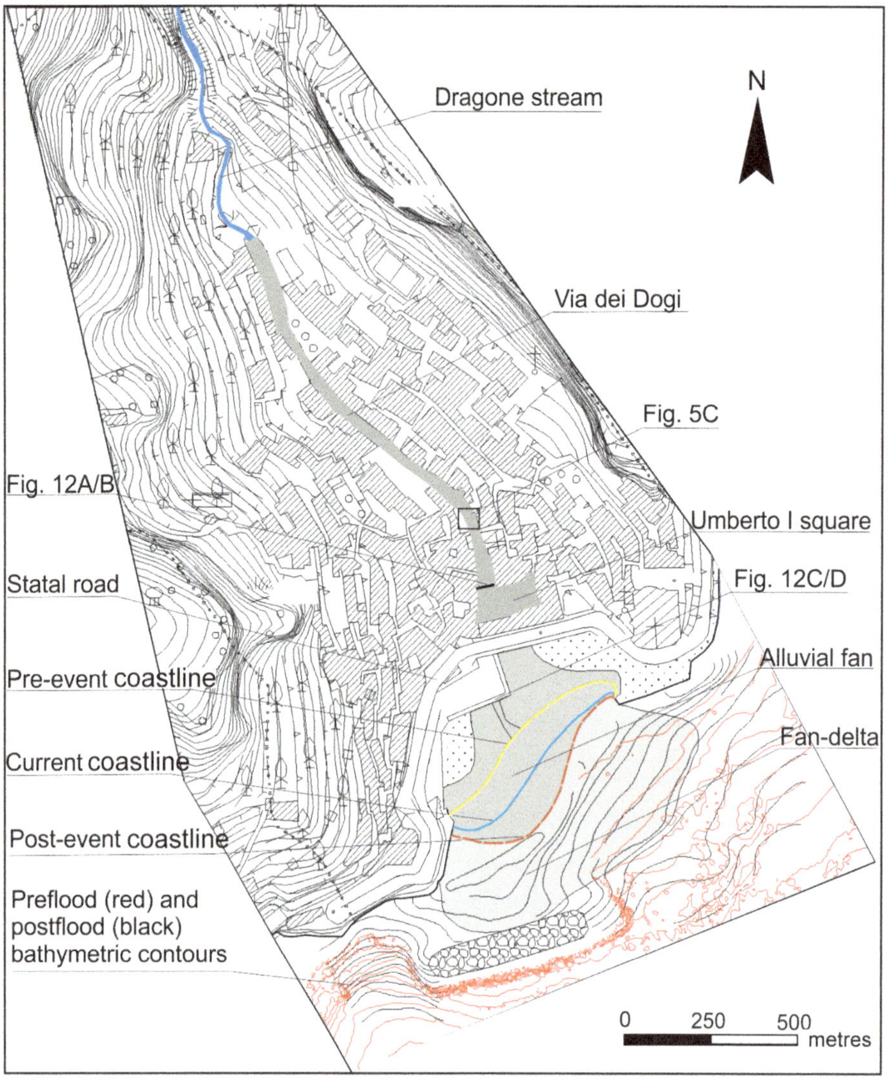

Figure 13. Map of the flooded area with indication of the alluvial fan and the submerged fan delta.

Amalphitana, Ignoti auctoris 1588, cited by Camera (1881): "at the end of the past month of August 1588 much lava fell down ... it destroyed the Seggio building ... the force of the lava removed trees, wood, earth and rocks and ... it filled the harbour and pushed the sea back seven roads (14 m) thereby enlarging the harbour".

- 20 January 1764: this flood caused much damage to the Santa Maria Acquabona church and to many flour mills as well as to some bridges. Extensive landslides coming from Scala hit Atrani, causing two deaths. The sediment transfer from Scala to Atrani indicates a mass transport along the main stream, likely produced by a flash flood.
- 17 January 1780: heavy rain hit the village of Atrani, causing 26 casualties, a huge landslide, and extensive damage (Greco, 1787). The elevated number of victims suggests a flash flood event.
- 18 August 1949: on 18 August, water masses flooded along the main street and square of Atrani, reaching a height of about 3 m. A widespread pattern of destruction characterized this event: boats, nets, and fishing gear were swept away or submerged by mud; there was also damage to buildings, destruction of roads, aqueducts, and sewer systems. Thousands of cubic metres of material and muddy debris were left in the Umberto I square as well as on the beach, producing a shoreline progradation of about 20 m.
- 1 October 1949: the effects of the past flood events were still evident when a second one occurred with greater violence. The breaking up of the main street occurred at several places, and water supply pipelines, recently refitted, were destroyed again. Huge amounts of debris and mud were transported all the way down to the coast in addition to the material recently transported by the 18

Table 4. Flood events occurred at Atrani from 1540 to August 2010 with indication of the induced effects. MF: major flash flood. Mf: minor flash flood.

Year	Month	Day	Damage	Victims	Geological effect	Flood type
1540	10	8	Severe	–	Slides	MF
1588	8	31	Major	Some	Shoreline progradation	MF
1764	1	20	Major	2		MF
1780	1	17	Severe	26		MF
1823	10	18	Minor			Mf
1824	10	3	Minor			Mf
1904	10	7	Minor			Mf
1924	3	27	Minor		Slides	Mf
1935	3	1	Minor			Mf
1949	8	18	Major		Shoreline progradation	MF
1949	10	1	Major		Shoreline progradation	MF
1953	9	11	Minor			Mf
1954	10	25	Minor			Mf
1969	3	15	Minor			Mf
1984	8	28	Minor	1		Mf
1987	10	6	Minor			Mf
1988	9	14	Minor			Mf
2007	9	20	Minor			Mf

August flood, so that a large beach developed at the foot of coastal cliffs, between the village of Atrani and the nearby locality of Castiglione.

8 Conclusions

Detailed field surveys and measurements, along with information from eyewitnesses and home-made videos, proved to be critical for reconstructing and modelling the flash flood that affected the village of Atrani on 9 September 2010. The collected data were combined with meteorological and historical analyses and marine geophysical surveys in order to reconstruct the physical features of the flood event and to evaluate the recurrence of flash floods in the study area. The main results can be summarized as follows.

- The rainstorm that generated the Atrani event lasted 4 h and was strongly conditioned by the local orography and positive thermic anomalies of the coastal waters during the warm season.

- In Atrani the rainfall event lasted about 1 h with cumulative rainfall of 80.8 mm and maximum rainfall intensity of nearly 120 mm h^{-1}. It was produced by a single storm cell elongated in a NE–SW direction, with a very flat elliptical shape and of limited areal extent (from 50 to 70 km^2) that can be ascribed to a β type mesoscale convective system.

- The estimated peak discharge of the clear water produced in the Dragone stream is about 65 $m^3 s^{-1}$ while the estimated total peak discharge (water + sediment) is 80 $m^3 s^{-1}$, leading to a sediment concentration of about 20 % in volume.

- Sediment removal mostly occurred through linear erosion that significantly engraved tributaries and the main stream. The displaced materials were mostly composed of pyroclastic deposits and solid wastes occurring at the channel beds and behind the hydraulic check dams. A reduced size of erodible sediment stored in channels may consequently reduce the mud-flow hazard and provide protection for the residential area on the alluvial fan in Atrani.

- The analysis of historical sources shows that 18 flood events occurred in the Dragone catchment in the last five centuries. Six events were classified as major flash floods and twelve as minor flash floods. The internal stratigraphic architecture of the Dragone fan delta confirms the recurrence of floods in the Early Medieval Cool Period (ca. AD 500–AD 800), in the Medieval Warm Period (ca. AD 900–AD 1100), and the Little Ice Age (ca. AD 1400–AD 1850).

The approach used in this work is representative of geomorphological and urban settings characterized by very small ungauged and rocky watersheds with ephemeral discharge, and local communities mostly living at the stream mouths. In these settings the importance of erosional processes claim the use of different data sources for predictive water models that include geological and hydrological analyses.

Acknowledgements. The authors wish to thank Crescenzo Minotta, geologist at Autorità di Bacino Campania Sud, and Maria Carla Sorrentino, archaeologist and member of the "Centro Universitario Europeo per i Beni Culturali" for support with post-event bathymetric survey and assistance in the field, respectively. We also thank Luigi Amato of the Cultural Association "Sos Dragone" that helped to reconstruct the timing of the flooding event. This work was funded by a research grant CUP: G23F11000790004 to Crescenzo Violante. Two anonymous referees are thanked for their constructive comments on the submitted manuscript.

References

Barnolas, M. and Llasat, M. C.: A flood geodatabase and its climatological applications: the case of Catalonia for the last century, Nat. Hazards Earth Syst. Sci., 7, 271–281, doi:10.5194/nhess-7-271, 2007.

Barriendos, M., Llasat, M. C., Barrera, A., and Rigo, T.: The study of flood events from documentary sources, Methodological guidelines for historical sources identification and flood characterization in the Iberian Peninsula, in: Palaeofloods, Historical Floods and Climatic Variability: Applications in Flood Risk Assessment (Proceedings of the Phefra Workshop, Barcelona, 16–19 October, 2002), edited by: Thorndycraft, V. R., Benito, G., Barriendos, M., and Llasat, M. C., 87–92, 2003.

Budillon, F., Violante, C., Conforti, A., Esposito, E., Insinga, D., Iorio, M., and Porfido, S.: Event beds in the recent prodelta stratigraphic record of the small flood-prone Bonea Stream (Amalfi Coast, Southern Italy), Marine Geology, 222–223, 419–441, 2005.

Calcaterra, D. and Parise, M.: The contribution of historical information in the assessment of the landslide hazard, in: The Use of Historical Data in Natural Hazard Assessments, edited by: Glade, T., Albini, P., and Frances, F., Adv. in Nat., and Tech. Hazards Res., 17, Kluwer Academic Publishers, 201–217, 2001.

Calcaterra, D., Parise, M., and Palma, B.: Combining historical and geological data for the assessment of the landslide hazard: a case study from Campania, Italy, Nat. Hazards and Earth System Sciences, 3 (1/2), 3–16, 2003.

Camera, M.: Memorie storico diplomatiche dell'antica città e ducato di Amalfi, (reprinted Centro Cultura e Storia amalfitana, Amalfi, 1999), 80 pp., 1881.

Casas Planes, A., Benito, G., Díez Herrero, A., and Barriendos, M.: Sphere-Gis: Implementation of an historical and palaeoflood geographical information system, in: Palaeofloods, Historical Floods and Climatic Variability: Applications in Flood Risk Assessment (Proceedings of the Phefra Workshop, Barcelona, 16th–19th October, 2002), edited by: Thorndycraft, V. R., Benito, G., Barriendos, M., and Llasat, M. C., 363–368, 2003.

Celico, P., De Vita, P., Monacelli, G. Scalise, A. R., and Tranfaglia, G.: Hydrogeological map of southern Italy. APAT, Università di Napoli "Federico II", Programma INTERREG IIC – Territory regulation and contrast to the drought, Sottoprogetto "Hydrological cycle analysis", 55 pp., 2005

CEMPID: Centro funzionale per la previsione meteorologica e il monitoraggio meteo pluvio idrometrico e delle frane: Rapporto dell'evento del 9 Settembre 2010 nel territorio dei comuni di Scala ed Atrani, Napoli, 38 pp., 2010.

Chow, V. T., Maidment, D. R., and Mays, L. W. (Eds): Applied Hydrology, McGraw Hill New York, 572 pp., 1988.

Ciervo F., Papa M. N., Medina V., and Bateman A.: Simulation of flash floods in ungauged basins using post-event surveys and numerical modelling, J. Flood Risk Manage., 1–13, 2014.

Cinque, A. and Robustelli, G.: Alluvial and coastal hazards caused by long-range effects of Plinian eruptions: the case of the Lattari Mts. After the AD 79 eruption of Vesuvius, in: Geohazard in Rocky Coastal Areas, edited by: Violante C., Geological Society of London, Special Publications, 322, 155–171, doi:10.1144/SP322.7, 2009.

Costa, J. E.: Rheologic, geomorphic, and sedimentologic differentiation of water floods, hyperconcentrated flows, and debris flows, in: Flood Geomorphology edited by: Baker, V. R., Kochel, R. C., and Patten, P. C., Wiley-Intersciences, New York, 113–122, 1988.

Esposito E., Porfido S., Violante C., and Alaia F.: Disaster induced by historical floods in a selected coastal area (southern Italy), in: Palaeofloods, Historical Floods and Climatic Variability: Applications in Flood Risk Assessment (Proceedings of the Phefra Workshop, Barcelona,16–19 October, 2002), edited by: Thorndycraft, V. R., Benito, G., Barriendos, M., and Llasat, M. C., 143–148, 2003.

Esposito, E., Porfido, S., Violante, C., Biscarini, C., Alaia, F., and Esposito, G.: Water events and historical flood recurrences in the Vietri sul Mare coastal area (Costiera Amalfitana, southern Italy), in: The Basis of Civilization – Water Science?, edited by: Rodda, G., and Ubertini, L., International Association of Hydrological Sciences Publication, 286, 1–12, 2004.

Fujita, T. T.: Mesoscale classifications: their history and their application to forecasting, in: Mesoscale Meteorology and Forecasting, edited by: Ray, P. S., American Meteorological Society, Boston, 18–35, 1986.

Greco, M.: Cronaca di Salerno, Reprinted Palladio Salerno, 36 pp., 1787.

Gusman, A. J., Teal, M. J., Todesco, D., and Bandurraga, M.: Estimating Sediment/Debris Bulking Factors, in: Proceeding of 33rd IAHR Congress, Vancouver, British Columbia, Canada, 9–14 August 2009, 83–91, ISBN: 978-1-61738-231-4, 2009.

Gutierrez, F., Parise, M., De Waele, J., and Jourde H.: A review on natural and human-induced geohazards and impacts in karst, Earth Science Rev., 138, 61–88, doi:10.1016/j.earscirev.2014.08.002, 2014.

Guzzetti, F., Cardinali, M., and Reichenbach, P.: The AVI Project: A bibliographical and archive inventory of landslides and floods in Italy, Environ. Manage., 18, 623–633, 1994.

Guzzetti, F. and Tonelli, G.: Information system on hydrological and geomorphological catastrophes in Italy (SICI): a tool for managing landslide and flood hazards, Nat. Hazards Earth Syst. Sci., 4, 213–232, doi:10.5194/nhess-4-213-2004, 2004.

Hungr, O., Evans, S. G., Bovis, M., and Hutchinson, J. N.: Review of the classification of landslides of the flow type, Environ. Eng. Geosci., 7, 221–238, 2001.

Jakob, M. and Hungr, O. (Eds.): Debris Flow Hazards and Related Phenomena, Springer-Praxis Book in Geophysical Science, Springer-Verlag Berlin Heidelberg NY, ISBN 3-540-20726-0, 733 pp., 2005.

Jourde, H., Roesch, A., Guinot, V., and Bailly-Comte, V.: Dynamics and contribution of karst groundwater to surface flow during Mediterranean flood, Environ. Geol., 51, 725–730, 2007.

Larsen, M. C., Wieczorek, G. F., Eaton, L. S., Morgan, B. A., and Torres-Sierra, H.: Natural hazards on alluvial fans; The Venezuela debris flow and flash flood disaster, US Geological Survey, Fact Sheet, 103-01, 2001.

Lionello, P., Bhend, J., Buzzi, A., Della-Marta, P. M., Krichak, S., Jansà, A., Maheras, P., Sanna, A., Trigo, I. F., and Trigo, R.: Cyclones in the Mediterranean region: climatology and effects on the environment, in: Mediterranean Climate Variability, edited by: Lionello, P., Malanotte-Rizzoli, P., Boscolo, R., Elsevier, 325–372, ISBN10-0-444-52170-4, 2006.

Llasat, M. C., Barriendos, M., Barrera, T., and Rigo, T.: Floods in Catalonia (NE Spain) since the 14th century. Climatological and meteorological aspects from historical documentary sources and old instrumental records, J. Hydrol., 313, 32–47, 2005.

Marchi, L. and Tecca, P. R.: Some observations on the use of data from historical documents in debris flow studies, Nat. Hazards, 38, 301–320, 2006.

Morel, C. and Senesi, S.: A climatology of mesoscale convective systems over Europe using satellite infrared imagery. II: Characteristics of European mesoscale convective systems, Q. J. Roy. Meteor. Soc., 128, 1973–1995, 2002.

Orlansky, I.: A rational subdivision of scales for atmospheric processes, B. Am. Meteor. Soc., 56, 527–530, 1975.

Passerini, G.: Intorno alle cause del disastro del marzo 1924 nella Penisola Sorrentina, Boll. Ist. Agrario di Scandicci, Ricerche ed esperienze istituite dal 1914 al 1925, 8, ISBN: 0366-1881, 249–269, 1925.

Penta, F., Lupino, R., Camozza, F., and Esu, F.: Effetti dell'alluvione del 26 ottobre 1954 nel Salernitano, Rivista Italiana di Geotecnica, 6, 245–257, 1954.

Perez, F. L.: Matrix granulometry of catastrophic debris flows (December 1999) in central coastal Venezuela, Catena, 45, 163–183, 2001.

Pierson, C.: Hyperconcentrated flow—transitional process between water flow and debris flow, in: Debris-flow Hazards and Related Phenomena, edited by: Jakob, M., and Hungr, O, Springer/Praxis, Chichester, UK, 159-202, 2005.

Plate, J: Flood risk and flood management, Journal of Hydrology, 267, 2–11, 2002.

Porfido S., Esposito E., Alaia F., Molisso F., and Sacchi M.: The use of documentary sources for reconstructing flood chronologies on the Amalfi rocky coast (southern Italy). in: Geohazard in Rocky Coastal Areas, edited by: Violante C., Geological Society of London, Special Publications, 322, 173–187, doi:10.1144/SP322.8, 2009.

Raffel, M., Willert, C. E., Wereley, S. T., and Kompenhans, J.: Particle Image Velocimetry, A Practical Guide, 2nd Ed., Springer Berlin. 447 pp. 2007.

Sacchi, M., Molisso, F., Violante, C., Esposito, E., Insinga, D., Lubritto, C., Porfido, S., and Toth, T.: Insights into flood dominated fan deltas: very high-resolution seismic examples off the Amalfi cliffed coasts, eastern Tyrrhenian Sea, in: Geohazard in Rocky Coastal Areas, edited by: Violante C., Geological Society of London, Special Publications, 322, 33–72, doi:10.1144/SP322.2, 2009.

Sigurdsson, H., Carey, S., Cornell, W., and Pescatore, T.: The eruption of Vesuvius in AD 79, Nat. Geogr. Res., 1, 332–387, 1985.

Singh, V. P. (Ed.): Hydrologic Systems. Watershed Modelling, Prentice Hall Englewood Cliffs, New Jersey, 320 pp., 1989.

Smith, K. and Ward, L.: Floods. Physical processes and human impacts, Chichester, John Wiley, 382 pp., 1998.

Thunis, P. and Bornstein, R: Hierarchy of Mesoscale Flow Assumptions and Equations, J. of Atmos. Sci., 53, 380–397, 1996.

Tropeano, D. and Turconi, L.: Using historical documents for landslide, debris flow and stream flood prevention, Applications in Northern Italy, Nat. Hazards, 31, 663–679, 2004.

USDA-SCS: Technical Release 55: Urban Hydrology for Small Watersheds, USDA (U.S. Department of Agriculture), http://www.cpesc.org/reference/tr55.pdf, 1986a.

USDA-SCS: National Engineering Handbook, Section 4 – Hydrology, Washington, D.C., 80 pp., 1986b.

USACE HEC: Hydrologic Modelling System version 3.5, User's Manual (CPD-74A), 306 pp., 2010.

Violante, C.: Rocky coast: geological constrains for hazard assessment, in: Geohazard in Rocky Coastal Areas, edited by: Violante C., Geological Society of London, Special Publications, 322, 1–31, doi:10.1144/SP322.1, 2009.

Violante, C., Biscarini, C., Esposito, E., Molisso, F., Porfido, S., and Sacchi, M.: The consequences of hydrologic events on steep coastal watersheds: the Costa d'Amalfi, eastern Tyrrhenian Sea, in: The Role of Hydrology in Water Resource Management, edited by: Liebscher, H. J., Clarke, R., Rodda, J., Schultz, G., Schumann, A., Ubertini, L., and Young, G., Capri Italy, 2008. IAHS Publication, 327, 102–113, 2009.

5

Spatial impact and triggering conditions of the exceptional hydro-geomorphological event of December 1909 in Iberia

S. Pereira[1], A. M. Ramos[2], J. L. Zêzere[1], R. M. Trigo[2], and J. M. Vaquero[3]

[1]Centre for Geographical Studies, Institute of Geography and Spatial Planning, Universidade de Lisboa, Lisboa, Portugal
[2]Instituto Dom Luiz (IDL), Faculdade de Ciências, Universidade de Lisboa, Lisboa, Portugal
[3]Departamento de Física, Universidad de Extremadura, Mérida, Spain

Correspondence to: S. Pereira (susana-pereira@campus.ul.pt)

Abstract. According to the DISASTER database the 20–28 December 1909 event was the hydro-geomorphologic event with the highest number of flood and landslide cases that occurred in Portugal in the period 1865–2010 (Zêzere et al., 2014). This event also caused important social impacts over the Spanish territory, especially in the Douro Basin, having triggered the highest floods in more than 100 years at the river's mouth in the city of Oporto.

This work has a dual purpose: (i) to characterize the spatial distribution and social impacts of the December 1909 hydro-geomorphologic DISASTER event over Portugal and Spain; (ii) to analyse the meteorological conditions that triggered the event and the spatial distribution of the precipitation anomalies.

Social impacts that occurred in Portugal were obtained from the Disaster database (Zêzere et al., 2014) whereas the data collection for Spain was supported by the systematic analysis of Spanish daily newspapers. In addition, the meteorological conditions that triggered the event are analysed using the 20th Century Reanalysis data set from NOAA and precipitation data from Iberian meteorological stations.

The Iberian Peninsula was spatially affected during this event along the SW–NE direction spanning from Lisbon, Santarém, Oporto, and Guarda (in Portugal), to Salamanca, Valladolid, Zamora, Orense, León, and Palencia (in Spain). In Iberia, 134 DISASTER cases were recorded (130 flood cases; 4 landslides cases) having caused 89 casualties (57 due to floods and 32 due to landslides) and a further total of 3876 affected people, including fatalities, injured, missing, evacuated, and homeless people.

This event was associated with outstanding precipitation registered at Guarda (Portugal) on 22 December 1909 and unusual meteorological conditions characterized by the presence of a deep low-pressure system located over the NW Iberian Peninsula with a stationary frontal system striking the western Iberian Peninsula. The presence of an upper-level jet (250 hPa) and low-level jet (900 hPa) located SW–NE oriented towards Iberia along with upper-level divergence and lower-level convergence favoured large-scale precipitation. Finally, associated with these features it is possible to state that this extreme event was clearly associated with the presence of an elongated Atmospheric River, crossing the entire northern Atlantic Basin and providing a continuous supply of moisture that contributed to enhance precipitation.

This work contributes to a comprehensive and systematic synoptic evaluation of the second most deadly hydro-geomorphologic DISASTER event that has occurred in Portugal since 1865 and will help to better understand the meteorological system that was responsible for triggering the event.

1 Introduction

In the Iberian Peninsula, extreme precipitation events that occur during winter (December-January-February-March) have been historically associated with progressive flooding events in the major rivers (Salgueiro et al., 2013), and exceptionally with flash floods that typically occur in small watersheds or urban areas (Llasat et al., 2005; Fragoso et al., 2012) and also landslide events (Zêzere and Trigo, 2011; Zêzere et al.,

2014, 2015). These events have been the source of major socio-economic impacts, with human and material damage (Barriendos and Rodrigo, 2006; García et al., 2007; Vicente-Serrano et al., 2011; Trigo et al., 2014, 2015).

Regularly, these extreme events have been evaluated based on case studies, like for instance the large flood in the Guadiana River in 1876 (Trigo et al., 2014), the November 1967 flash flood in the Lisbon region (Trigo et al., 2015), the 10 June 2000 flash flood in Catalonia (Llasat et al., 2003), and the 20 February 2010 flash floods in Madeira (Fragoso et al., 2012). Major floods that occurred in the Iberian Peninsula were extensively analysed in the Tagus Basin (Benito et al., 2003; Salgueiro et al., 2013; Machado et al., 2015) and to a lesser extent in the Douro Basin (e.g. Morán-Tejeda et al., 2012).

Recently, a long-term database of hydro-geomorphological cases was developed for Portugal covering the period 1865–2010 in the project DISASTER (Zêzere et al., 2014). This database accounts for 1621 flood cases and 281 landslide cases that were responsible for 1251 fatalities and more than 41 000 homeless people (Zêzere et al., 2014). This database includes social consequences (fatalities, injured, missing, evacuated, and homeless people) caused by flood and landslide cases referred to in Portuguese newspapers during the period of 1865–2010. The DISASTER database is the only database containing detailed data on the social impacts of hydro-geomorphologic disasters in Portugal as it accounts for any flood or landslide that, independently of the number of affected people, caused casualties or injuries, or missing, evacuated, or homeless people. Therefore, each DISASTER case is a unique hydro-geomorphologic occurrence, which is related to a unique geographic location and a specific period of time (e.g. the place where the flood or landslide harmful consequences occurred on a specific date). Within this database, a DISASTER event is a set of flood and landslide cases sharing the same trigger mechanism and thus clustered in the temporal domain, which may have a widespread spatial extension and a certain magnitude (Zêzere et al., 2014).

The December 1909 flood in Portugal is considered the most extraordinary flood of the 20th century along the Douro River since systematic records of the river flow began (Rodrigues et al., 2003). This particular flood also had a strong contribution from the Spanish area of the Douro Basin, at a time when the hydrologic regime was only conditioned by natural factors due to the absence of dams to control the river flow (Rodrigues et al., 2003). It is worth stressing that after the flash flood of November 1967 (with circa 500 deaths) the 1909 flood was the second deadliest Disaster event in Portugal in the period 1865–2010, and generated 35 fatalities (29 caused by floods and six caused by landslides). In addition, according to the DISASTER database the 20–28 December 1909 was the DISASTER event with the highest number of flood and landslide cases (67 flood cases and three landslide cases) occurring in Portugal in the period 1865–2010 (Zêzere et al., 2014). As we show later in the descriptive analysis there were an even larger number of fatalities in Spain (54), raising the total number of deaths to 89.

The societal impacts associated with the 1909 event were particularly high over the NW Iberian Peninsula (Fig. 1). Thus, while most observed floods were concentrated in Portuguese and Spanish provinces of the Douro Basin, other areas were equally affected, including the Minho Basin and Tagus valley (Santarém), Lisbon region and also the northern sector of the Tagus Basin in the Spanish territory (Fig. 1).

Besides the intrinsic relevance of this extreme episode from the impact perspective, this event was associated with some outstanding precipitation values and unusual meteorological conditions. However, to the best of our knowledge this extreme hydro-meteorological episode has never been evaluated in detail, namely by Portuguese and Spanish authorities or research teams. This lack of previous assessments resulted in our view from several distinct factors, including (1) lack of sufficient long-term time series of precipitation covering most of the affected territory, (2) lack of surface and upper-air meteorological fields capable of describing the most important atmospheric circulation features of the episode, and (3) the inexistence of a list of places with the affected people and socio-economic impacts for the complete event over Iberia. These three shortcomings result from the time that has passed since the event in the early 20th century. However, these caveats have been minimized to a large extent in recent years. Thus, a number of initiatives in Portugal and Spain have favoured the digitalization of time series from the late 19th century and beginning of 20th century, allowing the assessment of long-term changes in daily precipitation (Gallego et al., 2011; Cortesi et al., 2014), or related to major historical floods and extreme precipitation events (Trigo et al., 2014; Domínguez-Castro et al., 2015). Likewise, a new type of long-term reanalysis has been developed (Compo et al., 2011), and despite being based on surface data, it is capable of reproducing upper-air fields (e.g. Trigo et al., 2014). Finally, as stated previously, a comprehensive list of people affected (fatalities, injured, displaced, evacuated, and disappeared) has been obtained through the DISASTER database (Zêzere et al., 2014) although covering only the Portuguese territory.

In this context, this work has a dual purpose: (i) to characterize the spatial distribution and social impacts of the December 1909 hydro-geomorphologic DISASTER event over Portugal and Spain; (ii) to analyse the meteorological conditions that triggered the event and the spatial distribution of the precipitation anomalies. We aim at characterizing the meteorological synoptic forcing of a major hydro-geomorphologic event that occurred in the last 150 years and registered in the DISASTER database. However, besides this meteorological/climatological assessment we are equally interested in obtaining a full characterization of the associated human cost and the spatial extent of associated social im-

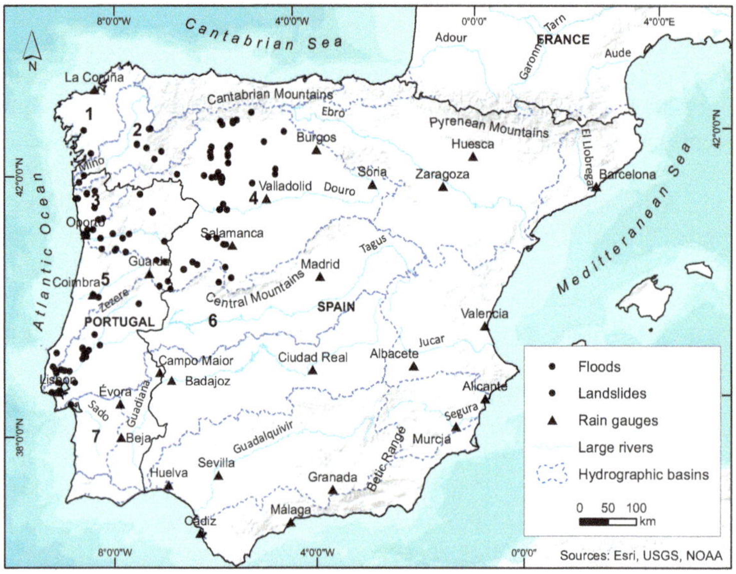

Figure 1. Study area and flood and landslide DISASTER cases from 20–28 December 1909 event over Iberia. Source: Disaster database for Portugal and Spanish newspapers (see Table 1). Numbers in the map correspond to the affected hydrographic basins: (1) Coruña basins; (2) Minho Basin; (3) Lima and Cávado basins; (4) Douro Basin; (5) Águeda, Mondego, and West basins; (6) Tagus Basin; (7) Sado Basin.

pacts, as such evaluation has never been done before. Overall, we expect to gain a better understanding of the climatic driving forces of hydro-geomorphologic disaster events and corresponding societal impacts over Iberia.

2 Study area and precipitation regime

The Iberian Peninsula is located southwest of Europe and includes the countries of Andorra, Portugal, and Spain. It is surrounded by the Mediterranean Sea (on the southeast and east), the Atlantic Ocean (on the west), and the Cantabrian Sea (on the north). The central part of the Iberian Peninsula is occupied by the Meseta, a vast plateau (Douro and Tagus Cainozoic basins) that is characterized by hilly and smooth terrain. The central Meseta is surrounded by mountains (e.g. Cantabrian Mountains, Pyrenean Mountains, the Betic Range, and Central Mountains) and is sculptured by the fluvial network (Santisteban and Schulte, 2007) (Fig. 1).

There are four large international Iberian rivers (Minho, Douro, Tagus, and Guadiana), all flowing predominantly along the NE–SW direction, with their lower courses in Portugal (Fig. 1). The seasonal precipitation regime in the Iberian Peninsula controls the major river flow regime (Trigo et al., 2004) like in other Mediterranean regions, where most of the river flow is registered in winter and spring time (Daveau, 1998).

In the Iberian Peninsula, there is a high spatial and temporal variability of precipitation (Serrano et al., 1999; Trigo et al., 2004; Cortesi et al., 2013) due to its location in the southern rim of the Atlantic zonal circulation (Trigo and Palutikof, 2001). The successive passage of frontal systems from west to east over the North Atlantic Basin (most frequent in the winter) produces continuous and persistent daily precipitation, mostly in the western Iberian Peninsula, which can cause major river floods (Salgueiro et al., 2013). Precipitation distribution over the Iberian Peninsula depends also on the wind trajectory (SW, W, or NW) and on the local orographic effects (Goodess and Jones, 2002; Salgueiro et al., 2013; Cortesi et al., 2014) that influence precipitation intensity and type (snow vs. hail or rainfall) at the regional scale.

Numerous studies have shown that the NAO has a strong impact on the occurrence of extreme flooding in the major rivers, namely along the Tagus River (Benito et al., 2003; Silva et al., 2012; Salgueiro et al., 2013; Machado et al., 2015) and the Guadiana River (Ortega and Garzón, 2004).

In addition, Ramos et al. (2015) show that the presence of the so-called Atmospheric River (AR) structure over the Iberian Peninsula can enhance the precipitation amount, as was the case of another outstanding historical flood event in southern Iberia in 1876 (Trigo et al., 2014). Therefore this will be also taken into account when analysing the 1909 hydro-geomorphologic DISASTER event. ARs are identified as narrow plumes of enhanced moisture transport that are usually present in the core of the warm conveyor belt and occur on the oceans along the warm sector of extra-tropical cyclones (Ramos et al., 2015). They are usually W–E (or SW–NE) oriented and steered by pre-frontal low-level jets along the trailing cold front and subsequently feed the precipitation in the extra-tropical cyclones. The large quantity of water vapour within the AR can generate heavy precipitation and floods when they make landfall (Dettinger, 2011; Lavers et al., 2011). Regarding the Iberian Peninsula, the

Table 1. Newspapers explored for the December 1909 event data collection.

Newspaper	Reference period	Category	Distribution	Spatial incidence
			Portugal	
Diário de Notícias	Since 1865	Daily	National	Mainly the metropolitan area of Lisbon and the Tagus valley region
Jornal de Notícias	Since 1888	Daily	Regional	North region (mainly the metropolitan area of Oporto)
Correio de Mirandela	1907–1937	Weekly	Local	North region (Trás-os-Montes)
			Spain	
La Época	1849–1936	Daily	National	Madrid
La Correspondencia de España	1859–1925	Daily	National	Madrid
El Imparcial	1867–1933	Daily	National	Madrid
El Siglo Futuro	1875–1936	Daily	National	Madrid
El Liberal	1879–1939	Daily	Regional	Madrid, Seville, Barcelona, Bilbao, Murcia

impact of ARs in extreme precipitation has been assessed systematically (Ramos et al., 2015) but also in relation to several well-known events of floods and extreme precipitation with major socio-economic consequences (Couto et al., 2012; Liberato et al., 2013). There are several methodologies to identify ARs (Gimeno et al., 2014), here we adopted the method requiring the existence of a long ($>$ 2000 km length) and narrow ($<$ 500 km width) band of water vapour at low levels from the subtropics ($>$ 5 g kg^{-1} of specific humidity at 900 hPa), while wind speed should be considered above 12.5 ms^{-1} (Lavers et al., 2011).

3 Materials and methods

3.1 Historical data sources

The data source used to analyse the event of December 1909 in Portugal was the DISASTER Database (Zêzere et al., 2014). The data collection process used to construct the DISASTER database was based on painstaking analysis of numerous daily and weekly newspapers, and is explained in full detail in Zêzere et al. (2014).

In 1909 there were available only four newspapers for data collection in Portugal (Zêzere et al., 2014). The national daily newspaper "Diário de Notícias" published since 1865 in Lisbon provided the longest time period to data collection. As well as this daily newspaper the "Jornal de Notícias", published in Oporto (North Portugal) since 1888, was also analysed. The remaining two newspapers are weekly local and regional newspapers, one located in the North ("Correio de Mirandela") and another one in the South of the country ("O Algarve: o semanário independente").

In Spain there is no equivalent to the DISASTER database built with similar criteria in order to allow data comparison between the two countries. As a consequence, we have adopted the same methodology and criteria used in the DISASTER database to ensure data integrity and to enable data comparison. The data collection process was supported by the systematic analysis of Spanish daily newspapers of the entire month of December 1909. Four daily national newspapers ("La Época", "La Correspondencia de España", "El Imparcial", "El Siglo Futuro") published in Madrid were used for data collection and a regional newspaper ("El Liberal") with spatial incidence in Madrid, Sevilla, Barcelona, Bilbao and Murcia was also explored (Table 1). In summary, we can state that, although not included in the original DISASTER database, all locations selected in Spain relative to this event fulfil the DISASTER requirements as described in Zêzere et al. (2014). In this regard they can be considered DISASTER-like cases and the use of this denomination will be applied indiscriminately throughout this work, to both cases and locations in Portugal and Spain.

Each DISASTER case includes details on process characteristics and associated damages. The first includes data on type (flood or landslide), subtype, date, location, triggering factor, and data source. In addition, the following damage data were recorded: number of casualties, injuries, missing, evacuated or homeless people, type of damages in buildings, number of affected buildings, type of damage in road and railroad, extent of interruptions in road and railroad circulation (Zêzere et al., 2014). Information about the gender of the fatalities was also obtained from the content analysis of the newspapers as well as the circumstances surrounding the fatalities (Pereira et al., 2015). The latter were classified into four classes according to the location where the fatalities occurred – namely, inside a building in a rural area; inside a building in an urban area; outdoors in a rural area; outdoors in an urban area.

Location of DISASTER cases in both countries was obtained using a point shapefile. The precision of the location depends on the detail of the news and was classified in the following classes: (i) exact coordinates (accuracy associated with scale 1 : 1000); (ii) local toponymy (accuracy associated with scale 1 : 10 000); (iii) local geomorphology (accuracy associated with scale 1 : 25 000 scale); (iv) centroid of the parish; and (v) centroid of the municipality (Zêzere et al., 2014).

Additionally, in order to evaluate the relative level of damages along the affected regions during this historical event, three damage indices (fatalities index, homeless index, and total affected index) were computed, taking into account the resident population per region obtained from the Census data in Portugal (1911) and Spain (1910):

$$\text{Fatalities index} = \frac{\text{No. of fatalities}}{\text{Resident population}} \cdot 100\,000 \quad (1)$$

$$\text{Homeless index} = \frac{\text{No. of homeless people}}{\text{Resident population}} \cdot 100\,000 \quad (2)$$

$$\text{Total affected index} = \frac{\text{No. of total affected people}}{\text{Resident population}} \cdot 100\,000. \quad (3)$$

The number of total affected people includes the total number of fatalities, injuries, missing people, evacuated people, and homeless people. Damage indices were mapped and classified into five classes (low, moderate, high, very high, extreme) according to natural breaks of index distribution. Natural breaks classes are based on natural groupings inherent in the data that best group similar values and maximize the differences between classes. With this methodology each affected region ("Distrito" in Portugal and "Provincia" in Spain) was classified according to the fatalities index, homeless index, and total affected index. These indices are important to compare the societal impacts of floods and landslides observed in different regions; in addition these indices can be applied to other historical events in other study areas.

3.2 Precipitation data

Precipitation data for 27 meteorological stations of Iberia (Fig. 1) used here is summarised in Table 2 including the name, elevation (in m) mean annual precipitation (MAP), mean monthly precipitation (MMP) in December, and total precipitation in 1909 (December, 22 December, and 20–28 December).

Seven meteorological stations are located in Portugal (Beja, Campo Maior, Coimbra, Évora, Guarda, Lisbon, and Porto) and were digitized by the Geophysical Institute Infante Dom Luiz (IDL) within the ERA-CLIIM project (Stickler et al., 2014). In addition, 20 meteorological stations are located in Spain (Albacete, Alicante, Badajoz, Barcelona, Burgos, Cádiz, Ciudad Real, Granada, Huelva, Huesca, La Coruña, Madrid, Málaga, Murcia, Salamanca, Sevilla, Soria, Valencia, Valladolid, and Zaragoza) and were digitized in the framework of the project EMULATE which are the most complete and reliable daily precipitation records extending back to the mid-19th century (Brunet et al., 2007). A large fraction of these stations were used in a long-term assessment of tendencies in the frequency of days characterised by low, medium, and intense precipitation (Gallego et al., 2011). Long-term daily precipitation data of Oporto (1906–2007), Guarda (1906–1994), Lisbon (1864–2013), and Salamanca (1894–2003) stations were further explored in detail considering the period from September to December. For the complete set of stations, climatological reference periods correspond to the 1971–2000 normal in Portugal (IPMA, Portuguese Meteorological Office) and to the 1981–2010 normal in Spain (AEMet, Spanish Meteorological Office).

3.3 Synoptic data

The atmospheric circulation during the December 1909 event and prior months was evaluated at different timescales. This was achieved with the recently developed 20th Century Reanalysis version 2 (Compo et al., 2011), hereafter 20 CR.

The 20 CR data set has been shown to be suitable for long-term assessment over Iberia, such as major floods event in 1876 (Trigo et al., 2014) or the decadal change of wind resources (Kirchner-Bossi et al., 2014). Several fields related to both surface and tropospheric levels were extracted. From the surface the 6 h sea level pressure field (hPa) and 6 h accumulated precipitation were used). In addition the following variables were also extracted and analysed: (i) for the lower levels (900 hPa), the 6 h fields of wind vector (m s^{-1}) and specific humidity (g kg^{-1}), and (ii) for the upper levels (500 and 250 hPa) the 6 h fields of wind vector (m s^{-1}), geopotential height, and temperature. The ensemble mean fields for all variables were assessed using a global grid of $2° \times 2°$.

3.4 NAO data

NAO data of monthly mean normalized difference of sea level pressures between Lisbon (Portugal) and Stykkisholmur/Reykjavik (Iceland) (Hurrell, 1995) were obtained from the National Centre of Atmospheric Research (NCAR) covering the period from 1865 onwards.

4 Results

4.1 Societal impacts of the December 1909 event

In Iberia, 134 DISASTER cases were recorded – 130 flood cases and four landslides cases (Fig. 1). Table 3 summarizes the social consequences of the December 1909 event in Iberia. Considering the data sources from both affected countries, floods that occurred during the December 1909 event caused 57 fatalities, 4 injured, 15 missing people, 1104 evacuated people, and 2564 homeless people (Table 3). Landslides (debris flows) caused 32 fatalities and 100 homeless people (Table 3). In total this event affected 3876 people in the Iberian Peninsula.

In this section the spatial distribution of the societal impacts will be explored per hydrographic basin whereas the results of the damage indices are presented per region. The temporal distribution of the DISASTER cases per day (20–

Table 2. Available precipitation stations for December 1909 event in Iberia.

Rain gauge	Elevation (m)	MAP (mm)	MMP in December (mm)	TP in December 1909 (mm)	TP (mm) in 22 December 1909	TP (mm) 20–28 December 1909
			Portugal			
Beja	246	571.8	100.6	132.9	0.0	58.9
Campo Maior	280	*	*	125.1	2.0	57.3
Coimbra	141	905.1	126.8	225.3	31.3	83.9
Évora	200	609.4	102.7	172.2	1.5	85.8
Guarda	1019	882.0	141.8	451.1	185.5	343.9
Lisbon	77	725.8	121.8	139.1	60.9	84.0
Oporto	93	1253.5	194.7	287.0	63.0	158.6
			Spain			
Albacete	699	353.0	31.0	14.5	0.0	0.0
Alicante	82	311.0	25.0	12.6	0.0	0.0
Badajoz	185	447.0	69.0	129.8	0.9	51.3
Barcelona	420	588.0	40.0	13.9	0.0	0.0
Burgos	881	546.0	63.0	119.2	23.2	53.0
Cádiz	30	523.0	92.0	81.7	9.1	27.5
Ciudad Real	627	402.0	59.0	70.0	6.0	22.6
Granada	685	365.0	56.0	45.8	10.0	17.0
Huelva	19	525.0	99.0	48.8	5.7	13.1
Huesca	541	480.0	44.0	52.6	0.0	16.9
La Coruña	67	1014.0	131.0	166.1	13.0	53.2
Madrid	679	421.0	51.0	88.8	0.5	32.0
Málaga	7	534.0	100.0	49.1	12.8	15.5
Murcia	57	297.0	29.0	11.6	0.0	0.0
Salamanca	790	372.0	42.0	80.7	20.6	36.4
Sevilla	31	539.0	99.0	69.3	19.2	36.3
Soria	1083	512.0	50.0	80.4	7.5	25.9
Valencia	11	475.0	48.0	33.3	0.0	0.0
Valladolid	691	433.0	53.0	137.0	18.0	51.5
Zaragoza	245	322.0	21.0	8.2	0.0	0.0

Note: MAP – mean annual precipitation; MMP – mean monthly precipitation; TP – total precipitation; * Campo Maior station was only in operation between December 1872 and December 1909.

28 December) will also be analysed as well as the surrounding circumstances of fatalities.

4.1.1 Spatial distribution of the impacts

The Iberian Peninsula was spatially affected during this event along the SW–NE direction spanning from Lisbon to the south slope of the Cantabrian Mountains (Fig. 1). The Douro Basin and the Tagus Basin registered 62 and 15 % of total DISASTER cases, respectively (Fig. 2a), but the highest DISASTER cases density (1.4/1000 km^2) was registered in the Lima and Cávado basins located in the NW sector of the Iberian Peninsula.

The locations of the DISASTER cases were also analysed taking into account their relative position in three main sectors of the corresponding hydrographic basin (headwater sector, middle sector, and lower sector). These sectors were identified within each hydrographic basin by dividing the corresponding DEM (pixel = 1 km) into three elevation classes of equal area. The lower sector of the affected hydrographic basins contains more than half of DISASTER cases considering the total hydrographic basins affected by the December 1909 event (Fig. 2b). In the western smallest hydrographic basins of Lima and Cávado basins and Sado Basin (3 and 7, Table 3) the DISASTER cases only occurred in the lower sector. Disaster cases located in the headwater sector were only registered in the major river basins of Douro and Tagus (10.6 and 4.8 %, of total cases, respectively).

The spatial distribution of fatalities generated by floods and landslides in Iberia during the 1909 event is presented in Fig. 3a. Floods were particularly disastrous along the Douro Basin, where 60 % of total fatalities associated with floods were registered (Table 3). However, the DISASTER case

Table 3. Social consequences of the December 1909 event per hydrographic basin.

Hydrographic basin	Area (km^2)	No. of disaster cases	% disaster cases	Disaster cases density (#/1000 km^2)	No. of fatalities	No. of injuries	No. of evacuated people	No. of homeless people	No. of missing people	Total affected people
				Floods						
1 – Coruña basins	12 663	2	1.5	0.16	0	0	3	84	0	87
2 – Minho Basin	16 993	8	6.2	0.47	9	0	18	196	0	223
3 – Lima and Cávado basins	6277	9	6.9	1.43	2	0	14	0	8	24
4 – Douro Basin	97 502	83	63.8	0.85	34	2	595	2104	6	2741
5 – Águeda, Mondego, and West basins	14 119	7	5.4	0.50	4	0	410	3	0	417
6 – Tagus Basin	81 115	20	15.4	0.25	8	2	61	177	1	249
7 – Sado Basin	10 074	1	0.8	0.10	0	0	3	0	0	3
Floods Total		*130*	*100*		*57*	*4*	*1104*	*2564*	*15*	*3744*
				Landslides						
2 – Minho Basin	16 993	1	25	0.06	26	0	0	100	0	126
4 – Douro Basin	97 502	2	50	0.02	3	0	0	0	0	3
6 – Tagus Basin	81 115	1	25	0.01	3	0	0	0	0	3
Landslides total		*4*	*100*		*32*	*0*	*0*	*100*	*0*	*132*
Total		*134*			*89*	*4*	*1104*	*2664*	*15*	*3876*

Note: hydrographic basins are shown in Fig. 1.

that generated more fatalities (26 fatalities) was produced by a single landslide (debris flow) that occurred in Orense province in the Minho Basin.

Floods generated the vast majority (96.2 %) of total homeless people, and among these, 82 % were located in the Douro Basin, mostly on the Spanish side of the basin (Fig. 3b, Table 3). In Spain, the provinces of Salamanca, Palencia, Zamora Orense, and León registered the highest number of homeless people due to floods along the Douro tributaries (Tormes, Carrión, Esla, Tera, and Luna rivers) (Fig. 4). In Portugal, the highest number of homeless people was registered at Santarém, in the lower sector of the Tagus hydrographic basin.

During the DISASTER event of December 1909 it was also necessary to evacuate people (Fig. 3c) rescued from the waters or flooded buildings. There were two major spots of evacuated people in Portugal, the first in the Lisbon district and the second at the mouth of the Douro River in Oporto City (Oporto district), where 400 and 105 people, respectively, were evacuated. In Spain, there was an important spot of evacuated people (101 people) near the confluence of the Carrión River with the Douro River in the province of Palencia (Fig. 4). The Douro Basin includes 53.9 % of total evacuated people registered during the December 1909 event.

Table 3 highlights the Douro Basin as the major zone impacted by floods. This basin registered 64% of total Disaster flood cases and 73 % of total affected people by floods. In the case of landslides, the Minho Basin had only a single DISASTER case, however, this single event was responsible for the highest number of fatalities (26) and homeless people (100) generated by landslides.

Figure 5 illustrates the spatial variability for different Iberian regions of the three damage indices considered (fatalities index, homeless index, and total affected index). The Fatalities Index was classified as Extreme in the provinces of Zamora and Orense (Minho Basin) and Very High in Bragança and Salamanca in the middle sector of the Douro Basin and Santarém in the lower sector of the Tagus Basin. In the provinces of Palencia, Zamora, and Salamanca, located in the Douro Basin headwater and middle sectors, the Homeless Index was Extreme (Fig. 5b). The same provinces registered an Extreme Total Affected Index (Fig. 5c), which was mainly justified by the very high number of homeless people located in the headwater and middle sectors of the Douro Basin. The total affected index was very high in the lower section of the Tagus Basin (Lisbon and Santarém districts) as well as in the headwater and middle sectors of the Douro Basin (León province), and the Minho Basin (Orense Province).

Despite the described social impacts, the December 1909 event also caused huge economic losses in Iberia. In Spain, several settlements were completely submerged by water and a great number of buildings were destroyed, especially in the areas drained by the main tributaries of the Douro River (e.g. in the provinces of Zamora, Salamanca, León, Palencia; Fig. 4). Also, in Portugal Peso da Régua and the riverside areas of Oporto and Vila Nova de Gaia were submerged by the Douro River.

Major cities of Spain (e.g. Salamanca) and Portugal (e.g. Oporto) were kept isolated, without communications (mail, telegraph) with the outside world due to widespread cuts in road and rail networks (submerged or destroyed roads

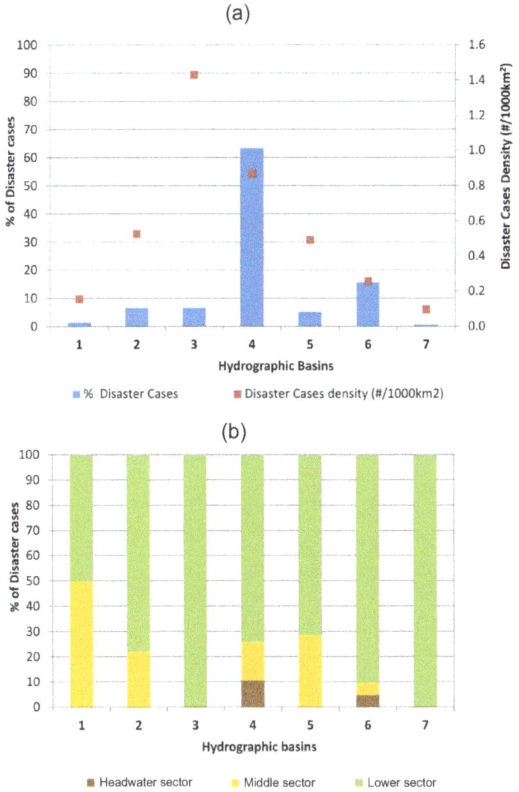

Figure 2. Percentage of Disaster cases and Disaster cases density per hydrographic basin **(a)** and percentage of Disaster cases per hydrographic basin sector **(b)**. Hydrographic basin code: 1 – Coruña basins; 2 – Minho Basin; 3 – Lima and Cávado basins; 4 - Douro Basin; 5 – Águeda, Mondego and West basins; 6 – Tagus Basin; 7 – Sado Basin.

and railroads, bridges collapsed). Rail connection between Portugal and Spain was also interrupted for several days.

Finally, it was also reported that some cattle herds were killed by the floods in the Spanish provinces affected by the event, as well as widespread destruction of cropped areas (e.g. vegetables, grain) that caused food scarcity after the floods. ("La Correspondencia de España", 22–28 December 1909).

4.1.2 Temporal distribution of the impacts

Daily evolution of fatalities and affected people generated during the December 1909 event is shown in Fig. 6. Between 20 and 22 December 48.8 % of the Disaster cases were registered (Fig. 6a). In addition, the highest daily number of DISASTER cases (58) and fatalities (52) were registered in 22 December. In terms of total affected people, it is possible to recognize two peaks, the major one on 22 December (1842 affected people) and a second on 25 December (1014 affected people).

In the Douro Basin the daily evolution of the social impacts (Fig. 6b) is quite similar to the complete set of the event. In contrast, the Tagus Basin only registered human impacts on 22, 24, and 27 December (Fig. 6c), with a peak of Disaster cases and affected people on 22 December.

The spatial distribution of the DISASTER cases per day is shown in Fig. 7. In general terms, DISASTER cases that occurred between 20 and 23 December arose predominantly in the western part of Portugal with some isolated spots in Spain (Salamanca, Zamora, León, and Orense). Disaster cases registered between 24 and 28 December were widespread in Spain in the Douro and Tagus headwater and middle sectors. Some isolated spots occurred in Portugal on 27 and 28 December in the lower sectors of the Minho, Lima and Cávado, and Tagus basins. Therefore, overall the temporal distribution of DISASTER cases is consistent with an atmospheric circulation in the Iberian Peninsula coming from SW to NE. However, we must keep in mind that information flow at the time was considerably slower than today, particularly during disruptive events such as major natural hazards. In this regard, it is possible that some of these later dates can correspond to cases that were reported a few days after their real occurrence without correction of the reported timing.

4.1.3 Circumstances surrounding fatalities

Circumstances surrounding the fatalities by gender during the December 1909 event are presented in Fig. 8. Fatalities were dominant in rural areas, both inside buildings and outdoors (87.6 % of total fatalities). For the hydro-geomorphological cases where the gender of the victim is known, it is clear that male fatalities were dominant (81 % of total fatalities). Also male fatalities were prevalent in rural areas. The predominance of male fatalities generated by floods and landslides is consistent with previous observations on mortality associated with hydro-geomorphologic disasters in Portugal (Pereira et al., 2015). The highest percentage of total fatalities with gender not reported that occurred inside buildings in rural areas (Fig. 8) correspond mostly to fatalities generated by landslides (81.3 %).

4.2 Precipitation event and hydrologic context

Daily long-term cumulative precipitation from 1 September to 31 December is shown in Fig. 9 (grey thin lines) for Oporto, Lisbon, Guarda, and Salamanca meteorological stations. In addition, the long-term mean (magenta line) and the 95th percentile (black line) for 1909 (red line) are also highlighted. It is clear that the December 1909 event was preceded by an intense precipitation in late autumn, especially after 15 November where it surpasses the mean in all locations and also a period of enhanced precipitation after the 17 December. These features are particularly notable at the Guarda station, where the cumulative precipitation between 17 and 31 December 1909 reached a total of 406.7 mm. Mean monthly precipitation for December at Guarda station is 141.8 mm, while the total precipitation registered in De-

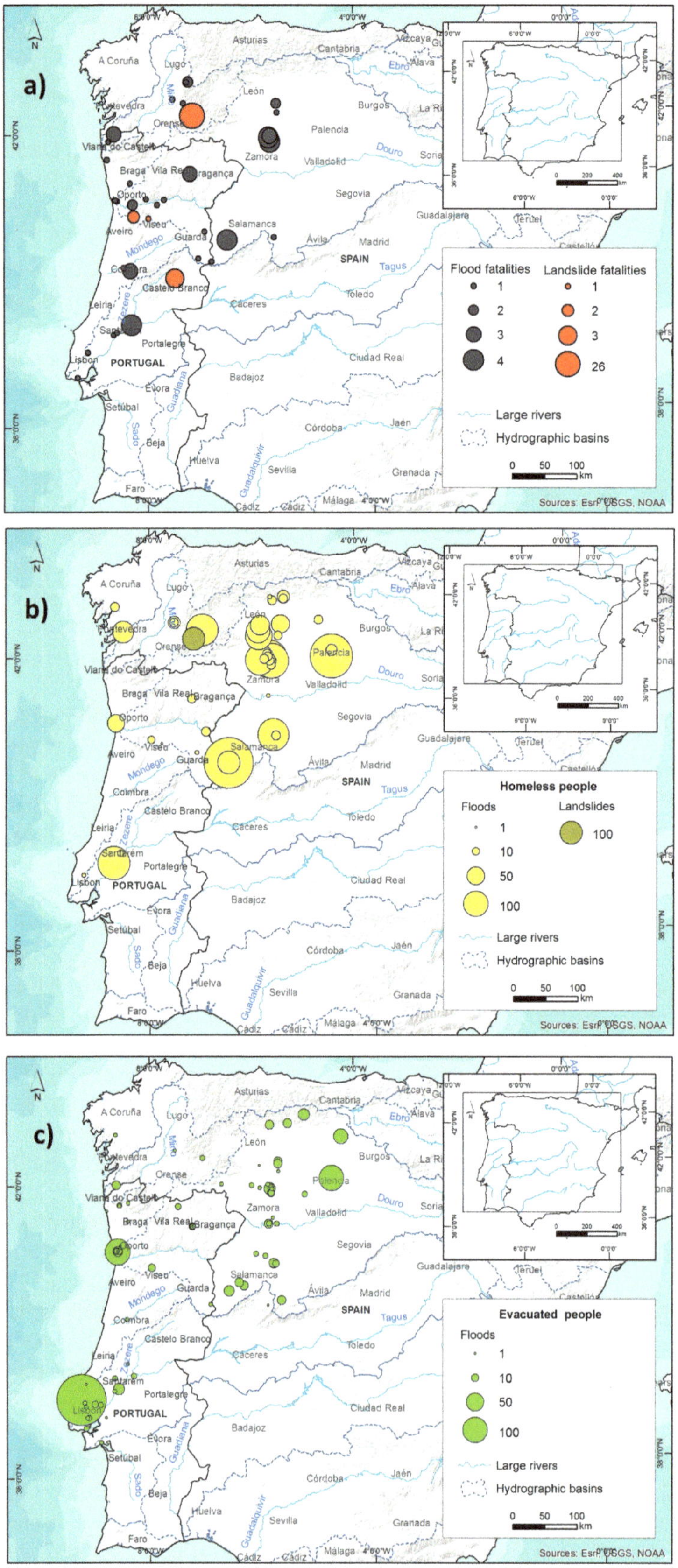

Figure 3. Social consequences of the DISASTER cases: fatalities (**a**), homeless people (**b**), and evacuated people (**c**).

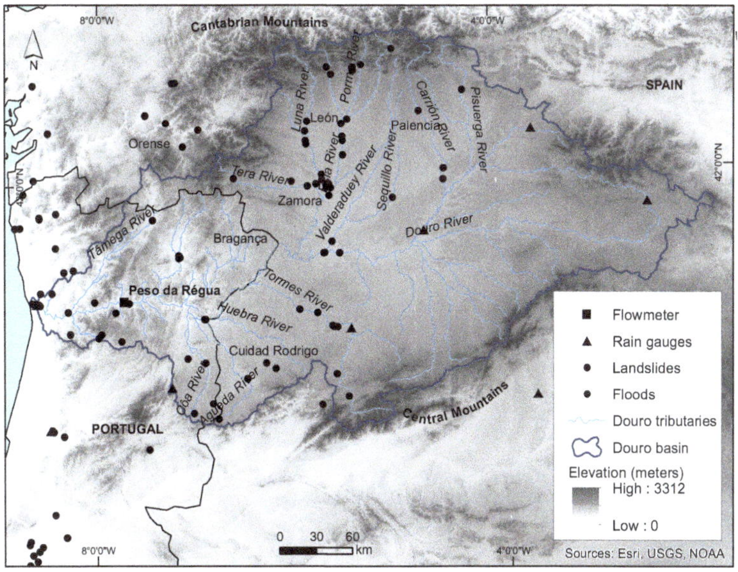

Figure 4. Douro Basin and DISASTER cases of the December 1909 event.

cember 1909 was 451.1 mm (Table 2). In December 1909 a total of 80.7 mm was registered at Salamanca (although much lower than the value observed at Guarda, it corresponds to twice the amount of the mean monthly precipitation for December) (Table 2). It is important to mention that in November and December 1909 strong negative values of NAO (−1.9 and −2.5, respectively) were registered. The NAO plays an important role in shaping European precipitation variability (Trigo et al., 2008) especially in the Iberian Peninsula, where negative values of NAO are associated with above-normal precipitation in the region. This significant impact results from the displacement of the storm track further south of its normal position during months with negative NAO index, leading to a higher frequency of Atlantic low-pressure systems (mid-latitude cyclones) that travel into the region (Trigo et al., 2004, 2008). The impacts of the negative values of the NAO help to put into context the above-average precipitation that occurred in November and December.

After 22 December 1909 the cumulative precipitation (red line) was higher than the 95th percentile (black line) at Guarda and Salamanca stations, both located in the Douro Basin. The extreme rainy period is very clear at the Guarda station where a daily maximum precipitation of 185.2 mm was recorded on 22 December 1909, corresponding to a return period of 141 years. Regarding Oporto and Lisbon, both stations recorded values of accumulated precipitation up to 22 December below the 95th percentile but still above the long-term mean. In addition, for Lisbon, two particular years with major floods are also highlighted for comparison: 1876 (blue line) and 1967 (green line).

The relevance of above-average precipitation before a major event depends on the temporal and spatial scales of the phenomenon being analysed. The December 1909 event along with the 1876 event (Trigo et al., 2014) highlights the key role played by previous accumulated precipitation in the occurrence of floods during events lasting several days in the major river basins of the western Iberian Peninsula, particularly for the period prior to the construction of major dams (Trigo et al., 2014). In contrast, floods occurring after intense bursts of precipitation over relatively confined areas of western Iberia can induce flash floods, independently of the precipitation that occurred in previous weeks or months. That was the case of the sub-daily 1967 event which occurred during the night between 25 and 26 November 1967, which was confined to the Lisbon metropolitan area, i.e. without affecting large river basins. The 1967 event was the deadliest storm affecting Portugal since at least the early 19th century with several flash floods in heavily constructed suburban areas around Lisbon causing more than 500 fatalities. In this case, the accumulated precipitation prior to 25 and 26 November 1967 was not relevant (being below average) and the flash floods occurred mainly as a result of very intense hourly precipitations, ranging in duration from 4 to 9 h, compatible with return periods of 100 years or more (Trigo et al., 2015).

As already mentioned, the precipitation record corresponding to the December 1909 event was particularly notable at the Guarda station. Therefore, in order to further evaluate the time duration and magnitude of the event, the accumulated precipitation values (from 1, 2, 5, 10, 15, 30, and 45 days) for the Guarda station (88 years) were computed and all values were ranked, with the top 10 events being represented in Fig. 10. A similar methodology was applied in a previous analysis of an outstanding event, the record floods of 7 December 1876 in southern Iberia (Trigo et al., 2014). Figure 10 shows on the x axis the 10 highest

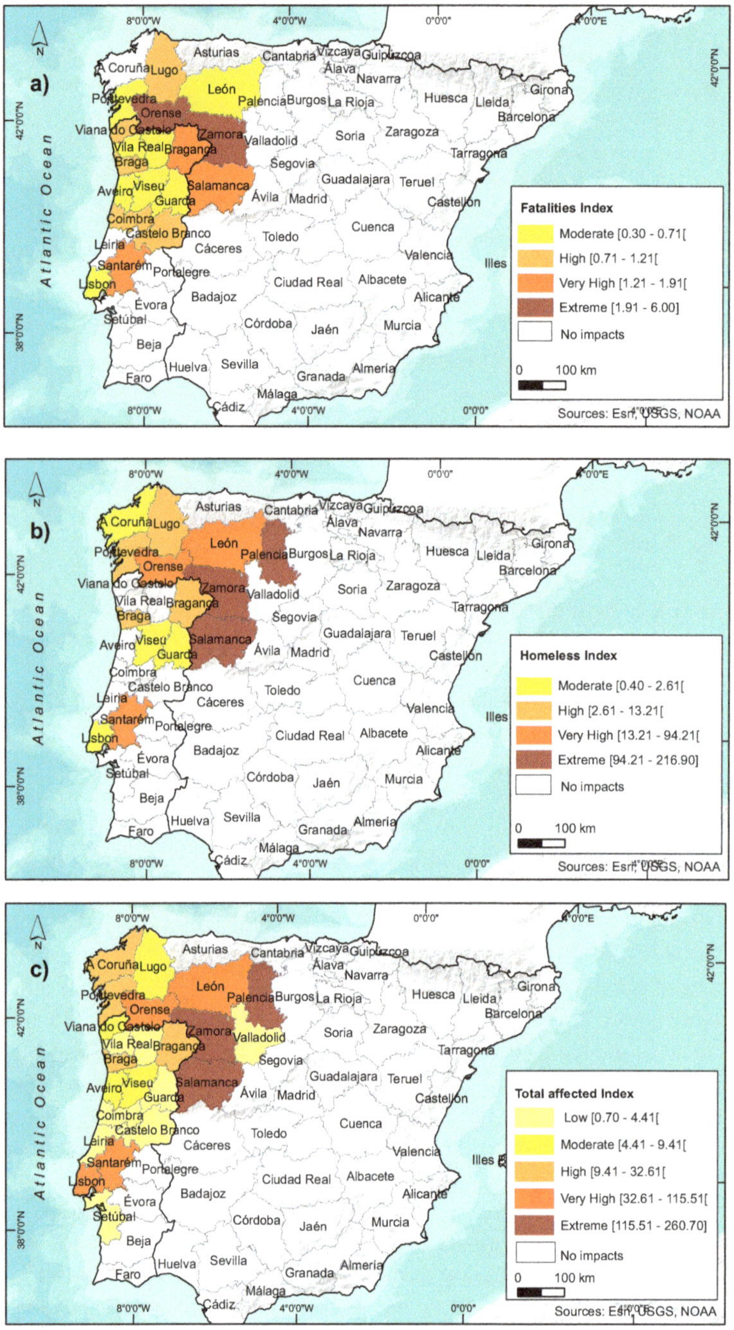

Figure 5. Fatalities index **(a)**, homeless index **(b)**, and total affected index **(c)** of the December 1909 event for affected regions over Iberia.

values of maximum accumulated precipitation for different temporal durations (*y* axis). The blue bars correspond to accumulated precipitation (*z* axis) for different durations of the December 1909 event. For the complete precipitation series of Guarda station, the December 1909 event occupies the first position in the cumulated precipitation for 1 day (185.2 mm) and the second position for 2 days (279.8 mm) (Fig. 10).

Interestingly, the extreme precipitation values observed in Guarda and to a less extent in Salamanca were not observed in most locations with pluviometers. Weather stations located near the Atlantic coast (Oporto and Lisbon) did not record extreme accumulated precipitation in the December 1909 event (Figs. 9 and 11) and the stations located in the eastern part of Spain registered null or low precipitation (below 20 mm). Estimated precipitation of 22 and 20–28 December 1909 was interpolated spatially using a co-kriging geostatistical technique considering the precipitation measured in the stations (points) and the Iberia SRTM 90 m DEM's

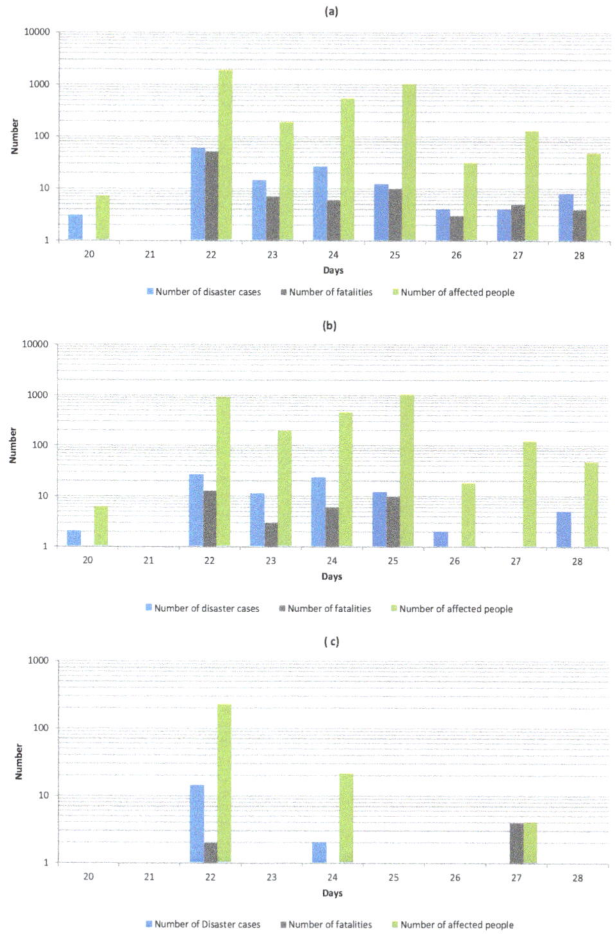

Figure 6. Daily number of DISASTER cases, fatalities, and affected people generated by the December 1909 event **(a)**, in the Douro Basin **(b)**, and in the Tagus Basin **(c)**. Notice the logarithmic scale used.

with 90 m resolution (Jarvis et al., 2008) (Fig. 11a and b). Estimated precipitation shows that the event entered in the Atlantic coast (south of Lisbon region and crossed the peninsula in a SW–NE direction. In addition, the extreme precipitation that occurred on 22 December in the Douro Basin, namely in the Guarda and Salamanca regions (Fig. 11a), is impressive.

According to Rodrigues et al. (2003) the flood of 1909 was referenced as the second historical flood measured at Peso da Régua, in the Portuguese sector of the Douro River (Fig. 4) in the last 300 years. In the historical flood of 1909 a flow rate of 16 700 $m^3\,s^{-1}$ was reached, which was only surpassed by the flood of 1739 (18 000 $m^3\,s^{-1}$).

A graph with the evolution of sub-daily flow rate at Peso da Régua for the period of 18–30 December 1909 is shown in Fig. 12, using data from Rodrigues et al. (2003). At the beginning of this period a constant rise in the flow rate was recorded, from 2000 $m^3\,s^{-1}$ on 18 December to 6000 $m^3\,s^{-1}$ in the early hours of 21 December. Then, the flow rate suddenly increased to 16 000 $m^3\,s^{-1}$ during the day of 23 December, as a consequence of the extreme precipitation that was registered along the Douro Basin during the previous days, especially on 22 December. The maximum flow rate of 16 700 $m^3\,s^{-1}$ corresponds to a return period of 80 years (Rodrigues et al., 2003). At Peso da Régua the river level rose 25 m during this flood ("Jornal de Notícias", 24 December 1909). It should be noticed that at Peso da Régua the flow rate of 6000 $m^3\,s^{-1}$ has a 16-year return period and implies a 13 m river level rise, which is considered an extraordinary flood in the region (Rodrigues et al., 2003).

The flow rate starts to decrease at the end of 23 December but a second peak was registered at the end of 24 December with similar magnitude (16 200 $m^3\,s^{-1}$), following a secondary precipitation peak that is not evident in precipitation daily records. Only afterwards, i.e. after 25 December, did the flow rate decrease progressively until the end of the month (2800 $m^3\,s^{-1}$ on 30 December). It is important to note that the flow rate was higher than 6000 $m^3\,s^{-1}$ between 21 and 28 December, coinciding with the Disaster event time interval.

4.3 Assessing atmospheric circulation

Here we intend to characterize the atmospheric circulation affecting western Iberia between 15 November and 28 December. However, the main focus of this section will be on 22 December, in order to understand the occurrence of highest precipitation values within this day.

The importance of the accumulated precipitation observed between mid-November and mid-December culminating with the intense precipitation on 22 December has already been shown in Sect. 4.2. The 6-hourly sequence of SLP and corresponding accumulated precipitation (both obtained from the 20 CR) between 15 November at 00:00 UTC and 28 December at 18:00 UTC can be seen in the Supplement video V1. The week between the 16 and 21 November was characterized by a stationary low-pressure system that affected the Iberian Peninsula with precipitation occurring during these 6 days. The most intense precipitation over these days was recorded on 16 November at 18:00 UTC with high precipitation rates. According to the specific humidity and wind vectors at 900 hPa (shown in video V2 in the Supplement also between 15 November to 28 December) high moisture availability was present on 16 November at 18:00 UTC due to high moisture advection from the western African Atlantic coast as a consequence of intense southwest winds.

Between 22 November and 12 December the atmospheric circulation in the Iberian Peninsula was mainly of anticyclonic nature with sporadic precipitation in the north Iberian Peninsula due to the passage of some low-pressure systems north of the domain (video V1 in the Supplement). On 13 December a low-pressure system had its cyclogenesis west of the Azores and become stationary near the Azores until 21 December. From 17 December (video V1 in the Supplement)

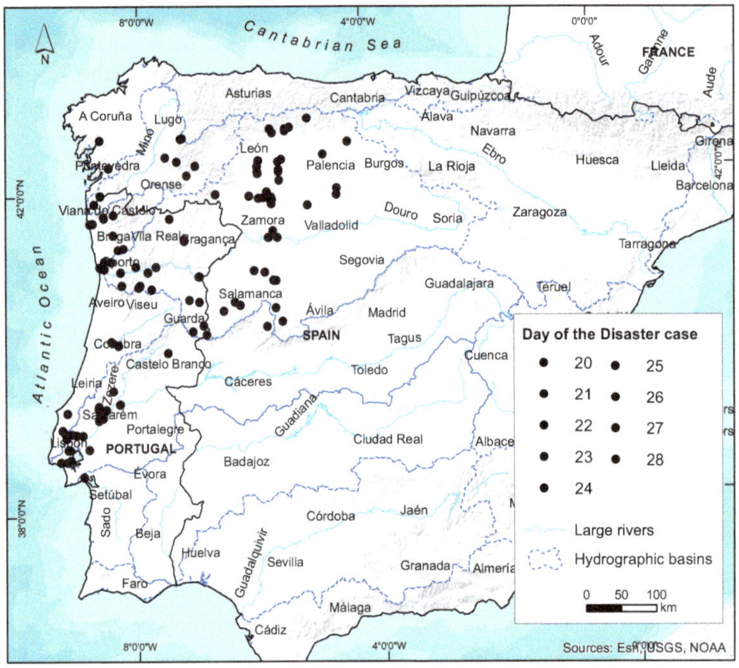

Figure 7. DISASTER cases of the December 1909 event per day.

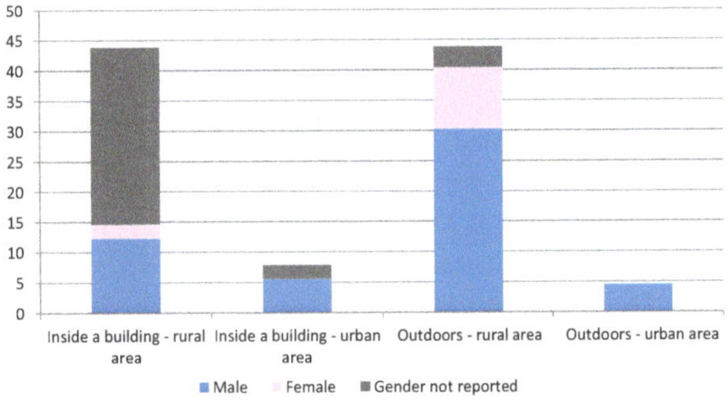

Figure 8. Circumstances surrounding the fatalities by gender during the December 1909 event in Iberia.

onwards it affected the western Iberian Peninsula with almost continuous precipitation every day until 22 December, when the most intense precipitation occurred (as seen in Fig. 13).

A thorough analysis is now provided covering the days of most intense precipitation, as seen by the sequential panels every 6 h of SLP and the corresponding accumulated precipitation (both from the 20 CR) represented from 22 December at 00:00 UTC to 23 December at 00:00 UTC (Fig. 13). This figure reveals the presence of a large deep low-pressure system (966 hPa) located over the Atlantic Ocean, northwest of the Iberian Peninsula and extending from Iceland to southern Portugal and to Central France. This low-pressure system remained quasi-stationary from 22 December 00:00 until 12:00 UTC when it started to move east towards the British Isles and its central core began to increase pressure (Fig. 13d and e). In addition, the frontal system associated with this low pressure struck the western Iberian Peninsula from a SW–NE direction. Although the 20th Century Reanalysis does not provide explicit front analysis, the precipitation pattern around the cyclone centre corresponds clearly to a frontal system when compared with a conceptual model of an extratropical cyclone (Shapiro and Keyser, 1990). In fact we must emphasize that the vast majority of the Atlantic mid-latitude cyclones that struck Iberia are characterized by the presence of frontal systems (sometimes already in the occlusion phase), that extend the precipitation impacts considerably to the south of the low-pressure system centre (Trigo et al., 2004).

With the aim of further analysing the upper- and lower-level dynamics of this event we computed the divergence fields simultaneously with the wind speeds for both the 900 and 250 hPa (Fig. 14). The wind speed and associated divergence at the 250 hPa level is shown in the upper panels of Fig. 14, for 00:00 and 12:00 UTC respectively. The most interesting feature for this event is associated with the presence of an upper-level jet branch, oriented along a SW–NE direction towards the Iberian Peninsula but with its origin in the eastern coast of USA. This feature presents an area with strongest upper-level divergence over Portugal at 00:00 and 12:00 UTC, with values above $3 \times 10^{-5}\,s^{-1}$ favouring deep convection. In addition, at lower levels (900 hPa), a very intense low-level jet (winds up to $30\,m\,s^{-1}$ close to Portugal) associated with the frontal system embedded within the southern flank of the low-pressure system is also present with a similar SW–NE orientation. This intense advection of mass

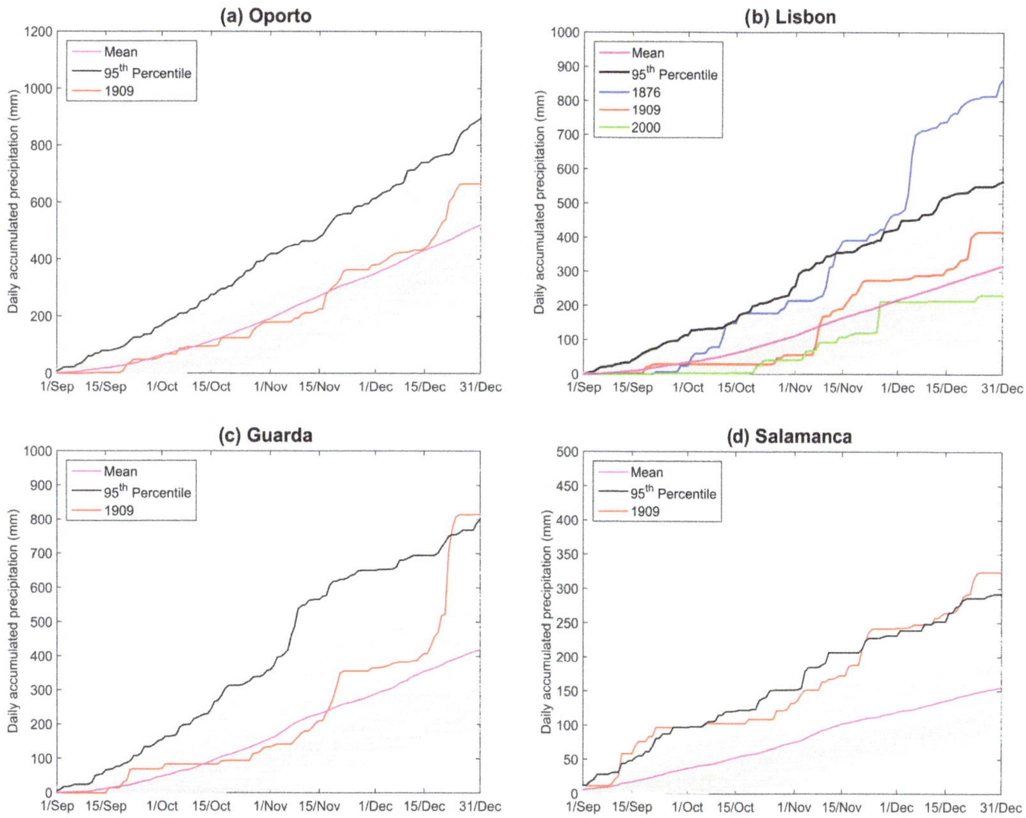

Figure 9. Cumulative precipitation from 1 September to 31 December using daily precipitation data from **(a)** Oporto, **(b)** Lisbon, **(c)** Guarda, and **(d)** Salamanca. Each year of cumulative precipitation is represented in grey. The mean and the 95th percentile of precipitation are represented in magenta and black respectively and accumulated precipitation in the year of 1909 is represented by the red line. In addition, the historical storm years of 1876 (blue) and 1967 (green) in Lisbon are also highlighted.

implies strong low-level convergence at the western coast of the Iberian Peninsula ($-3 \times 10^{-5}\,s^{-1}$). In summary, we can state that the large-scale conditions in the area of the western Iberian Peninsula were optimal for intense precipitation, i.e. frontal uplift associated with upper-level divergence and low-level convergence. The displacement of the jet stream so far south was also responsible for the stationary position of the low-pressure system for so long.

However, one should bear in mind that the presence of the low-pressure system and its associated frontal system and the favourable dynamical mechanisms for intense precipitation explained before cannot alone explain the intense precipitation that occurred on 22 December.

The analysis of Fig. 15 (upper panel, geopotential height and temperature at 500 hPa) shows that the geopotential height is characterized by a strong trough located south of the Azores thus favouring the southerly displaced position of the jet in the middle of the Atlantic (Fig. 14). Along with this trough there is a negative temperature gradient between 60 and 30° N with dry and cold air aloft near the Iberian Peninsula which also could enhance the formation of convection within the frontal system.

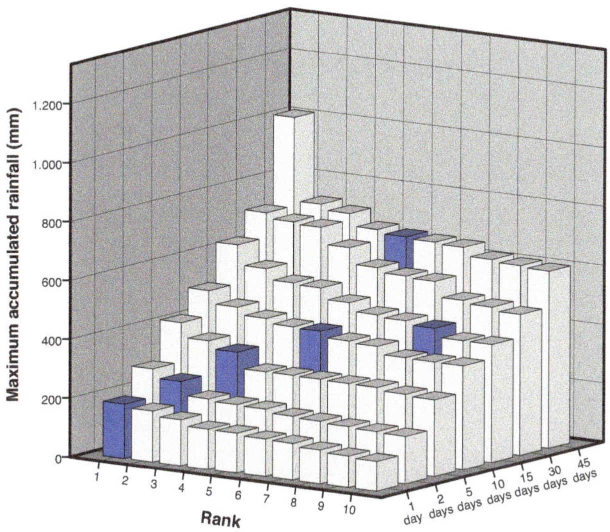

Figure 10. The 10 highest values (rank) of maximum accumulated precipitation from 1, 2, 5, 10, 15, 30, and 45 days in Guarda for the period 1906–1994. Blue bars correspond to the December 1909 event.

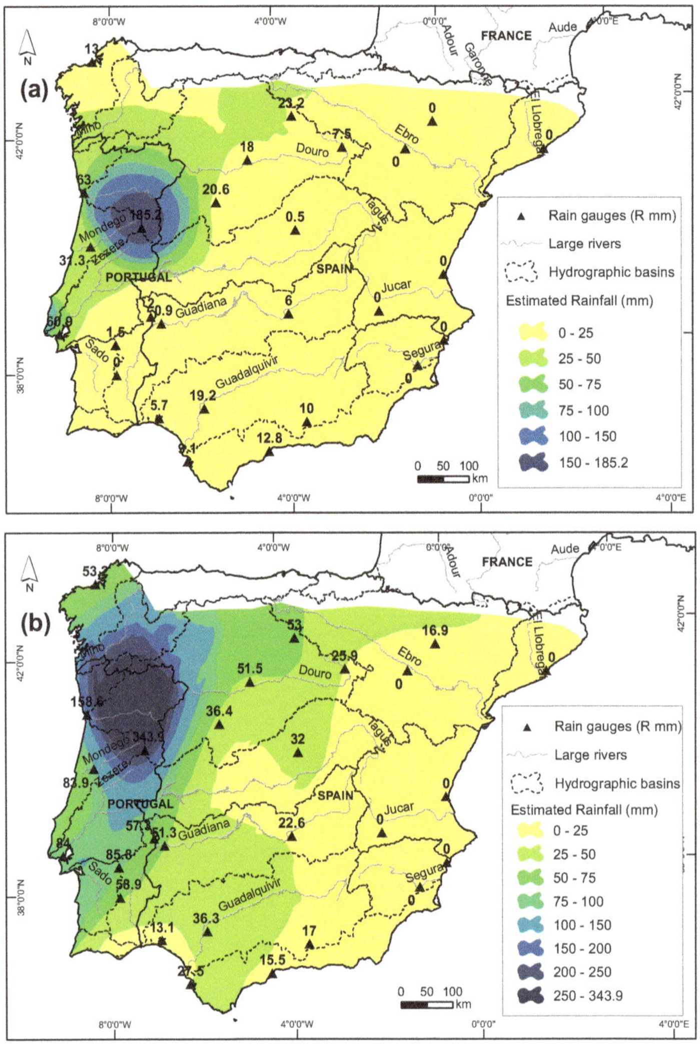

Figure 11. Co-kriging interpolation of total precipitation of 22 December **(a)** and event cumulated precipitation in 20–28 December 1909 **(b)** in Iberia.

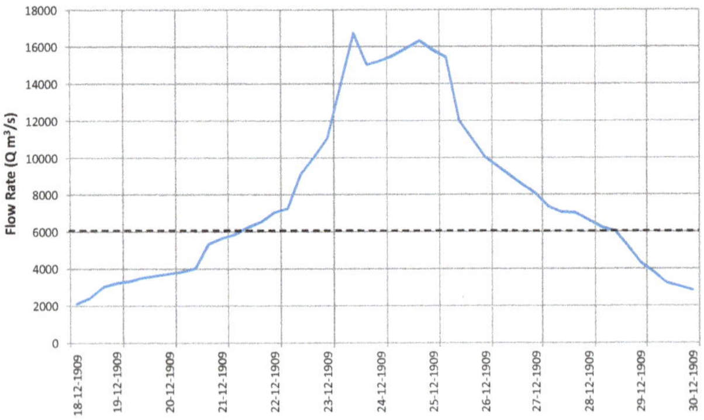

Figure 12. Flow rate ($m^3\,s^{-1}$) at Peso da Régua from 18–30 December 1909. Source: SNIHR (Rodrigues et al., 2003). The dashed line corresponds to an extraordinary flood at Peso da Régua (Portugal).

To analyse the moisture availability, which can enhance the precipitation intensity, the specific humidity and wind vectors at 900 hPa are also shown for the 00:00 and 12:00 UTC (Fig. 15, lower panels). In both time steps, a narrow and prolonged band of moisture being transported from sub-tropical latitudes can be seen, with origin close to the Caribbean islands. This narrow band presents an intense core with values above $9\,g\,kg^{-1}$, over the Atlantic Ocean, and moves towards extratropical latitudes, by a southwestern low-level jet with wind speeds clearly above $24\,m\,s^{-1}$ (Fig. 14). Therefore the two main criteria for the identification of AR are fulfilled (see Sect. 2) on 22 December, enhancing the precipitation intensity during this particular day.

On 23 December, the low-pressure system moves NW towards the British Isles and the Iberian Peninsula becomes less under the influence of this low pressure. Despite this,

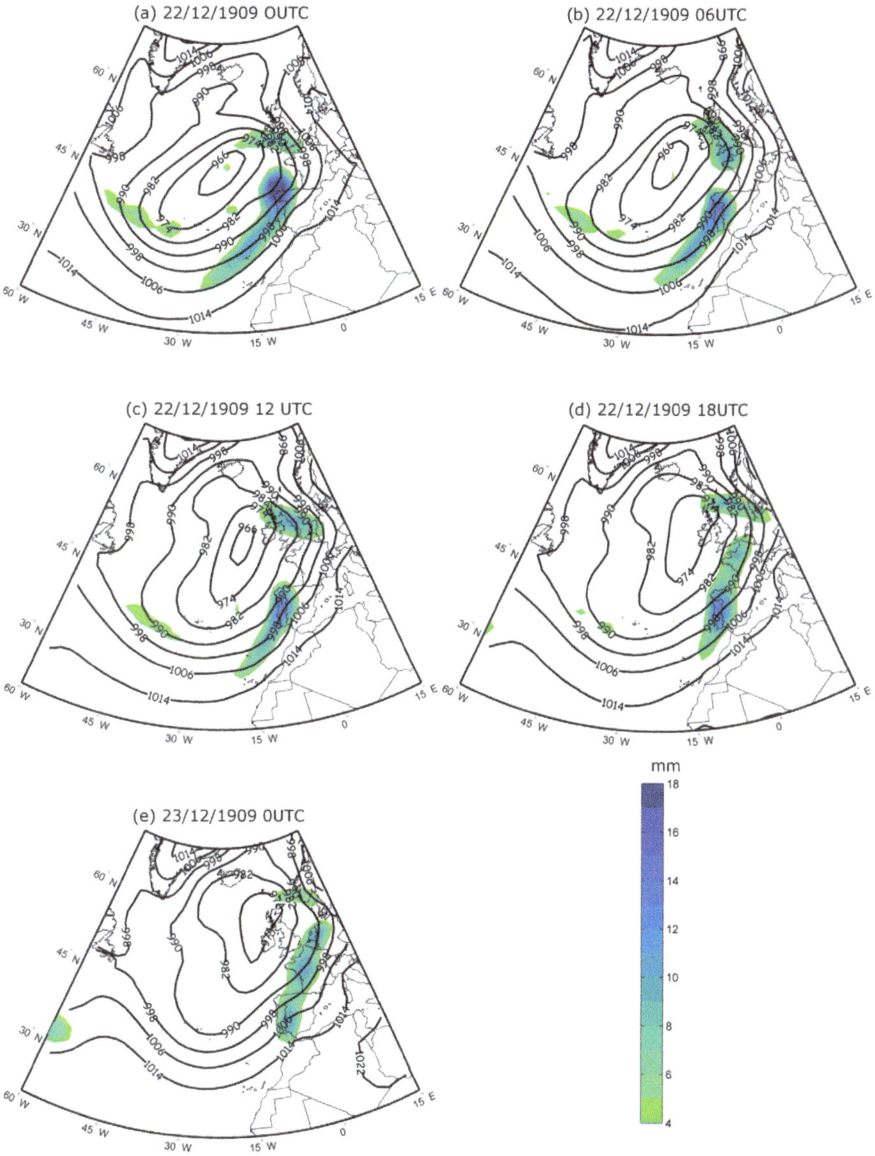

Figure 13. The sea level pressure field (contour, hPa) and 6 h accumulated precipitation (shaded, mm) for 22 December 1909 at **(a)** 00:00 UTC, **(b)** 06:00 UTC, **(c)** 12:00 UTC, **(d)** 18:00 UTC, and **(e)** 23 December at 00:00 UTC.

some light precipitation continues to occur between 23 and 25 December as shown in video V1 in the Supplement.

Summarizing, the synoptic analysis shows that 22 December was characterized by the presence of a deep low-pressure system located over the NW Iberian Peninsula with a stationary frontal system striking the western Iberian Peninsula. The presence of an upper-level jet (250 hPa) and low-level jet (900 hPa) located on SW–NE oriented towards the Iberian Peninsula along with upper-level divergence and lower-level convergence favoured large-scale precipitation. Along with these two features, the presence of an AR (with high content of moisture) contributed to a large extent to enhanced precipitation.

5 Discussion and conclusions

This study provides the first comprehensive assessment of a major hydro-geomorphologic event that occurred in Iberia in the last 150 years, including: (i) the spatial and temporal characteristics of the societal impacts based on newspapers analysis; (ii) the corresponding spatial distribution of precipitation anomalies using recently digitized daily precipitation data; and (iii) the meteorological conditions that triggered the event using the 20 CR.

Data collection of human and societal disaster impacts on newspapers was very useful to obtain the spatial distribution of the hydro-geomorphologic DISASTER cases in both countries and provided an important historical complement

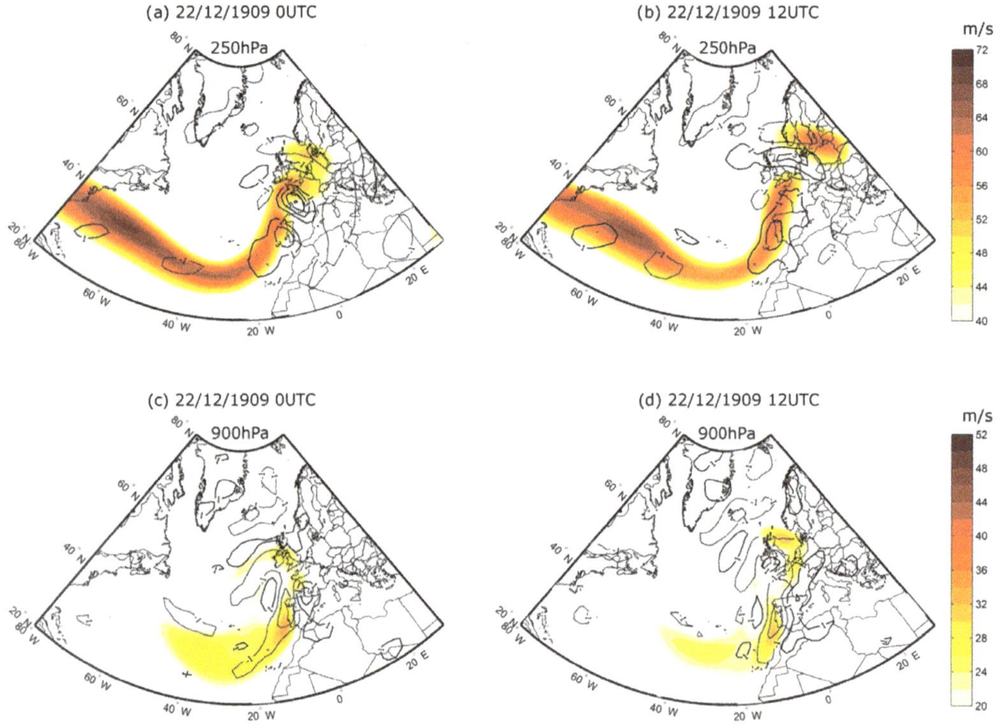

Figure 14. Divergence (contours; $10^{-5}\ s^{-1}$) at 22 December 1909 at 00:00 and 12:00 UTC for **(a, b)** the 250 Pa level, and **(c, d)** at the 900 Pa level. Along with the divergence field, the respective wind speed (shaded, $m\ s^{-1}$) is also shown.

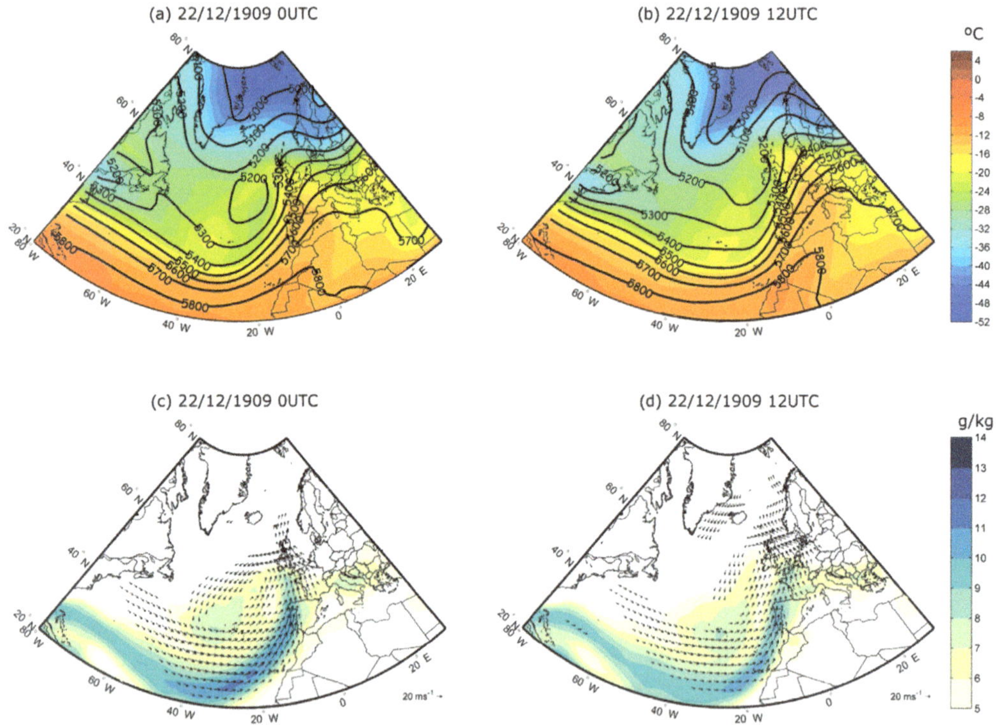

Figure 15. The geopotential height (contour, m) and air temperature (shaded; °C) at 500 hPa level for 22 December: **(a)** 00:00 UTC and **(b)** 12:00 UTC. The wind vector ($m\ s^{-1}$) and specific humidity (shaded, $g\ kg^{-1}$) at 900 hPa level for the 22 December 1909 at **(c)** 00:00 UTC and **(d)** 12:00 UTC are also shown. Only wind vectors with wind speeds above 12.5 $m\ s^{-1}$ are shown.

useful to offset the lack of precipitation data in some of the affected areas. In this work new information concerning the event impacts in Spain was gathered with respect to the already available DISASTER database for Portugal. According to newspapers, the December 1909 event generated 89 fatalities, 4 injured, 1104 evacuated, and 2664 homeless people. However, the number of injured, evacuated, and homeless people is certainly underreported because the newspapers clearly highlighted flood and landslide cases that generated fatalities.

Hydro-geomorphological event damage assessment is particularly difficult to quantify due to the diversity of damage (human, economic, structural, or functional) and the scarce information reported in contemporary newspapers. This problem is more accentuated in the study of historical events like the December 1909 event because it is virtually impossible to reconstruct the level of loss verified in the different affected regions. To help in tackling this caveat we have introduced three damage indices that allowed us to compare societal damages in the affected regions, because these indices are normalized by the resident population at the time of the event. In general terms damage indices were higher in Spanish regions than in Portuguese. Fatalities associated with this event were more abundant in rural areas. In addition, this study demonstrates that males were more exposed to flood risk, which is explained by the more active role of males in agriculture activities and pasturing at that time.

The length of this event extending from 20 to 28 December 1909 represents an extreme episode affecting most of the western and northwestern sector of the Iberian Peninsula from the impact perspective, particularly along the Douro Basin. In addition, the temporal distribution of DISASTER cases is consistent with a frontal system that crossed the Iberian Peninsula travelling on a SW to NE orientation and associated with a low pressure system located near the British Isles.

The December 1909 event was characterized by a period of enhanced precipitation after 17 December and was associated with some outstanding precipitation records particularly notable at the Guarda station, where the cumulative precipitation between 17 and 31 December 1909 reached approximately three times more than the mean monthly precipitation for December. Since 22 December 1909 the cumulative precipitation was higher than the 95th percentile at Guarda and Salamanca stations, both located in the Douro Basin.

The extreme precipitation was not observed in most locations of the Iberian Peninsula where pluviometers were available. Weather stations of Oporto and Lisbon located near the Atlantic coast in the mouth of the Douro and Tagus rivers, respectively, did not record extreme accumulated precipitation in the December 1909 event. Nevertheless, these areas were also affected by floods as a result of the intense precipitation that occurred in the upper and middle sectors of the Douro and Tagus hydrographic basins.

The limited available daily precipitation data do not show the precipitation amount that may have occurred in the affected area of the event especially on higher mountain ranges such as Gerês or Cantabrian Mountains. Since the ARs are rich in water vapour and are associated with strong winds that force the water vapour up mountain sides, the atmospheric conditions at that time did not inhibit upward motions which were responsible for higher precipitation values in mountain areas than over flat coastal areas. This can only be seen at Guarda rain gauge (altitude of 1019 m) that registered a daily precipitation record, but we believe that equally extremely high precipitation values must have occurred also in other mountain ranges especially in the Cantabrian Mountains due to the nature of the floods that occurred in northern Spain.

The assessment of the large-scale atmospheric conditions suggests that the extreme precipitation event verified in December 1909 resulted from the merging of unusual meteorological conditions. The presence of atmospheric instability due to the existence of a low-pressure system and its associated frontal and post-frontal sectors and also the presence of an AR structure (with particularly high content of humidity) responsible for the advection of moisture towards the Iberian Peninsula contributed to enhance precipitation. It was shown by Ramos et al. (2015) that 60 % of the most anomalous precipitation days are associated with the presence of an AR in the Iberian Peninsula. Since the average frequency of ARs during the extended winter (ONDJFM) over the Peninsula is around 10 % (although at different latitudes) we believe that more historical Douro floods (Rodrigues et al., 2003; Fragoso et al., 2015) could be associated also with the presence of ARs and deserve a specific assessment.

The DISASTER event of December 1909 took place at a time when the relationship between the precipitation and the river flow was more straightforward in the absence of dams. Nowadays, Douro and Tagus rivers have several dams that control flow rates and reduce flood peaks, thus reducing also the risk of flooding even when higher than normal precipitation occurs (Trigo et al., 2004). Nevertheless, there are some lessons to be learnt from the reconstruction of this major historical hydro-geomorphologic event.

First, data collection of societal disaster impacts based on local and national newspapers can be considered a good source of information to achieve temporal and spatial distribution of impacts, characteristics of circumstances surrounding fatalities and gender of affected people. However, such a document-based disaster inventory is typically incomplete and, as a rule, the reliability of information is higher for the temporal distribution than for the spatial location of disaster cases.

Second, the characterization of the meteorological synoptic forcing of the DISASTER event is crucial to understand the rainfall distribution in space and time, which is of paramount importance to explain the observed social impacts. In addition, in the case of floods the complete fluvial system has to be considered, as the flooding on the lower part of the basins (e.g. the Douro and the Tagus river basins) is related to rainfall occurring in the middle and upper sectors of the basins.

Third, the detailed reconstruction of past hydro-geomorphological disastrous events such as the one presented here is very useful to improve our knowledge on extreme events which may occur again and may have a high impact on the territory. Therefore, this work is also relevant for assessing the exceptionality of past hydro-geomorphologic disasters.

Lastly, the correct understanding of the meteorological drivers of major flood and landslide events like the December 1909 event may improve the early warning systems for floods and landslides in Iberia as the critical meteorological conditions can be nowadays anticipated some days prior to the event.

Acknowledgements. Susana Pereira and Alexandre M. Ramos were supported by the Portuguese Foundation for Science and Technology (FCT) through grants SFRH/BPD/69002/2010 and SFRH/BPD/84328/2012, respectively. José M. Vaquero was supported by the Government of Extremadura. This research was supported by the Portuguese Foundation for Science and Technology (FCT) through the project FORLAND – Hydro-geomorphologic Risk in Portugal: Driving Forces and Application for Land Use Planning (PTDC/ATPGEO/1660/2014) and Junta de Extremadura (Research Group grants 15137). We thank the anonymous reviewers for helpful comments.

References

Barriendos, M. and Rodrigo, F. S.: Study of historical flood events on Spanish rivers using documentary data, Hydrolog. Sci. J., 51, 765–783, doi:10.1623/hysj.51.5.765, 2006.

Benito, G., Diéz-Herrero, A., and Villalta, M.: Magnitude and frequency of flooding in the Tagus basin (Central Spain) over the last millennium, Climatic Change, 58, 171–192, 2003.

Brunet, M., Sigró, J., Saladie, O., Aguilar, E., Moberg, A., Lister, D., and Walther, A.: Long-term changes in extreme temperatures and precipitation in Spain, Contrib. Sci., 3, 331–342, doi:10.2436/20.7010.01.11, 2007.

Compo, G. P., Whitaker, J. S., Sardeshmukh, P. D., Matsui, N., Allan, R. J., Yin, X., Gleason, B. E., Vose, R. S., Rutledge, G., Bessemoulin, P., Brönnimann, S., Brunet, M., Crouthamel, R. I., Grant, A. N., Groisman, P. Y., Jones, P. D., Kruk, M. C., Kruger, A. C., Marshall, G. J., Maugeri, M., Mok, H. Y., Nordli, Ø., Ross, T. F., Trigo, R. M., Wang, X. L., Woodruff, S. D., and Worley, S. J.: The Twentieth Century Reanalysis Project, Q. J. Roy. Meteorol. Soc., 137, 1–28, doi:10.1002/qj.776, 2011.

Cortesi, N., Trigo, R. M., Gonzalez-Hidalgo, J. C., and Ramos, A. M.: Modelling monthly precipitation with circulation weather types for a dense network of stations over Iberia, Hydrol. Earth Syst. Sci., 17, 665–678, doi:10.5194/hess-17-665-2013, 2013.

Cortesi, N., Gonzalez-Hidalgo, J. C., Trigo, R. M., and Ramos, A. M.: Weather types and spatial variability of precipitation in the Iberian Peninsula, Int. J. Climatol., 34, 2661–2677, doi:10.1002/joc.3866, 2014.

Couto, F. T., Salgado, R., and Costa, M. J.: Analysis of intense rainfall events on Madeira Island during the 2009/2010 winter, Nat. Hazards Earth Syst. Sci., 12, 2225–2240, doi:10.5194/nhess-12-2225-2012, 2012.

Daveau, S.: Geografia de Portugal: O Ritmo Climático e a Paisagem, João Sá da Costa, Lisbon, 1998.

Dettinger, M.: Climate Change, Atmospheric Rivers, and Floods in California – A Multimodel Analysis of Storm Frequency and Magnitude Changes, J. Am. Water Resour. Assoc., 47, 514–523, doi:10.1111/j.1752-1688.2011.00546.x, 2011.

Domínguez-Castro, F., Ramos, A. M., García-Herrera, R., and Trigo, R. M.: Iberian extreme precipitation 1855/1856: an analysis from early instrumental observations and documentary sources, Int. J. Climatol., 35, 142–153, doi:10.1002/joc.3973, 2015.

Fragoso, M., Trigo, R. M., Pinto, J. G., Lopes, S., Lopes, A., Ulbrich, S., and Magro, C.: The 20 February 2010 Madeira flash-floods: synoptic analysis and extreme rainfall assessment, Nat. Hazards Earth Syst. Sci., 12, 715–730, doi:10.5194/nhess-12-715-2012, 2012.

Fragoso, M., Marques, D., Santos, J., Alcoforado, M., Amorim, I., Garcia, J., Silva, L., and Nunes, M.: Climatic extremes in Portugal in the 1780s based on documentary and instrumental records, Clim. Res., 66, 141–159, doi:10.3354/cr01337, 2015.

Gallego, M. C., Trigo, R. M., Vaquero, J. M., Brunet, M., García, J. A., Sigró, J., and Valente, M. A.: Trends in frequency indices of daily precipitation over the Iberian Peninsula during the last century, J. Geophys. Res., 116, D02109, doi:10.1029/2010JD014255, 2011.

García, J. A., Gallego, M. C., Serrano, A., and Vaquero, J. M.: Trends in Block-Seasonal Extreme Rainfall over the Iberian Peninsula in the Second Half of the Twentieth Century, J. Climate, 20, 113–130, doi:10.1175/JCLI3995.1, 2007.

Gimeno, L., Nieto, R., Vázquez, M., and Lavers, D. A.: Atmospheric rivers: a mini-review, Front. Earth Sci., 2, 1–6, doi:10.3389/feart.2014.00002, 2014.

Goodess, C. M. and Jones, P. D.: Links between circulation and changes in the characteristics of Iberian rainfall, Int. J. Climatol., 22, 1593–1615, doi:10.1002/joc.810, 2002.

Hurrel, J. W.: Decadal trends in the North Atlantic Oscilation: regional temperatures and precipitation, Science, 269, 676–679, 1995.

Jarvis, A., Reuter, H. I., Nelson, E., and Guevara, E.: Hole-filled SRTM for the globe Version 4, available from the CGIAR-CSI SRTM 90 m Database, available at: http://srtm.csi.cgiar.org (last access: 9 November 2015), 2008.

Kirchner-Bossi, N., García-Herrera, R., Prieto, L., and Trigo, R. M.: A long-term perspective of wind power output variability, Int. J. Climatol., 35, 2635–2646, doi:10.1002/joc.4161, 2014.

Lavers, D. A., Allan, R. P., Wood, E. F., Villarini, G., Brayshaw, D. J., and Wade, A. J.: Winter floods in Britain are connected to atmospheric rivers, Geophys. Res. Lett., 38, L23803, doi:10.1029/2011GL049783, 2011.

Liberato, M. L., Ramos, A. M., Trigo, R. M., Trigo, I. F., Durán-Qesada, A. M., Nieto, R., Gimeno, L., Durán-Quesada, A. M., Nieto, R., and Gimeno, L.: Moisture Sources and Large-Scale Dynamics Associated With a Flash Flood Event, in: Lagrangian Modeling of the Atmosphere, edited by: Lin, J., Brunner, D., Gerbig, C., Stohl, A., Luhar, A., and Webley, P., American Geophysical Union, Washington, D.C., 111–126, 2013.

Llasat, M. D. C., Rigo, T., and Barriendos, M.: The "Montserrat-2000" flash-flood event: a comparison with the floods that have occurred in the northeastern Iberian Peninsula since the 14th century, Int. J. Climatol., 23, 453–469, doi:10.1002/joc.888, 2003.

Llasat, M.-C., Barriendos, M., Barrera, A., and Rigo, T.: Floods in Catalonia (NE Spain) since the 14th century. Climatological and meteorological aspects from historical documentary sources and old instrumental records, J. Hydrol., 313, 32–47, doi:10.1016/j.jhydrol.2005.02.004, 2005.

Machado, M. J., Botero, B. A., López, J., Francés, F., Díez-Herrero, A., and Benito, G.: Flood frequency analysis of historical flood data under stationary and non-stationary modelling, Hydrol. Earth Syst. Sci., 19, 2561–2576, doi:10.5194/hess-19-2561-2015, 2015.

Morán-Tejeda, E., Lorenzo-Lacruz, J., López-Moreno, J. I., Ceballos-Barbancho, A., Zabalza, J., and Vicente-Serrano, S. M.: Reservoir Management in the Duero Basin (Spain): Impact on River Regimes and the Response to Environmental Change, Water Resour. Manage., 26, 2125–2146, doi:10.1007/s11269-012-0004-6, 2012.

Ortega, J. A. and Garzón, A.: Influencia de la oscilación del Atlántico norte en las inundaciones del Río Guadiana, in: Riesgos naturales y antrópicos en Geomorfología, edited by: Benito, G. and Diez-Herrero, A., CSIC, Madrid, 117–126, 2004.

Pereira, S., Zêzere, J. L., Quaresma, I., Santos, P. P., and Santos, M.: Mortality patterns of hydro-geomorphologic disasters, Risk Anal., 22, doi:10.1111/risa.12516, 2015.

Ramos, A. M., Trigo, R. M., Liberato, M. L. R., and Tomé, R.: Daily Precipitation Extreme Events in the Iberian Peninsula and Its Association with Atmospheric Rivers, J. Hydrometeorol., 16, 579–597, doi:10.1175/JHM-D-14-0103.1, 2015.

Rodrigues, R., Brandão, C., and Costa, J. P.: As cheias no Douro ontem, hoje e amanhã, SNIRH, Instituto da Água, Lisbon, 2003.

Salgueiro, A. R., Machado, M. J., Barriendos, M., Pereira, H. G., and Benito, G.: Flood magnitudes in the Tagus River (Iberian Peninsula) and its stochastic relationship with daily North Atlantic Oscillation since mid-19th Century, J. Hydrol., 502, 191–201, doi:10.1016/j.jhydrol.2013.08.008, 2013.

Santisteban, J. I. and Schulte, L.: Fluvial networks of the Iberian Peninsula: a chronological framework, Quaternary Sci. Rev., 26, 2738–2757, doi:10.1016/j.quascirev.2006.12.019, 2007.

Serrano, J. A., Garcia, J. A., Mateos, V. L., Cancillo, M. L., and Garrido, J.: Monthly Modes of Variation of Precipitation over the Iberian Peninsula, J. Climate, 12, 2894–2919, 1999.

Shapiro, M. A. and Keyser, D.: Fronts, jet streams and the tropopause, in: Extratropical Cyclones: The Erik Palmen Memorial Volume, edited by: Newton, C. W. and Holopainen, E. O., American Meteorological Society, 167–191, 1990.

Silva, A. T., Portela, M. M., and Naghettini, M.: Nonstationarities in the occurrence rates of flood events in Portuguese watersheds, Hydrol. Earth Syst. Sci., 16, 241–254, doi:10.5194/hess-16-241-2012, 2012.

Stickler, A., Brönnimann, S., Valente, M. A., Bethke, J., Sterin, A., Jourdain, S., Roucaute, E., Vasquez, M. V., Reyes, D. A., Allan, R., and Dee, D.: ERA-CLIM: Historical Surface and Upper-Air Data for Future Reanalyses, B. Am. Meteorol. Soc., 95, 1419–1430, doi:10.1175/BAMS-D-13-00147.1, 2014.

Trigo, R. M. and Palutikof, J. P.: Precipitation Scenarios over Iberia?: A Comparison between Direct GCM Output and Different Downscaling Techniques, J. Climate, 14, 4422–4446, 2001.

Trigo, R. M., Pozo-Vázquez, D., Osborn, T. J., Castro-Díez, Y., Gámiz-Fortis, S., and Esteban-Parra, M. J.: North Atlantic oscillation influence on precipitation, river flow and water resources in the Iberian Peninsula, Int. J. Climatol., 24, 925–944, doi:10.1002/joc.1048, 2004.

Trigo, R. M., Valente, M. A., Trigo, I. F., Miranda, P. M. A., Ramos, A. M., Paredes, D., and García-Herrera, R.: The impact of North Atlantic wind and cyclone trends on European precipitation and significant wave height in the Atlantic, Ann. N.Y. Acad. Sci., 1146, 212–234, doi:10.1196/annals.1446.014, 2008.

Trigo, R. M., Varino, F., Ramos, A. M., Valente, M., Zêzere, J. L., Vaquero, J. M., Gouveia, C. M., and Russo, A.: The record precipitation and flood event in Iberia in December 1876: description and synoptic analysis, Front. Earth Sci., 2, 1–15, doi:10.3389/feart.2014.00003, 2014.

Trigo, R. M., Ramos, C., Pereira, S., Ramos, A. M., Zêzere, J. L., and Liberato, M. L.: The deadliest storm of the 20th century striking Portugal; flood impacts and atmospheric circulation, J. Hydrol., doi:10.1016/j.jhydrol.2015.10.036, in press, 2015.

Vicente-Serrano, S., Trigo, R. M., López-Moreno, J., Liberato, M. L., Lorenzo-Lacruz, J., Beguería, S., Morán-Tejeda, E., and El Kenawy, A.: Extreme winter precipitation in the Iberian Peninsula in 2010: anomalies, driving mechanisms and future projections, Clim. Res., 46, 51–65, doi:10.3354/cr00977, 2011.

Zêzere, J. L. and Trigo, R. M.: Impacts of the North Atlantic Oscillation on Landslides, in: Hydrological, socioeconomic and ecological impacts of the North Atlantic Oscillation in the Mediterranean Region, advances in global change research, edited by: Vicente-Serrano, S. and Trigo, R. M., Springer, Dordrecht, Heidelberg, London, New York, 199–212, doi:10.1007/978-94-007-1372-7, 2011.

Zêzere, J. L., Pereira, S., Tavares, A. O., Bateira, C., Trigo, R. M., Quaresma, I., Santos, P. P., Santos, M. and Verde, J.: DISASTER: a GIS database on hydro-geomorphologic disasters in Portugal, Nat. Hazards, 72, 503–532, doi:10.1007/s11069-013-1018-y, 2014.

Zêzere, J. L., Vaz, T., Pereira, S., Oliveira, S. C., Marques, R., and Garcia, R. A. C.: Rainfall thresholds for landslide activity in Portugal: a state of the art, Environ. Earth Sci., 73, 2917–2936, doi:10.1007/s12665-014-3672-0, 2015.

6

An approach to build an event set of European windstorms based on ECMWF EPS

R. Osinski[2,1], **P. Lorenz**[3,1], **T. Kruschke**[4,1], **M. Voigt**[5,1], **U. Ulbrich**[1], **G. C. Leckebusch**[6,1], **E. Faust**[7], **T. Hofherr**[7], **and D. Majewski**[8]

[1]Institute of Meteorology, Freie Universität Berlin, Berlin, Germany
[2]CNRM/GAME, Météo-France and CNRS, Toulouse, France
[3]Meteo Service Weather Research GmbH, Berlin, Germany
[4]GEOMAR Helmholtz Centre for Ocean Research Kiel, Kiel, Germany
[5]GFZ German Research Centre for Geosciences, Potsdam, Germany
[6]School of Geography, Earth and Environmental Sciences, University of Birmingham, Birmingham, UK
[7]Munich Re, Munich, Germany
[8]German Weather Service, Offenbach, Germany

Correspondence to: U. Ulbrich (ulbrich@met.fu-berlin.de)

Abstract. The properties of European windstorms under present climate conditions are estimated on the basis of surface wind forecasts from the European Centre for Medium-Range Weather Forecast (ECMWF) Ensemble Prediction System (EPS). While the EPS is designed to provide forecast information of the range of possible weather developments starting from the observed state of weather, we use its archive in a climatological context. It provides a large number of modifications of observed storm events and includes storms that did not occur in reality. Thus it is possible to create a large sample of storm events, which entirely originate from a physically consistent model, whose ensemble spread represents feasible alternative storm realizations of the covered period. This paper shows that the huge amount of identifiable events in the EPS is applicable to reduce uncertainties in a wide range of fields of research focusing on winter storms. Windstorms are identified and tracked in this study over their lifetime using an algorithm based on the local exceedance of the 98th percentile of instantaneous 10 m wind speed, which is associated with a storm severity measure. After removing inhomogeneities in the data set arising from major modifications of the operational system, the distributions of storm severity, storm size, and storm duration are computed. The overall principal properties of the homogenized EPS storm data set are in good agreement with storms from the ERA-Interim data set, making it suitable for climatological investigations of these extreme events. A demonstrated benefit in the climatological context by the EPS is presented. It gives clear evidence of a linear increase of maximum storm intensity and wind field size with storm duration. This relation is not recognizable from a sparse ERA-Interim sample for long-lasting events, as the number of events in the reanalysis is not sufficient to represent these characteristics.

1 Introduction

According to the records of insurance and re-insurance companies, windstorms are the most costly natural hazards in Europe (Münchener Rückversicherungs-Gesellschaft, 2011). Fortunately, the most extreme events occur very rarely, but this makes it difficult to estimate their recurrence periods and other statistical characteristics, which can only be estimated with large error bars assigned to them (cf. Della-Marta et al., 2009). Studies estimating these parameters make use of reanalysis and station data (e.g., Della-Marta et al., 2009 or Hofherr and Kunz, 2010) or climate simulations (e.g., Leckebusch et al., 2006). Most recently, a catalogue of damaging European windstorms was produced by Roberts et al. (2014), based on the European Centre for Medium-Range Weather

Forecast (ECMWF) ERA-Interim reanalysis. As one implication of this study, it can be said that the Ensemble Prediction System (EPS) provides a reasonable opportunity to enlarge such a catalogue substantially. So far, statistical models like the random walk or Markov Chain Monte Carlo models are often used to extend samples for the estimation of the recurrence of severe storm events or extreme wind speed with return periods of over 1000 years (e.g., Dukes and Palutikof, 1995). We use the EPS for the same purpose of extending the sample size with the distinction that all EPS events are fully based on a physical model, which has the big advantage of a good consistency and coverage of the potential storm-related risk. In a statistical sense, observations represent the realized reality. Ensemble forecasts as part of the regular weather forecasts demonstrate that individual weather events could have developed differently, starting from basically the same initial weather conditions. In this sense, observations do not provide information on potential alternative developments that could have been reality with a similar probability.

Studies on the EPS are mainly focused on the quality of the prediction. An example of such a study related to European winter storms can be found in Buizza and Hollingsworth (2002), where the focus lies on the predictability of the heavily impacting winter storms of the year 1999. Froude (2006, 2009) has analyzed the predictability of storm tracks and extratropical cyclones using a cyclone-tracking algorithm by Hodges (1994). Froude and Gurney (2010) focused on the application of the EPS for the oil and gas industry. The output of the ECMWF EPS in an impact-based study was used for estimating the range of potential storm surge events at the German bight (Koziar and Renner, 2005). The small area investigated in this study is, however, not representative of winter storms in Europe. The current study aims at assessing climatological properties of European winter storms, produced by the operational ECMWF Ensemble Prediction System. Such an approach requires minimizing the effects from inhomogeneities in the EPS introduced by the regular updates of this operational system. They could potentially produce systematic deviations from observed storms, the latter being represented by the ERA-Interim reanalysis in our study. Beyond these changes, there could be systematic forecast-lead-time-dependent trends in the EPS data set, affecting storm characteristics like severity, duration, or the affected areas. Possible breaks (e.g., between different model cycles), trends (e.g., a model drift), and biases (e.g., different wind speed distributions according to different model resolutions) caused by the EPS inhomogeneities in the detected storm properties must be initially addressed in order to carry out climatological investigations.

The paper aims at demonstrating that it is possible to produce statistics of storms under observed climate conditions based on EPS forecasts, leading to more reliable results than traditional approaches based on reanalysis data. Our aim is a representation of the recent climate, which distinguishes our approach from others based for example on climate projections. To summarize, our study intends to describe a possibility of producing more reliable storm statistics which are still very close to the observed climate. The final event set is comparable to those which are stochastically generated based on a fixed historical sample, with the distinction that the stochastics is replaced by the application of a physical model in our case.

2 Data

Instantaneous 10 m wind speed data at different archiving time steps as mentioned farther below are considered. The area of investigation covers the Atlantic–European region spanning from 40° W to 40° E and 25 to 80° N. For part of the studies in this paper (explicitly mentioned in the respective sections), the entire Northern Hemisphere was used in order to avoid boundary effects. An extended winter season is used from September to May.

2.1 ERA-Interim

An archive of 6-hourly ERA-Interim reanalysis data (Dee et al., 2011) is used. At the time the current study was performed, data before the year 1989 were not available, so the period considered is 1989 to 2010. ERA-Interim uses the 4D-Var assimilation scheme and the Integrated Forecast System (IFS) release Cy31r2 at a horizontal spectral resolution of T_L255. The same system release was operational for the EPS from 12 December 2006 until 5 June 2007, but with horizontal resolution of T_L399 (for details refer to Palmer et al., 2007).

2.2 ECMWF Ensemble Prediction System

This section provides some relevant aspects about the ECMWF EPS. A more detailed description of the EPS can be found in Palmer et al. (1992, 2007) and Molteni et al. (1996). The Ensemble Prediction System of the ECMWF became operational in December 1992 (see Table 1 for an overview). Initially, 32 perturbed forecast members (based on the method of singular vectors; in the following abbreviated as "pf") plus one control forecast (not perturbed against the original analysis, but using the EPS model system instead of its deterministic counterpart; in the following abbreviated as "cf") were produced. The number of perturbed ensemble members was increased to 50 in December 1996. Since October 1998, some of the EPS runs have been produced including perturbations in the model physics. With increasing computing power, continuous upgrades of the system lead to improvements in the forecast skill (cf. Palmer et al., 2007). The horizontal resolution was increased from T63 as follows: T_L159 (December 1996), T_L255 (November 2000), and T_L399 (February 2006) to eventually (not used in this study) T_L639 (January 2010). The resolution of the singular vectors was changed from T21L31 to T42L31 (March 1995),

T42L40 (October 1999), and eventually T42L62 (February 2008) (Palmer et al., 2007). Changes in the data assimilation scheme (Rabier et al., 2000; Mahfouf and Rabier, 2000; Klinker et al., 2000) from 3D-Var to 4D-Var were introduced in November 1997 (cf. Bouttier and Rabier, 1997). The EPS integration time is 15 days, but after 10 days of forecast the horizontal resolution is decreased. Since March 2003, the system has been initialized twice a day, at 12:00 and 00:00 UTC. In order to take the major changes into account, the data set was split into periods with constant horizontal resolution (Table 1). Data used in this study cover the period until 25 January 2010, thus excluding the latest period with T_L639 resolution. Depending on the period, the EPS data are available at 12-, 6-, and 3-hourly temporal resolution. As ERA-Interim is only available at 6-hourly resolution, the EPS data with a 3 h resolution were used in subsets of 6-hourly resolution. For the 12-hourly data, ERA-Interim was also used at this temporal resolution (time steps at 00:00 and 12:00 UTC).

3 Methods

3.1 Identification and characterization of storms at midlatitudes – wind tracking

For the identification and characterization of European winter windstorms, an impact-related wind-tracking algorithm is used. It was introduced by Leckebusch et al. (2008) and has been further developed since then. An overview of the actual scheme is provided by Kruschke (2015). It identifies grid points belonging to windstorms by searching for spatial clusters of grid points (extending over an area of at least 1.6×10^5 km^2) where the local 98th percentile of wind speed is exceeded. The choice of the 98th percentile is motivated by the relevance of this threshold for storm damages (Klawa and Ulbrich, 2003). The identified clusters are connected to a track using a nearest-neighbor criterion. The maximum distance allowed to connect two clusters to a windstorm track is limited by an assumed maximum wind field propagation velocity of 120 km h^{-1}. In the present study a minimum lifetime of 24 h of an identified windstorm must be fulfilled, equivalent to three archived time steps for the 12 h temporal resolution and five time steps for 6 h resolution periods (Table 1). By summing the cube of the 98th-percentile exceedances belonging to a track, an objective storm severity measure is determined. This measure, called the storm severity index (SSI), is calculated for each storm over all time steps t and grid points k affected by exceedances of the 98th percentile assigned to a storm. It is meant to characterize the severity of storms, taking intensity, size, and duration of the storms into account, as is shown in Eq. (1):

$$\mathrm{SSI} = \sum_{t}^{T}\sum_{k}^{K}\left[\left(\max\left(1, \frac{v_{k,t}}{v_{\mathrm{perc},k}}\right) - 1\right)^{3} \cdot A_k\right]. \qquad (1)$$

$v_{k,t}$ is the wind velocity in grid cell k at time instance t, $v_{\mathrm{perc},k}$ the 98th percentile in grid cell k, and A_k the area of grid cell k. The SSI values are normalized to a grid cell of unit size. This is done using the grid cell area A_k to reduce the resolution dependence when applying different models and to eliminate a latitude dependence. A resolution dependence can still remain, as models of different resolutions can produce different wind speed distributions. This will be discussed in the following section. The algorithm was originally developed for the application with reanalysis and climate data. The medium-range ensemble EPS consists of single forecasts from which we use the first 10 days. For each day up to twice a day (12:00 and 00:00 UTC initializations) 50 perturbed forecasts and an additional control forecast were produced and archived. The algorithm is applied on each individual forecast. This means that, when combining the members and forecasts with different lead times, a single day is represented by up to (50pf+1cf) $\times$ 2 initializations $\times$ 10 days = 1020 equivalent days. To avoid boundary effects at the beginning and end of the forecasts, the period had to be reduced to be able to generate representative samples. We restricted our sample to EPS runs initialized at 12:00 UTC with storms starting inside a window of 6 forecast days (to be discussed in Sect. 4.3). This results in an enlargement of the sample deducible from reanalysis by a factor of 300 using perturbed forecasts.

3.2 Homogenization of the EPS

The improvements introduced into the operational EPS system mentioned above will affect the results of the tracking procedure in different ways, but a main impact is due to the changes in spatial and temporal resolution. Hence, we subdivide the data into subperiods of the same spatiotemporal resolution and apply a two-step procedure to homogenize windstorm identification and SSI calculation across these subperiods: first, the 98 % quantiles of each subperiod are scaled towards a common basis, using the ERA-Interim data set as a reference. We call this the "climatological scaling" of the threshold used for windstorm identification (see Sect. 3.2.1). Second, a quantile–quantile mapping approach (cf. Boé et al., 2007; Maraun, 2013) is used for exceedances of the 98th percentile to provide matching shapes of the upper tail of the wind speed distribution, which is a requirement of SSI calculations, homogenous across all subperiods. This second step is called "scaling of exceedance" in the context of this study (see Sect. 3.2.2).

3.2.1 Climatological scaling

Subdividing the EPS data set into periods which are homogeneous in terms of the horizontal resolution of the model system (see Table 1) reflects the finding that different resolutions of the EPS system produce different wind speed biases and, as a consequence, biases in SSIs, storm duration, and size.

Table 1. Overview of general characteristics of the EPS (*used temporal resolution*) pf: perturbed forecast; cf: control forecast.

Time frame	Spatial resolution resolution	Temporal (h)	Number of member	Initialisations at
21 Nov 1992–9 Dec 1996*	T63	12	32pf+1cf	12:00 UTC
10 Dec 1996–12 Jan 2000	T_L159	12	50pf+1cf	12:00 UTC
13 Jan 2000–20 Nov 2000	T_L159	6 (12)	50pf+1cf	12:00 UTC
21 Nov 2000–24 Mar 2003	T_L255	6	50pf+1cf	12:00 UTC
25 Mar 2003–27 Jun 2005	T_L255	6	50pf+1cf	00:00 and 12:00 UTC
28 Jun 2005–31 Jan 2006	T_L255	3 until 126/144 h and 6 (6)	50pf+1cf	00:00 and 12:00 UTC
1 Feb 2006–25 Jan 2010	T_L399	3 until 144 h (6)	50pf+1cf	00:00 and 12:00 UTC

* Before January 1994 only three forecasts per week available, major change on the system with introduction of the IFS in March 1994 and the introduction of IFS cycle 12r1, which led to a significant reduction in the model bias of 10m wind speed (http://www.ecmwf.int/en/forecasts/documentation-and-support/evolution-ifs/cycle-archived/1994-summary-changes).

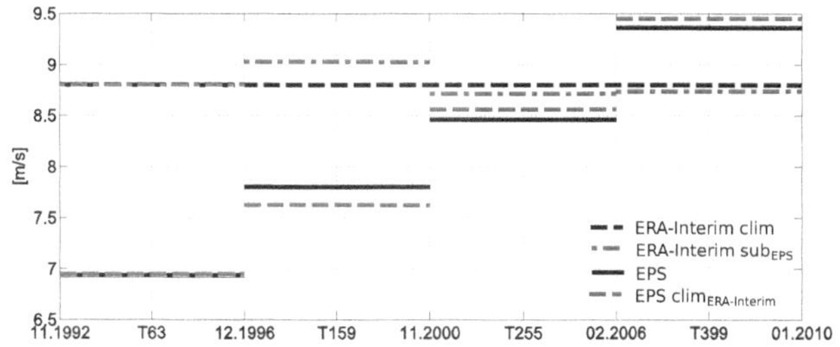

Figure 1. 98th percentile as average over all land boxes (according to land-sea masks of the data set) in the domain from 40° W to 40° E and 25 to 80° N for different EPS subperiods with corresponding 98th ERA-Interim land box percentile: climatological ERA-Interim percentile based on ERA-Interim for period 1989–2010 (ERA-Interim clim), ERA-Interim percentiles for the periods of the corresponding EPS periods (ERA-Interim sub$_{EPS}$), percentiles of raw EPS data (EPS), and climatologically adjusted EPS percentiles (EPS clim$_{ERA\text{-}Interim}$).

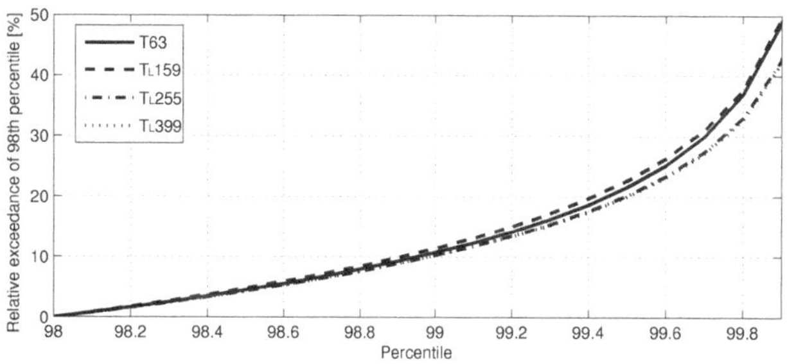

Figure 2. Visualization of tail differences in the wind speed distribution of the four subperiods (see Table 1) of the EPS. Shown: relative exceedance of 98th EPS percentile as land average. Internal climate variability of the disjunct periods is excluded by utilization of the climatological scaling (for details see text).

A model with a coarse resolution represents an average of a larger grid cell area than a model with a fine resolution. The finer the model resolution, the better the orographic effects that can be captured. This influences the wind speed distributions, but differences in wind speed characteristics for the periods considered can also originate from climate variability. The latter becomes evident when the ERA-Interim data are used for estimating this threshold for the whole period and for the same subperiods: Fig. 1 shows 98th ERA-Interim percentiles using all land grid points in the Atlantic–European area chosen. Land grid points are shown, as the major interest is related to storm damages over land, but the method is applied on all individual cells of the entire grid. The estimates for the four subperiods vary from the percentile computed for the complete period 1989 to 2010. The percentiles of the EPS versions with coarser horizontal resolution are found to be lower than those with higher resolution. The effect from T_L159 to T_L255 is much stronger than from T_L255 to T_L399. Note that for this intercomparison an interpolation towards the ERA-Interim grid had to be performed. The correction factor for the 98 % quantile of each grid cell is computed taking the factor due to climate variation (as estimated from ERA-Interim) into account.

3.2.2 Scaling of exceedance

After climatological scaling of the identification threshold, the wind speeds exceeding this 98th percentile still differ between the subperiods, as shown in Fig. 2. The presented differences in the tail seem to be very small, but as the cubic of these values is used and summed over a larger quantity of grid cells for the SSI calculation, cf. Eq. (1), they are impacting the results. For this reason, a quantile–quantile mapping is used. It is a standard method used for a bias correction; see, e.g., Maraun (2013). The method chosen estimates empirically percentiles in equidistant steps (0.1 %) for both EPS and ERA-Interim. A wind value in the EPS, which corresponds to the ith percentile of the EPS wind speed distribution, is corrected in the way that it has afterwards the value of the ith percentile of the ERA-Interim wind distribution. After both climatological scaling and quantile–quantile mapping, the ERA-Interim 98th percentile and the exceeding wind speeds mapped on the ERA-Interim distribution can be

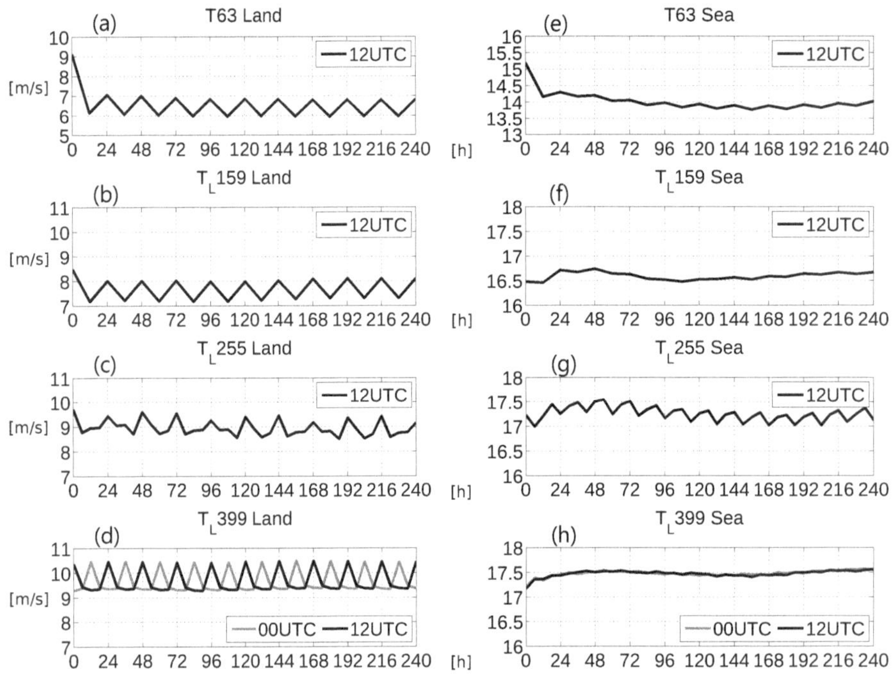

Figure 3. Average of 98th percentile ($m\,s^{-1}$) for different forecast lead times (right axis, h after initialization): for T63 12-hourly **(a, e)**, T_L159 12-hourly **(b, f)**, T_L255 6-hourly **(c, g)**, and T_L399 6-hourly **(d, h)**. **(a)–(d)** for land grid boxes, **(e)–(h)** for sea grid boxes.

used for the SSI calculation in every subperiod. A quantile–quantile mapping for the different periods without previous climatological scaling is not suitable, as it would completely remove the (real) climate variations.

4 EPS storm validation

4.1 Spin-up effects, threshold, and diurnal cycle

Even though spin-up effects in numerical simulations are well known, their magnitudes in the ECMWF EPS have not been a major issue in the scientific literature. An exception is the report by Lamquin et al. (2009) focusing on humidity in the upper troposphere. Results of an analysis on systematic variations of 98 % quantiles of wind speed are given in Fig. 3 for the T63, T_L159, T_L255, and T_L399 resolutions. Average values over all land and all sea boxes in the area considered have been computed for archiving steps of the forecasts. For both land and sea grid points a small initialization effect in the first 6 to 12 h of the forecasts becomes visible. The percentile value in the T_L159 resolution over land, for example, is about $0.5\,m\,s^{-1}$ higher during the first one to two archiving time steps than subsequently. Over sea, there seems to be an effect with opposite signature (lower initial values) in the first 12 to 18 forecast hours. The data for T_L399 over sea show the same initialization effect. The dominant feature in Fig. 3 is, however, a diurnal cycle with an amplitude of about $1\,m\,s^{-1}$ over land. Maxima occur at the forecast time steps valid at noon (12:00 UTC). Note that a corresponding cycle is also found in the ERA-Interim data, with about the same amplitude (not shown). Conventional observations confirm that the daily cycle in the 10 m wind speed over land is a realistic feature (Lapworth, 2008, 2012). The EPS with T_L255 is characterized by an interfering daily periodicity and an 18 h periodicity. As the daily cycle is small over sea, the 18 h periodicity is clearly visible in Fig. 3g. The irregular behavior of the EPS with T_L255 resolution is apparently related to the stochastic perturbations of the model physics used during the respective period (A. Beljaars, personal communication, November 2012) as the unperturbed control forecast produces a regular daily cycle (figure not shown). A more thorough investigation of the 18 h cycle is beyond the scope of the present paper. We have not attempted to remove it from the investigation, but in comparing the windstorm statistics for this EPS resolution with the other periods we found no evidence for a systematic effect.

4.2 Modifications of observed storms in the EPS: storm "Emma"

Different EPS members started at different lead times will produce modifications of observed storm events in terms of their genesis time, track, and intensity. Before considering the respective statistics for the whole time series, we consider the storm event named[1] Emma (28 February 2008) as an example in more detail. At a lead time of 6 h, all of the

[1]Names are given by the Institute of Meteorology of the Freie Universität Berlin.

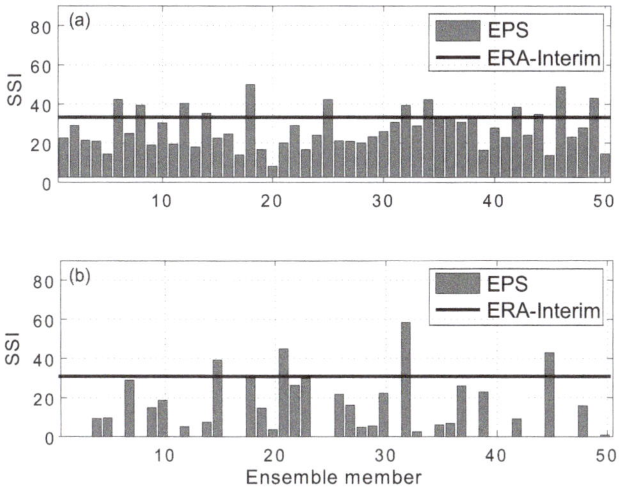

Figure 4. SSIs for representations of the storm Emma (28 February 2008, 18:00 UTC, detected in ERA-Interim) in 50 EPS members: **(a)** 6 h lead time, initialized 28 February 2008 at 12:00 UTC and **(b)** 90 h lead time, initialized 25 February 2008 at 00:00 UTC.

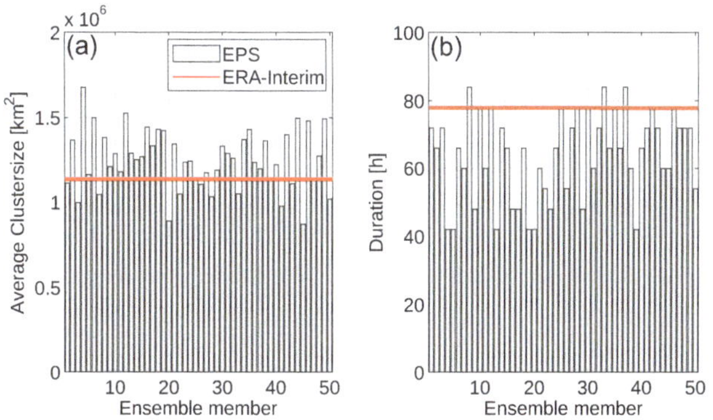

Figure 5. Representation of the storm Emma (28 February 2008, 18:00 UTC, detected in ERA-Interim) in 50 EPS members initialized 28 February 2008 at 12:00 UTC. **(a)** Average cluster size (km^2), **(b)** duration (h).

50 EPS runs produce a storm fulfilling our criteria that can be assigned to the observed one (Fig. 4a). The majority of the simulated events are weaker than the intensity computed from ERA-Interim, but for 12 members the simulated storm is stronger than observed. At a lead time of 90 h, taken as a second example (Fig. 4b), in several runs no storm is found. One member, however, produces a storm of about double the observational SSI. The variations in SSI originate from variations in the intensity at individual grid points, in area and in storm lifetime, as depicted in Fig. 5 for the 6 h lead time. The track of Emma in ERA-Interim and in the individual EPS members (Fig. 6) is found by identifying a storm core from the weighted local SSI contributions of all storm grid points at a time step, and connecting the centers from different time steps (Leckebusch et al., 2008). While in many other cases the observed storm is found close to the center of the EPS

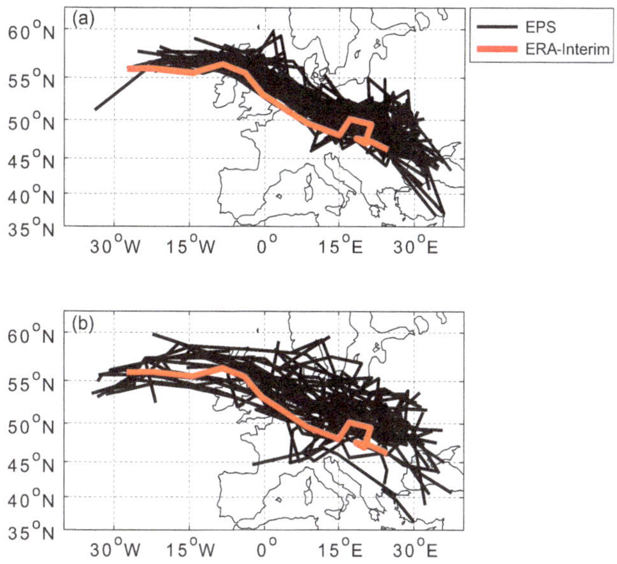

Figure 6. Tracks for representations of the storm Emma (28 February 2008, 18:00 UTC, detected in ERA-Interim) in 50 EPS members initialized on **(a)** 28 February 2008 at 12:00 UTC and 50 members initialized on **(b)** 25 February 2008 at 00:00 UTC.

ensemble member storms, all EPS tracks of Emma at this lead time are located northward of the ERA-Interim storm (Fig. 6a). For the 90 h lead time (Fig. 6b), the spread between the modified Emma tracks is larger. A notable feature of Emma is the fact that the observed Emma tends to be at the border of the EPS ensemble also for the long lead time. This example demonstrates that extreme EPS events can be feasible representations, but the northward shift is not systematic in the EPS. SSI values for all events detected in ERA-Interim and the EPS (starting inside a 6-day window; see Sect. 4.3) over the period 2001 to 2010 are shown in Fig. 7. Over the entire period, the range of SSI in the EPS is much larger than in ERA-Interim. A larger range of SSI values was expected, as the EPS can include forecasts with slightly higher wind velocities. The definition of the SSI, using cubic exceedances, enlarges the range of values. As the motivation for the SSI definition is damage potential, the additional events help to better estimate potential storm risks for Europe, in particular with respect to the occurrence of the most extreme storms.

4.3 Comparison of storm properties in the EPS and ERA-Interim

In order to compare the entire ensemble of storms in the EPS with those detected in the ERA-Interim data set, events not entirely captured in a forecast must be excluded. They would erroneously be taken as short(er)-lived storm events. This situation may be present if a storm is detected at the initialization time. In this case, it may have existed before but could not be completely tracked on the basis of the driving data. Removing all storms existing at the start of the fore-

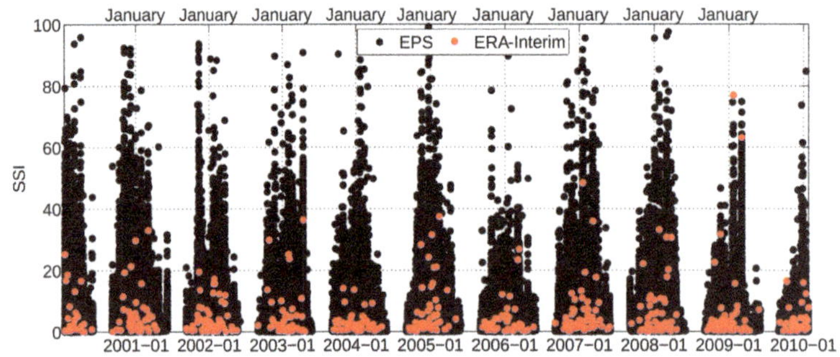

Figure 7. SSIs for all storms in the period 13 January 2000 (10 m wind available at 6-hourly resolution for the EPS) to 25 January 2010 for ERA-Interim and for the EPS with initializations at 12:00 UTC. The months June, July, and August are excluded.

Table 2. Average storm properties of EPS (ERA-Interim).

Resolution	No. per year	Size [10^6 km^2]	Duration [h]
T63* (12-hourly)	50.9 (54.0)	0.70 (0.65)	41.2 (43.8)
T_L159 (12-hourly)	45.9 (49.3)	0.75 (0.79)	49.2 (50.4)
T_L255 (6-hourly)	47.6 (45.0)	0.71 (0.76)	41.4 (42.0)
T_L399 (6-hourly)	47.9 (50.5)	0.74 (0.74)	42.0 (42.6)

* Based on data from 1 January 1995 to 9 December 1996.

cast, however, allows the full range of storm durations to enter the statistics without a bias. A similar kind of problem would occur with storms existing at the end of the 10-day forecast time. Here, the same solution cannot be applied as it would prefer short-duration storms for genesis occurring rather late in the forecast period. We decided to restrict the evaluated storms to those generated a maximum of 6 days after forecast initialization, leaving 4 days as a maximum duration. There is still a problem with storms lasting 4 days or longer. According to ERA-Interim, only 0.8 % of storms are this long-lasting, and only some of them (namely, those generated at one of the time steps just before the 6-day limit) are affected. We expect the impact on the results to be small. Also, the choice of 6 days is motivated in the fact that it leads to an equal frequency of evaluated time steps at 0, 6, 12, and 18 h forecast time, thus ameliorating the effects of the 18 h periodicity in intensities mentioned earlier.

Initializations at 00:00 UTC are only available after March 2003, as is shown in Table 1. We wanted to be sure to avoid an overrepresentation of the period 2003 to 2010 in the statistics and thus use only the 12:00 UTC initializations. Nevertheless we looked into the forecasts initialized at 00:00 UTC and found no systematic difference compared with the runs starting at 12:00 UTC. Using the 6-day window, one initialization per day, and 50 perturbed forecasts for the period 2000 to 2010 yields a storm sample 300 times larger than available from reanalysis data for the same period.

Storm properties in the EPS compared to ERA-Interim

The average number, size, and duration of storm events per year found in the four different time periods characterized by the specific EPS resolutions are given in Table 2, both for the EPS and ERA-Interim. The number of events in the EPS is the ensemble average over all available ensemble members, initializations per day, and over the forecast length limited to storms lying inside the described 6-day window (cf. Sect. 4.3). This number can thus be directly compared to the ERA-Interim values given in the same table. The respective values are similar between the two data sets, meaning that the storm properties in the EPS ensemble average are in good agreement with ERA-Interim. In order to compare the severity distributions of the EPS and ERA-Interim events, seven severity classes were formed making sure that there is a reasonable number of events in each of the classes to permit statistical tests. Subperiods with constant horizontal resolution of the EPS are again distinguished. Note that the SSI values calculated from data with 12-hourly resolutions (T63 and T_L159) are expected to be lower than those from 6-hourly resolutions (T_L255 and T_L399) (Fig. 8) due to the additional time steps included for the latter. It can be seen how the results of the wind tracking differ for the EPS without using any scaling technique, using only the climatological scaling, the scaling of exceedance, or both together. When both scaling techniques are used together, the severity distributions of the EPS and ERA-Interim are comparable for all subperiods except for EPS T63. For the latter, the scaling corrects for an overestimation of severity, resulting in a good agreement in the highest four severity classes. The larger number of weak events has its origin in model biases of 10 m wind speed[2] during the early years (1992 to 1994) of the data period of the T63 EPS. As it is difficult to evaluate the benefit of the scaling techniques visually, a normal distribution was fitted to the logarithm of the SSI. The Anderson–Darling test (Thode, 2002) indicates that the logarithm of the SSI is normally dis-

[2]http://www.ecmwf.int/en/forecasts/documentation-and-support/evolution-ifs/cycle-archived/1994-summary-changes

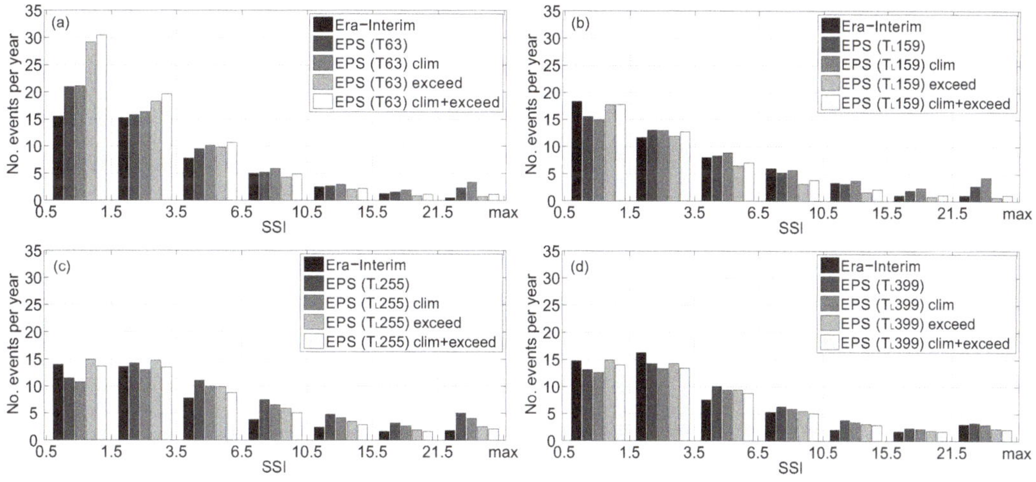

Figure 8. No. of storm events per year subdivided according to the severity, for the four individual subperiods with constant horizontal resolution: **(a)** T63, **(b)** T_L159, **(c)** T_L255, and **(d)** T_L399 of the EPS (T63 and T_L159 at 12-hourly resolution for EPS and ERA-Interim). First bar is for ERA-Interim, the other for the EPS (bars from left to right): second – EPS raw data; third – processed by climatological scaling; fourth – processed by scaling of exceedance; and fifth – applying both scaling techniques on the data.

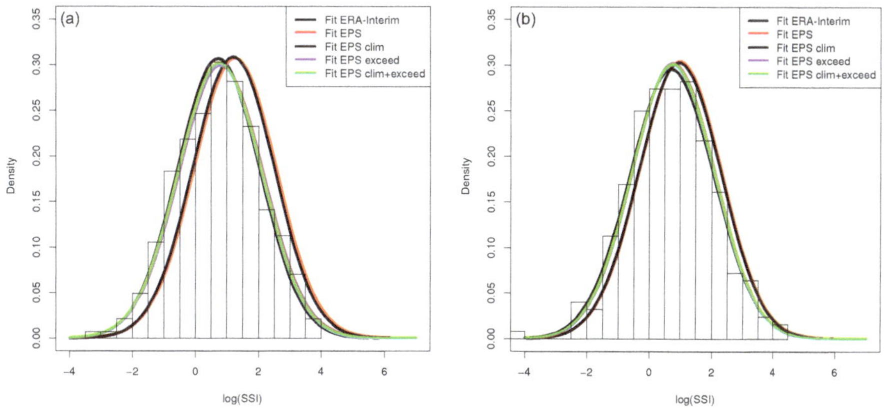

Figure 9. Fit of normal distributions to logarithm of SSI for **(a)** EPS in T_L255 and **(b)** EPS in T_L399 without scaling techniques, with climatological scaling, with exceedance scaling, and with both scaling techniques together. Raw EPS data and climatologically scaled data are similar and differ more greatly from the observation than using the exceedance scaling or both together.

tributed. The benefit from the scaling techniques is illustrated in Fig. 9. Looking for the raw EPS data at T_L399 resolution, one sees that they concur better with ERA-Interim than the data in T_L255. The effect of the climatological scaling is relatively small. Using both scaling techniques together, the distributions between the EPS and ERA-Interim look very similar. The fit parameters are shown in Table 3. Fit parameters were estimated using maximum likelihood. The exact standard errors of the parameters are very small in the EPS case due to its very large sample. The mean and standard deviation lies in between the error resulting from ERA-Interim. This means that the EPS ensemble mean represents well the storm climate which can be found in ERA-Interim. Storm representations in the EPS and ERA-Interim with comparable SSI values show, on average, comparable storm duration as well as the storm size (not shown).

5 Spatiotemporal EPS storm properties

5.1 Pure and modified EPS storms

Most considerations in this paper are based on the assumption that the EPS produces modifications of storms in the real world (subsequently called "modified EPS storms"), or, for some ensemble members, low wind speeds and thus no storm at all. However, the EPS can produce storm events that have no real-world counterpart. As for statistical investigations independent and identically distributed (iid) random variables are necessary; such pure events are particularly interesting, because they can increase the sample of independent events. Figure 10 shows a sketch of the definition of pure and modified storms in this study. To identify pure EPS storm events, events are sought for which no simultaneous counterpart can

Table 3. Parameters and their errors of fitted normal distribution to logarithm of SSI for the EPS using both scaling techniques together and ERA-Interim.

Model	Mean	SD	Error mean	Error SD
ERA-Interim (period of EPS T_L255)	0.73	1.30	0.077	0.055
EPS T_L255	0.77	1.32	0.003	0.002
ERA-Interim (period of EPS T_L399)	0.72	1.35	0.086	0.061
EPS T_L399	0.76	1.33	0.005	0.004

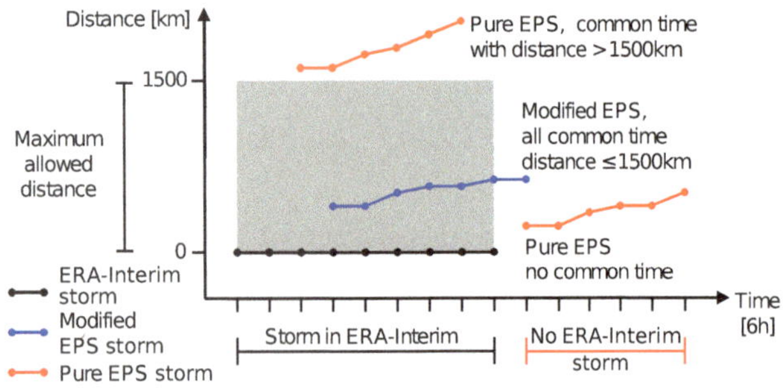

Figure 10. Sketch of definition for pure and modified EPS storms.

be found in ERA-Interim. We also regard events as pure if there is a spatial distance of more than 1500 km between contemporaneous events, as this is a typical synoptic scale of the investigated phenomena.

5.2 EPS storms during the forecast time

Using the aforementioned method to separate pure and modified storms, it can be assumed that close to the initialization time almost only modified storms can be found in the EPS (Fig. 11). All ensemble members are likely to produce the storm that actually occurred, even if properties like size and duration as well as severity vary between the different realizations. For long lead times, however, there is an increased number of pure EPS storms (grey lines in Fig. 11). The example of the storm Emma illustrates that for longer lead times a number of ensemble members do not show the storm at all, and a larger variability can be found in the intensities. Note that the average number of all storms in the EPS is nearly constant over the forecast time in spite of the small variation in the percentile values (Fig. 3) over forecast time. This number is similar to its ERA-Interim counterpart, supporting our approach to use the individual period's own percentile for storm identification. A diurnal variation in the number of storms related to the diurnal variation in the 98th percentiles (Fig. 3) is reflected in Fig. 11. As the percentile values used for the wind tracking are based on all data, their values lie between the minimum and maximum value of the 6-hourly or 12-hourly resolution. As at 12:00 UTC the 98th-percentile value is above the 98th percentile of the entire data set, the probability of an exceedance at this time of the day is larger than for the other times. For this reason the number of both first and final storm track detections is larger at 12:00 UTC than for the other times.

5.3 Spatial distribution of storms

In order to investigate whether there is a difference in the spatial distribution of European winter storms between ERA-Interim and the EPS, the effect of each grid cell by all detected storms per EPS subperiod is computed. The footprint (region of grid cells which is affected by a storm) of each detected storm is analyzed, and for each grid cell the number of footprints affecting this particular grid cell is counted. Figure 12 shows the results, and for comparability the area effects for ERA-Interim are calculated for the same time frames as the EPS subperiods. The results with the ERA-Interim and EPS T_L255 resolutions have identical grid points and are thus comparable without interpolation. For the comparison for the EPS with T_L399 resolution, the result for ERA-Interim was interpolated to this resolution. For this specific analysis, the entire Northern Hemisphere was used for the tracking to avoid boundary effects caused by a limitation of the area. The basic distribution of the effects is similar in ERA-Interim and the EPS. The lower number (300 times; EPS with 50 members lasting over 6 days) of events available in the observational data causes a much noisier distribution than what is obtained from the EPS. There are local maxima in ERA-Interim for example over north Africa and the Mediterranean which the forecast model is not able to reproduce.

5.4 Modified vs. pure EPS storms

The interest in pure EPS storms originates from the wish to find events that are independent of modifications of ERA-Interim storms. Using the same procedure as in the section before to determine the spatial effects, but only for footprints of pure EPS storms, defined after the method explained in Fig. 10, the results are shown in Fig. 13 for the EPS with T_L255. Over the Atlantic the number for the pure EPS storms is lower than over north Africa and eastern Europe. The major pathway of the storm systems is not so strongly affected by pure EPS storms as the regions where storms appear less frequently. The absolute number of pure events can be seen by combining Fig. 12 with Fig. 13. Then we have about 1 pure event over the Atlantic and about 1.5 to 2 over central

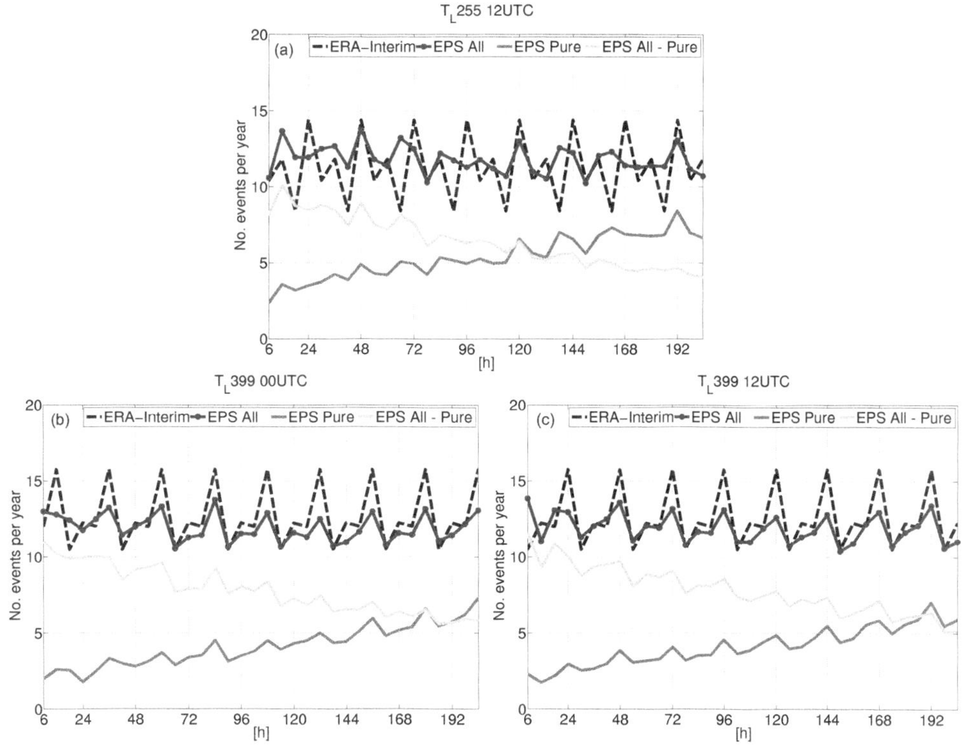

Figure 11. Temporal evolution of the number of first storm detections during the integration time (h) after initialization **(a)** T_L255 – 12:00 UTC; **(b)** T_L399 – 00:00 UTC; and **(c)** T_L399 – 12:00 UTC. Values for ERA-Interim at 00:00, 06:00, 12:00, and 18:00 UTC are repeated.

Europe. This has the consequence that the use of pure EPS storms as a supplemental amount of events for increasing an independent sample of modified storms leads to a bias in the spatial distribution of storms. Using the presented method, the dependency between events to create an iid sample is defined by a comparison to ERA-Interim. Another feasible approach is to use a matching criterion in between all of the EPS events or a bootstrap-like sampling of alternative realizations of the past. Such approaches using the ECMWF EPS were successfully applied for estimations of return periods of European winter storms by Osinski (2015).

5.5 Storm intensity vs. duration

A benefit of storm statistics based on the EPS instead of reanalysis is the larger number of storms available for statistical studies of typical midlatitude storms. Figure 14 shows a clear correlation between the storm duration and the maximum wind field size, which is the maximum of the area of exceedance of the 98th percentile that is assigned to the storm at each particular time step. For storms with durations of up to 54 h, ERA-Interim shows a comparable picture to the EPS. This can be explained by the fact that the number of observed storms of this timescale is large enough to provide reliable statistics. The EPS indicates that the average growth rate of storms is independent of their duration, while the duration determines the maximum size of the wind field. For long-lasting events there seems to be an asymmetry between the growth and the decline, where the growth seems to be faster than the decline. With respect to storm severity, a similar interdependence is found (Fig. 15). Again, the intensification rate of storms on average is nearly independent of storm duration.

6 Conclusions

Atlantic–European windstorms were identified in the archived data set of the ECMWF Ensemble Prediction System forecasts in the period December 1992 to January 2010. The identification of potentially damaging windstorms was based on the excess over the local 98th percentile of wind speeds (Leckebusch et al., 2008; Kruschke, 2015), only taking into account events which have a minimum area at a single archived time step and a minimum duration of 24 h (with fulfillment of the minimum area criterion in each of them). The fact that the operational EPS changed its characteristics during the data period led to changes in the value of the 98th percentile of wind speed. Hence a homogenization procedure was applied to four subperiods characterized by different spatial resolutions of the system. Temporal variation of the percentile due to climatic variations and variations with

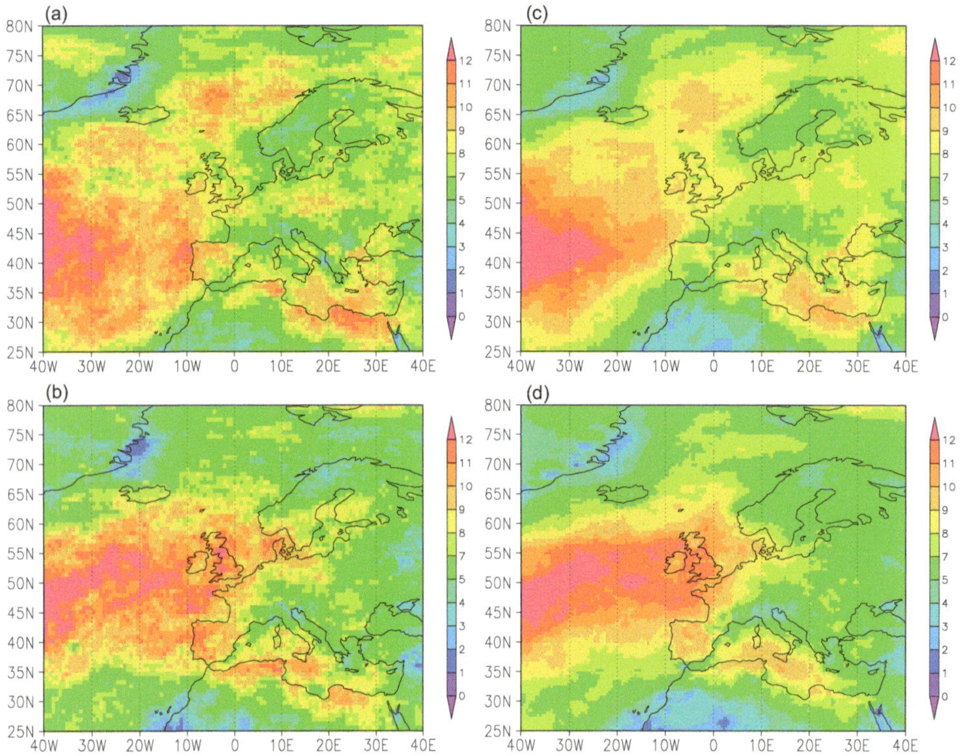

Figure 12. Accumulated yearly number of detected storms (sum of footprints per year) for time frame of the EPS resolution T_L255 **(a, c)** and T_L399 **(b, d)**, ERA-Interim **(a, b)**, and EPS **(c, d)** normalized by ensemble size by dividing by 50 members and 6 forecast days.

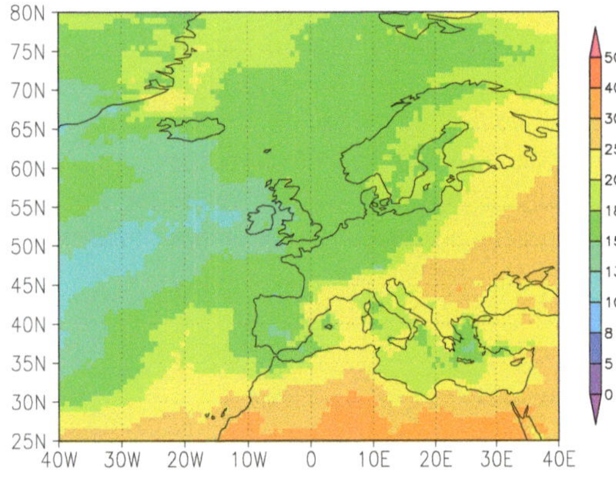

Figure 13. Percentage of number of EPS storms affecting the grid cell and being pure in the EPS with T_L255 initialized at 12:00 UTC.

respect to the cubic excess over the percentile (assumed to be model version specific) were taken into account. A diurnal cycle in the 98th percentile of the 10 m wind speed was observed in the EPS, which is also present in ERA-Interim. These diurnal variations comprise a systematically higher value of the threshold percentile for 12:00 UTC only, which is about $1\,\mathrm{m\,s^{-1}}$ larger than the respective values at the other 6 hourly time steps. This effect also leads to a diurnal variation in the number of storm initiations and ends, as detected by the here-applied storm identification scheme. Averaged over a large number of storms, this diurnal variation can be seen in the severity at different times of day. This behavior is, however, partly hidden in the EPS with T_L255 resolution, as these forecasts additionally exhibit an 18 h periodicity in the threshold for individual time steps presumably assigned to the specific stochastical perturbations imposed in the ensemble-generating process during the respective period. None of these effects had an apparent strong impact on the subsequent evaluations of the EPS as all forecast time steps inside a 6-day window were taken into account. The overall EPS storm properties were found to be similar to ERA-Interim storm properties. On average the EPS produces the same number of storm days as ERA-Interim. There is no systematic tendency over lead time in the total number of storms. The EPS produces developments of storms which have no observational counterpart. While the principal statistical properties are the same as for modifications of modified representatives of real storms, their share in the total number increases with increasing lead time. They have a spatial distribution of occurrence that is different from the observed and modified storms, with a focus on the Mediterranean and eastern Europe.

As the spatial distribution and the number, the size, and duration of events of same severity are in good agreement with "real" storm events, the EPS can be used to increase the

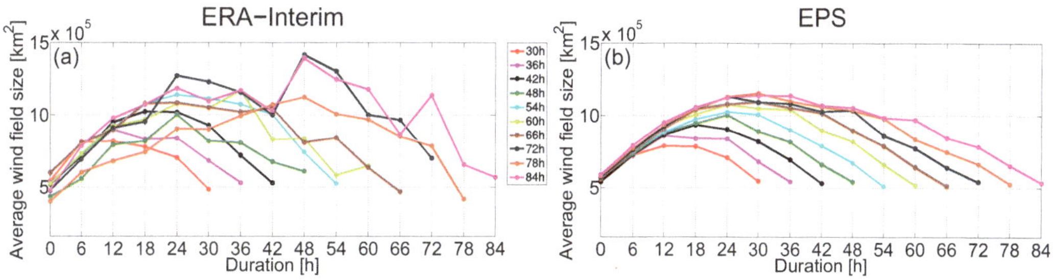

Figure 14. Wind field size during storm duration for storm duration between 30 and 84 h; **(a)** for ERA-Interim and **(b)** for the EPS.

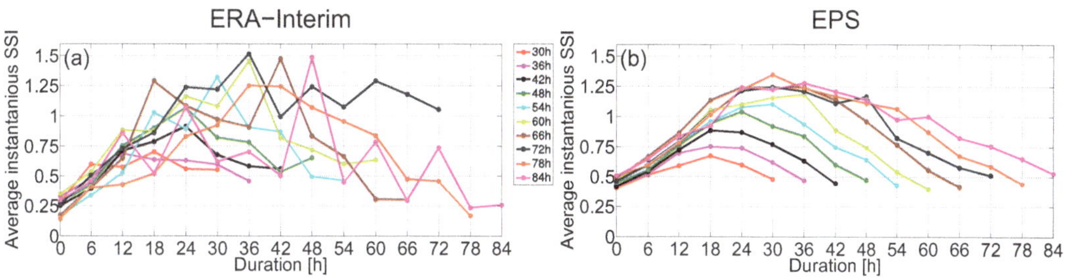

Figure 15. Storm severity during storm duration for storm duration between 30 and 84 h; **(a)** for ERA-Interim and **(b)** for the EPS.

sample size for European winter storm studies by a factor up to the number of ensemble members, initializations per day, and forecast time. As we used 50 perturbed members and storms starting inside a 6-day window, we get a sample size increase of 300 times. The statistics of the storms indicate a clear increase of maximum intensity and extension of Atlantic–European storms with their duration. This result from the EPS cannot be obtained easily from reanalysis as the number of very strong events is too low to provide stable statistics. Another example of analyses possible by using the huge sample of storm events deducted from the EPS is the estimation of return periods of specific storms and intensities. Such return periods will naturally be associated with smaller uncertainties than those in other studies (e.g., Della-Marta et al., 2009). However, for such a study, it has to be taken into account that storm representations are not statistically independent; see Osinski (2015). They are also limited to climate conditions (e.g., SSTs) during the 10 year period considered. Still, the consideration of EPS storms enables us to estimate the potential for an occurrence of storms more extreme than observed based on a physical modeling approach.

The range of severity in the EPS is much larger than in ERA-Interim. Model biases resulting from different model versions and/or resolutions were eliminated using the quantile–quantile mapping approach. Spatiotemporal properties of the storms are realistic compared to ERA-Interim, and also the range of wind velocity is realistic. For this reason, also the SSI values are realistic. The range of SSI values is larger, because the EPS contains a wide range of storm modifications, including those with higher wind speeds. Modifications to stronger winds are additionally amplified when calculating the SSI by utilizing the cubic threshold exceedance.

The climatology based on the EPS is intended to be close to the observed development of climate conditions, and it must be distinguished from alternative approaches such as climate simulations for present-day greenhouse gas and solar forcing, for example, which allow the models to produce windstorms largely independent from the observed development of weather and climate in the time period considered. If independence from observations is a requirement, coupled general circulation model (CGCM) runs may be the better choice. In the sense of an event set, we do not expect complete independency but just variations of storms, as is done, e.g., for stochastic event sets out of a fixed historical sample. Finally, the way that events are selected for construction of an event set will be dependent on the specific purpose of that event set, and so approaches are not discussed further in this paper.

To sum up, the EPS shows realistic storm properties with a wide range of modifications in the storm properties, where storms can be found with a higher possible impact than appeared as in reality; thus the ability to use this data set for statistical studies is given.

Acknowledgements. For the possibility to carry out this study we would like to thank the Munich Re for their financial support and the open-access publication fund of the Freie Universität Berlin for financing this publication. We are also grateful to the German Weather Servive (DWD) and the ECMWF for providing access to the EPS data and ERA-Interim reanalysis. The authors thank the two anonymous referees for their constructive comments.

References

Boé, J., Terray, L., Habets, F., and Martin, E.: Statistical and dynamical downscaling of the Seine basin climate for hydro-meteorological studies, Int. J. Climatol., 27, 1643–1655, doi:10.1002/joc.1602, 2007.

Bouttier, F. and Rabier, F.: The operational implementation of 4D-Var, ECMWF Newsletter Number 78, European Centre for Medium-Range Weather Forecasts, Shinfield Park, Reading, UK, http://www.ecmwf.int/sites/default/files/elibrary/1997/14646-newsletter-no78-winter-199798.pdf (last access: 20 January 2016), 1997.

Buizza, R. and Hollingsworth, A.: Storm prediction over Europe using ECMWF Ensemble Prediction System, Meteorol. Appl., 9, 289–305, doi:10.1017/S1350482702003031, 2002.

Dee, D. P., Uppala, S. M., Simmons, A. J., Berrisford, P., Poli, P., Kobayashi, S., Andrae, U., Balmaseda, M. A., Balsamo, G., Bauer, P., Bechtold, P., Beljaars, A. C. M., van de Berg, L., Bidlot, J., Bormann, N., Delsol, C., Dragani, R., Fuentes, M., Geer, A. J., Haimberger, L., Healy, S. B., Hersbach, H., Holm, E. V., Isaksen, L., Kallberg, P., Köhler, M., Matricardi, M., McNally, A. P., Monge-Sanz, B. M., Morcrette, J.-J., Park, B.-K., Peubey, C., de Rosnay, P., Tavolato, C., Thepaut, J.-N., and Vitart, F.: The ERA-Interim reanalysis: configuration and performance of the data assimilation system, Q. J. Roy. Meteorol. Soc., 137, 553–597, doi:10.1002/qj.828, 2011.

Della-Marta, P. M., Mathis, H., Frei, C., Liniger, M. A., Kleinn, J., and Appenzeller, C.: The return period of wind storms over Europe, Int. J. Climatol., 29, 437–459, doi:10.1002/joc.1794, 2009.

Dukes, M. D. G. and Palutikof, J. P.: Estimation of Extreme Wind Speeds with Very Long Return Periods, J. Appl. Meteorol., 34, 1950–1961, doi:10.1175/1520-0450(1995)034<1950:EOEWSW>2.0.CO;2, 1995.

Froude, L. S. R.: Regional Differences in the Prediction of Extratropical Cyclones by the ECMWF Ensemble Prediction System, Mon. Weather Rev., 137, 893–911, 2009.

Froude, L. S. R.: The Predictability of Storm Tracks, PhD thesis, Environmental Systems Science Center, University of Reading, Reading, http://ethos.bl.uk/OrderDetails.do?uin=uk.bl.ethos.440078 (last access: 2 April 2012), 2006.

Froude, L. S. R. and Gurney, R.: Storm Prediction Research and its Application to the Oil/Gas Industry, in: NATO Science for Peace and Security Series C: Environmental Security, edited by: Troccoli, A., Springer Netherlands, 241–252, doi:10.1007/978-90-481-3692-6_16, 2010.

Hodges, K. I.: A General Method for Tracking Analysis and Its Application to Meteorological Data, Mon. Weather Rev., 122, 2573–2586, doi:10.1175/1520-0493(1994)122<2573:AGMFTA>2.0.CO;2, 1994.

Hofherr, T. and Kunz, M.: Extreme wind climatology of winter storms in Germany, Clim. Res., 41, 105–123, 2010.

Klawa, M. and Ulbrich, U.: A model for the estimation of storm losses and the identification of severe winter storms in Germany, Nat. Hazards Earth Syst. Sci., 3, 725–732, doi:10.5194/nhess-3-725-2003, 2003.

Klinker, E., Rabier, F., Kelly, G., and Mahfouf, J.-F.: The ecmwf operational implementation of four-dimensional variational assimilation. III: Experimental results and diagnostics with operational configuration, Q. J. Roy. Meteorol. Soc., 126, 1191–1215, doi:10.1002/qj.49712656417, 2000.

Koziar, C. and Renner, V.: MUSE Modellgestützte Untersuchungen zu Sturmfluten mit sehr geringen Eintrittswahrscheinlichkeiten – Teilprojekt: Numerische Berechnung physikalisch konsistenter Wetterlagen mit Atmosphärenmodellen, Tech. rep., DWD, www.bsh.de/de/Meeresdaten/Projekte/MUSE/MUSE_Abschlussbericht.pdf (last access: 22 April 2012), 2005.

Kruschke, T.: Winter wind storms: Identification, verification of decadal predictions, and regionalization, PhD thesis, Freie Universität Berlin, Dept. of Earth Sciences, Institute of Meteorology, Berlin, Germany, http://www.diss.fu-berlin.de/diss/receive/FUDISS_thesis_000000099397, last access: 19 August 2015.

Lamquin, N., Gierens, K., Stubenrauch, C. J., and Chatterjee, R.: Evaluation of upper tropospheric humidity forecasts from ECMWF using AIRS and CALIPSO data, Atmos. Chem. Phys., 9, 1779–1793, doi:10.5194/acp-9-1779-2009, 2009.

Lapworth, A.: The evening wind, Weather, 63, 12–14, doi:10.1002/wea.167, 2008.

Lapworth, A.: Forecasting the wind at sunset, Weather, 67, 65–67, doi:10.1002/wea.880, 2012.

Leckebusch, G. C., Koffi, B., Ulbrich, U., Pinto, J. G., Spangehl, T., and Zacharias, S.: Analysis of frequency and intensity of European winter storm events from a multi-model perspective, at synoptic and regional scales, Clim. Res., 31, 59–74, 2006.

Leckebusch, G. C., Renggli, D., and Ulbrich, U.: Development and application of an objective storm severity measure for Northeast Atlantic region, Meteorol. Z., 17, 575–587, 2008.

Mahfouf, J.-F. and Rabier, F.: The ECMWF operational implementation of four-dimensional variational assimilation. II: Experimental results with improved physics, Q. J. Roy. Meteorol. Soc., 126, 1171–1190, doi:10.1002/qj.49712656416, 2000.

Maraun, D.: Bias Correction, Quantile Mapping, and Downscaling: Revisiting the Inflation Issue, J. Climate, 26, 2137–2143, doi:10.1175/JCLI-D-12-00821.1, 2013.

Molteni, F., Buizza, R., Palmer, T. N., and Petroliagis, T.: The ECMWF Ensemble Prediction System: Methodology and validation, Q. J. Roy. Meteorol. Soc., 122, 73–119, doi:10.1002/qj.49712252905, 1996.

Münchener Rückversicherungs-Gesellschaft, Geo Risks Research, N.: Significant natural catastrophes 1980–2011; 10 costliest winter storms in Europe ordered by overall losses, http://www.munichre.com/app_pages/www/@res/pdf/NatCatService/significant_natural_catastrophes/2011/NatCatSERVICE_significant_winter_storms_eco_june2011_en.pdf (last access: 5 February 2015), 2011.

Osinski, R.: Untersuchungen zu Atlantisch-Europäischen Winterstürmen anhand des operationellen Ensemblevorhersagesystems des EZMW, Ph.D. thesis, Freie Universität Berlin, Dept. of Earth Sciences, Institute of Meteorology, Berlin, Germany, http://www.diss.fu-berlin.de/diss/receive/FUDISS_thesis_000000098264, last access: 12 January 2015.

Palmer, T., Molteni, F., Mureau, R., Buizza, R., Chapelet, P., and Tribbia, J.: Ensemble prediction, Technical Memorandum 188, European Centre for Medium Range Weather Forecasts, http://www.ecmwf.int/sites/default/files/elibrary/1992/

11560-ensemble-prediction.pdf (last access: 12 April 2012), 1992.

Palmer, T. N., Buizza, R., Leutbecher, M., Hagedorn, R., Jung, T., Rodwell, M., Vitart, F., Berner, J., Hagel, E., Lawrence, A., Pappenberger, F., Park, Y.-Y., von Bremen, L., and Gilmour, I.: The Ensemble Prediction System – Recent and Ongoing Developments, Technical Memorandum 540, European Centre for Medium Range Weather Forecasts, paper presented to the 36th Session of the SAC, 8–10 Oct 2007, http://www.aosc.umd.edu/~ekalnay/syllabi/AOSC614/PalmerENStm540.pdf (last access: 5 July 2011), 2007.

Rabier, F., Järvinen, H., Klinker, E., Mahfouf, J.-F., and Simmons, A.: The ECMWF operational implementation of four-dimensional variational assimilation. I: Experimental results with simplified physics, Q. J. Roy. Meteorol. Soc., 126, 1143–1170, doi:10.1002/qj.49712656415, 2000.

Roberts, J. F., Champion, A. J., Dawkins, L. C., Hodges, K. I., Shaffrey, L. C., Stephenson, D. B., Stringer, M. A., Thornton, H. E., and Youngman, B. D.: The XWS open access catalogue of extreme European windstorms from 1979 to 2012, Nat. Hazards Earth Syst. Sci., 14, 2487–2501, doi:10.5194/nhess-14-2487-2014, 2014.

Thode, H.: Testing For Normality, Statistics, textbooks and monographs, Taylor & Francis, Marcel Dekker Inc., New York, 2002.

7

Debris flow susceptibility mapping using a qualitative heuristic method and Flow-R along the Yukon Alaska Highway Corridor, Canada

A. Blais-Stevens and P. Behnia

Geological Survey of Canada, 601 Booth Street, Ottawa, K1A0E8, Canada

Correspondence to: A. Blais-Stevens (andree.blais-stevens@canada.ca)

Abstract. This research activity aimed at reducing risk to infrastructure, such as a proposed pipeline route roughly parallel to the Yukon Alaska Highway Corridor (YAHC), by filling geoscience knowledge gaps in geohazards. Hence, the Geological Survey of Canada compiled an inventory of landslides including debris flow deposits, which were subsequently used to validate two different debris flow susceptibility models. A qualitative heuristic debris flow susceptibility model was produced for the northern region of the YAHC, from Kluane Lake to the Alaska border, by integrating data layers with assigned weights and class ratings. These were slope angle, slope aspect, surficial geology, plan curvature, and proximity to drainage system. Validation of the model was carried out by calculating a success rate curve which revealed a good correlation with the susceptibility model and the debris flow deposit inventory compiled from air photos, high-resolution satellite imagery, and field verification. In addition, the quantitative Flow-R method was tested in order to define the potential source and debris flow susceptibility for the southern region of Kluane Lake, an area where documented debris flow events have blocked the highway in the past (e.g. 1988). Trial and error calculations were required for this method because there was not detailed information on the debris flows for the YAHC to allow us to define threshold values for some parameters when calculating source areas, spreading, and runout distance. Nevertheless, correlation with known documented events helped define these parameters and produce a map that captures most of the known events and displays debris flow susceptibility in other, usually smaller, steep channels that had not been previously documented.

1 Introduction

Debris flow modelling has been carried out by several researchers – e.g. Costa (1984), Hungr et al. (1984), Berti and Simoni (2007), Beguería et al. (2009), Hussin et al. (2012), and Fischer et al. (2012), among many others. Over the past decades, there have been ongoing discussions about building a pipeline roughly parallel to the Yukon Alaska Highway Corridor, in northern Canada (YAHC; Fig. 1a). Portions of the proposed pipeline route intersect critical geological regions and areas prone to geological hazards, such as landslides including debris flows, earthquakes due to active faults, subsidence from thermo-karstic erosion, and permafrost degradation. One of the roles of the Geological Survey of Canada (GSC), as part of Natural Resources Canada, is to reduce risk of geohazards to critical linear infrastructure, such as a proposed pipeline project by providing baseline geoscience information to decision makers.

The YAHC is roughly 875 km long and about 40 km wide and covers roughly 22 000 km^2 as it does not include the portions that extend into British Columbia (Fig. 1a). As a first step in assessing the types of geohazards, the GSC in collaboration with the Yukon Geological Survey (Huscroft et al., 2004), compiled an inventory of geological hazards, particularly landslides (Blais-Stevens et al., 2010a). Several landslide types were observed including debris flows, debris slides, rockfalls, rock slides, active layer detachment slides, retrogressive thaw slumps, among others (Blais-Stevens et al., 2011). Once the baseline landslide information was compiled, the next step was to provide an overview of the landslide distribution and landslide "hot spots" by plotting distribution maps and producing regional qualitative landslide

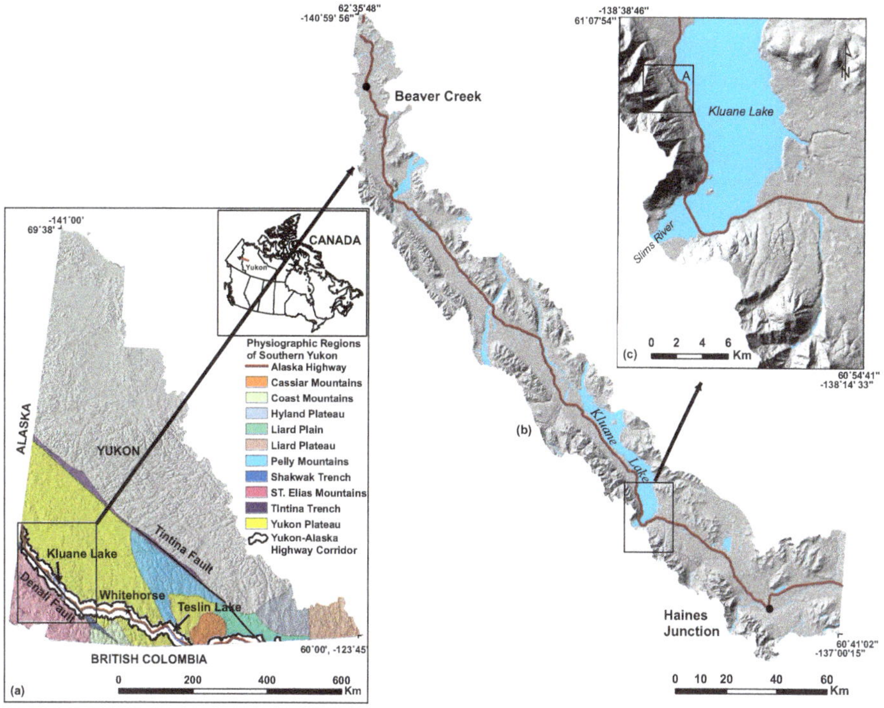

Figure 1. **(a)** Physiographic setting of southern Yukon (modified from Mathews, 1986 and Huscroft et al., 2004) and **(b)** northern portion of YAHC where regional debris flow susceptibility mapping was carried out. **(c)** South end of Kluane Lake area where the quantitative susceptibility assessment was performed. Box A shows the location of Fig. 4.

susceptibility maps (Blais-Stevens et al., 2010b, 2011, 2012, 2014; Couture et al., 2010). Different susceptibility models were created for different types of landslides, namely rockfall/rock slides, active layer detachment slides, retrogressive thaw slumps, and debris flows.

Choosing which landslide susceptibility method to apply in our study area depended on several factors. Some of the factors were scale, availability of geological and geomorphological data in digital format, software and hardware capabilities, and technical experience with GIS platform (Soeters and van Westen, 1996; van Westen et al., 2008). For the entire YAHC, we chose the simplest approach – a qualitative heuristic method – as the study area was vast and uniform geological information was available at a broad scale. We modified and adapted the models first used for landslides triggered in permafrost along the Mackenzie Valley (Riopel et al., 2006) and other types of landslides (including debris flows) for the Sea to Sky Highway in southern British Columbia (Blais-Stevens et al., 2012). Furthermore, previous studies revealed that debris flows along the Alaska Highway had occurred previously and blocked the highway, particularly in the south end of Kluane Lake where the highway was built close to steep slopes (Evans and Clague, 1989; Lipovsky, 2005; Koch et al., 2014; Fig. 1c). Such an event occurred in 1988 following a torrential downpour; the highway was blocked by debris for 7 days (Evans and Clague, 1989; Lipovsky, 2005). Therefore, with a higher resolution Digital Elevation Model (DEM; 10 × 10 m), we used a quantitative method using Flow-R to assess the potential source areas in steep channels and corresponding debris flow susceptibility (Horton et al., 2013) in the area where documented debris flows had occurred. Hence, the objectives of the paper are to (1) assess the regional debris flow susceptibility using a qualitative heuristic method for the northern portion of the corridor (Fig. 1b), a method used to highlight the potential zones of initiation; (2) for the south end of Kluane Lake (Fig. 1c), define the local potential source areas for debris flows; and (3) model the debris flow susceptibility by calculating propagation and spreading, using the quantitative Flow-R method.

2 Physiographic setting and geology

The Yukon Alaska Highway Corridor traverses 10 physiographic regions. From northwest to east in a broad sense, the relief varies from mountainous to rounded hills, to lowlands. The maximum elevation within the study area is 2799 m, located roughly southeast of the southernmost point of Kluane Lake and the minimum elevation, and occurs about 15 km west of Haines Junction (Fig. 1b). Two major faults cut across the YAHC, namely the Denali and Tintina faults (Fig. 1a). For a detailed description of the physiographic regions, the reader is referred to Mathews (1986) and Huscroft et al. (2004).

The YAHC tectonic setting is largely the result of 190 million years of accretion of exotic terranes and intervening sediments onto the ancient North American continental plate (Gordey and Makepeace, 2003; Huscroft et al., 2004). Along the YAHC, from west to east, the Kluane ranges are composed of young sedimentary rocks (< 150 Ma old). Further east, felsic intrusions dominate within the Yukon Plateaus. Towards Whitehorse, Upper Cretaceous volcanic rocks and valley-filling Quaternary basalts are exposed. Within the Liard Lowland eastward, there are outcrops of Paleozoic sedimentary rocks (Gordey and Makepeace, 2003; Huscroft et al., 2004).

The abundance of surficial sediments along the YAHC is the result of several Quaternary glaciations. Hillslopes are covered by till blankets and veneer. Coarser ice-marginal meltwater deposits are commonly flanked along these hillslopes. In the steep mountainous terrain, weathered bedrock and reworked till form colluvium along valley sides. These are a major source for mass movement within the YAHC (Huscroft et al., 2004).

2.1 Climate and vegetation

The study area is characterized by a sub-arctic continental climate with long, cold winters, and short mild summers with low relative humidity and low to moderate precipitation averaging 340 mm per year. The mean daily temperatures vary from −27 °C in January to 27 °C in July (Huscroft et al., 2004). The YAHC lies within the Boreal Forest Ecozone, which is dominated by white and black spruce. Trembling aspen grows in meadows and open forests. The treeline varies between 1400 m in the east and 1200 m in the northwest corridor (Huscroft et al., 2004).

3 Previous work

Landslide characterization and inventory compilation in the Yukon was carried out by Huscroft et al. (2004). Blais-Stevens et al. (2010a) published a regional landslide inventory based solely on aerial photo interpretation (1960–1990 aerial photos from National Air-Photo Library) that also included the contribution by Huscroft et al. (2004). Interpretation of high-resolution satellite imagery (WorldView-2) collected in 2010 helped fine tune and update the inventory that was used to validate regional landslide susceptibility models as well as field verification (Blais-Stevens et al., 2010b, 2011, 2012).

Horton et al. (2008, 2013) used a distributed empirical model that can be transposed to a small area within YAHC where an inventory of past events exist. In addition to the qualitative heuristic method, the south portion of the Kluane Lake, an area adjacent to the Kluane Ranges (Fig. 1a, b, and c) was a suitable area to test this method because (1) a high-resolution DEM was available for this area, (2) the area has witnessed extreme debris flow events (July 1988), and (3) it contained several mapped debris flow deposits to validate the model.

4 Regional qualitative heuristic susceptibility mapping

The qualitative approach used for landslide susceptibility mapping is a simple heuristic approach modified from Soeters and van Westen (1996), Riopel et al. (2006), and Blais-Stevens et al. (2012). Previous publications by Blais-Stevens et al. (2010b, 2011, 2012) demonstrated landslide susceptibility mapping for the entire corridor using a 30 × 30 m DEM (22 000 km^2; Fig. 1a). In this study, we focused on the northern portion, the more mountainous region of the corridor (7600 km^2 Fig. 1b) and used a similar approach, but with a higher-resolution DEM (5 × 5 m generated by Natural Resources Canada's Canadian Centre for Geospatial Information, 2014, using Canvec products and Topo to Raster tool in ArcGIS). The simple equation is defined by a susceptibility index for debris flows in which the instability factors served as the variables. Each variable was assigned a percent weight relative to each other based on expert knowledge according to its importance in slope instability. The factors of instability were derived from the surficial geology map or generated from DEM as data layers and consequently reclassified. Classes within each parameter were also rated based on their potential contribution to slope failure. The resulting parametric equation is the sum of all variables and represents a susceptibility index (SI) ranging between 0 and 1 for each pixel (5 × 5 m). The heuristic modelling was carried out independently of the debris flow deposit inventory where it was later used to validate the susceptibility map. The criteria considered relevant to initiation of debris flows in the study area include proximity to drainage system, slope angle, sediment type, plan curvature, and slope aspect. The drainage was considered important as debris flows usually are triggered in steep streams. This data layer was generated by creating five buffer zones around the drainage system. Each zone was rated based on its proximity to the drainage system (Table 1). Slope angle was considered an equally important factor in initiation of a debris flow based on expert knowledge and the literature. According to Ortigao and Kanji (2004), the minimum slope angles reported in the literature are above 20 to 25°. There is an increase in the number of debris flows as the slope angle increases. At slope angles higher than 34 to 37°, there is a decrease because of the existence of rock scarps (Ortigao and Kanji, 2004). Similarly, Dai and Lee (2001) noted that at slope gradients $\geq 40°$, the slope forming material of the terrain is composed of weathered rock which is stronger and less prone to failure than colluvium. Thus, the slope map was generated from a 5 × 5 m DEM and was reclassified into five classes and assigned the highest rating to slope angles ranging from 25–45° (Table 1). Sediment availability is another

Table 1. Debris flow parameters and class ratings.

Surficial geology (G) unit	Rating	Slope angle (°) (S1)	Rating	Slope aspect (°) (S2)	Rating
Colluvium, tuff	1.0	0–15	0.1	0–45	0.1
Eolian, lacustrine (vC, C), glacio-lacustrine (vC, C)	0.7	15–25	0.5	45–135	0.5
Till, drift, glacio-lacustrine (M)	0.5	25–45	1.0	135–225	1.0
Glacio-lacustrine(F)	0.4	45–55	0.5	225–315	0.5
Glaciofluvial	0.3	56–90	0.1	315–360	0.1
Organics, alluvium, rock, anthropogenic	0.1				
Proximity to drainage (m) (D)	**Rating**	**Plan curvature (-1 / $100\,m^{-1}$) (C)**	**Rating**		
0–50	1.0	≤ -2	1.0		
50–100	0.75	-2 to -0.5	0.8		
100–150	0.5	-0.5 to -0.1	0.5		
150–200	0.25	> -0.1	0.1		
> 200	0.1				

vC = very coarse, C = coarse, M = medium, F = fine.

important factor in initiation of debris flows, but to a slightly lesser extent relative to drainage and slope angle due to the broad scale of the surficial geology map relative to the DEM. The surficial geology map (Yukon Geological Survey, 2015) was thus reclassified into six classes based on the favourability of the sediments in developing debris and were rated accordingly (Table 1). Plan curvature was also a factor included in the equation as it reflects areas where debris can accumulate. The more concave the slope curvature, the higher the rating (Table 1). We also considered the direction of slope (i.e. the slope aspect) as a potential contributing factor, but to a lesser degree than the other data layers. The south-facing slope is more exposed to solar radiation. This, in turn, could contribute to either increased snow melt or permafrost thaw and subsequently to the drainage system. The slope aspect map, also generated from the 5 × 5 m DEM, was reclassified into five classes, with the highest ratings assigned to slopes facing SE to SW (Table 1). All geological information was compiled and processed on an ArcGIS platform (v. 10.2.2). Once the parameter maps were reclassified and rated, they were integrated using

$$\mathrm{SI} = 0.3\,D + 0.3\,S1 + 0.2\,G + 0.15\,C + 0.05\,S2, \qquad (1)$$

where D is proximity to drainage system, $S1$ is the slope angle, G is the type of surficial geology deposit, C is the plan curvature, and $S2$ is the slope aspect. The resulting debris flow susceptibility map (Fig. 2) shows a susceptibility index (SI) ranging from 0 to 1 in which the high values are the areas of high susceptibility (shown in red). Although the highly susceptible areas are dominant in the most western region of the corridor, there are areas of high susceptibility found close to the highway, near Kluane Lake.

We used the mapped debris flow inventory consisting of 306 deposits to evaluate the debris flow susceptibility map. Given that the inventory only includes the deposits and not the initiation zones, we made the assumption that the source area likely occurs within at least 500 m distance uphill from the apex of the deposit. There are potentially other source areas uphill from these minimum distances (as shown in the Flow-R method below), but our assumption is based on what we consider a realistic minimum limit for most of the steep streams. To delineate the potential source area (initiation zone), the stream network was extracted from the flow accumulation map and catchment basins were created from the apex of each debris flow deposit. Areas 100 m around the streams were selected and were intersected with the catchment basins. The areas covered by both catchment basins and streams within 500 m upstream from the debris flow deposits were selected as potential source areas (Fig. 5) where the maximum susceptibility value occurring within each polygon was extracted. A success rate of prediction curve was generated by plotting the cumulative percentage area from high to low susceptibility against the cumulative percentage of debris flow sources predicted (Chung and Fabbri, 2003). The qualitative heuristic susceptibility map shows a high prediction rate with respect to debris flow source areas (Fig. 3). About 80 and 90 % of the debris flow sources are predicted within the top 5.5 % and 10.5 % of high susceptibility areas, respectively.

5 Quantitative susceptibility mapping – Flow-R application to the southern end of Kluane Lake, YAHC

We used the Flow-R model developed by Horton et al. (2008, 2013) to assess debris flow susceptibility and focus on a smaller area of the YAHC at the south end of Kluane Lake. Flow-R, developed under Matlab® by Horton et al. (2013) stands for *Flow path assessment of gravitational hazards at*

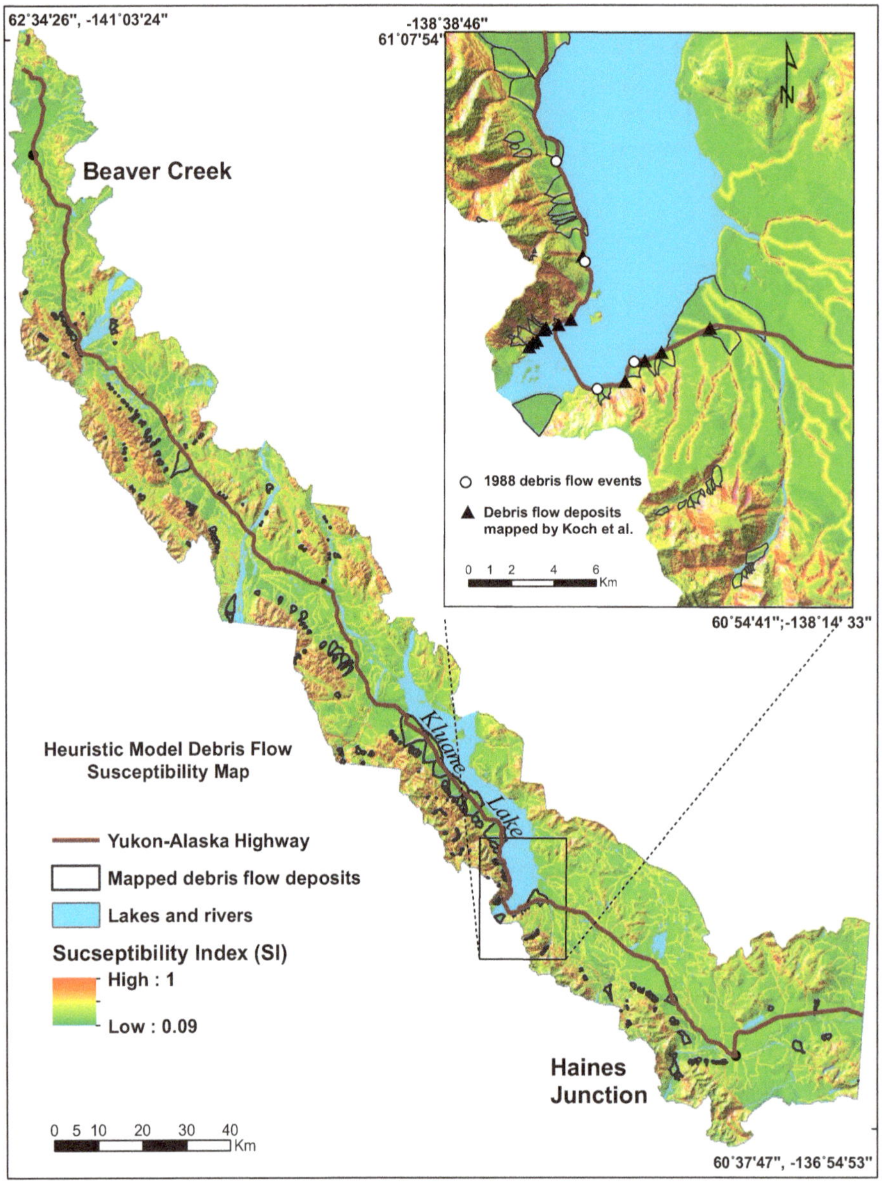

Figure 2. Debris flow susceptibility map created using the qualitative heuristic method for the northern part of YAHC (Fig. 1b). Inset map shows the southern end of the Kluane Lake area (Fig. 1c).

a Regional scale and was downloaded from www.flow-r.org. The main data set required for susceptibility assessment in Flow-R is a grid-based DEM. The quality of DEM is of great importance for the accuracy of the results. We used a 5 × 5 m DEM and resampled it to 10 m in order to reduce the roughness and avoid the effect of channelization that can occur with very high resolution data so that a wider propagation could be captured within the fans. The following concepts of Flow-R and the implemented models are summarized from Horton et al. (2013). Susceptibility assessment using Flow-R involves two stages:

Step 1 – Delineation of debris flow source areas based on the geological, morphological, and hydrological criteria critical in debris flow occurrence. These controlling parameters are used in grid format and are classified according to their favourability in debris flow initiation. The data are classified as favourable if initiation is possible, excluded if the initiation is unlikely, and ignored if there is not enough evidence in favourability of the class. The classified input parameters are integrated based on the following rule: a grid cell is considered a source area if it was classified as favourable in at least one of the parameter maps, but was never classified as excluded (Horton et al., 2013).

Step 2 – Propagation of the source areas. The potential source areas are propagated using two types of algorithms: (1) spreading algorithms which determine the path and the

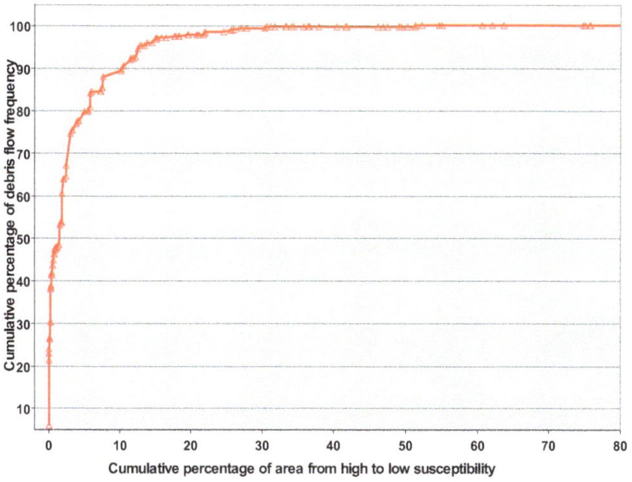

Figure 3. Prediction rate of qualitative heuristic debris flow susceptibility map represents a good correlation between debris flow susceptibility and inventory.

way debris flows spread and (2) algorithms which are based on the friction laws and control the runout distance of debris (Horton et al., 2013).

The spreading algorithms address flow direction algorithms and persistence functions and describe downslope movement of material. We used the flow direction algorithm of Holmgren (1994) and applied the height factor added by Horton et al. (2013) in order to smooth the roughness of the DEM and obtain a more consistent spreading. Another influencing factor which is implemented in Flow-R is inertial parameter. The flow direction is weighted based on the change in direction according to the persistence function. The weights resulted from the persistence function and those from flow direction are combined to provide the overall susceptibility (see Gamma, 2000 and Horton et al., 2013 for equations).

The runout distance algorithms control the distance which can be reached by debris flows. Two types of algorithms are available in Flow-R to assess the friction loss: a two-parameter friction model based on Perla et al. (1980) and a simplified friction model based on maximum possible runout distance. The Perla model, used in this study, is based on a non-linear friction law and calculates the velocity of the flow at the end of segment i. The model requires the value of friction coefficient μ and mass-to-drag ratio ω to be provided. In order to keep the energy within reasonable values, a maximum threshold is introduced to avoid achieving unrealistic velocities.

5.1 Source area delineation

In order to delineate the source areas in the study area, we considered slope angle and curvature as the morphological factors, upslope contributing area as the hydrological factor, and surficial geology as the lithological factor.

Slope angles above 20 to 25° are reported by Ortigao and Kanji (2004) for initiation of most debris flows in the literature. According to Takahashi (1981) and Rickenmann and Zimmermann (1993), most debris flows occur at slope angles higher than 15° (Horton et al., 2013). We considered this value as the lower initiation slope angle in this study area for this method.

The plan curvature was used to identify the concave curvatures as possible source areas based on the assumption that debris flows tend to occur in gullies where the curvature is concave. There is no established threshold value for curvature in the literature. Horton et al. (2013) used a curvature of $-2/100\,\mathrm{m}^{-1}$ based on a 10 m DEM, and analysis of orthophotographs in Switzerland. Fischer et al. (2012) used values from $-1.5/100\,\mathrm{m}^{-1}$ to $-0.5/100\,\mathrm{m}^{-1}$ for debris flow susceptibility mapping in Norway and Park et al. (2013) used curvature values from $-2/100\,\mathrm{m}^{-1}$ to $-1/100\,\mathrm{m}^{-1}$ for debris flow hazard zonation analysis in Korea. In order to establish a proper curvature threshold for the study area, we tested two threshold values of $-1.5/100\,\mathrm{m}^{-1}$ and $-1.0/100\,\mathrm{m}^{-1}$ to select the favourable source areas. The selected source areas and their resulting propagation areas were examined using the mapped debris flow deposits of the study area. Most of the mapped debris flow deposits were captured by the propagation area in both cases, but the threshold value of $-1.0/100\,\mathrm{m}^{-1}$ provided a better agreement and could capture more debris flow deposits (Fig. 4).

Water, commonly as infiltration of rainfall or convergence of ground water into the soil, contributes to instability of steep slopes by increasing the soil weight and decreasing the soil strength (Savage and Baun, 2005). To estimate the water input, the flow accumulation which represents the upslope contributing area was used. Based on the analysis of past debris flow events, a relationship could be established between the upslope contributing area and slope. Horton et al. (2008, 2013) defined two curves for identifying the lower limit of debris flow initiation in the central Alps. The first curve is based on the work of Heinimann (1998; in Horton, 2013); and is considered as rare events. The second curve is based on the work of Rickenmann and Zimmermann (1993) on the extraordinary 1987 event in Switzerland. The threshold values for rare and extreme events are given in Eqs. (2) and (3), respectively (Horton et al., 2008, 2013):

$$\begin{cases} \tan\beta_{\mathrm{thres}} = 0.32 S_{\mathrm{uca}}^{-0.2} & \text{if} \quad S_{\mathrm{uca}} < 2.5\,\mathrm{km}^2 \\ \tan\beta_{\mathrm{thres}} = 0.26 & \text{if} \quad S_{\mathrm{uca}} \geq 2.5\,\mathrm{km}^2 \end{cases} \tag{2}$$

$$\begin{cases} \tan\beta_{\mathrm{thres}} = 0.31 S_{\mathrm{uca}}^{-0.15} & \text{if} \quad S_{\mathrm{uca}} < 2.5\,\mathrm{km}^2 \\ \tan\beta_{\mathrm{thres}} = 0.26 & \text{if} \quad S_{\mathrm{uca}} \geq 2.5\,\mathrm{km}^2, \end{cases} \tag{3}$$

where $\tan\beta_{\mathrm{thres}}$ is the slope threshold, and S_{uca} is the surface of the upslope contributing area. The main difference between the rare and extreme events occurs for the small catchments with areas between 1 and 10 ha. Thus, using the

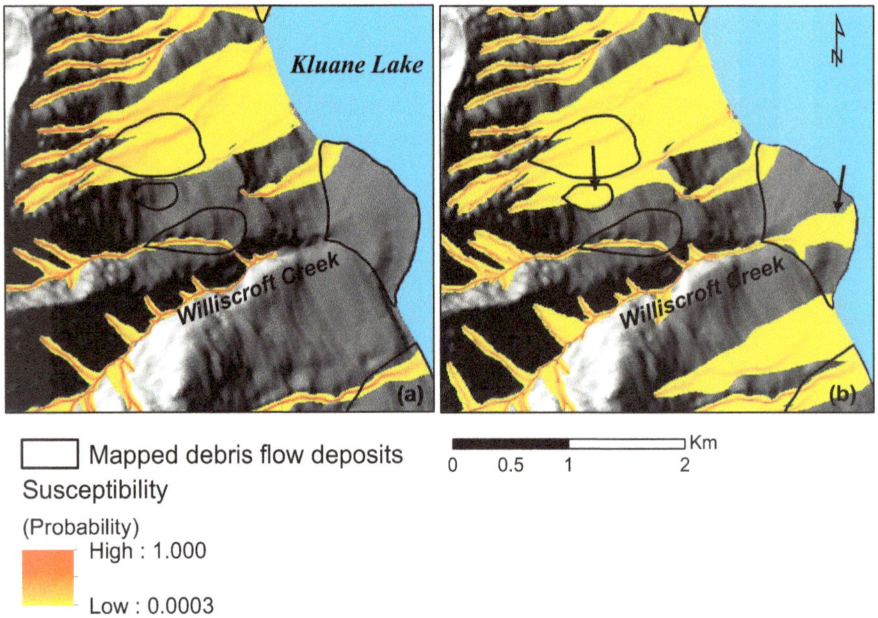

Figure 4. a) The propagation of source areas in and around Williscroft Creek area (box A in Fig. 1): **(a)** based on the threshold value of $-1.5\,/\,100\,\mathrm{m}^{-1}$, and **(b)** based on the threshold value of $-1\,/\,100\,\mathrm{m}^{-1}$. Note the two debris flow deposits that are captured only in **(b)** are shown with black arrows.

rare events limit would cause the omission of some potential sources, which are small but still can contribute to a debris flow event (Horton et al., 2008). In order not to miss any potential source area, we decided to use the extreme events limit. Furthermore, comparing the mapped debris flow deposits in the study area with the source areas created using both threshold values (Eqs. 2 and 3) showed that the extreme events limit is more consistent with the past debris flow events along the Yukon Alaska Highway area.

The surficial geology deposits were classified into favourable, excluded, and ignored, based on the type of surficial unit and suitability as being a source for sediments and debris material. The colluvium, till, and the drift units were classified as favourable, the organic material as excluded, and the rest of units as ignored. Once the parameter maps were classified, they were integrated to create the final source areas (Fig. 5). We also compared the estimated catchment source areas for the heuristic method to the calculated source areas from Flow-R. Hence, many of the calculated source areas overlap with the estimated source areas (Fig. 5).

5.2 Susceptibility assessment

In order to calculate the potential spread from source areas, we used the Holmgren (1994) method and its modified version as the flow direction algorithms (Horton et al., 2013). Using Δh equal to 1 and 2 m provided somewhat more lateral spreading; except for a few debris flow fans, the effect was not very significant. This was expected for a 10 m DEM because the change of Δh affects mainly the finer resolution DEMs (Horton et al., 2013). Holmgren (1994) proposed the value of exponent x in the flow direction algorithm in the range of 4 to 6 and Claessens et al. (2005) suggested the value of 4 for debris flows. We applied the value of 4 in most of our simulations and compared the spreading results with the mapped debris flow deposits in the study area. Exponent values equal to 5 and 6 were also tested. These propagation results were not as satisfactory as with x equal to 4. Therefore, we selected the value of 4 for the rest of the simulations. To set the inertial algorithm which weights the flow proportion based on the persistence function for each direction change (w), we selected the proportional approach which assigns the weights of 1, 0.8, 0.4, 0, and 0, for w_0, w_{45}, w_{90}, w_{135}, and w_{180}, respectively.

Perla's (1980) two-parameter friction model was used for the assessment of runout distance. Because there was no information on the mass-to-drag ratio ω and friction coefficient μ for the debris flow deposits in the study area, it was decided to calibrate these values by comparing the propagation areas resulting from different sets of ω and μ and the mapped debris flow deposits, while keeping the spreading parameters fixed. The tested values range from 10 to 1000 for ω and 0.01 to 0.3 for μ. The resulting simulations using varying sets of ω and μ yield almost similar propagation for most of the debris flow deposits except for the relatively larger ones (Figs. 6 and 7). With ω equal to 10 and μ equal to 0.3, only one very small debris flow deposit was completely reached in runout distance. About 69 % of the mapped debris flow deposits were completely or partly reached when ω equal to 20 and μ equal to 0.02 were used. Most of the debris flow

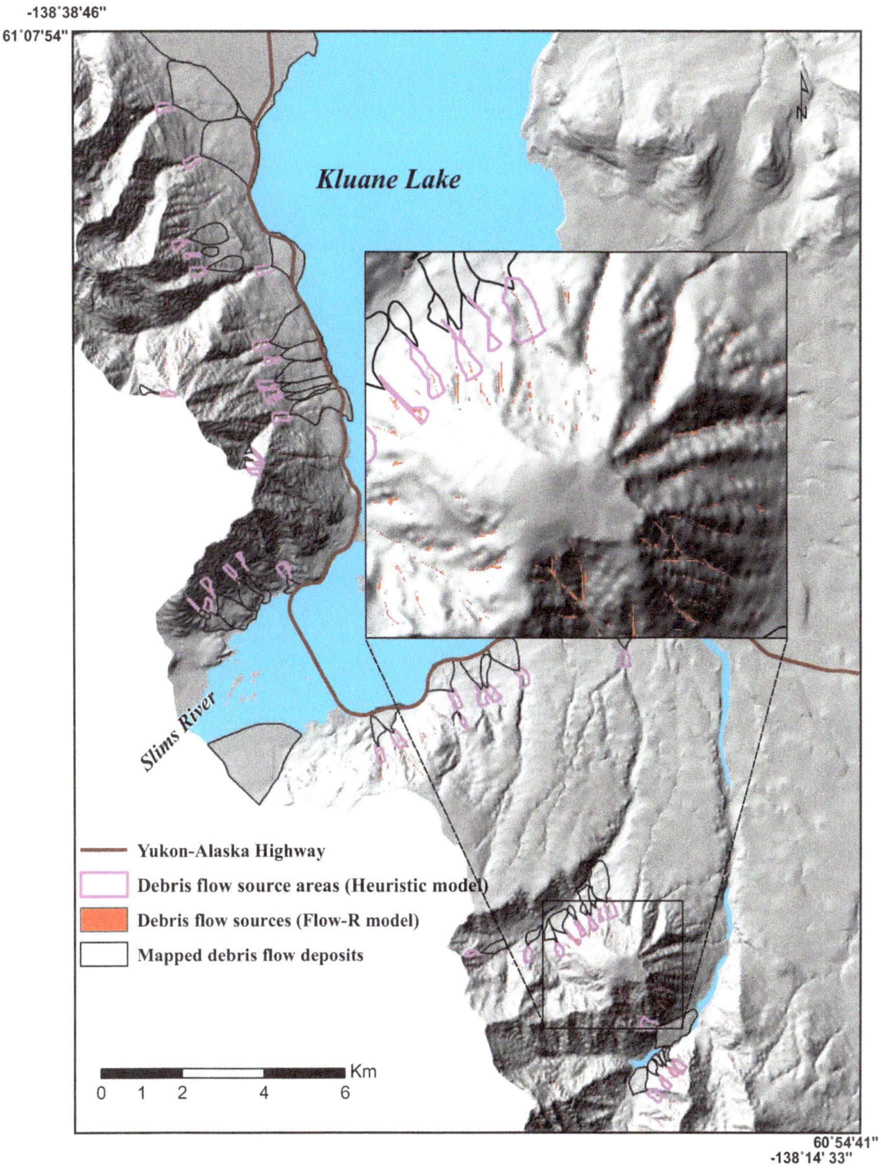

Figure 5. Potential debris flow source areas for the south end of Kluane Lake derived by integrating controlling factors. Inset map displays an enlargement of the calculated source areas (red areas; Flow-R) and the assumed potential source areas (purple lines; Heuristic method). Refer to Fig. 1c for location.

deposits with the exception of two very large and three relatively large fans were reached with $\omega = 200$ and $\mu = 0.06$. The same result was also achieved with other sets of ω and μ (e.g. $\omega = 150$ and $\mu = 0.05$, $\omega = 100$ and $\mu = 0.04$). The best fit for the three relatively larger fans (A, B, and C in Fig. 6) were achieved with $\omega = 150$ and $\mu = 0.02$ because it provided a better fit for these fans, but did not change the propagation pattern for the rest of debris flow deposits (Fig. 6). Of the two very large fans, one is part of a larger fan located in the northwestern edge of the study area (A in Fig. 6) with only a very small portion reached in one area. The source area for this fan is likely located outside of the limit of the study area. The other very large fan, located southeast of Kluane Lake (E in Figs. 6 and 7), is just partly reached when ω greater than 600 and μ equal to 0.02 (or ω greater than 100 at $\mu = 0.01$) are used (Fig. 7). Further increasing the mass-to-drag ratio mainly affects the runout distance for fans C, D, and E and a few channels in the east and northeast of the study area and does not change the propagation pattern for the rest of the area (Figs. 6 and 7). The calculated propagation for fan E is likely too short in this case because the first source area is located about 6 km upstream from the apex of this fan. Figure 7 displays correlation with the mapped debris flow deposits (Blais-Stevens, 2010a), but also the ones

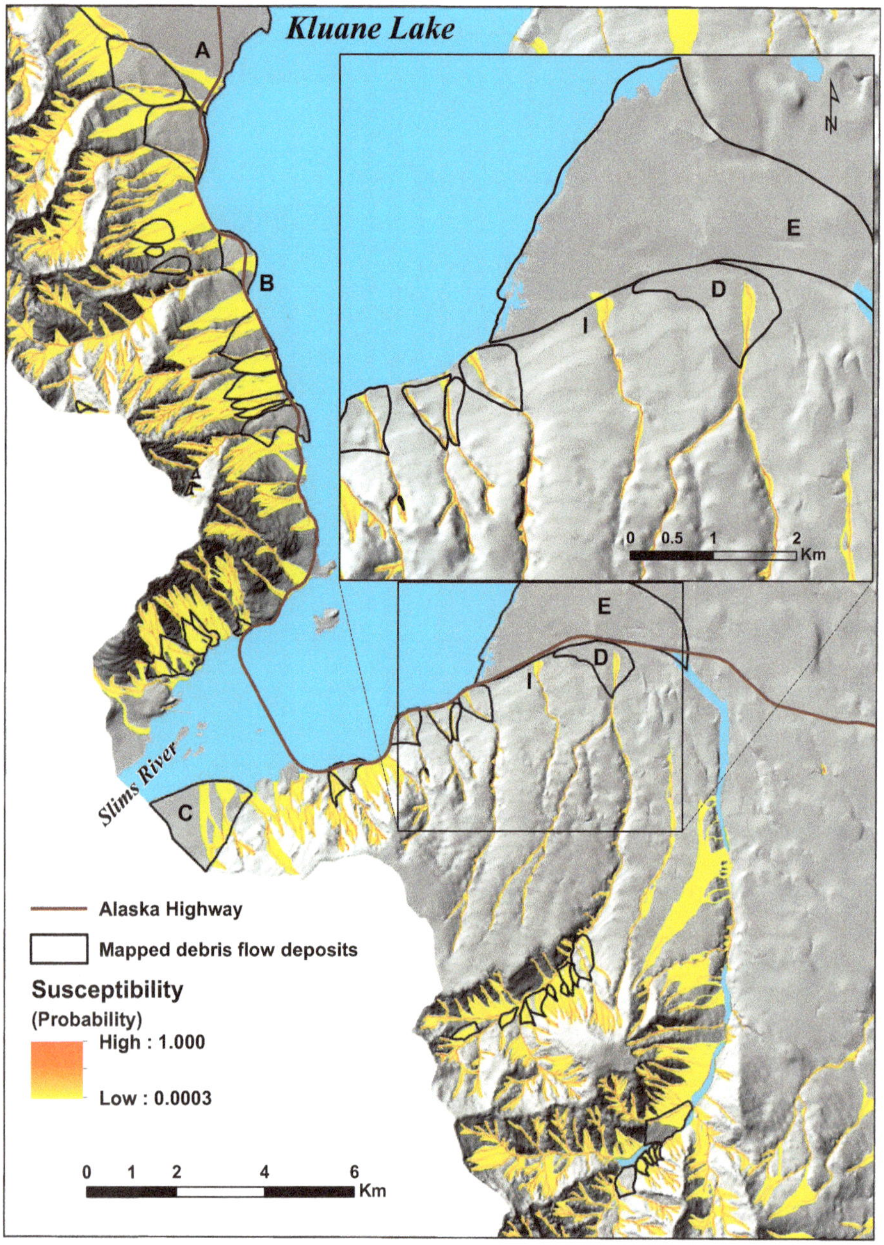

Figure 6. Debris flow susceptibility map derived from propagation of source areas using $\omega = 150$ and $\mu = 0.02$. Note no propagation reaches debris flow deposit E.

studied in detail by Koch et al. (2014) and those that were documented after the 1988 event (Evans and Clague, 1989; Lipovsky, 2005). Testing various sets of ω and μ showed that the runout distance was more sensitive to the friction coefficient rather than mass-to-drag ratio changes, particularly at higher values of μ.

5.3 Evaluation

The debris flow susceptibility map created using the propagation of source areas represents the present-day conditions. The mapped debris flow deposits were used to evaluate the susceptibility map although these deposits likely have been accumulating since deglaciation (Koch et al., 2014) spreading into large fans. Therefore it is expected that the spreading would not cover all of the fan but it should cover at least a part of it in present-day conditions and delineate areas within the fan that are more susceptible to deposition (e.g. red zones). Out of 49 debris flow deposits, most are reached in terms of runout distance, but some of them are not completely covered in terms of lateral spreading. Nevertheless, debris can

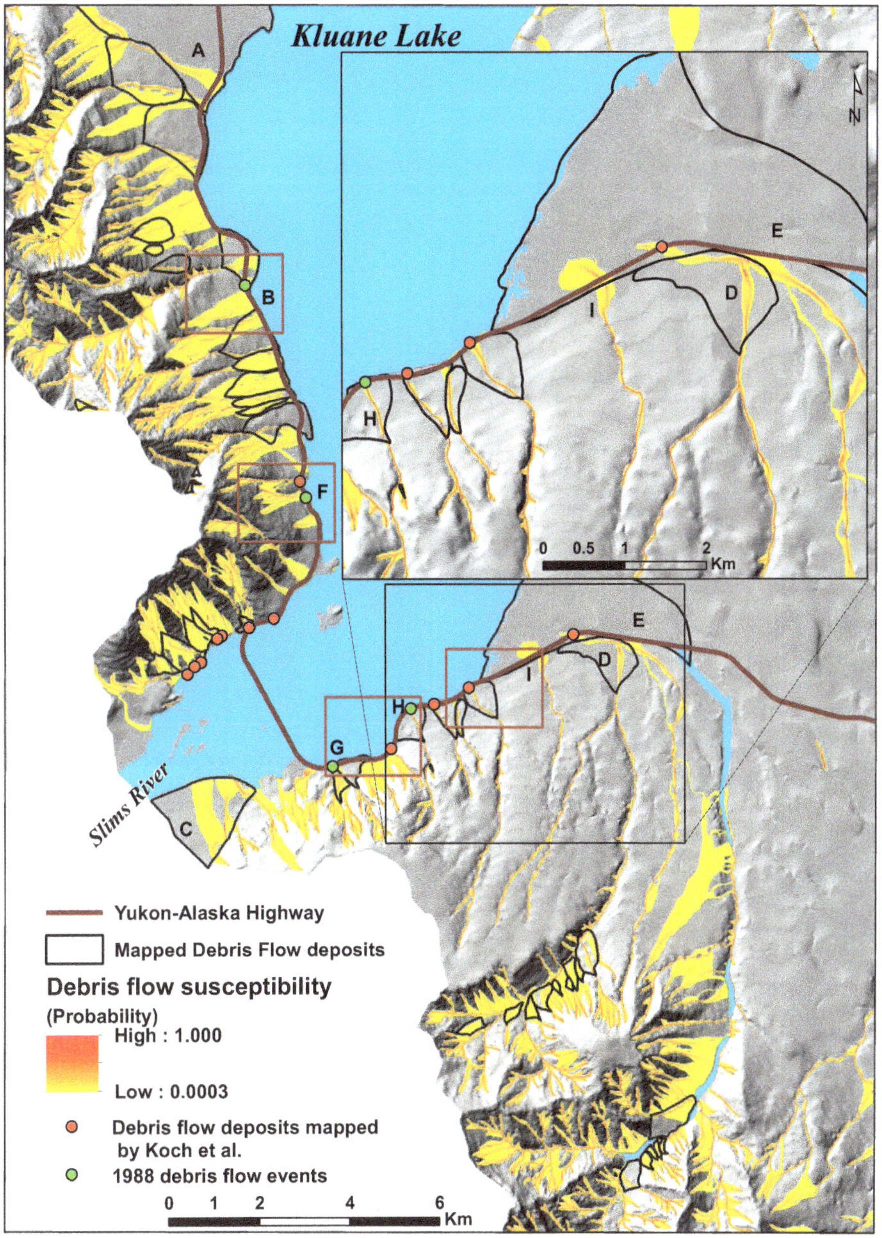

Figure 7. Debris flow susceptibility map derived from propagation of source areas using $\omega = 600$ and $\mu = 0.02$. Minor propagation is seen reaching the southwest edge of debris flow deposit E. Also note the propagation pattern in C, D, and I. Debris flow susceptibility displays a good correlation with the mapped debris flow deposits (black lines from Blais-Stevens, 2010a), the documented 1988 event from Evans and Clague (1989; green dots), and from the detailed study from Koch et al. (2014; red dots).

still flow, propagate, and block the highway even as relatively small fans (Evans and Clague, 1989).

Debris flows triggered by heavy rains in 1988 and 1967 (locations B, F, G, and H in Figs. 7 and 8) are also used to evaluate the debris flow susceptibility map. Figure 8 displays WorldView-2 images overlaid by the susceptibility map at the location of these events. Williscroft Creek watershed (Figs. 7 and 8a) is a typical active fan along the foot of the Kluane Range (Evans and Clague, 1989) in which the highway was severed at B during heavy rains in July 1988. About 3.5 km north of Slims River (Fig. 7), two debris flows occurred in Holocene colluvium, one of which blocked the highway at F (Figs. 7, 8b and 9). Another debris flow occurred about 1.5 km east of Slims River (G in Figs. 7 and 8c) and covered more than 500 m of the highway (Evans and Clague, 1989). A debris flow triggered by heavy rains during the summer of 1967 (location H in Figs. 7 and 8c) also blocked the highway. All four locations where the high-

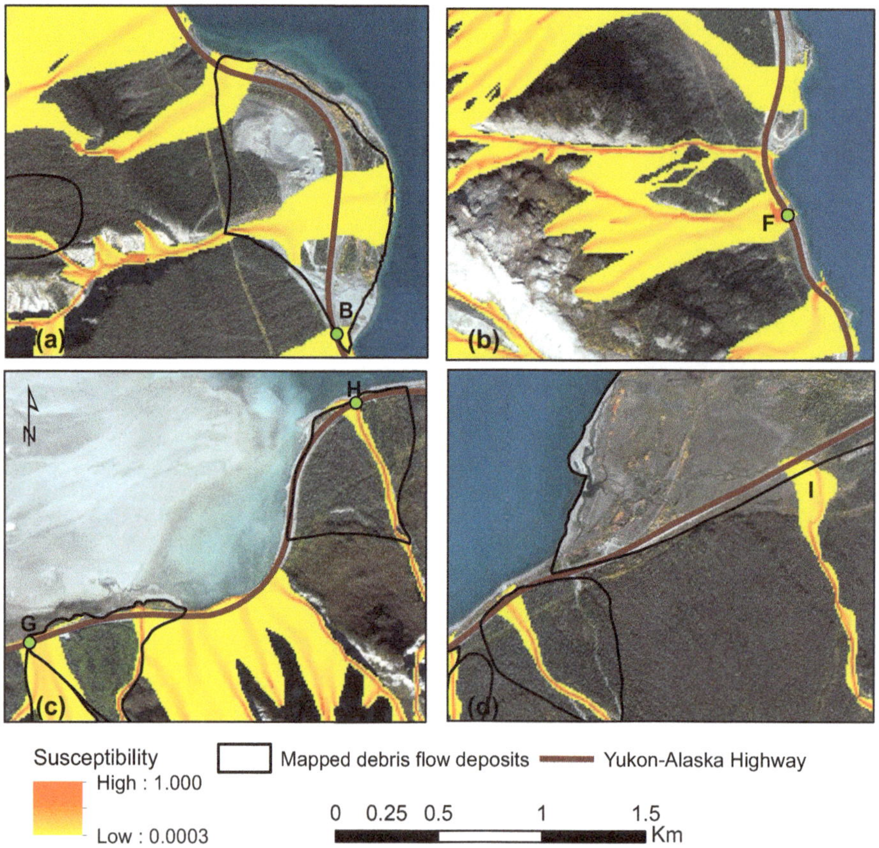

Figure 8. **(a)**, **(b)**, **(c)**. The susceptibility map overlaid on WorldView2 satellite imagery (2010) shows the location of debris flow deposits which blocked the highway in 1988 (B, F, and G) and in 1967 (H). **(d)** A potential debris flow (I) that could reach the highway.

way was severed are covered by the propagation pattern in the debris flow susceptibility map created using Flow-R. The Williscroft Creek watershed is mainly covered in the middle part, but also is reached at B where the highway was blocked.

Koch et al. (2014) studied 30 large debris flow deposits along the YAHC between Beaver Creek and the south end of Kluane Lake using dendrochronology and tephrochronology to evaluate the potential hazard of debris flows on the highway. A number of 15 debris flow deposits lie in the south end of Kluane Lake (red dots in Fig. 7). Three of them coincide with the 1988 debris flow events documented by Evans and Clague (1989). All except one (Fig. 7) of these fans are captured by the propagation of source areas established using Flow-R.

Nevertheless, the susceptibility map also shows some areas for debris flow deposition that were not outlined as debris lobes in the inventory or other sources. These in some cases extend past the highway (e.g. debris flow I in Figs. 7 and 8d).

6 Conclusion

Two separate methods were used to assess the debris flow susceptibility in the YAHC focusing on the northern portion of the corridor for the qualitative heuristic method and the south end of Kluane Lake for the quantitative method using Flow-R.

The qualitative method was based on expert knowledge using geological and geomorphological data layers with assigned weights and class ratings. The resulting susceptibility map displayed the steep channels as high-susceptibility zones for potentially initiating debris flows. For the most part, downstream from the high-susceptibility zones, where the lower-susceptibility zones are identified, is where the mapped debris flow deposits are found. Validation was carried out using 306 debris flow deposits and their source area estimated within 500 m upstream of the apex of the deposits. A success rate curve demonstrated that about 80 and 90 % of the debris flow sources were predicted within 5.5 and 10.5 % of high-susceptibility areas, respectively. In the southern portion of Kluane Lake, where 1988 debris flow deposits were documented and studied in detail, the susceptibility zones upstream from the location of these deposits are high.

A quantitative approach for debris flow susceptibility focused on a smaller area within the YAHC, at the southern end of Kluane Lake, where historical and pre-historic events had been documented. The Flow-R method allowed us to

Figure 9. Oblique photo of debris flow channels at Horseshoe Bay, southwestern end of Kluane Lake (photo by C. Huscroft). During the 1988 storm event, debris flow deposits blocked the Alaska Highway (e.g. red arrow) for up to 7 days (Evans and Clague, 1989). (Site F in Figs. 7 and 8b.)

quantitatively determine the potential source areas and also calculate potential spreading and runout. Trial-and-error calculations were needed during the modelling because there was not enough detailed information on the debris flow deposits to calculate threshold values for parameters controlling source area delineation, spreading, and runout distance assessment. Nevertheless, correlation with known documented events helped us define these parameters and produce a map that captures most of the known events and displays debris flow susceptibility in other steep channels. Applying Flow-R method for a large area with varied topography, i.e. using the same threshold values for the whole area, may result in overgeneralization in some areas. However, except the area southeast of Kluane Lake, which is characterized by a flatter topography, the rest of the study area has a similar topography. Besides, using a wide range of values for parameters controlling spreading and runout distance did not change the propagation pattern dramatically, except for a few large and relatively large debris flow deposits.

The debris flow susceptibility map correlates well with known events and has also displayed other potential channels that could be susceptible to debris flows. The correlation with known events may or may not exactly match with the size of the mapped debris flow deposits. This is expected, given that some of these debris flow fans have been building since deglaciation time, while the Flow-R method reflects present-day conditions.

The two debris flow susceptibility regional models were able to delineate areas that have high susceptibility either from the initiation zone (heuristic) or calculated sources and propagation (Flow-R). These are meant to serve a first step in assessing the debris flow hazard. However, for further quantitative hazard assessment, site-specific investigations on magnitude and frequency of events would be needed.

Acknowledgements. This project was funded by Natural Resources Canada's Program of Energy and Research and Development. The authors would like to thank Sharon Smith (Geological Survey of Canada project leader), Panya Lipovsky (Yukon Geological Survey), and John Clague (Simon Fraser University) for their continual input. P. Horton (Terr@num Ltd) is acknowledged for his advice, comments and suggestions. Crystal Huscroft (Thompson River University) is thanked for the photo of Horseshoe Bay. Amaris Page, Marian Kremer, Ariane Castagner, and Joe Koch are acknowledged for their technical assistance. Jeff Harris (Geological Survey of Canada) is also thanked for his comments and suggestions. Pierre Gravel and his colleagues (Canadian Centre for Geospatial Information) are thanked for providing a new high-resolution DEM.

References

Beguería, S., Van Asch, Th. W. J., Malet, J.-P., and Gröndahl, S.: A GIS-based numerical model for simulating the kinematics of mud and debris flows over complex terrain, Nat. Hazards Earth Syst. Sci., 9, 1897–1909, doi:10.5194/nhess-9-1897-2009, 2009.

Berti, M. and Simoni, A.: Prediction of debris flow inundation areas using empirical mobility relationships, Geomorphology, 90, 144–161, 2007.

Blais-Stevens, A., Couture, R., and Page, A.: Landslide inventory along the Alaska Highway Corridor, Yukon, Geol. Surv. of Can., Ottawa, Ont., Open File 6654, DVD, 2010a.

Blais-Stevens, A., Couture, R., Page, A., Koch, J., Clague, J. J., and Lipovsky, P.: Landslide susceptibility, hazard and risk assessments along pipeline corridors in Canada, in: Proceedings of the 63rd Canadian geotechnical conference and 6th Canadian permafrost conference, Calgary (AB), 878–885, 2010b.

Blais-Stevens, A. Lipovsky, P., Kremer, M., Couture, R., and Page, A.: Landslide inventory and susceptibility mapping for a proposed pipeline route, Yukon Alaska Highway Corridor, in: Proceedings of the Second World Landslide Forum, Rome, Italy, 362, (ESS Cont. # 20110101), 2011.

Blais-Stevens, A. Lipovsky, P., Kremer, M., Couture, R., and Smith, S.: Landslide inventory and susceptibility mapping for the Yukon Alaska Highway Corridor, in: Proceedings of the 11th International and 2nd North American Symposium on landslides and engineered slopes, Banff, Alberta, 1, 777–782, 2012.

Blais-Stevens, A., Kremer, M., Bonnaventure, P. P., Smith, S. L., Lipovsky, P. and Lewkowicz, A. G.: Active layer detachment slides and retrogressive thaw slumps susceptibility mapping for present-day and projected climate conditions along the Yukon Alaska Highway Corridor – A qualitative heuristic approach, in: Proceedings of the Engineering Geology for Society and territory, IAEG conference, Torino, Springer, 1, 449–453, (GSC Contribution: 20130369), 2014.

Chung, C.-J. F. and Fabbri, A. E.: Validation of spatial prediction models for landslide hazard mapping, Nat. Hazards, 30, 451–472, doi:10.1023/B:NHAZ.0000007172.62651.2b, 2003.

Claessens, L., Heuvelink, G. B. M., Schoorl, J. M., and Veldkamp, A.: DEM resolution effects on shallow landslide hazard and soil redistribution modelling, Earth Surf. Proc. Land., 30, 461–477, doi:10.1002/esp.1155, 2005.

Costa, J. E.: Physical geomorphology of debris flows, in: Developments and Applications of Geomorphology, edited by:Costa, J. E. and Fleisher, P. J., Berlin, Springer-Verlag, 268–317, 1984.

Couture, R., Blais-Stevens, A., Page, A., Koch, J., Clague, J. J., and Lipovsky, P. S.: Landslide susceptibility, hazard and risk assessment along pipelines in Canada, in: Proceedings of the 11th International Association for Engineering Geology and the Environment Congress, edited by: Williams, A. L., Pinches, G. M., Chin, C. Y., McMorran, T. J., and Massey, C. I., Auckland, New Zealand, 5–10 September, 1023–1031, 2010.

Dai, F. C. and Lee, C. F.: Terrain-based mapping of landslide susceptibility using a geographical information system: a case study, Can. Geotech. J., 38, 911–923, doi:10.1139/cgi-38-5-911, 2001.

Evans, S. G. and Clague, J. J.: Rain-induced landslides in the Canadian Cordillera, July 1988, Geos. Canada, 16, 193–200, 1989.

Fischer, L., Rubensdotter, L., Sletten, K., Stalsberg, K., Melchiorre, C., Horton, P., and Jaboyedoff, M.: Debris flow modeling for susceptibility mapping at regional to national scale in Norway, in: Proceedings of the 11th International and 2nd North American Symposium on Landslides, Banff, Alberta, Canada, 3–8 June 2012, 723–729, 2012.

Gamma, P.: dfwalk – Ein Murgang-Simulationsprogramm zur Gefahrenzonierung, Geographisches Institut der Universität Bern, 2000.

Gordey S. P. and Makepeace, A. J.: Yukon digital geology, version 2.0 Geol. Surv. of Can., Ottawa, Ont., Open File 1749 and Yukon Geol. Surv., Whitehorse, Yukon, Open File 2003-9(D), 2 CD-ROMS, 2003.

Heinimann, H. R.: Methoden zur Analyse und Bewertung von Naturgefahren, Bundesamt für Umwelt, Wald und Landschaft, (BUWAL), 85, 247 pp., 1998.

Holmgren, P.: Multiple flow direction algorithms for runoff modelling in grid based elevation models: An empirical evaluation, Hydrol. Process., 8, 327–334, doi:10.1002/hyp.3360080405, 1994.

Horton, P., Jaboyedoff, M., and Bardou, E.: Debris flow susceptibility mapping at a regional scale, in: Proceedings of the 4th Canadian Conference on Geohazards, Québec, Canada, 2008, 399–406, 2008.

Horton, P., Jaboyedoff, M., Rudaz, B., and Zimmermann, M.: Flow-R, a model for susceptibility mapping of debris flows and other gravitational hazards at a regional scale, Nat. Hazards Earth Syst. Sci., 13, 869–885, doi:10.5194/nhess-13-869-2013, 2013.

Hungr, O., Morgan, G. C., and Kellerhals, R.: Quantitative analysis of debris torrent hazards for design of remedial measures, Can. Geotech. J., 21, 663–677, 1984.

Huscroft, C. A., Lipovsky, P. S., and Bond, J. D.: A regional characterization of landslides in the Alaska Highway corridor, Yukon, Yukon Geol. Surv., Whitehorse, Yukon, Open File Rep. 2004-18, 65 pp., 2004.

Hussin, H. Y., Quan Luna, B., van Westen, C. J., Christen, M., Malet, J. P., and van Asch, Th. W. J.: Parameterization of a numerical 2-D debris flow model with entrainment: a case study of the Faucon catchment, Southern French Alps, Nat. Hazards Earth Syst. Sci., 12, 3075–3090, 2012.

Koch, J., Clague, J. J., and Blais-Stevens, A.: Debris flow chronology and potential hazard along the Alaska Highway in southwest Yukon Territory, Environ. Eng. Geosci., 25–43, 2014.

Lipovsky, P. S.: Rapid landscape change and human response in the Arctic and sub-Arctic, an interdisciplinary conference, Field guide, Yukon College, Whitehorse, Yukon, 43–101, 2005.

Mathews, W. H.: Physiographic map of the Canadian Cordillera, Geol. Surv. Can., Ottawa, Ont., 1 : 5 000 000 scale, 1986.

Ortigao, J. A. R. and Kanji, M. A.: Landslide classification and risk management, in: Handbook of Slope Stabilization, edited by: Oritgao, J. A. R. and Sayao, A. S. F. J., Springer Verlag, Heidelberg, 27–66, 2004.

Park, D., Lee, S., Nikhil, N. V., Kang, S., and Park, J.: Debris flow hazard zonation by probabilistic analysis (MT. Woomyeon, Seoul, Korea), IJIRSET, 2, 2381–2390, 2013.

Perla, R., Cheng, T. T., and McClung, D. M.: A two-parameter model of snow-avalanche motion, J. Glaciol., 26, 197–207, 1980.

Rickenmann, D. and Zimmermann, M.: The 1987 debris flows in Switzerland: documentation and analysis, Geomorphology, 8, 175–189, 1993.

Riopel, S., Couture, R., and Tewari, K.: Mapping susceptibility to landslides in a permafrost environment: case study in the Mackenzie Valley, Northwest Territories, GeoTech Event, Ottawa, Canada, 18–21 June, 13 pp., 2006.

Savage, W. and Baun, R.: Instability of steep slopes, in: Debris-flow hazards and related phenomena, edited by: Matthias, J. and Oldrich, H., Springer, Berlin, Heidelberg, 53–70, 2005.

Soeters, R. and van Westen, C. J.: Slope instability recognition, analysis, and zonation, in: Landslides, investigation and mitigation, edited by: Turner, A. K. and Schuster, R. L., Transportation Research Board, National Research Council, Special Report 247, National Academy Press, Washington, DC, 129–177, 1996.

Takahashi, T.: Estimation of potential debris flows and their hazardous zones: Soft countermeasures for a disaster, Nat. Disaster Sci., 3, 57–89, 1981.

van Westen, C. J., Castellanos, E., and Kuriakose, S. L.: Spatial data for landslide susceptibility, hazard, and vulnerability assessment: An overview, Eng. Geol., 102, 112–131, 2008.

Yukon Geological Survey, http://www.geology.gov.yk.ca/digital_surficial_data.html, last access: 21 April 2015.

8

A review and critical analysis of the efforts towards urban flood risk management in the Lagos region of Nigeria

U. C. Nkwunonwo[1,2], **M. Whitworth**[2], **and B. Baily**[3]

[1]Department of Geoinformatics and Surveying, University of Nigeria, Enugu Campus, Nigeria
[2]School of Earth and Environmental Sciences, University of Portsmouth, Portsmouth, UK
[3]Department of Geography, University of Portsmouth, Portsmouth, UK

Correspondence to: U. C. Nkwunonwo (ugonna.nkwunonwo@port.ac.uk)

Abstract. Urban flooding has been and will continue to be a significant problem for many cities across the developed and developing world. Crucial to the amelioration of the effects of these floods is the need to formulate a sound flood management policy, which is driven by knowledge of the frequency and magnitude of impacts of these floods. Within the area of flood research, attempts are being made to gain a better understanding of the causes, impacts, and pattern of urban flooding. According to the United Nations office for disaster reduction (UNISDR), flood risk is conceptualized on the basis of three integral components which are frequently adopted during flood damage estimation. These components are: probability of flood hazard, the level of exposure, and vulnerabilities of elements at risk. Reducing the severity of each of these components is the objective of flood risk management under the UNISDR guideline and idea of "living with floods". On the basis of this framework, the present research reviews flood risk within the Lagos area of Nigeria over the period 1968–2012. During this period, floods have caused harm to millions of people physically, emotionally, and economically. Arguably over this period the efforts of stakeholders to address the challenges appear to have been limited by, amongst other things, a lack of reliable data, a lack of awareness amongst the population affected, and a lack of knowledge of flood risk mitigation. It is the aim of this research to assess the current understanding of flood risk and management in Lagos and to offer recommendations towards future guidance.

1 Introduction

Flood events and impacts in recent times have arguably been unprecedented and affected the lives of hundreds of millions of people across the world. These impacts have been shared by both developed and developing countries (DCs) with rapid urban expansion taking place in many flood-prone areas. Concerns for flooding and the associated human impacts are clearly of global significance, especially when allied with the fears of climatic change and associated changes in rainfall events and sea level rise (Kundzewicz et al. 2014). The rapidly growing urban environments in many areas correspond with a lack of urban planning strategies, the deterioration and lack of capacity of urban drainage infrastructure and an increased rate of development on floodplains (Gill 2004; CII, 2001). Additionally, the increasing densities of populations (particularly in the urban areas of most DCs such as Nigeria), alongside the poor level of awareness and the limited efforts of many stakeholders towards flood risk reduction are critical issues undermining possible efforts towards addressing the hazard (Action Aid, 2006; McMichael et al., 2006; Raaijmakers et al., 2008). The present research attempts to clarify these issues using a synthesis and analysis of available historical flood data in the Lagos area of Nigeria, spanning the period 1968 to 2012. This research also reviews the current status and awareness of pluvial flood risk within Lagos and makes recommendations to help improve the management of these events.

2 Description of the Lagos metropolis of Nigeria

The Lagos metropolis is a densely populated low-lying coastal area in the southwestern part of Nigeria, West Africa. The city is located within geographical coordinates of 3.1 to 3.4° E longitude and 6.5 to 6.8° N latitude and covers a land area of approximately 1100 km^2 (or 425 sq. miles). It is bordered in the south by the Atlantic Ocean (see Fig. 1). With a dense network of roads and buildings, and several inland waterways including the Lagos Lagoon which empties into the Atlantic, the conurbation serves as a major hub for transportation, tourism, and economic activities in Nigeria. With a population of over 20 million people (LSG, 2012), the Lagos metropolis is the biggest city in Nigeria, (although the smallest land area), the second largest city in Africa, and the seventh largest city in the world. The population growth rate in the Lagos metropolis is estimated at 3.2 % (World Bank, 2013). Against this background, the United Nations predicts that Nigeria will be one of the eight countries expected to account collectively for half of the total population increase in the world from 2005 to 2050, and will by 2100, have a population of between 505 million and 1.03 billion (United Nations, 2004). High population density is a major impasse in the Lagos region, subjecting the area to lack of space for the myriad of human activities, which often manifests itself in muddled human settlements, overcrowding, slum envelopments, pollution, illegal structures, and other social and environmental disorder. These factors, either singularly or in combination, have the potential to increase exposure to hazard, especially the risks associated with pluvial urban flooding events.

3 Flooding in Lagos – frequency, causes, and impacts

Flooding and flood risk management are issues of grave significance in Lagos (Aderogba, 2012a, b). It is clear from previous studies (for example, Ajibade et al. 2013, 2014; Adelekan, 2013), that flooding in the area can be devastating, affecting hundreds of thousands of people and causing considerable economic damage. A typical example is the July 2011 flooding event, which affected approximately 5 thousand people and resulted in about 25 deaths. The direct economic losses resulting from the event totalled about 50 billion Nigerian naira (i.e. USD 250 million). Public utilities including road networks, bridges, and schools were destroyed. In addition, houses collapsed, private homes were submerged, and several cars were swept away by flood water (IFRC, 2011; Oladunjoye, 2011).

3.1 Frequency of occurrence

From a historical perspective, Lagos has always been susceptible to various types of flooding which are well documented from the 1960s onwards (Odunuga, 2008; Oyebande, 1974; Etuonovbe, 2011). However, in recent years pluvial flooding events (rainfall-related), have arguably been more widespread (Olajuyigbe et al., 2012). With the exception of 1973, the drought year, flooding in Lagos has occurred annually (usually between July and October rainy season) with an increasing intensity and an increased severity of impacts from 1960 onwards (Oyebande, 1974). According to FME (2012), Lagos is one of the few locations in Nigeria with more frequent flooding events (see Fig. 2). A number of floods have occurred in the Lagos area, although keeping track of events in the Nigerian context is challenging due partly to lack of relevant data collection capacities. As a result, data relating to hydrodynamics and historical flooding events are often lacking (Ajibade et al., 2013). According to previous studies, more severe flooding has been recorded in select areas of Lagos including Lagos Island, Apapa, Ikeja, Mushin, Surulere, and parts of Ikorodu (Oyebande, 1974; Odunuga, 2008). There is no clear explanation for this peculiar circumstance, but it seems to follow a pattern of spatial and temporal distribution of rainfall within and around Lagos.

Table 1 shows a summary of major flooding events and associated consequences in the Lagos metropolis of Nigeria from 1968 to 2012. These data which appear to represent generalized flooding situations were obtained from a wide range of sources including EM-DAT and the Nigerian FME (Federal Ministry of Environment). It is argued that the conclusions that can be drawn about flooding in Lagos from these data sets relate to events of higher magnitudes and return periods (Guha-Sapir et al., 2013). Only journalistic and non-quantitative evidence is available for lesser impacts and more frequent flooding events (see for example, IFRC, 2012). The problem with these forms of evidence is that they often do not have ethical and empirical groundings. For most of the events considered, data relating to flood duration and impacts in terms of number of people displaced, mortality, and economic losses were not available. On the basis of this inconsistency, the effectiveness of flood management policies is being queried (Adelekan, 2015). In many cases, different types of flood damage were aggregated. Indeed, this situation adversely affects accurate flood damage estimation since a critical understanding of Lagos flood risk in the context of flood damage typology is difficult.

To qualitatively identify risk levels in this area, an approach used in the 2005 World Bank Hotspot project was adopted (Dilley et al., 2005). In this approach, records of flood events and affected areas were coupled with the population density of the local enumeration areas in Lagos and mapped in a GIS. The result of this analysis as shown in Fig. 3, indicates that Ajeromi-Ifelodun and Mushin areas are at a higher risk of flooding than other areas. This approach, regardless of its simplicity, offers a potentially valuable insight into flooding patterns in Lagos. However, much uncertainty lies within the results and this suggests the need for more detailed research that will investigate the flood risk lev-

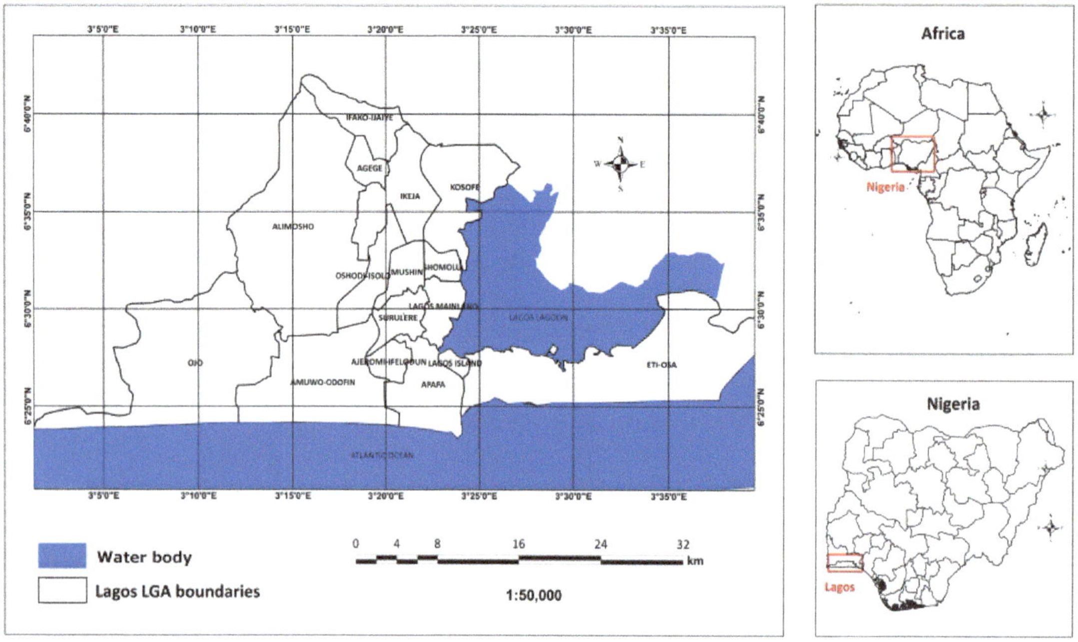

Figure 1. The Lagos metropolis of Nigeria. Inset shows the location of Lagos State in Nigeria.

els at Lagos using a more detailed quantitative or qualitative data set.

3.2 Causes of flooding

Over the past 2 decades, the causes of flooding in Lagos have received significant attention in the literature (Ayoade and Akintola, 1980; Action Aid, 2006; Adeloye and Rustum, 2011; Adelekan, 2013; Aderogba, 2012a; Oshodi, 2013; Ajibade et al., 2013, 2014; Soneye, 2014). Debates arising from the literature indicate that Lagos floods are mainly the consequences of climate-change-induced short-duration-high-intensity or long-duration-low-intensity rainfall (Houston et al., 2011). This is unsurprising considering that climate change has arguably influenced regional precipitation patterns in recent history. Odjugo (2006) concluded that there are now more high-intensity short-duration rainfall events and more low-intensity long-duration rainfall events than there were 3 decades ago. Other factors have also been investigated with reference to the causes of these floods in Lagos. These include the topography of the area, land use (LU) and land cover (LC) modifications, and influence of canals, lagoons, and beaches (Aderogba, 2012a; Aderogba et al., 2012; Odunuga, 2008). Other factors considered are urbanization and population growth, poor urban planning, and poor environmental management and the indiscriminate disposal of solid waste (Lamond et al., 2012; Adeloye and Rustum, 2011). It is also suggested that tidal and co-tidal influences and frequent incursion from the Atlantic into the lowlands during heavy storms also play important roles (Ojinnaka, 2013).

These factors (schematized in Fig. 4), seem to influence the occurrence of the hazard and the exposure of elements at risk. However, in relation to the vulnerabilities of social systems to flooding in the area, the development of slum settlements and poor perception of flooding among local communities, urban residents and the general public are critical factors (Agbola and Agunbiade, 2007; Nkwunonwo, 2013; Ayoade and Akintola, 1980; Odunuga et al., 2012; Oloke et al., 2013; BNRCC, 2008).

3.3 Impacts of flooding

To date, flooding in Lagos is characterized by severe consequences, which raise concerns about a lack of early warning and evacuation systems. Impacts from flooding (as illustrated in Fig. 5) are compounded by population density and the upward trends of urban growth in the area (Aderogba, 2012b). Also, flood water depth, inundation extent and duration as well as water flow velocity play contributory roles. The general impacts (such as displacement from homes, mortality, physical injuries, disruption of economic activities, destruction of urban infrastructure, and submergence of buildings) that relate to social systems directly have been extensively considered in the literature (Ugwu and Ugwu, 2013; Adigun et al., 2013; Ajibola et al., 2012; Aderogba, 2012b; Olajuyigbe et al., 2012). However, there are reports that Lagos flooding causes severe additional impacts including the loss of social values, spread of vector-borne diseases, as well as air and water pollution (Adelekan, 2010; Olajuyigbe et al., 2012; Bashir et al., 2012).

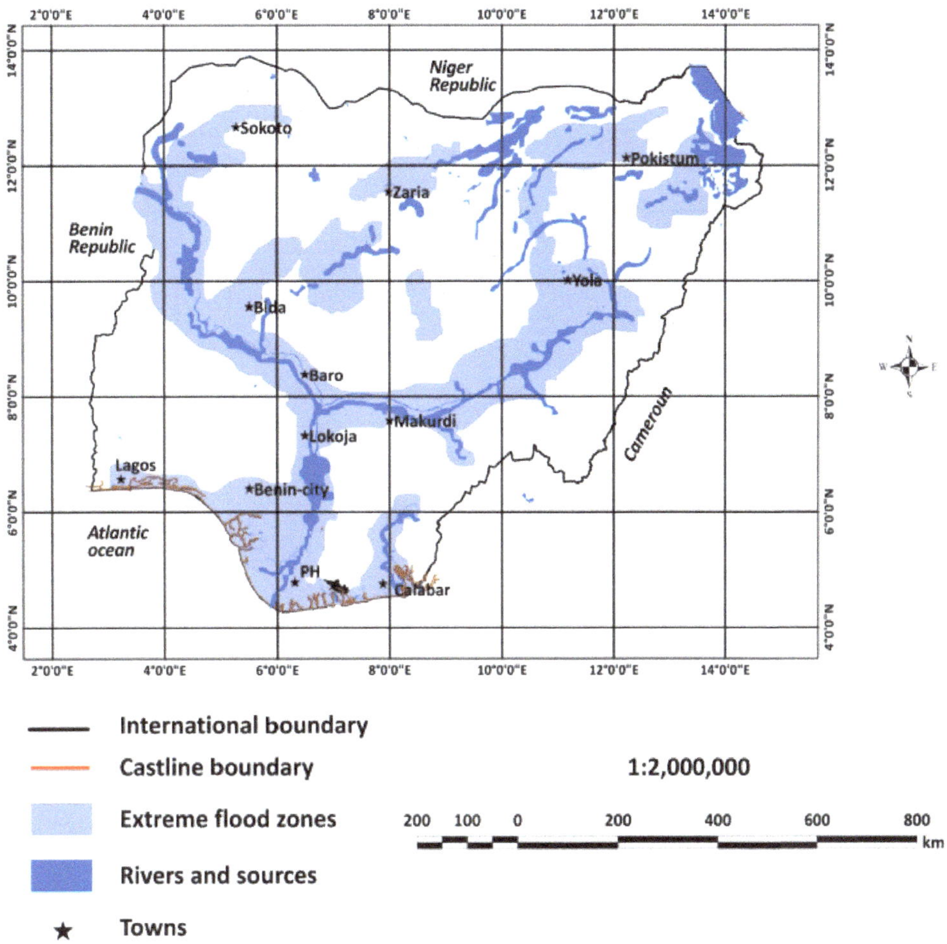

Figure 2. Spatial distribution of areas affected by extreme floods in Nigeria between 2000 and 2012. Adapted from the Federal Ministry of Environment (2012).

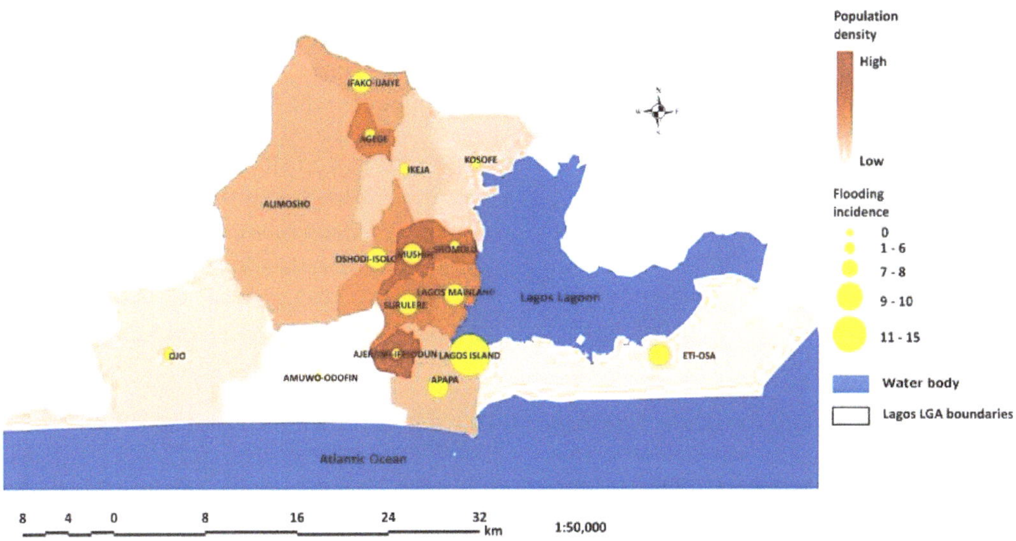

Figure 3. Flood risk levels in the Lagos area qualitatively determined by coupling population density with list of flooding events and locations.

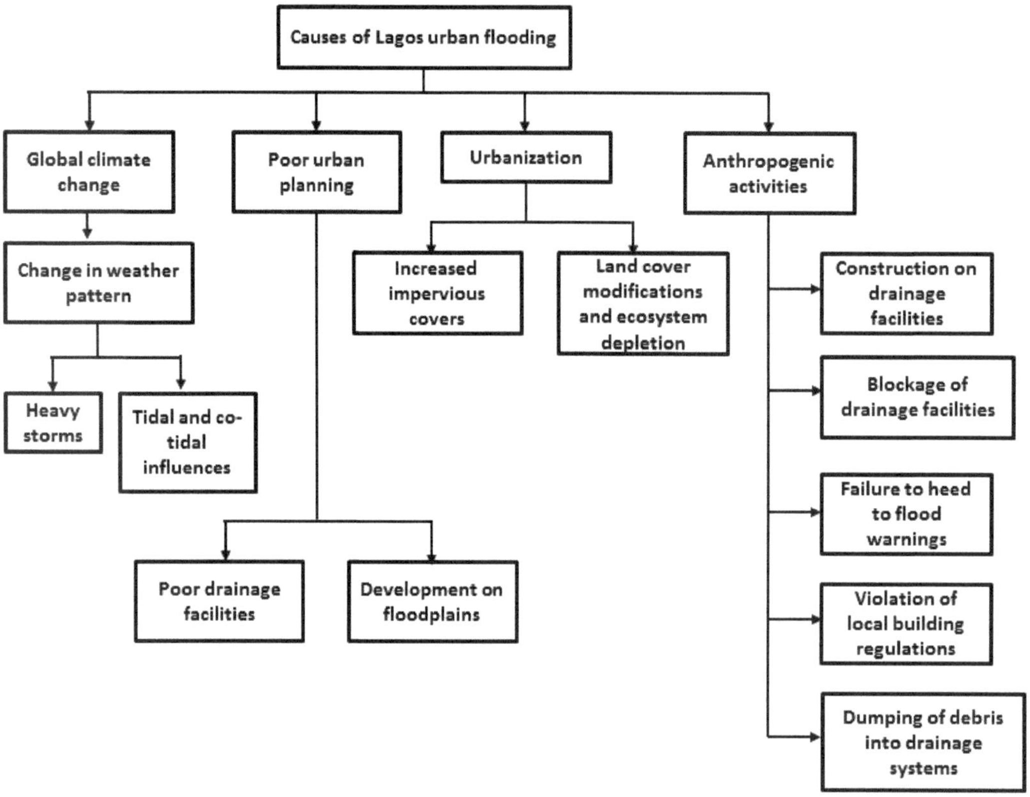

Figure 4. Main causes of urban flooding in the Lagos area of Nigeria showing global climate change, poor urban planning, urbanization, and anthropogenic activities.

Olajuyigbe et al. (2012) report that flood hazard increases city-wide poverty as a result of the farmlands which are destroyed and essential services which are often interrupted. Adelekan (2010) investigated these impacts using four poor urban communities in Lagos as case studies and identified three significant scales: individual, household, and community. At the individual scale, the reluctance of friends and family to visit one another while in flooded houses affects social relationships. This has broad adverse implications on community lifestyle and further compounds depression among flood victims in Lagos. Food insecurity is equally an important issue at this scale as food items stored in individual homes are often lost during flooding. In addition, there can be numerous health impacts including chronic skin infections from exposure to contaminated environmental systems and increased effects on those with an already poor health history.

Household and community scales of impacts are mainly indicated by the secondary effects of flooding in Lagos. Household impacts include deterioration of building quality, intrusion of contaminated water into apartments, lack of good drinking water, and loss or damage to household properties including sanitation facilities. The community impacts include an unclean environment, disruption of movement, and damage to public utilities. Urgent needs arise where community schools were flooded and schooling for children has been interrupted. This is an important issue within the context of human development. In many other DCs where it is also applicable, community leaders and the local authorities have often instigated measures to ensure that children's schooling is not interrupted despite the magnitude of flooding. In Bangladesh for example, a strategy known as "floating schools" in which classrooms are constructed on boats is being put in place during flooding (Huq and Aslam, 2003). This enables provision of uninterrupted education for children who have been impacted and whose education has been disrupted by flood catastrophes.

The impacts of flooding in Lagos also trigger concerns for environmental management, urban sustainability, and the development, governance, and the vulnerability of urban residents and local communities. Other factors of concern are humanitarian needs and services, especially primary health delivery (Soneye, 2014; Ajibade, 2013; Lamond et al., 2012). Needless to say, concerns for solid waste management are crucial as long as the indiscriminate dumping of wastes into drainage systems remains prevalent within Lagos. One example of this is the water sold in polythene sachets which is the major source of drinking water for residents. It is perceived that many residents dispose of the containers which end up in drainage facilities. Being a non-degradable waste,

Figure 5. Some flooding scenes examples in the Lagos metropolis of Nigeria: **(a)** living room submerged by flood water, **(b)** residential building submerged, **(c)** local community affected by flood waters, and **(d)** expressway overwhelmed by flood water. Source: Online images of flooding in Lagos, Nigeria.

it accumulates over time and blocks these drainage facilities. Unfortunately, little has been done to address such an important issue in flood control management.

4 Present research and gaps in knowledge

Despite the frequency of occurrence, causes, and severity of impacts of Lagos urban flooding alongside the understanding of a potentially increased risk in the future a better understanding of flooding in Lagos is lacking. Whilst most of the data relating to these impacts are at local levels and limited to small geographical areas, wide-ranging cross-disciplinary intervention measures are constrained. Lack of data and poor data scale present an important limitation to comprehensive urban flood impact assessment and management. Flood risk management efforts in Lagos considered in the light of global current flood risk management practices (see Sect. 5) reveals significant gaps. A number of non-structural approaches which promote flood risk assessment and management are frequently ignored. Efforts to manage and assess flood risk in Lagos are aimed at both prevention and control of flooding. Moreover, such efforts have been unprecedented in Nigeria and clearly demonstrate a practical commitment to fighting the flood hazard (Njoku and Udeagha, 2013; Obeta, 2014). However, results from various studies which highlighted the increasing vulnerabilities to flooding indicate that such efforts have so far been of little assistance to the victims of flooding.

The problems of a lack of data and poor data scale in Lagos are exacerbated by the general lack of funds and inadequate access to improved technology alongside a lack of political will (Nkwunonwo et al., 2014; Adeloye and Rustum, 2011). Unfortunately, the level of existing knowledge regarding the state of affairs is unsatisfactory and fails to assist in providing a potential solution to ways of reducing the hazard or its impacts on human populations. The more critical and disturbing scenario is that the Lagos area is a fast-growing city within which a great deal of the population currently lives within areas prone to flooding. Locally, the majority of these areas are slums which provide provisional dwelling places to poor urban residents. Adelekan (2010) argued that these residents often do not have the capacity to cope with flood hazard or the ability to quickly recover from losses. Moreover, the vast spatial distribution of such flood-prone areas in Lagos increases the exposure of the human population to flooding. A clearer picture taken from a global and regional perspective shows that Lagos is among the top 20 cities with

Table 1. A summary of major flooding events and associated threats in the Lagos metropolis of Nigeria from 1968 to 2012. Source: EM-DAT (2014), FME (2012) and published works.

S/No.	DATE	LGA(S) AFFECTED	DURATION (DAYS)	CAUSE (S)	NO OF PEOPLE DISPLACED	MORTALITY	ECONOMIC LOSS (N)	AFFECTED HOUSES/ OTHERS
1	Oct 2012	Lagos city*	Many days, unspecified	Heavy rain	Thousands	>50	Millions, unspecified	Many*, including interruption of traffic and other activities
2	Jul 2011	Lagos island, Mainland, Mushin	2 days	Heavy rain	10 000	100	Millions, unspecified	Many*
3	Oct 2010	Lagos island, Apapa, Kosofe,	Many days, unspecified	Heavy rain	Thousands	20	Millions, unspecified	Many* including interruption of traffic and other activities
4	Jul 2009	Lagos city*	Many days	Heavy rain	Many	Nil	Millions, unspecified	Many*
5	Oct 2008	Lagos city*	N/A	Heavy rain	Not specified	No data	Millions, unspecified	Many* including interruption of traffic and other activities
6	Aug 2007	Ikorodu, Kosofe and Abeokuta	15	Heavy rain	5000	17	Millions, unspecified	5000
7	Jul 2005	Lagos city	5	Heavy storm	3000	25	Millions	N/A
8	Jun 2004	Lagos city	2	Heavy rain	1000	Nil	Millions	Drainages
9	Jul 2002	Lagos city	3	Heavy rain	200	2	Millions	Many*
10	Jun, Jul, Sep 2000	Victoria Island and Ikoyi	2	Brief Torrential Rain	500	Nil	Millions, unspecified	Tens of thousands
11	May, Jun, Jul 1999	Mushin and Idiaraba	N/A				70 000 000	
12	Jul 1990	Lagos city	2	Heavy rain	3000	5	Thousands	Many*, not specified
13	Jul 1990	Lagos city	2	Heavy rain	500	Nil	N/A	Hundreds of inhabitants
14	Jun 1974	Idiaraba, Ikorodu, Surulere, and Yaba	Many days, unspecified	Heavy rain	Thousands	Nil	N/A	
15	Jun 1972	Lagos Island	N/A	Heavy rainfall	Not specified	Nil	N/A	Traffic was disrupted, few houses
16	Jul 1971	Lagos Island	5	Heavy rainfall	Not specified	Nil	N/A	Traffic was disrupted, few houses
17	Jul 1970	Lagos Island	N/A	Winds, accompanied by short-duration, high-intensity rain	Nil	Nil	5000	Few
18	Jun 1969	Surulere and Yaba	10	Short-duration, high-intensity rain	Nil	Nil	N/A	Many*, not specified
19	Jun 1968	Lagos Island and Ijora.	N/A	Heavy storm	Nil	Nil	6000	Traffic was disrupted, Few houses

* Grouped instead of treating as separate variables due to lack of data. N/A: Not available. Nil: No information.

increasing numbers of the present and future population exposed to flooding (see Table 1). Within such an environment, destruction of biodiversity and depletion of ecosystem functions to earn a livelihood are common practices. Given the issues of coping capacity and resilience which remain unresolved for the burgeoning vulnerable population, sustainable development under such circumstances is difficult and almost impossible.

The present research attempts to address the challenges of flooding within and around Lagos through a review of literature and flood information covering the hazard in the area and how it has so far been managed. Available historical flood data in the area, spanning the period 1968 to 2012, were synthesized and analysed. These data were obtained from various online databases (for example: NEST, 1991; Guha-Sapir et al., 2013) and published data on historical flooding in Lagos, Nigeria. Sourcing data from a wide range of databases is an attempt to clarify the issue of lack of flood data which creates significant gaps in knowledge relating to flood management efforts in Lagos. The research discusses the challenges faced by Lagos in managing and reducing flooding impacts. It contextualizes the current situation and puts forward relevant recommendations for more effective remedies to alleviate the threats of flooding in the area. Ultimately, it is argued that the lack of more robust techniques such as flood modelling and assessment of vulnerability accounts for a stark limitation in the efforts towards addressing the challenges of flooding in the area. In particular, the importance of flood modelling in flood risk reduction and the need for it to be included in the present and future efforts at reducing the impacts of flooding are emphasized. This research in general and the recommendations in particular are driven by three key aims: firstly, to understand the unique situation which exists in Lagos in relation to flooding; secondly, to align the focus of flood risk reduction in the Lagos area with that of more developed countries such as the US, the Netherlands, and the United Kingdom; and finally, to suggest measures for possible improvements.

5 Flood risk management: the general concept

In recent times, a shift from flood hazard control to flood risk reduction has been found necessary (Samuels et al., 2006). This is due to the inexorable nature of the hazard and the general perception that climate change, demographic, and economic activities would influence regional and global occurrences and consequences of flooding in the future (Kundzewicz et al., 2010, 2014). The importance of this new framework lies in the way in which risk is contextualized, which enables strategic and adaptable policies to be easily formulated.

There are various interpretations of the concept of risk in the literature (see for example Renn, 2008) and this complicates a cross-disciplinary approach to risk reduction. An

Table 2. Top 20 countries ranked in terms of population exposed to coastal flooding in the 2070s, including both climate change and socio-economic change and showing present-day exposure. (Source: Nicholls et al., 2008; OECD, Paris.) (*Highlight of Nigeria is by the authors.)

Rank	Country	Urban agglomeration	Exposed population (current)	Exposed population (future)
1	India	Calcutta	1 929 000	14 014 000
2	India	Mumbai	2 787 000	11 418 000
3	Bangladesh	Dhaka	844 000	11 135 000
4	China	Guangzhou	2 718 000	10 333 000
5	Vietnam	Ho Chi Minh City	1 931 000	9 216 000
6	China	Shanghai	2 353 000	5 451 000
7	Thailand	Bangkok	907 000	5 138 000
8	Myanmar	Rangoon	510 000	4 965 000
9	USA	Miami	2 003 000	4 795 000
10	Vietnam	Hai Phòng	794 000	4 711 000
11	Egypt	Alexandria	1 330 000	4 375 000
12	China	Tianjin	956 000	3 790 000
13	Bangladesh	Khulna	441 000	3 641 000
14	China	Ningbo	299 000	3 305 000
15*	Nigeria	Lagos	357 000	3 229 000*
16	Cote d'ivoire	Abidjan	519 000	3 110 000
17	USA	New York	1 540 000	2 931 000
18	Bangladesh	Chittagong	255 000	2 866 000
19	Japan	Tokyo	1 110 000	2 521 000
20	Indonesia	Jakarta	513 000	2 248 000

agreed perception of risk is still lacking despite decades of research. In an attempt to address this conceptual conflict, the United Nations in 1992 proposed a seminal definition: "risk is the expected loss (of lives, persons injured, property damaged, and economic activity disrupted) due to a particular hazard for a given area and reference period". In addition, the United Nations International Strategy for Disaster Reduction defined risk as the combination of probability of a hazard and its negative consequences (UNISDR, 2004). Within these frameworks, hazard-specific and place-based concepts of risk are highlighted (see for example Penning-Rowsell and Chatterton, 1997).

Regardless of conceptual disparity, on the basis of hazard, risk is too often hypothesized as the product of hazard and its consequences (Samuels, 2000; Samuels et al., 2006; De Wrachien et al., 2011; Solín and Skubincan, 2013). This concept of risk generally embodies three components which are considered fundamental to various methods used today in flood risk assessment and management and they include: "probability of hazard occurrence", "exposure", and "vulnerabilities" of elements at risk. Global current flood risk management practices or "best practices" in flood risk management can be identified in a wide range of sources (see for example Ashley et al., 2007; Fratini et al., 2012; Sayers et al., 2013) which aim to reduce the likelihood and/or the consequences of floods. The idea of "living with floods rather than fighting them" is the underlying framework. This philosophy tends towards a policy whereby societies adapt to floods by being prepared and having the right attitude towards damage reduction (van Ogtrop et al., 2005). It stems from three key considerations: (1) the limitations or failures of traditional flood control structural measures; (2) the need for a people-friendly means of tackling flooding; and (3) the goal to lessen all impacts of extreme floods while at the same time exploiting all benefits of ordinary floods (UNISDR, 2004; Di Baldassarre and Uhlenbrook, 2012). These considerations are often perceived in the context of a holistic management cycle which covers prevention, protection, preparedness, emergency response, recovery, and lessons learned (Samuels, 2000). Within this framework, a sound understanding of flooding, accurate and actionable assessment of flood hazard and flood risk, knowledge-based decision making, and strong political leadership are indispensable (UNISDR, 2004). As a result, integrated approaches which combine structural and non-structural measures are extensively applied (EC, 2004; Pitt, 2008; Kazmierczak and Carter, 2010; Sayers et al., 2013).

Structural measures which aim at reducing the occurrence of flood hazard are technically based and involve channelization and the use of natural and man-made barriers to contain waters in rivers and seas. Non-structural measures have the aim of reducing vulnerabilities and developing the capacities of ecological systems to cope with floods through multidisciplinary approaches (Few, 2003; Schanze, 2006; Miceli et al., 2008; Sayers et al., 2013). Non-structural measures include flood risk mapping, land use zoning and planning, flood vulnerability assessment, flood proofing, flood modelling, institutionalization of policies, flood awareness campaign, flood insurance, flood forecasting, relocation of properties, resettlement of human population, and green infrastructure planning (Merz et al., 2007; Jha et al., 2012; Smith, 2013). Fundamental to these measures is the assessment of flood risk which is a procedure to estimate the damage or negative impacts resulting from flooding. Flood negative impacts are often classified in terms of tangible and intangible damage. Tangible damage such as damage to properties and infrastructure are those that can be associated with monetary values. Intangible damage is difficult to express in monetary terms and may include depression, anxiety, and loss of life. Categorization of flooding impacts as direct and indirect damage is also applicable. Direct damage includes those which result from direct contact with flood water. When direct flood damage is tangible, a broader classification, direct tangible flood damage, applies. Indirect tangible flood damage is a secondary consequence of direct flood impacts and may include economic disruption, individual misfortune, and an increase in water-borne diseases (Samuels et al., 2006; Hammond et al., 2015).

The assessment of flood damage on the basis of damage typology is characterized by significant limitations. By means of heterogeneous methodologies, assessment of direct tangible flood damage has received significant attention in the literature (Hammond et al., 2015; Samuels et al., 2006). Damage functions, especially stage–damage curves which have been applied in many case studies around the world, remain potential tools for assessing direct and indirect tangible flood damage (Smith, 1994; Merz et al., 2004). A major limitation of stage–damage curves arises from the neglect of water flow velocity which also influences flood damage. To date, the difficulty in assessing intangible flood damage remains unresolved. Methods are being proposed in the current literature, for example the concept of "anxiety–productivity–income" which assumes that anxiety is a function of flood water depth and duration (Lekuthai and Vongvisessomjai, 2001; Price and Vojinovic, 2008). The "Risk-to-life" model, which uses flood characteristics and an estimate of the number of people exposed to flooding to assess the possible mortality, was proposed by Jonkman et al. (2008). A comprehensive review of other current methods to assess intangible flood losses can be found in Hammond et al. (2015).

Despite the progress made so far, uncertainties and the lack of time series stage data and good-quality damage data are major factors constraining the ubiquitous assessment of flood risk. In line with predictions of worsened future risk scenarios, these constraints underscore the importance of developing capacities for estimating flood hazard (in terms of water depth, velocity, etc.), exposure, and vulnerabilities of elements at risk as discrete components. The review presented in this research considers various aspects of Lagos flood risk assessment and management in light of the general context of flood risk assessment. With regard to flood risk assessment, procedures to estimate flood hazard are reviewed. Fundamental to the research is knowledge about the extent to which current and future ecosystem exposure to flood hazard have been influenced. Given reported trends in urbanization and population growth (summarized in Sect. 8.1), it can be assumed that exposure of systems to flooding in Lagos is at a critical level.

6 The management and reduction of flood risk in Lagos

6.1 General measures

General measures to tackle the challenges of flooding in Lagos have been discussed by Oshodi (2013). Recent practices have included:

1. The expansion of drainage infrastructure within the city heartland.
2. The annual debris removal from principal drainage facilities within the city heartland.
3. Providing advice to the inhabitants of flood plains and wetlands to relocate.
4. The demolition of homes in flood-prone areas.
5. Proposed resettlement scheme for the residents of Ogun River catchment areas.

Oshodi (2013) claimed that these practices are being carried out, although it could be argued that they are politically influenced. One example of this is the expansion and upgrading of the primary drainage facilities mainly being carried out in Bariga, Surulere, and Gbagada. The annual clearing of primary and secondary channels by Lagos State Government through the Ministry of the Environment is carried out principally in the metropolitan areas. Evacuation and resettlement is carried out for residents who live in flood-prone areas. A proposed resettlement scheme for residents of Ajegunle community near Ikorodu was undertaken between October 2011 and January 2012 (Oshodi, 2013). The move was necessitated based on the belief that the current location of the community in the Lagos urban Master Plan was originally zoned as wetland for agricultural use. The area was a major catchment of Ogun River. It could be argued that the lack of clear implementation policy for the plan with enormous housing shortfall in the city to cope with rapid population growth and urbanization has led to the conversion of the area into residential use. Significant environmental impacts are often associated with the use of land for purposes it was not originally allotted for. Flooding can be the case when such inappropriate use affects ecological equilibrium. However, a sound investigation is needed to validate this claim, especially within the context of Lagos.

Major failures of these general measures arise from the issues of continuity and the scope of application. These measures are often limited to the core urban areas of the city, excluding the majority of the outskirts (Oshodi, 2013). Many areas within Lagos did not benefit from the expansion of drainage facilities; moreover, the project was not completed because of a change in administration. These projects often suffer when new governments abandon uncompleted projects of their predecessors. This shows the potential for urban mismanagement and the degree of limitation in urban development given that there are many abandoned projects while new ones are being considered. There are occasional inconsistencies in the annual cleaning of the channels. Sometimes this operation is delayed, which allows debris to accumulate in the channels. Coupled with careless attention to the channels after cleaning, accumulation of debris has led to early collapse of the channels. This causes potholes on motorways and retention surfaces for water during flooding.

The problems of where to relocate to and the sparcity of support for relocation are overwhelming. Given the vast distribution of flood-prone areas and the limited financial resources to facilitate resettlement, the choice of which slum locations to resettle is complicated. Communities which cannot be resettled often face the risk of having individual homes demolished. Examples of this include the Agege and Ijeshatedo demolitions in August 2011 and that of Ijora-Badia in 2010, 2012, and 2013. The demolition of homes is understandably a distressing measure to those affected. It could be argued that a particularly controversial aspect is the failure and neglect of the state government to provide any form of alternative housing arrangements to those whose houses have been demolished. There have been a few exceptional cases, for example the Ijora-Badia (World Bank assisted) drainage channels project, which necessitated the demolition and burning of homes. Following a community-led protest the affected families were awarded relocation assistance costs. On the basis of these potential government inadequacies, Ahonshi (2002), Kamunyori (2007), and Basinsksi (2009), cited in Lawanson (2015), argued that the achievement of government's urban sustainability goals (which include general flood management measures) in Lagos are often without regard to the needs of the poor residents of the area. At the same time, demolition and eviction frequently leads to forced social disconnection amongst families, and compensation and legal process on behalf of those affected can take a significant amount of time to complete. This is in part an aspect of social vulnerability which needs to be investigated within the context of Lagos urban flooding. Under the circumstances of eviction and resettlement uncertainties, most people occupying the flood-prone areas seem to ignore government's flood risk mitigation proposals and flood warnings. This influences large-scale impacts of urban flooding in Lagos. However, despite numerous tensions, this issue has only received limited attention in the literature and in governance.

6.2 Institutional efforts

One critical aspect of flood response is the institutional efforts which have been undertaken by local authorities and stakeholders. Odunuga (2008) recognized several flood preventive and curative initiatives ranging from community self-assistance actions to World Bank assisted programmes. Recently, key initiatives, which include the Drain Dock and The Emergency Flood Abatement Gang (EFAG), were launched by the government of Lagos state to improve current efforts towards addressing the challenges of flooding. The ministries of Environment, Works, and Health as well as the Lagos Metropolitan Development and Governance Project (LMDGP) have a number of initiatives aimed at controlling flood hazard in the area, and these include shoreline protection, low carbon emissions, the school advocacy programme, and the climate change club. Lagos is also the first region in Nigeria to carry out a detailed topographic mapping of the area with LiDAR (Light Detection and Ranging) data acquisition and GIS-based analysis aimed at addressing the challenges of flooding. In addition to these efforts, the Nigerian government and international community have been active with measures to address the challenges of flooding at various locations within the country including the Lagos area (Olorunfemi, 2011; NIHSA, 2013). As well as engineering works such as dams, bridges, and sustainable urban drainage systems there has also been financial assistance to victims of flooding and this appears to be a common practice. These are undertaken by the National Emergency Management Agency (NEMA), the Nigeria Hydrological Services Agency (NIHSA), the Nigerian Meteorological Agency (NIMET), and the National Environmental Standards and Regulations Enforcement Agency (NESREA) which in 2009 superseded the Federal Environmental Protection Agency (FEPA). It is not intended to discuss the structure, specific roles, and the unique position of these agencies with regard to flood management in Lagos. These aspects have been comprehensively discussed by Obeta (2014). However, it is important to mention that the activities of these institutions with regards to disaster management are generally coordinated by NEMA.

Although detailed data are not available, a key point will be highlighted in terms of the historical perspective of disaster management institutions, to give clarity to a synthetic description of the temporal evolution of flood awareness in Lagos. The institutional framework in Nigeria can be traced back more than four decades in history. The federal government of Nigeria has since the first, second, and third National Development Plans of 1962–1968, 1970–1974, and 1975–1980 initiated plans for the management of all disasters including flooding. This was organized through the federal and state ministries of works. On the basis of flooding and associated hazards, the primary aim of these initiatives was to create awareness among the citizenry and to develop sound response strategies. This development has evolved to the present time into what is now known as institutional approaches to managing disaster. Lately, the institutional framework has incorporated operations such as flood warning through the NISHA, improving general flood awareness through the National Orientation Agency (NOA), and integration of local state and federal disaster emergency management agencies.

Despite the recent initiatives, these developments have been criticised as weak, while the roles of the institutions are not clearly defined (Adeaga et al., 2005; Oshodi, 2013; Soneye, 2014; Nkwunonwo et al., 2014; Adelekan, 2015). Critically, current measures undertaken by these agencies appear aimed to control flood rather than mitigate its impacts on ecological systems. Efforts are being made to facilitate evacuation and provide flood victims with urgent humanitarian needs. Environmental sustainability and policy, social responses, physical intervention, and environmental management are also critical issues requiring attention (Aderogba et al., 2012; Olajuyigbe et al., 2012; Aderogba, 2012b; Adeaga, 2008, Ilesanmi, 2010). However, despite the increasing number of people affected, the effectiveness of these efforts is challenged. Whilst it is unreasonable to claim that the weakness of these flood mitigation measures probably leads to more frequent flooding in the area, in can be argued that such measures have improved the experience of the general population with regard to flooding.

6.3 Recent research efforts

From the literature, researchers have suggested several options in relation to possible flood hazard mitigation and adaption responses. Adedeji et al. (2012) highlighted the importance of building the capacity for flood preparedness through spatial planning and land management. Ogunsote et al. (2011) suggested combating environmental degradation through sustainable landscaping. The need for sustainable management of solid waste which was emphasized in section 3 of this paper was recommended by Folorunsho and Awosika (2001). Komolafe et al. (2014) argued for the adoption of proactive measures to risk management and adaptation whilst constant geophysical and hydrological evaluation of rising groundwater levels was emphasized by Oyedele et al. (2009). Adelekan (2013) reiterated the UNISDR recommendation which calls for the participation of the private sector in risk management through investment decision making in building and construction. Other factors besides flood prevention are also important to reduce the potential impacts of flood events. The humanitarian relief supply chain for victims of flooding in the Lagos area was investigated by Soneye (2014). This study identified the need for more empirical investigation into such crucial components of flood risk management in Lagos. In relation to the planning framework, sustainable housing development and functionality of planning laws and regulations as well as the role of governance in flood management in the Lagos area and indeed in Nigeria have been examined by a number of authors including Aluko (2011) and Oshodi (2013).

As well as the general discussions in the literature, there is other research which focuses on integral components of flood risk such as probability of flooding, and exposure and vulnerability to flooding. But these are insufficient, and leave significant gaps in knowledge. Insufficient knowledge of the vulnerabilities to flooding of local communities, urban residents, and the general public constrains effective flood risk management in Lagos. Urban flooding in particular has arguably not received the attention it deserves. A general critique, which should provide a nuanced understanding of the strengths and limitations of present efforts to address the threats of flooding in the Lagos area, is lacking. The lack of flood data and other ancillary data, as discussed in Sects. 3 and 5, is a major setback towards containing these threats and has not been fully addressed. Importantly, the focus of these studies which solely rested on general knowledge of the causes, impacts, and remedies of flooding suggests that the global view of the situation has been imperfect. More scientific approaches such as the flood modelling and flood vulnerability assessment which drives recent approaches to flood risk management in more developed countries are generally lacking. Little action has been undertaken to raise public awareness of flood risk or to address gendered vulnerability, as highlighted by Odunuga et al. (2012), Ajibade et al. (2013), and Adelekan (2010). As an unprecedented measure, the Lagos state government has made significant efforts at providing high-resolution air-borne LiDAR data and topographic maps which promote research towards flood risk in the area. However, since many of these data sets are produced and sold commercially, the limited access of researchers to them arguably undermines their usefulness. A possible solution to such a limited access to LiDAR data is to apply global data sets such as ASTERGDEM (Advanced Spaceborne Thermal Emission and Reflection Radiometer Global Digital Elevation Model) and STRM (Shuttle Radar Topographic Mission). However, the resolution of spatial data is a crucial factor in flood modelling and flood risk assessment. Due to the geomorphology of urban features which significantly influence hydrodynamics, global data sets often do not provide realistic results when used to model flood hazard in urban areas (van de Sande et al., 2012). Flood modelling research is still looking into possible ways of simulating accurate flood variables on the basis of low-scale global data sets.

7 Flood risk assessment in Lagos

There has been a great deal of research about the assessment of flood risk in Lagos in general. However, it can be argued that none of these studies explicitly assessed the level of exposure to flooding of social and environmental systems. This aspect of flood risk assessment can be derived from studies that consider LU and LC change analyses, and population expansion and urbanization (see Table 3). From the literature, it is clear that vulnerability is being assessed under varying contexts and objectives. No study known to the authors has considered flood hazard estimation, although Odjugo (2006) and Adelekan (2010) investigated rainfall patterns over spatial and temporal scales. Odjugo (2006) investigated the changing rainfall pattern in Lagos over 3 decades. Adelekan (2010) on the basis of qualitative survey and secondary data analyses attributed increasing flood risks in Lagos Island mainly to changes in the frequency and intensity of rainstorms between 1971 and 2005. Other studies that estimate flood hazard were embodied within the general framework of climate change research (see for example Aderogba, 2012a). The lack of an explicit study on flood hazard estimation makes it difficult to appreciate how the probability of flood hazard occurrence is being determined in Lagos. Moreover, a clear understanding of the evolutionary trend of the hazard is denied. A consensus view is that the hazard has been increasing over the last 5 decades (Odunuga, 2008; Adelekan, 2015). It is expected that future research should be directed towards this crucial aspect of flood risk assessment.

7.1 Exposure

LU and LC analyses are procedures to classify land use/cover and to identify rates and patterns of change. However, the outcome of these analyses to Lagos has continuously suggested expansion in the urban areas. LU and LC analyses provides a basis for researchers and policy formulators to analyse the current and future urban exposure to environmental hazards especially flooding. Flood risk is heavily influenced by human and economic developments which evolve within spatial and temporal scales (Samuels et al., 2006). These developments are known to influence the hydrological cycle particularly through the development of more impervious surfaces which reduce soil infiltration capacity and lead to increased surface water runoffs. The spatial and temporal dimensions of flood risk can in part be directly related to LU and LC modifications. Barredo and Engelen (2010) and Gupta and Nair (2010) have shown that modelling LU and LC scenarios is integral to flood risk management. Identification of the population and urban growth pattern is also fundamental (Barredo and Demicheli, 2003). Many studies covering this crucial aspect of flood risk assessment within the context of the Lagos area exist in the literature (Table 3). Some of the major conclusions emerging from these studies show that significant sprawl and development in residential and industrial or commercial land classes are progressively taking place. There has been also been increasing pressure on arable lands (Akpomrere and Nyorere, 2012) and depletion of wetlands, mangroves, and swamps are significant issues (Obiefuna et al., 2013). The associated migration of people into Lagos city from other rural communities and states in Nigeria escalates residential needs and challenges. Barredo and Demicheli (2003) predicts that up to 27 million people will inhabit Lagos by 2020. Based on these studies, it can be shown that the level of human and economic resources exposed to urban flooding in Lagos is undoubtedly high. Figure 6 shows the 2012 LU and LC of the Lagos metropolis of Nigeria. Despite this, Lagos is relatively low on a global scale when compared to Calcutta, Dhaka, and Miami (Nicholls et al., 2007). However, the need to identify and address the vulnerabilities of exposed systems should be prioritized.

7.2 Vulnerability

There are numerous studies that have considered the vulnerabilities of social and environmental systems to flooding within the Lagos context (for example, Action Aid, 2006; Douglas et al., 2008; Adelekan, 2010; Ajibade et al., 2013, 2014; Nkwunonwo et al., 2015; Nsorfon, 2015; Olokesusi et al., 2015). Action Aid (2006) investigated vulnerability but tied Lagos with five other African cities. The study was carried out on the basis of key management criteria including local people's perceptions of the causes of flooding, adaptation, and the community's social coping capacity. Some of the limitations in the study were addressed by Douglas et al. (2008), who considered the vulnerability on the basis of climate change and adaptation strategies for the urban poor in Africa. Adelekan (2010) investigated the vulnerability of coastal communities in Lagos and responses to changing climatic conditions. The patterns of flood vulnerability and resilience amongst women were investigated by Ajibade et al. (2013). On the basis of political ecology, Ajibade et al. (2014) argued that the two crucial factors responsible for vulnerabilities of social systems in Lagos are limited access to housing and weak housing rights. The sources of social vulnerability to floods in informal settlements of Lagos was investigated in Nsorfon (2015) while Nkwunonwo et al. (2015) highlighted the relevance of assessing such vulnerability for the whole of Lagos. Olokesusi et al. (2015) investigated the influence of awareness of and responses to flood warnings on physical vulnerabilities of the affected communities in Lagos.

Although these studies are major contributions to knowledge which provide some evidence to suggest that vulnerability is an important issue of flood risk in Lagos, significant gaps still exist in the literature with regard to accurate assessment of vulnerability in the area. Most of these studies were based on a limited random sample of data which is insufficient to make accurate generalization. Critical issues of vulnerability analyses such as choice and measurement of vulnerability indices were not addressed. This contradicts a widely accepted philosophy that accurate assessment of vulnerability is related to these critical issues (Adger, 2006). Classical analyses of vulnerability should be able to consider every available factor that undermines the chances of resistance available to social and environmental systems.

Table 3. Studies on exposure to urban flooding in the Lagos area of Nigeria.

S/No.	Author(s)	Study	Context	Major findings
1	Nwafor (1986)	Physical environment, decision-making and land use development in metropolitan Lagos	Urban growth	Rapid changes in land use classes were detected. Major upward trend in urban growth since 1960. Urban renewal and highway development radiating from the inner city to the hinterland were the major catalysts.
2	Abiodun (1997)	The challenges of growth and development in metropolitan Lagos.	Urban growth	Extensive and persistent urban growth. Key growth factors are the city's economic vitality and pivotal position in Nigeria's economy. Major growth challenges are survival of urban residents and city's sustainable development.
3	Barredo and Demicheli (2003)	Urban sustainability in developing countries' megacities: modelling and predicting future urban growth in Lagos.	Urban growth	By 2020, Lagos will experience astronomical spatial growth as a direct consequence of population expansion. Up to 27 million people will inhabit Lagos by 2020.
4	Adepoju et al. (2006)	Land use/land cover change detection in metropolitan Lagos (Nigeria):	LU and LC change detection	Lagos urban growth is phenomenal. Between 1984 and 2002, about 35 % increase in urban areas was recorded
5	Sunday and Ajewole (2006)	Spatial determinants of urban land use change in Lagos, Nigeria.	LU and LC change detection	Changing pattern of LU and LC in Lagos is characterized by significant socio-economic and environmental implications. Flooding and other implications are expected to worsen in the future given this present trend.
6	Braimoh and Onishi (2007)	Implications of the changing pattern of LC of the Lagos coastal area of Nigeria.	LU and LC change detection	Remote sensing was used to investigate LU changes, while binary logistic regression was used to model the probability of observing urban development as a function of spatially explicit independent variables.
7	Odunuga (2008)	Urban LU change and the flooding patterns in Ashimowu Watershed, Lagos, Nigeria	LU scenario and flooding (Alimosho)	A progressive increase in built-up area at the rate of $28\,\text{ha}\,\text{yr}^{-1}$ between 1965 and 2003.
8	Olaleye et al. (2009)	LU change detection and analysis using remotely sensed data in Lekki Peninsula area of Lagos, Nigeria.	LU and LC change detection	Between 1964 and 2003 built-up areas grew from 40.93 to 7271.19 ha. Evolution in new classes of LU was observed, which include industrial, commercial and recreational LU class.
9	Nwokoro and Dekolo (2012)	LU change and environmental sustainability: the case of Lagos metropolis.	LU change and environmental sustainability	Between 1990 and 2006, built-up areas increased by approximately 17 %. There was an obvious loss of forest resources and agricultural land to urban development.
10	Akpomrere and Nyorere (2012)	LU patterns and economic development of Ikeja in Lagos state, Nigeria: the GIS approach.	LU patterns and economic development	From 1962 to 1994, the built-up area in Lagos rose from 6.55 to 63.90 and from 63.90 to 67.99 % between 1994 and 2004. Decreases in vegetation cover and undeveloped areas were recorded.
11	Adebayo (2012)	Impact of urban LU changes on property values in metropolitan Lagos.	Urban LU changes and property	Significant changes in LU pattern from residential to commercial. Significant implication in property values in the area. These created social and environmental problems such as traffic congestion and noise pollution. The need for adequate LU planning was identified.
12	Nkwunonwo (2013)	LU/LC mapping of the Lagos metropolis of Nigeria using 2012 SLC off Landsat ETM+ Satellite Images	LU and LC mapping	By 2012, four major LU/LC themes were identified in Lagos: water body, vegetation, residential, and industrial areas. Urban areas, namely residential and industrial areas, account for more than 54 % of the whole Lagos metropolis.
13	Obiefuna et al. (2013)	Spatial changes in the wetlands of Lagos/Lekki Lagoons of Lagos, Nigeria.	Wetland changes	Mangrove wetlands decreased from 88.51 to $19.95\,\text{km}^2$ between 1984 and 2006 at $-3.12\,\text{km}^2\,\text{yr}^{-1}$. Swamps decreased from 344.75 to $165.37\,\text{km}^2$ between 1984 and 2006 at $-8.15\,\text{km}^2\,\text{yr}^{-1}$. Built-up areas increased from 48.97 to $282.78\,\text{km}^2$ at $10.61\,\text{km}^2\,\text{yr}^{-1}$. Water body decreased from 685.58 to $654.98\,\text{km}^2$ at $-0.16\,\text{km}^2\,\text{yr}^{-1}$. Bare land increased from 24.32 to $72.73\,\text{km}^2$ at $2.2\,\text{km}^2\,\text{yr}^{-1}$. Vegetation decreased marginally from 1369.15 to $1361.08\,\text{km}^2$ at $-0.37\,\text{km}^2\,\text{yr}^{-1}$. Most of the increase in built-up area occurred in the Eti-Osa Local Government Area (LGA) and then in the Kosofe LGA. The key implication is of flood risk on affected areas.

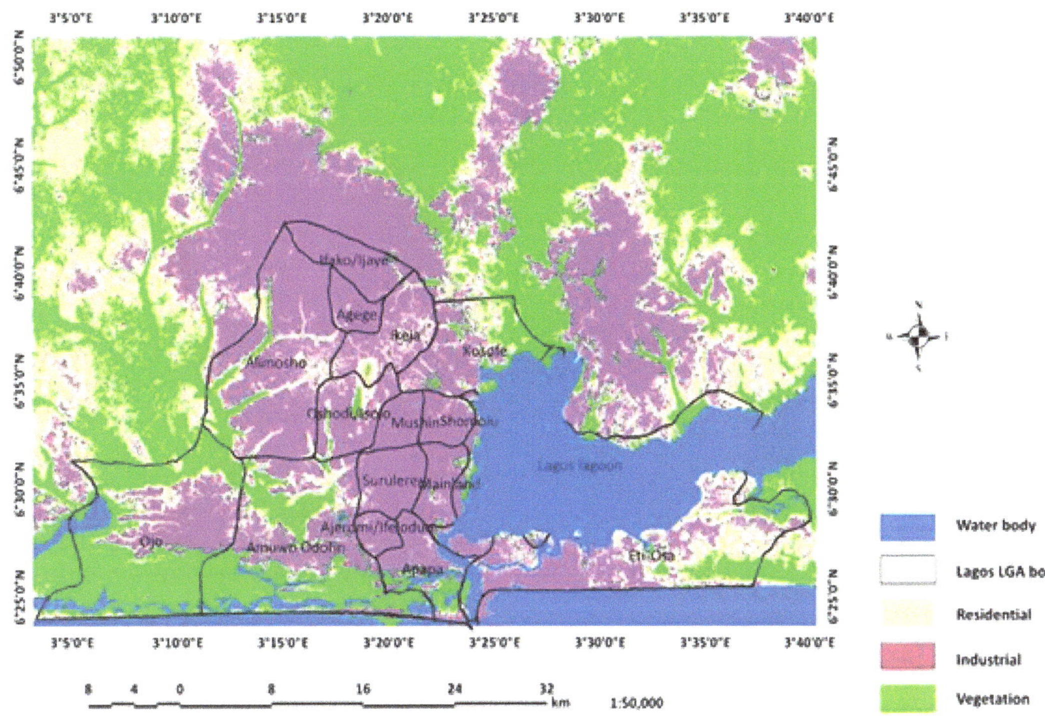

Figure 6. 2012 Land use (LU) and Land Cover (LC) map of the Lagos metropolis of Nigeria. Source: Nkwunonwo (2013).

8 The case for flood modelling in Lagos

Based on the issues identified by this research with regard to Lagos, three significant issues emerge which are: (1) lack of data relating to flooding and its consequence, (2) lack of accurate flood risk assessment, and (3) the "unidentifiable method" of flood hazard estimation. To address these issues, flood modelling which is widely being applied in hazard research (see for example Samuels et al., 2006; Merz et al., 2010) to estimate flood hazard should be considered. Attention should be given to the specific objectives that can be achieved in Lagos. It is also important to relate these objectives to specific factors which account for the reasons flood modelling has been largely ignored in the area.

Whilst it is widely acknowledged that stage–damage functions (presented in Sect. 5) are viable tools for flood risk assessment (Samuels et al., 2006; Merz et al., 2010; Hammond et al., 2015), the benefits which Lagos flood risk assessment can derive from such a tool should be investigated. This depends on the availability of various data relating to water depth and extent of flooding events. Even the use of probability concepts to estimate flood hazard requires data relating to a large sample of historical flooding events. As these data are not readily available for Lagos, a flood modelling approach can be used to reconstruct particular historical events (such as 1 in 50, 1 in 100, 1 in 200, 1 in 500, and 1 in 1000 flood return periods) in terms of inundation depth, extent, and water flow velocity, and on the basis of simulated data predict future occurrences. This will largely improve warning systems currently in place. However, when flooding occurs, and a predicted return period is the case, field survey should be carried out to determine properties and assets that are exposed. If the present market values of the exposed elements are known, damage functions on the basis of ex post flood hazard estimate can be used to assess the level of risk following the flooding event.

Theoretically, flood modelling is a procedure to investigate how rainfall or any other hydrological input into a drainage area is transformed to a streamflow (Chow et al., 1988). Since many of the floods in Lagos are pluvial based, accurate flood modelling results will assist in delineating areas likely to get increased runoff volumes and flooding. Flood modelling will play an important role in producing Lagos flood hazard and flood risk maps which do not currently exist. Needless to say, a comprehensive flood risk and flood hazard map for Lagos will increase public awareness of flood risks and inform stakeholder decision about flood mitigation options and policies in terms of structural and/or non-structural measures. From the point of view of cost benefit analysis (CBA), Lagos area will derive maximum benefits in the long run if flood modelling is the basis of any flood management operation.

Flood modelling is used to promote flood risk reduction in the US, the Netherlands, and the United Kingdom, but in Nigeria it is arguably too often ignored. This has continued to raise the question of how actionable flood risk assessment can be achieved. A number of structural measures are being put in place. However, an issue that has been less considered in the Nigerian context is the reliability of structural measures in the absence of accurate flood data and scientific means of acquiring such data. Flood modelling has been largely ignored in Lagos for reasons such as data requirement and availability, lack of specialist technology, funding, and the applied skills required. Academic research seems to be the most likely option in terms of the responsibility to develop flood models. The present research has been unable to refer to specific contributions of academic research towards flood modelling. However, the lack of bespoke or generalized flood models and poor application in flood risk management in the area seem to suggest that flood modelling procedures considered by academic curricula in Lagos are mainly theoretical and on the periphery. As NIHSA (2013) pointed out, such an intensive research should be funded by the government and other interested bodies. To date, the researchers are not aware of any such development. Moreover, the uncertainties associated with flood modelling can discourage investment of resources into it. To the best of the authors' knowledge, no public agency in Lagos undertakes the procedure as a specific role.

It can be argued that relatively few studies have highlighted the relevance of flood modelling and its implications with the paucity of topographic data (Nkwunonwo et al., 2014; van de Sande et al., 2012), although Adeaga (2008) implemented a flood hazard mapping and risk management in the northeastern part of Lagos. Although flood modelling was mentioned in these studies, solutions to the problems raised remain largely unanswered. Critically, current measures undertaken by flood management agencies appear to control flood rather than mitigate its impacts on human populations and urban infrastructure.

9 Flooding in Lagos – the way forward

The review of flood management covered in this research prompts several recommendations for reducing flood risk in Lagos. These recommendations are based on three key considerations: first, is the understanding and demonstration of the roles more scientific approaches (such as flood modelling,) can play in flood risk reduction within the context of Lagos; second, is the need to align the focus of flood risk reduction in the Lagos area to the objectives of similar measures in more developed countries such as the US, the UK, and the Netherlands. This should be linked to improving collaboration between Lagos and indeed Nigeria and the developed countries in terms of promoting a more effective flood risk management philosophy. In hazard research, the need for collaboration to strengthen learning and information exchange is increasingly being identified (Thomalla et al., 2006). Moreover, similar to the European Union flood risk directive, regional collaborations are being advocated for countries in Africa (Holloway, 2003). Thirdly, there is a need to promote awareness of flooding among local communities, urban residents and the general public and to delineate more suitable locations for relocation of human populations during flooding events.

Further specific recommendations include:

1. The government of Lagos state should as a matter of urgency prioritize legislation and provision of resources towards flood hazard and flood risk mapping for the whole of Lagos state. This is the basis of flood risk mitigation within the European Union framework, which requires all constituting states to prepare flood hazard and flood risk maps to promote the concept of living with floods (EC, 2004).

2. Flood risk reduction under the "living with floods" idea is multi-disciplinary, indicating that various industries can assist in reducing the impacts of flooding. This is the case in the UK in particular (EA, 2010). In view of the widening of public awareness of flooding, there is a need for improved collaboration between the Lagos state government and federal ministries, departments, and agencies such as NEMA, NESREA, and NIHSA.

3. Flood alert and flood early warning systems should be improved within the Lagos area. Globally, there is a growing concern regarding the dissemination of and response to flood warnings (Moore et al., 2005; Parker et al., 2009). On the basis of this, researchers in the UK and other European countries are anxious that flood warning systems are failing to yield the expected scale of damage savings (Penning-Rowsell et al., 2000; Parker et al., 2009). A review of 20 years of the progress of hazard warning systems in the US was presented by Sorensen (2000) although this research shows that no warning system is perfect for any hazard.

4. A series of issues need to be addressed with regard to the design of flood warning systems, and improved forecasting and response systems which correspond to the needs of the public and are often influenced by social, economic, and environmental variability. These issues should underlie improvements in any flood warning systems in Lagos. The cultural, social, and religious values of the people (which are important to many residents of Lagos) may also need to be taken into consideration when disseminating flood warnings. The reluctance of many people living in flood-prone areas to heed flood warnings should also be thoroughly investigated.

5. Flood insurance is a non-structural approach which many property owners have benefitted from in developed countries following flood disasters. It is argued that the benefits of flood insurance far outweigh the limitations which exist (Treby et al., 2006; Crichton, 2008). Flood insurance is still being applied in places like the

UK, the Netherlands, and Indonesia, although the coverage is at a relatively low scale (Jha et al., 2012). The need to enable flood victims to recover more quickly underscores flood insurance (Lamond and Penning-Rowsell, 2014). Purchase of insurance is highly dependent on a number of factors, including its availability and cost, the level of the provision of disaster relief, general risk awareness, and attitudes to collective and individual risk (Lamond et al., 2009). For the Lagos area, poor awareness of flood insurance is a major issue. Urban residents need to know that a means of assisting them recover from flood losses is possible. They must be given the opportunity to try flood insurance based on individual volition. To support the roles of flood insurance in Lagos, it is recommended that the role of FEMA in this regard should be extended to the state, whilst encouraging insurance companies to commence an awareness campaign for property owners to take positive steps towards purchasing insurance cover.

6. The enforcement of environmental standards and laws is often a key factor towards containing adverse effects of climate change including flooding (UN/ISDR, 2004). Indiscriminate waste disposal, construction along the flood plain, and the blockage of drainage facilities, among other anthropogenic activities influence flooding in Lagos and are often illegal. It can be argued that NESREA should embark on enforcement measures to address these issues.

7. The reaction to the 1953 floods in the Netherlands has arguably made the Dutch an exemplar in terms of flood management (Vis et al., 2003). Invariably, the success of flood risk reduction in the Netherlands is built on a strong commitment to avoid a repeat occurrence. The population in general and the Dutch government are committed to the collective efforts which underlie success towards addressing the challenges of flooding. Collective flood management costs each Dutch adult about USD 110 annually (Kazmierczak and Carter, 2010). For this reason, we recommend that urban residents and the general public in Lagos need to engage more fully in flood management and control. This would include raising awareness of the risks and educating the wider public about potential solutions. This would include encouraging adherence to environmental laws and a commitment to engage with flood alerts and early warning systems. To support this, more education is required to make the public more aware about flooding and its consequences. Qualitative research such as CBA, involving the local community should also be encouraged.

8. Globally, it appears that research is proportional to success towards addressing the threats of flooding. From scoping investigations it is clear that literature relating to flooding in Lagos appears embryonic compared to literature in the US, the Netherlands, China, and the UK for example. This is a strong indicator that more research is required for Lagos. Equally, the universities and research agencies should be encouraged to include flood awareness and management in their curricula and programmes. More research should be directed towards developing bespoke hydrologic and hydraulic flood models for simulating flood hazard and other hydrological parameters in the area.

9. Globally, accurate data play a key role in flood risk mitigation and the government of Lagos state has made an unprecedented attempt to improve its data by the acquisition of LiDAR data sets, although access to these data sets has been limited due to funds. Given the importance of these data sets and the need to optimize their usefulness for the Lagos area, the state government should consider subsidizing these data sets for universities and research institutions. Additionally, we also recommend further improvement on data acquisition such as SAR (Synthetic Aperture Radar) for flood modelling within the Lagos area. It is argued that the key to wider applied research requires easy access to data and a collaborative approach between researchers and government agencies.

10 Conclusion

Flooding is a global phenomenon, but the impacts in many urban areas in the developing countries can be overwhelming. It is easier to understand the threats of flooding in Lagos by the attention generated both in the media and in academic literature relating to social and environmental science. Flood waters have impacted upon the local population, destroyed critical infrastructure, and disrupted economic activities. However, based on "current practice" in flood management and flood risk reduction in the context of "living with floods", the actions of the state government and other stakeholders towards addressing the challenge of flooding in the Lagos area have arguably been limited. Unfortunately, relevant data on flood events are not readily available and the means of building a community resilient to flood threats have continued to elude present efforts.

This research is an attempt towards addressing the challenges of flooding in and around Lagos. Looking to the future, this research argues that flood modelling and assessment of vulnerability are necessary requisites for more effective results towards addressing the challenges of flooding in the Lagos area and indeed in Nigeria. The wider availability of high-quality data sets (including surface modelling and detailed demographic data) is a crucial step towards addressing this problem. It is recognized that a major limitation of this study and many other studies is the level of information available with regard to flooding in the Lagos area. The majority of the research offers a generalized overview of flooding rather than scaling the hazard to the local level. If flood data had been available for LGAs, it would have offered a better understanding of the spatial distribution of flooding over the epoch considered. In addition, there are places within various LGAs which may never have flooded over the period considered. It is important to investigate these locations with a view to finding out how they defend themselves from flooding. It was not intended within this study to address flood modelling and vulnerability assessment. However, the authors recommend future research that focuses on developing bespoke flood models for simulating flood hazard in the Lagos area of Nigeria.

Flooding is and will continue to be a hazard for the people of Lagos. The key to ameliorating the effects of flooding on the population is a more integrated, informed flood management philosophy which is linked to improved data collection and provision. This alongside greater public engagement and education as well as improved government coordination will greatly reduce the potential impacts of flood events on the population of Lagos.

Author contributions. This work was carried out in collaboration between all authors. Nkwunonwo Ugonna designed the study, performed the statistical analysis, managed literature searches, wrote the protocol, and wrote the first draft of the paper. Malcolm Whitworth and Brian Baily reviewed the first draft and made academic contributions. All authors read and approved the final paper.

Acknowledgements. This publication derives from a PhD research which is being funded by the Tertiary Institutions Education Trust Fund (TETFund) programme for staff of university of Nigeria. The Surveyors Council of Nigeria (SURCON) is acknowledged for providing some emergency grants. Data relating to flooding were sourced mainly from the Center for Research in Epidemiology of Disasters (CRED), Nigeria's Ministry of the Environment and the Nigeria Environmental Study Action/Team (NEST). Previous works in this area are equally acknowledged and so are the anonymous reviewers of this paper.

References

Abiodun, J. O.: The challenges of growth and development in metropolitan Lagos. The urban challenge in Africa: growth and management of its large cities, 192–222, 1997.

Action Aid: Climate Change, Urban Flooding and the Rights of the Urban Poor in Africa: Key Findings From Six African Cities, Action Aid International, London, 2006.

Adeaga, O.: Flood Hazard Mapping and Risk Management in Parts of Lagos N. E., available at: http://www.gsdi.org/gsdiconf/gsdi10/papers/TS13.3paper.pdf (last access: 10 March 2015), 2008.

Adeaga, O., Savic, D. A., Bertoni, J. C., Mariño, M. A., and Savenije, H. H. G.: A sustainable flood management plan for the Lagos environs, in: Sustainable Water Management Solutions for Large Cities: The Proceedings of the International Symposium on Sustainable Water Management for Large Cities (S2): Held During the Seventh Scientific Assembly of the International Association of Hydrological Sciences (IAHS) at Foz Do Iguaçu, Brazil, 3–9 April 2005, No. 293, International Assn of Hydrological Sciences, 226–229, 2005.

Adebayo, M. A.: Impact of urban land use changes on property values in Metropolitan Lagos, The Social Sciences, 4, 111–117, 2009.

Adedeji, O. H., Odufuwa, B. O., and Adebayo, O. H.: Building capabilities for flood disaster and hazard preparedness and risk reduction in Nigeria: need for spatial planning and land management, Journal of Sustainable Development in Africa, 14, 45–58, 2012.

Adeloye, A. J. and Rustum, R.: Lagos (Nigeria) flooding and influence of urban planning, Urban Design and Planning, 164, 175–187, 2011.

Adelekan, I.: Vulnerability of poor urban coastal communities to flooding in Lagos, Nigeria, Environ. Urban., 22, 433–450, 2010.

Adelekan, I. O.: Flood risk management in the coastal city of Lagos, Nigeria, Journal of Flood Risk Management, doi:10.1111/jfr3.12179, 2015.

Adelekan, I.: Private Sector Investment Decision in Building and Construction: Increasing, Managing and Transferring Risks: Case Study of Lagos, Nigeria, The United Nations Office for Disaster Risk Reduction, Global Assessment Report on Disaster Risk Reduction, 1–9, 2013.

Adepoju, M. O., Millington, A. C., and Tansey, K. T.: Land use/land cover change detection in metropolitan Lagos (Nigeria): 1984–2002, ASPRS 2006 Annual Conference Reno, Nevada May, 1–5, 2006.

Aderogba, K.: Qualitative studies of recent flood and sustainable growth and development of cities and towns in Nigeria, International Journal of Academic Research in Economics and Management Science, 1, 1–25, 2012a.

Aderogba, K.: Global warming and challenges of floods in Lagos metropolis, Nigeria, Academic Research International, 2, 448–468, 2012b.

Aderogba, K., Martins, O. M., Oderinde, S., and Afelumo, T.: Challenges of poor drainage systems and floods in Lagos metropolis, Nigeria, International Journal of Social Science and Education, 2, 412–427, 2012.

Adger, W. N.: Vulnerability, Global Environ. Chang., 16, 268–281, 2006.

Adigun, F., Abolade, O., and Yusuf, A. A.: Incidence of flood and its impacts: empirical evidence from Ajeromi-Ifeledun, Lagos State, Nigeria, International Journal of Innovative Research and Studies, 2, 239–254, 2013.

Agbola, T. and Agunbiade, E.: Urbanization, slum development and security of tenure: the challenges of meeting millennium development goal (MDG 7) in metropolitan Lagos, Nigeria, in: PRIPODE Workshop, Nairobi, Kenya, 11–13 June 2007.

Ahonsi, B. A.: Popular shaping of metropolitan forms and processes in Nigeria: glimpses and interpretations from an informed Lagosian, in: Documenta11_Platform4, edited by: Enwezor, O., Basualdo, C., Bauer, U. M., Leifeld, B., Ghez, S., Maharaj, S.,

Nash, M., Zaya, O., Müller, M., and Nollert, A., Under Siege: Four African Cities, , Freetown, Johannesburg, Kinshasa, Lagos and Ostfildern-Ruit, Hatje Cantz, 129–151, 2002.

Ajibade, I., McBean, G., and Bezner-Kerr, R.: Urban flooding in Lagos, Nigeria: patterns of vulnerability and resilience among women, Global Environ. Change, 23, 1714–1725, 2013.

Ajibade, I., Armah, F. A., Kuuire, V. Z., Luginaah, I., McBean, G., and Tenkorang, E. Y.: Assessing the bio-psychosocial correlates of flood impacts in coastal areas of Lagos, Nigeria, J. Environ. Plann. Man., 58, 445–463, doi:10.1080/09640568.2013.861811, 2014.

Ajibola, M. O., Izunwanne, E. M., and Ogungbemi, A. O.: Assessing the effects of flooding on residential property values In Lekki Phase I, Lagos, Nigeria, International Journal of Asian Social Science, 2, 271–282, 2012.

Akpomrere, O. R. and Nyorere, O.: Land Use Patterns and Economic Development of Ikeja in Lagos State, Nigeria: The Geographic Information System Approach, International Journal of Economic Development Research and Investment, 3, 39–47, 2012.

Aluko, O.: Sustainable housing development and functionality of planning laws in Nigeria: the case of cosmopolitan Lagos, Journal of Sustainable Development, 4, 139–150, 2011.

Ashley, R., Garvin, S., Pasche, E., Vassilopoulos, A., and Zevenbergen, C. (Eds.): Advances in urban flood management, CRC Press, London, UK, 2007.

Ayoade, J. O. and Akintola, F. O.: Public perception of flood hazard in two Nigerian cities, Environ. Int., 4, 277–280, 1980.

Bashir, O., Oludare, H., Johnson, O., and Aloysius, B.: Floods of fury in Nigerian cities, Journal of Sustainable Development, 5, 69–79, 2012.

Basinkski, S.: All Fingers Are Not Equal: A Report on Street Vendors in Lagos, Nigeria. Lagos: CLEEN Foundation, avialable at: http://www.cleen.org/allfingersarenotequalreport.pdf (last access: 10 March 2015). 2009.

Barredo, J. I. and Demicheli, L.: Urban sustainability in developing countries' megacities: modelling and predicting future urban growth in Lagos, Cities, 20, 297–310, 2003.

Barredo, J. I. and Engelen, G.: Land use scenario modeling for flood risk mitigation, Sustainability, 2, 1327–1344, 2010.

Braimoh, A. K. and Onishi, T.: Spatial determinants of urban land use change in Lagos, Nigeria, Land Use Policy, 24, 502–515, 2007.

BNRCC (Building Nigeria's Response to Climate Change): Vulnerability, Impacts and Adaptation, Climate Change in Nigeria, available at: www.infoo@climatechange.org (last access: 10 March 2015), 2008.

Chow, V. T., Maidment, D. R., and Mays, L. W.: Applied Hydrology, McGraw-Hill, Amsterdam, the Netherlands, 1988.

CII (Climate Change and Insurance): Chartered Institute for Insurance, Research report of the Chartered Institute for Insurance, London, 2001.

Crichton, D.: Role of insurance in reducing flood risk, Geneva Pap. R. I.-Iss. P., 33, 117–132, 2008.

De Wrachien, D., Mambretti, S., and Schultz, B.: Flood management and risk assessment in flood-prone areas: Measures and solutions, Irrig. Drain., 60, 229–240, 2011.

Di Baldassarre, G. and Uhlenbrook, S.: Is the current flood of data enough? A treatise on research needs for the improvement of flood modelling, Hydrol. Process., 26, 153–158, 2012.

Dilley, M., Chen, R. S., Deichmann, U., Lerner-Lam, A., Arnold, M., Agwe, J., Buys, P., Kjekstad, O., Lyon, B., and Yetman, G.: Natural Disaster Hotspots, A Global Risk Analysis, Washington, D.C., The World Bank, 2005.

Douglas, I., Alam, K., Maghenda, M., Mcdonnell, Y., McLean, L., and Campbell, J.: Unjust waters: climate change, flooding and the urban poor in Africa, Environ. Urban., 20, 187–205, 2008.

EA (Environment Agency): Working With Natural Processes to Manage Flood and Coastal Erosion Risk, Environment Agency, London, 2010.

EC (European Commission): Flood Risk Management – Flood Prevention, Protection and Mitigation, Communication From the Commission to the Council, the European Parliament, the European Economic and Social Committee and the Committee of the Regions, Brussels, 2004.

EM-DAT: The International Disaster Database – Centre for Research on the Epidemiology of Disasters – CRED: Flooding Data for Nigeria, available at: www.emdat.be/ (last access: 10 March 2015), 2014.

Etuonovbe, A. K.: The devastating effect of flooding in Nigeria, in: FIG Working Week, May 2011, available at: http://www.fig.net/pub/fig2011/papers/ts06j/ts06j_etuonovbe_5002.pdf (last access: March 2015), 2011.

Few, R.: Flooding, vulnerability and coping strategies: local responses to a global threat, Progress in Development Studies, 3, 43–58, 2003.

FME (Federal Ministry of Environment): Bulletin on Ecological Disasters, FME, Abuja, Nigeria, 25, 2012.

Folorunsho, R. and Awosika, L.: Flood Mitigation in Lagos, Nigeria through Wise Management of Solid Waste: a case of Ikoyi and Victoria Islands, Nigerian, UNESCO-CSI workshop, 19–23 November 2001, Maputo, 2001.

Fratini, C. F., Geldof, G. D., Kluck, J., and Mikkelsen, P. S.: Three Points Approach (3PA) for urban flood risk management: A tool to support climate change adaptation through transdisciplinarity and multifunctionality, Urban Water J., 9, 317–331, 2012.

Gill, S.: Literature Review: Impacts of Climate Change on Urban Environments – Draft 30 Copy, with contributions from: Pauleit, S., Ennos, R., Lindley, S., Handley, J., Gwilliam, J., and Ueberjahn-Tritta, A., The Impact of Flooding on Urban and Rural Communities, Centre for Urban and Regional Ecology, University of Manchester, UK, 2004.

Guha-Sapir, D., Hoyois, P., and Below, R.: Annual Disaster Statistical Review 2012: The Numbers and Trends, CRED, Brussels, 2013.

Gupta, A. K. and Nair, S. S.: Flood risk and context of land-uses: Chennai city case, Journal of Geography and Regional Planning, 3, 365–372, 2010.

Hammond, M. J., Chen, A. S., Djordjević, S., Butler, D., and Mark, O.: Urban flood impact assessment: A state-of-the-art review, Urban Water J., 12, 14–29, 2015.

Holloway, A.: Disaster risk reduction in Southern Africa: hot rhetoric-cold reality, African Security Studies, 12, 29–38, 2003.

Houston, D., Werritty, A., Bassett, D., Geddes, A., Hoolachan, A., and McMillan, M.: Pluvial (Rain-Related) Flooding in Urban Areas: the Invincible Hazard, Joseph Rowntree Foundation, York, UK, 2011.

Huq, S. and Alam, M.: Flood management and vulnerability of Dhaka City. Building Safer Cities: The Future of Disaster Risk, Washington, D.C., 121–135, 2003.

IFRC (International Federation of Red Cross and Red Crescent): Nigeria: Floods – July, available at: http://reliefweb.int/disaster/fl-2012-000138-nga (last access: 10 March 2015), 2012.

Ilesanmi, A. O.: Urban sustainability in the context of Lagos mega-city, Journal of Geography and Regional Planning, 3, 240–252, 2010.

Isong, M., Eyo, E., Eyoh, A., Nwanekezie, O., Olayinka, D. N., Udoudo, D. O., and Ofem, B.: GIS cellular automata using artificial neural network for land use change simulation of Lagos, Nigeria, Journal of Geography and Geology, 4, p. 94, 2012.

Jha, A. K., Bloch, R., and Lamond, J.: Cities and flooding: a guide to integrated urban flood risk management for the 21st century, World Bank Publications, Washington, D.C., 2012.

Jonkman, S. N., Vrijling, J. K., and Vrouwenvelder, A.: Methods for the estimation of loss of life due to floods: a literature review and a proposal for a new method, Nat. Hazards, 46, 353–389, 2008.

Kamunyori, S.: A growing space for dialogue: the case of street vending in Nairobi's Central business district, Unpublished MCP thesis, Massachusetts Institute of Technology, Boston, MA, 2007.

Kazmierczak, A. and Carter, J.: Adaptation to Climate Change Using Green and Blue Infrastructure, a Database of Case Studies, GRaBS project, University of Manchester, Manchester, UK, 2010.

Komolafe, A. A., Adegboyega, S. A. A., Anifowose, A. Y., Akinluyi, F. O., and Awoniran, D. R.: Air pollution and climate change in Lagos, Nigeria: needs for proactive approaches to risk management and adaptation, American Journal of Environmental Sciences, 10, 412–423, 2014.

Kundzewicz, Z. W., Hirabayashi, Y., and Kanae, S.: River floods in the changing climate – observations and projections, Water Resour. Manag., 24, 2633–2646, 2010.

Kundzewicz, Z. W., Kanae, S., Seneviratne, S. I., Handmer, J., Nicholls, N., Peduzzi, P., Mechler, R., Bouwer, L.M., Arnell, N., Mach, K., Muir-Wood, R., Brakenridge, G. R., Kron, W., Benito, G., Honda, Y., Takahashi, K., and Sherstyukov, B.: Flood risk and climate change: global and regional perspectives, Hydrolog. Sci. J., 59, 1–28, 2014.

Lagos State Government (LSG): Abstract of Local Government Statistics, Lagos: Lagos Bureau of Statistics, Ministry of Economic Planning and Budget Secretariat, Alausa, Ikeja, 2012.

Lamond, J. and Penning-Rowsell, E.: The robustness of flood insurance regimes given changing risk resulting from climate change, Climate Risk Management, 2, 1–10, 2014.

Lamond, J., Bhattacharya, N., and Bloch, R.: The role of solid waste management as a response to urban flood risk in developing countries, a case study analysis, in: Flood Recovery Innovation and Response, edited by: Proverbs, D., Mambretti, S., Brebbia, C., and de Wrachien, D., WIT Press, Southampton, 193–205, 2012.

Lamond, J. E., Proverbs, D. G., and Hammond, F. N.: Accessibility of flood risk insurance in the UK: confusion, competition and complacency, J. Risk Res., 12, 825–841, 2009.

Lawanson, T.: Potentials of the urban poor in shaping a sustainable Lagos metropolis, in: Untamed Urbanism, edited by: Allen, A., Lampis, A.,and Swilling, M., Routledge, New York, 108–118, 2015.

Lekuthai, A. and Vongvisessomjai, S.: Intangible flood damage quantification, Water Resour. Manag., 15, 343–362, 2001.

McMichael, A. J., Woodru, R. E., and Hales, S.: Climate change and human health: present and future risks, Lancet, 367, 859–869, 2006.

Merz, B., Kreibich, H., Thieken, A., and Schmidtke, R.: Estimation uncertainty of direct monetary flood damage to buildings. Natural Hazards and Earth System Sciences, 4,153–163, 2004.

Merz, B., Thieken, A. H., and Gocht, M.: Flood risk mapping at the local scale: concepts and challenges, in: Flood risk management in Europe, Springer, the Netherlands, 231–251, 2007.

Merz, B., Kreibich, H., Schwarze, R., and Thieken, A.: Review article "Assessment of economic flood damage", Nat. Hazards Earth Syst. Sci., 10, 1697–1724, doi:10.5194/nhess-10-1697-2010, 2010.

Miceli, R., Sotgiu, I., and Settanni, M.: Disaster preparedness and perception of flood risk: A study in an alpine valley in Italy, J. Environ. Psychol., 28, 164–173, 2008.

Moore, R. J., Bell, V. A., and Jones, D. A.: Forecasting for flood warning, C. R. Geosci., 337, 203–217, 2005.

NEST (Nigeria Environmental Study/Action Team): Nigeria's Threatened Environment: a National Profile, Ibadan, Nigeria: NEST, IRC Main, Abuja, Nigeria, 824 NG91, 9582, 1991.

Nicholls, R., Hanson, J. S., Herweijer, C., Patmore, N., Hallegatte, S., Corfee-Morlot, J., Château, J., and Muir-Wood, R.: Ranking of Port Cities With High Exposure and Vulnerabilities to Climate Extremes, OECD Environment Working Papers, No. 1, OECD Publishers, University of Southampton, UK, 2008.

NIHSA (Nigeria Hydrological Services Agency).: Flood Outlook, available at: http://nahs.org.ng/wp-content/uploads/2014/10/Nigeria-Flood-Outlook-for-2014.pdf (last access: 10 March 2015), 2013.

NIMET: 5 Nigerian Meteorological Agency: NIMET Weather Data Pack-1, available at: http://nimet.gov.ng/content/nimet-weather-data-pack-1 (last access: 10 March 2015), 2012.

Njoku, J. D. and Udeagha, M.: Assessing the flooding potentials of Oguta Lake Watershed using remote sensing technology, Paper presented at 5th Annual National Conference, Abuja, organized by the Nigerian Association of Hydrological Sciences (NAHS) at University of Nigeria, Nsukka, 21–30, 2013.

Nkwunonwo, U. C.: Land use/land cover mapping of the Lagos metropolis of Nigeria using 2012 SLC-off Landsat ETM+ Satellite Images, International Journal of Scientific and Engineering Research, 4, 1217–1223, 2013.

Nkwunonwo, U. C., Whitworth, M., Baily, B., and Inkpen, R.: The development of a simplified model for urban flood risk mitigation in developing countries, in: Vulnerability, Uncertainty, and Risk Quantification, Mitigation, and Management, ASCE, USA, 1116–1127, 2014.

Nkwunonwo, U. C., Whitworth, M., and Baily, B.: Relevance of social vulnerability assessment to flood risk reduction in the Lagos metropolis of Nigeria, British Journal of Applied Science & Technology, 8, 366–382, 2015.

Nsorfon, I. F.: Exploring Social Vulnerability to Natural Disasters in Urban Informal Settlements-Perspectives from Flooding in the Slums of Lagos, Nigeria, Doctoral dissertation, Universität zu Köln, Köln, 2015.

Nwafor, J. C.: Physical environment, decision-making and land use development in Metropolitan Lagos, GeoJournal, 12, 433–442, 1986.

Nwokoro, I. I. C. and Dekolo, S. O.: Land use change and environmental sustainability: The case of Lagos Metropolis, The Sustainable City VII: Urban Regeneration and Sustainability, 1, 157–167. 2012.

Obeta, C. M.: Institutional approach to flood disaster management in Nigeria: need for a preparedness plan, British Journal of Applied Science & Technology, 4, 4575–4590, 2014.

Obiefuna, J. N., Nwilo, P. C., Atagbaza, A. O., and Okolie, C. J.: Spatial changes in the wetlands of Lagos/Lekki Lagoons of Lagos, Nigeria, Journal of Sustainable Development, 6, p. 123, 2013.

Odjugo, P. A. O.: An analysis of rainfall pattern in Nigeria, Global Journal of Environmental Sciences, 4, 139–145, 2006.

Odunuga, S.: Urban Land Use Change and the Flooding in Ashimowu Watershed, Lagos, 20 Nigeria, PhD thesis, University of Lagos, Nigeria, 2008.

Odunuga, S., Oyebande, L., and Omojola, A. S.: Social-economic indicators and public perception on urban flooding in Lagos, Hydrology for Disaster Management: Special Publication of the Nigerian Association of Hydrological Sciences, NAHS, Abuja, 82–96, 2012.

Ogunsote, O. O., Adedeji, Y. M. D., and Prucnal-Ogunsote, B.: Combating environmental degradation through sustainable landscaping in emerging mega cities: a case study of Lagos, Nigeria, in: Proceedings of the 24th World Congress of Architecture "UIA2011 TOKYO", 25 September to 1 October, Tokyo, Japan, 16–21, 2011.

Ojinnaka, O.: Hydrography in Nigeria and Research Challenges, FIG Working Week 2013, Environment for Sustainability, TS05E – Hydrographic Education and Standards – 6439 (1-11), Abuja, available at: http://www.fig.net/pub/fig2013/papers/ts05e/TS05E_ojinnaka_6439.pdf (last access: 10 March 2015), 2013.

Oladunjoye, M.: Nigeria: July 10 Flooding – Lagos Gives Relief Materials to Victims, Daily Champion Newspaper, available at: http://allafrica.com/stories/201109080792.html (last access: 8 February 2015), 2011.

Olajuyigbe, A. E., Rotowa, O. O., and Durojaye, E.: An assessment of flood hazard in Nigeria: the case of mile 12, Lagos, Mediterranean Journal of Social Sciences, 3, 367–375, 2012.

Olaleye, J. B. and Abiodun, O. E.: Land Use Change Detection and Analysis Using Remotely Sensed Data in Lekki Peninsula Area of Lagos, Nigeria, Eliat Isreal: FIG Working Week, 1–15, 2009.

Oloke, O. C., Ijasan, K. C., Ogunde, A. O., Amusan, L. M., and Tunji-Olayeni, P. F.: Improving urban residents' awareness of the impact of household activities on climate change in Lagos State, Nigeria, Journal of Sustainable Development, 6, 56–64, 2013.

Olokesusi, F., Olorunfemi, F. B., Onwuemele, A., and Oke, M. O.: Awareness of and responses to the 2011 flood warnings among vulnerable communities in Lagos, Nigeria, in: Global Sustainability, Springer International Publishing, Switzerland, 203–223, 2015.

Olorunfemi, F. B.: Managing flood disasters under a changing climate: lessons from Nigeria and South Africa, Paper presented at NISER Research Seminar Series, NISER, Ibadan, 3 May 2011, 1–44, 2011.

Oshodi, L.: Flood management and governance structure in Lagos, Nigeria, Regions Magazine, 292, 22–24, 2013.

Oyebande, L.: Drainage protection to urban lands: an environmental challenge, Nigerian Geographical Association Conference, 16–21 December 1974, University of Nigeria, Nsukka, Enugu, 1–7, 1974.

Oyedele, K. F., Ayolabi, E. A., Adeoti, L., and Adegbola, R. B.: Geophysical and hydrogeological evaluation of rising groundwater level in the coastal areas of Lagos, Nigeria, B. Eng. Geol. Environ., 68, 137–143, 2009.

Parker, D. J., Priest, S. J., and Tapsell, S. M. Understanding and enhancing the public's behavioural response to flood warning information, Meteorol. Appl., 16, 103–114, 2009.

Penning-Rowsell, E. C. and Chatterton, J. B.: The Benefits of Flood Alleviation: A Manual of Assessment Techniques, Gower Publishing Company Limited, Aldershot, UK, 1977.

Penning-Rowsell, E. C., Tunstall, S. M., Tapsell, S. M., and Parker, D. J.: The benefits of flood warnings: real but elusive, and politically significant, Water Environ. J., 14, 7–14, 2000.

Pitt, M.: Lessons from the 2007 Floods, Pitt Review, London, 2008.

Price, R. K., and Vojinovic, Z.: Urban flood disaster management, Urban Water J., 5, 259–276, 2008.

Raaijmakers, R., Krywkow, J., and van der Veen, A.: Flood risk perceptions and spatial multicriteria analysis: an exploratory research for hazard mitigation, Nat. Hazards, 46, 307–322, 2008.

Renn, O.: Concepts of Risk: An Interdisciplinary Review Part 1: Disciplinary Risk Concepts, GAIA, 17, 50–66, 2008.

Samuels, P., Klijn, F., and Dijkman, J.: An analysis of the current practice of policies on river flood risk management in different countries, Irrig. Drain., 55, S141–S150, 2006.

Samuels, P. G.: An overview of flood estimation and flood prevention. In invited paper presented at the first international symposium on flood defense, Kassel Reports of Hydraulic Engineering, 9, G1–G11, 2000.

Sayers, P., Li, Y., Galloway, G., Penning-Rowsell, E., Shen, F., Wen, K., Chen, Y., and Le Quesne, T.: Flood Risk Management: a Strategic Approach, UNESCO, Paris, 2013.

Sayers, P. B., Hall, J.W., and Meadowcroft, I. C.: Towards risk-based flood hazard management 25 in the UK, Proceedings of ICE Civil Engineering, 150, 1 May 2002, 36–42, 2002.

Schanze, J.: Flood risk management – a basic framework. In Flood risk management: Hazards, vulnerability and mitigation measures, Springer, the Netherlands, 1–20, 2006.

Smith, K.: Environmental hazards: assessing risk and reducing disaster, Routledge, London, 2013.

Smith, D. I.: Flood damage estimation – a review of urban stage-damage curves and loss functions, Water SA, 20, 231–238, 1994.

Solín, L'. and Skubincan, P.: Flood risk assessment and management: review of concepts, definitions and methods, Geogr. J., 65, 23–44, 2013.

Soneye, A.: An overview of humanitarian relief supply chains for victims of perennial flood disasters in Lagos, Nigeria (2010–2012), Journal of Humanitarian Logistics and Supply Chain Management, 4, 179–197, 2014.

Sorensen, J. H.: Hazard warning systems: Review of 20 years of progress, Natural Hazards Review, 1, 119–125, 2000.
Sunday, O. A. and Ajewole, A. I.: Implications of the changing pattern of land cover of the Lagos Coastal Area of Nigeria, American-Eurasian Journal of Scientific Research, 1, 31–37, 2006.
Thomalla, F., Downing, T., Spanger-Siegfried, E., Han, G., and Rockström, J.: Reducing hazard vulnerability: towards a common approach between disaster risk reduction and climate adaptation, Disasters, 30, 39–48, 2006.
Treby, E. J., Clark, M. J., and Priest, S. J.: Confronting flood risk: implications for insurance and risk transfer. J. Environ. Manage., 81, 351–359, 2006.
Ugwu, L. I. and Ugwu, D. I.: Gender, floods and mental health: the way forward, International Journal of Asian Social Science, 3, 1030–1042, 2013.
UN (United Nations): World Population to 2300, United Nations Department of Economic and Social Affairs, Population Division, New York, 2004.
UN/ISDR: United Nations International Strategy for Disaster Reduction: Living with Risks: a global Review of Disaster Reduction Initiatives, 2004 Version Volume 1, available at: http://www.unisdr.org/files/657_lwr1.pdf (last access: 10 March 2015), 2004.
van de Sande, B., Joost, L., and Hoyng, C.: Sensitivity of coastal flood risk assessments to digital elevation models, Water, 4, 568–579, 2012.
van Ogtrop, F. F., Hoekstra, A. Y., and van der Meulen, F.: Flood management in the lower Incomati River Basin, Mozambique: two alternatives, J. Am Water Resour. As., 41, 607–619, 2005.
Vis, M., Klijn, F., De Bruijn, K. M., and Van Buuren, M.: Resilience strategies for flood risk management in the Netherlands, International Journal of River Basin Management, 1, 33–39, 2003.
World Bank: World Bank Population Growth Rate 2009–2013, available at: http://data.worldbank.org/indicator/SP.POP.GROW (last access: 10 March 2015), 2013.

9

Monitoring and modelling to guide coastal adaptation to extreme storm events in a changing climate

J. M. Brown[1], P. Ciavola[2], G. Masselink[3], R. McCall[3,4], and A. J. Plater[5]

[1]The National Oceanography Centre, Liverpool, UK
[2]University of Ferrara, Ferrara, Italy
[3]Plymouth University, Plymouth, UK
[4]Deltares, Delft, the Netherlands
[5]University of Liverpool, Liverpool, UK

Correspondence to: J. M. Brown (jebro@noc.ac.uk)

Abstract. Storms across the globe and their associated consequences in coastal zones (flooding and erosion), combined with the long-term geomorphic evolution of our coastlines, are a threat to life and assets, both socioeconomic and environmental. In a changing climate, with a rising global sea level, potentially changing patterns in storm tracks and storminess, and rising population density and pressures on the coastal zone, the future risk of coastal storm impacts is likely to increase. Coastal managers and policy makers therefore need to make effective and timely decisions on the use of resources for the immediate and longer term.

Research focused on "monitoring and modelling to guide coastal adaptation to extreme storm events in a changing climate" is becoming more common; its goal is to provide science-based decision support for effective adaptation to the consequences of storm impacts, both now and under future climate scenarios at the coast. The growing transfer of information between the science community and end-users is enabling leading research to have a greater impact on the socioeconomic resilience of coastal communities. This special issue covers recent research activities relating to coastal hazard mapping in response to extreme events, economic impacts of long-term change, coastal processes influencing management decisions and the development of online decision support tools.

1 Introduction

A clearer picture of how coastal hazards may change in the future will enable coastal communities and industries to adapt effectively to a changing coastal environment to achieve resilience over the long term. Two current aims of the research community to help attain this goal are the following:

- to assess and provide levels of probability on the impacts of climate change at the coast;
- to increase our understanding of the consequences of extreme storm impacts to help provide evidence-based policy advice.

To enable regional-scale and long-term study of the changing patterns within a coastal system, modelling approaches are often adopted to simulate waves (e.g. De winter et al., 2012), water levels (e.g. Howard et al., 2014) and morphological evolution (e.g. Ashton et al., 2001). However, for the projections to be credible the monitoring of different coastal types is required to determine their validity, constrain the uncertainty in model predictions, and identify important processes and interactions that need to be captured within numerical or conceptual models.

Multidisciplinary research and stakeholder engagement are necessary to ensure improved understanding of coastal storm impacts is translated into effective planning for societal benefit (Pine, 2011). Multiple projects across Europe and the international science community are therefore currently focused on monitoring and/or modelling extreme events from

Figure 1. An example of a UK site (Crosby, North West England) where the dune system has built up around an engineered defence (the pathway runs along the top of a seawall and revetment) due to its position relative to the spring high-tide level, thereby changing the management issues in this location from flood risk to safe access.

all fields of study to improve future coastal resilience through working with the end-user community. More recently, the need for data assimilation into coastal impact models is being addressed in order to improve the capability of early warning systems (Van Dongeren et al., 2013; Sembiring et al., 2015). An example of how research can guide management decisions is in the improved understanding of a systems response to intervention through the development of transferable tools that incorporate leading scientific developments to reduce the uncertainty in alternative solutions and designs for sustainable long-term defence schemes (e.g. Fig. 1). Within this special issue, the risk posed by natural hazards in coastal environments is studied. Coastal vulnerability is shown to be the balance of a community's susceptibility and resilience. The research focuses on different communities vulnerable to physical processes during extreme events and increased flood risk under future sea level scenarios. Two key research areas are (1) early warning systems that run operationally to enable rapid response to imminently forecast storms; and (2) geographic information system (GIS) decision-support tools that are used to visualise erosion and flood risk to inform risk management frameworks and test mitigation measures. By understanding regional vulnerability appropriate land-use planning can facilitate evacuation pathways and improve the positioning of critical infrastructure.

In Sect. 2 of this preface there is a brief overview of each contribution to the special issue. This is followed by a summary of the research in Sect. 3.

2 Research contributions

The research in this special issue is not only focused on storm events themselves, but also the long-term changes and adaptation to flood hazard. We first introduce the impact studies followed by those looking at long-term trends in both physical coastal process and socioeconomic development.

2.1 Storm impact studies at the event scale

Sea state and surge forecasts are well developed, while capabilities to consider coastal response are still improving. This section focuses on the coastal impact of nearshore storms in a changing climate. The need for monitoring in conjunction with early warning systems to calibrate morphological response and assimilate real-time offshore conditions is demonstrated to improve hazard forecasting (Harley et al., 2016). The ability to respond to a storm forecast can greatly reduce loss of life as well as storm damage and, therefore, economic costs. To capture uncertainty in forecasts, modelling systems are often run in ensemble mode (Flowerdew et al., 2010). Methods to reduce the uncertainty in morphological forecasts are now also being developed (Harley et al., 2016). Early warning is not only about the ability to accurately forecast a storm event in a timely manner, but also about the behaviour of society in response to such a warning (Pescaroli and Magni, 2015). To implement effective warning systems, there must be an understanding of what the hazard is and where vulnerable areas are (Perini et al., 2016; Lapidez et al., 2015). Such information can be assessed using cost–distance analysis and knowledge of land use (Perini et al., 2016). Developing these tools requires high resolution monitoring to produce quality maps for end-users to identify regional vulnerability. Continued monitoring enables capability assessment for more extreme and less frequent events when they occur (Perini et al., 2016). This highlights the need to catalogue storm levels as they occur (as done by Wadey et al., 2015), which also provides a basis for establishing any antecedent conditions that contribute to increased coastal vulnerability (see also Dissanayake et al., 2015). The development of risk maps based on cost–distance analysis can be used to identify the cost–benefit ratio of different coastal defence systems (Perini et al., 2016). Incorporating such maps into decision-support tools (see Knight et al., 2015) enables assessment of the time variation of the flood hazard during storms, and in response to climate change, sea-level rise and future management intervention. Improved knowledge of pathways for flood water also improves evacuation planning (Perini et al., 2016; Lapidez et al., 2015) in preparation for early warning.

Both social and behavioural patterns influence the effectiveness of an early warning, not only due to the response, but also through the preparedness and communication of information. Improvements in education, training, and accountability are more critical than further development in technology (Pescaroli and Magni, 2015). To improve evacuation plans, vulnerability assessments of coastal communities are required. To carry out such assessment, plausible extremes have to be considered to enable a worst-case impact assessment from possible storm tracks that influence a region. This typically requires the use of large-scale modelling studies (Lapidez et al., 2015). Many models are often limited by the availability of high resolution and timely input data. The efficiency of models can also limit the amount of detail that can be considered to allow multiple ensembles of simulations. Methods must therefore be considered to ensure critical features that are important for accurate impact modelling, are appropriately incorporated and are not smoothed during data processing (Hodges, 2015). Such techniques are important for inundation modelling, as defences and infrastructure that limits the flow of water must be accurately resolved to generate an accurate flood map (Hodges, 2015). The accuracy of the mapping is not only important for knowledge of flood depths and velocities to identify the hazard rating of events, but also to enable managers to consider cost–benefit analysis of varied solutions to coastal flood and erosion risk. To support managers in planning more strategically, the development of decision-support tools, which incorporate a range of flood maps to cover plausible hazard scenarios, is becoming more common (Knight et al., 2015). Such tools are designed to meet the needs of the end-user, but incorporate the latest advances in scientific models to provide evidence to underpin management decisions. One difficulty in generating such tools beyond the scale of a local region is the lack of standardized monitoring to enable identical model applications to populate the tool in a uniform way at the national level (Knight et al., 2015).

To aid the event selection incorporated into decision-support tools and to test early warning systems, there needs to be better categorization of the extreme events that impact a region (Wadey et al., 2015) and better understanding of how repeated shocks to a system modify the vulnerability (Dissanayake et al., 2015). Authorities are more frequently collecting long-term observations to understand the coastal processes to inform policy making. Following an extreme event managers need to categorise the severity relative to historical conditions to enable impact assessments of both engineered and natural defence systems (Wadey et al., 2015). This enables better understanding of the critical conditions that trigger a change in a system's resilience. In the most extreme cases this change may be permanent, requiring a change in management policy (Knight et al., 2015). Early warning of when a system may reach a tipping point is thus required to plan long-term management strategies. Long-term national monitoring also enables better analysis of the occurrence (Wadey et al., 2015) and impacts (Dissanayake et al., 2015) of storm "clustering" in both time and space. Spatial impact is of great importance for a nation due to the socioeconomic impacts, while for local communities a continued impact within a short time frame is a more immediate concern as repeated shock interspersed with short recovery time can increase vulnerability to later events. In terms of natural defence, the repeat impact of storm waves can modify the beach-dune profile, changing the system's vulnerability (Dissanayake et al., 2015). However, the cumulative erosive impact is found to be insensitive to the event chronology.

2.2 Long-term influence on coastal resilience

Over the long term, other factors in addition to storm conditions also need consideration to accurately assess changes concerning flood and erosion hazard. Rising sea level needs to be considered in assessments of extreme events (Wadey et al., 2015), while changes in coastal land use and infrastructure need consideration in assessment of the impact of events (Fakhruddin et al., 2015; Stevens et al., 2015; Sekovski et al., 2015). In support of proactive long-term adaptive management, guidance on suitable frameworks to apply when considering climate change is also required (Armstrong et al., 2015).

Coastal urbanization (Stevens et al., 2015; Sekovski et al., 2015) and modern life style, such as commuting to work, are placing people in a hazard zone, even if temporarily. The infrastructure that supports these coastal communities is also at risk (Fakhruddin et al., 2015). Analysis of the temporal variability in floodplain population and infrastructure vulnerability is therefore required to inform adaptive management. Transferable methods can be used to assess changes in risk in response to residential population growth relative to sea-level rise (Stevens et al., 2015), and to assess the change in hazard for chosen storm levels in response to different types of urban growth (Sekovski et al., 2015). It is suggested that population growth is a greater contributor to increased flood risk than sea-level rise (Stevens et al., 2015) and that a "compact" growth type increases the flood hazard over other modes of urbanization (Sekovski et al., 2015). To assess the susceptibility of a community's infrastructure, both the physical exposure and the repair capacity need to be considered in vulnerability indicator models as developed by Fakhruddin et al. (2015).

With urbanization and rising sea levels, coastal squeeze is becoming a greater threat to natural habitats. Ecosystem adaptation is therefore being considered more often as a mechanism to increase a system's resilience (Brown et al., 2011). One example of such a solution is coastal wetlands, which are recognized for their ability to lessen storm impacts and absorb flood water (Smolders et al., 2015). In response to economic growth within a changing coastal climate alternative solutions to flood and erosion risk management are emerging. At the same time, the environmental impact of an approach is now recognized as important and "softer" solutions are being considered (Smolders et al., 2015).

To manage the long-term coastal adaptation a hybrid of three principal approaches (Scenario-Led, Vulnerability-Led and Decision–Centric) is often applied (Armstrong et al., 2015). As understanding of climate change increases, practical guidance on how to respond needs to be addressed to ensure effective long-term management. This is particularly important at sites of coastal industry that have long life expectancies, e.g. the nuclear energy sector (see also Knight et al., 2015). Conflicting coastal interests often complicate coastal management, preventing a uniform approach across a region, although numerous interactions and feedbacks exist (Armstrong et al., 2015).

3 Summary

With rising sea levels, increasing economic costs, expanding populations within low-lying coastal regions, and defence degradation, coastal flood and erosion risk are increasing. It is not only lives, livelihoods and human health that are at risk, but also infrastructure, business and ecosystems. To enable sustained development socioeconomic analysis must be incorporated with oceanographic and geomorphic process studies to improve risk management frameworks and understand the fragility of natural, engineered and human-altered systems. Although there is great uncertainty associated with projected changes in storminess, it is known higher sea levels will increase the impact of present-day storms as elevated water levels enable ageing defences to be more easily overtopped and dune retreat to increase. This special issue demonstrates a good understanding of the joint probability of water levels and waves that cause coastal impacts are therefore needed. The impacts of waves, surge and their interactions are becoming more hazardous for many coastal systems, whether engineered, gravel, sand, estuarine, wetland or cliff coastlines. By combining modelling with monitoring schemes, a robust and reliable understanding of coastal dynamics can provide credible evidence for decision support in response to coastal hazards. The research presented within this special issue demonstrates recent developments and limitations in inundation modelling, morphological response modelling, storm cluster analysis and development of socioeconomic decision-support tools to better communicate risk and vulnerability to coastal communities. Through the improved understanding and awareness of storm impacts these studies aim to enable timely adaptation measures and resource allocations to be made strategically within management plans. However, practical guidance on best-practise approaches for long-term adaptive management is still needed.

Acknowledgements. The work described in this publication was brought together under the efforts of the EPSRC funded ARCoES project (grant no. EP/1035390/1), NERC National Capability funding to the NOC Marine Physics and Ocean Climate research directorate and the EPSRC Impact Accelerator Account funding to Liverpool University. P. Ciavola acknowledges support from the RISC-KIT project (EU contract 603458) and R. McCall would like to acknowledge funding by the EPSRC thought the NUPSIG project (grant no. EP/H040056/1) and Deltares (Strategic Research Programme 1209342).

References

Armstrong, J., Wilby, R., and Nicholls, R. J.: Climate change adaptation frameworks: an evaluation of plans for coastal Suffolk, UK, Nat. Hazards Earth Syst. Sci., 15, 2511–2524, doi:10.5194/nhess-15-2511-2015, 2015.

Ashton, A., Murray, B., and Arnauly, O.: Formation of coastline features by large-scale instabilities induced by high angle waves, Nature, 414, 296–299, 2001.

Brown, S., Nicholls, R. J., Vafeidism A., Hinkel, J., and Watkiss, P.: The Impacts and Economic Costs of Sea-Level Rise in Europe and the Costs and Benefits of Adaptation. Summary of Results from the EC RTD ClimateCost Project, in: The ClimateCost Project, Final Report, Vol. 1: Europe, edited by: Watkiss, P., Stockholm Environment Institute, Sweden, 44 pp., 2011.

De Winter, R. C., Sterl, A., de Vries, J. W., Weber, S. L., and Ruessink, B. G.: The effect of climate change on extreme waves in front of the Dutch coast, Ocean Dynam., 62, 1139–1152, 2012.

Dissanayake, P., Brown, J., and Karunarathna, H.: Impacts of storm chronology on the morphological changes of the Formby beach and dune system, UK, Nat. Hazards Earth Syst. Sci., 15, 1533–1543, doi:10.5194/nhess-15-1533-2015, 2015.

Fakhruddin, S. H. M., Babel, M. S., and Kawasaki, A.: Assessing the vulnerability of infrastructure to climate change on the Islands of Samoa, Nat. Hazards Earth Syst. Sci., 15, 1343–1356, doi:10.5194/nhess-15-1343-2015, 2015.

Flowerdew, J., Horsburgh, K., Wilson, C., and Mylne K.: Development and evaluation of an ensemble forecasting system for coastal storm surges, Q. J. R. Meteorol. Soc., 136, 1444–1456, 2010.

Harley, M. D., Valentini, A., Armaroli, C., Perini, L., Calabrese, L., and Ciavola, P.: Can an early-warning system help minimize the impacts of coastal storms? A case study of the 2012 Halloween storm, northern Italy, Nat. Hazards Earth Syst. Sci., 16, 209–222, doi:10.5194/nhess-16-209-2016, 2016.

Hodges, B. R.: Representing hydrodynamically important blocking features in coastal or riverine lidar topography, Nat. Hazards Earth Syst. Sci., 15, 1011–1023, doi:10.5194/nhess-15-1011-2015, 2015.

Howard, T., Pardaens, A. K., Bamber, J. L., Ridley, J., Spada, G., Hurkmans, R. T. W. L., Lowe, J. A., and Vaughan, D.: Sources of 21st century regional sea-level rise along the coast of northwest Europe, Ocean Sci., 10, 473–483, doi:10.5194/os-10-473-2014, 2014.

Knight, P. J., Prime, T., Brown, J. M., Morrissey, K., and Plater, A. J.: Application of flood risk modelling in a web-based geospatial decision support tool for coastal adaptation to climate change, Nat. Hazards Earth Syst. Sci., 15, 1457–1471, doi:10.5194/nhess-15-1457-2015, 2015.

Lapidez, J. P., Tablazon, J., Dasallas, L., Gonzalo, L. A., Cabacaba, K. M., Ramos, M. M. A., Suarez, J. K., Santiago, J., Lagmay, A. M. F., and Malano, V.: Identification of storm surge vulnerable areas in the Philippines through the simulation of Typhoon Haiyan-induced storm surge levels over historical storm tracks, Nat. Hazards Earth Syst. Sci., 15, 1473–1481, doi:10.5194/nhess-15-1473-2015, 2015.

Pescaroli, G. and Magni, M.: Flood warnings in coastal areas: how do experience and information influence responses to alert services?, Nat. Hazards Earth Syst. Sci., 15, 703–714, doi:10.5194/nhess-15-703-2015, 2015.

Perini, L., Calabrese, L., Salerno, G., Ciavola, P., and Armaroli, C.: Evaluation of coastal vulnerability to flooding: comparison of two different methodologies adopted by the Emilia-Romagna region (Italy), Nat. Hazards Earth Syst. Sci., 16, 181–194, doi:10.5194/nhess-16-181-2016, 2016.

Pine, J. C.: Enhancing the Resilience of Coastal Communities: Dealing with Immediate and Long-Term Impacts of Natural Hazards, in: Chapter 12.13 In Treatise on Estuarine and Coastal Science, edited by: Wolanski, E. and McLusky, D., Academic Press, Waltham, 271–288, 2011.

Sekovski, I., Armaroli, C., Calabrese, L., Mancini, F., Stecchi, F., and Perini, L.: Coupling scenarios of urban growth and flood hazards along the Emilia-Romagna coast (Italy), Nat. Hazards Earth Syst. Sci., 15, 2331–2346, doi:10.5194/nhess-15-2331-2015, 2015.

Sembiring, L., van Ormondt, M., van Dongeren, A., and Roelvink, D.: A validation of an operational wave and surge prediction system for the Dutch coast, Nat. Hazards Earth Syst. Sci., 15, 1231–1242, doi:10.5194/nhess-15-1231-2015, 2015.

Smolders, S., Plancke, Y., Ides, S., Meire, P., and Temmerman, S.: Role of intertidal wetlands for tidal and storm tide attenuation along a confined estuary: a model study, Nat. Hazards Earth Syst. Sci., 15, 1659–1675, doi:10.5194/nhess-15-1659-2015, 2015.

Stevens, A. J., Clarke, D., Nicholls, R. J., and Wadey, M. P.: Estimating the long-term historic evolution of exposure to flooding of coastal populations, Nat. Hazards Earth Syst. Sci., 15, 1215–1229, doi:10.5194/nhess-15-1215-2015, 2015.

Van Dongeren, A., Van Ormondt, M., Sembiring, L., Sasso, R., Austin, M., Briere, C., Swinkels, C., Roelvink, D., and Van Thiel de Vries, J.: Rip current predictions through model data assimilation on two distinct beaches, Proceedings of the 7th International Conference on Coastal Dynamics, Bordeaux, France, 1775–1786, 2013.

Wadey, M. P., Brown, J. M., Haigh, I. D., Dolphin, T., and Wisse, P.: Assessment and comparison of extreme sea levels and waves during the 2013/14 storm season in two UK coastal regions, Nat. Hazards Earth Syst. Sci., 15, 2209–2225, doi:10.5194/nhess-15-2209-2015, 2015.

10

The quantitative estimation of the vulnerability of brick and concrete wall impacted by an experimental boulder

J. Zhang[1,2], Z. X. Guo[2], D. Wang[2,3], and H. Qian[4]

[1]School of Energy and Power Engineering, Key Laboratory of Fluid and Power Machinery, Ministry of Education, Xihua University, Chengdu, 610039, China
[2]State Key Laboratory of Hydraulics and Mountain River Engineering, Sichuan University, Chengdu, 610065, China
[3]Highway Planning Survey & Design Institute, Sichuan Communications Department, Chengdu, 610041, China
[4]Chengdu Municipal Engineering Design and Research Institute, Chengdu, 610015, China

Correspondence to: Z. X. Guo (scugzx@scu.edu.cn)

Abstract. There is little historic data about the vulnerability of damaged elements due to debris flow events in China. Therefore, it is difficult to quantitatively estimate the vulnerable elements suffered by debris flows. This paper is devoted to the research of the vulnerability of brick and concrete walls impacted by debris flows. An experimental boulder (an iron sphere) was applied to be the substitute of debris flow since it can produce similar shape impulse load on elements as debris flow. Several walls made of brick and concrete were constructed in prototype dimensions to physically simulate the damaged structures in debris flows. The maximum impact force was measured, and the damage conditions of the elements (including cracks and displacements) were collected, described and compared. The failure criterion of brick and concrete wall was proposed with reference to the structure characteristics as well as the damage pattern caused by debris flows. The quantitative estimation of the vulnerability of brick and concrete wall was finally established based on fuzzy mathematics and the proposed failure criterion. Momentum, maximum impact force and maximum impact bending moment were compared to be the best candidate for disaster intensity index. The results show that the maximum impact bending moment seems to be most suitable for the disaster intensity index in establishing vulnerability curve and formula.

1 Introduction

After Wenchuan Earthquake on 12 May 2008, several catastrophic earthquake events in high magnitude (>6.5) occurred in China recently. For example, Yushu earthquake in Qinghan on 14 April 2010; Lushan earthquake in Ya'an, Sichuan on 20 April 2013; Ludian earthquake in Zhaotong, Yunnan on 3 August 2014 (Earthquake in China, http://www.ceic.ac.cn/). Huge volumes of sediment deposits and debris sources have been induced by earthquakes and contributed to new debris flow in higher frequency magnitude leading to much loss, both in life and economics (Tang et al., 2011b). Risk management is popularly applied to relieve or prevent debris flow disaster. Quantitative vulnerability estimation is a necessary element of risk estimation. During decades, however, the vulnerability research of debris flow had a slow development for several reasons. First, it is widely accepted that the behavior of debris flow is quite complicated and has not been clearly revealed yet (Rickenmann, 1999; Cui, 2009). Second, there are many kinds of vulnerable elements potentially affected by debris flow, and the indicator system involves many influence factors not only in natural vulnerability but also in social vulnerability (for example, economics, environment and human life) (Liu et al., 2012). Third, historic data are of much use for estimating vulnerability with a statistical approach, or for improving the accuracy of results (Totsching et al., 2011; Jakcob et al., 2012). Unfortunately, most of the databases and reports are about characteristics of debris flows but not on the vulnerability of the elements at risk (Tang et al., 2011a; Zhang et al., 2013). Fourth, the lack of data in vulnerability research is also relevant to the little attention from both engineers and scientists in past years. Unlike earthquakes and cyclones, local structural protection measures are quite applicable to decrease the damages caused by debris flows (Douglas, 2007).

There are mainly four evaluation methods for the vulnerability of the elements in debris flow: multi-index assessment, element value accounting, empirical vulnerability curve and physical experiment. Among these, the first two methods were widely applied in China for the regional vulnerability estimation based on the category and market value of affected elements. These methods are applicable for the regions that have a vast territory or high density of vulnerable elements. However, they fail to connect with debris flow intensity (Liu et al., 2001; Tang et al., 2005; Tie, 2009). Empirical vulnerability curves are applied mainly by European researchers. The curves are fitted with empirical data including the deposit height of debris flows and the vulnerability of investigated elements (Papathoma et al., 2011, 2012). At first, vulnerability assessment had a purely qualitative character, but after soon people realized that only quantitative data could lead effective evaluation (Bell and Glade, 2004; Romang, 2004; Michael et al., 2003). The first curve was provided by Fuchs et al. (2007) according to the debris flow event occurred in the Wartschenbach catchment in the eastern Alps, next to the city of Lienz, Austria. Totschnig et al. (2011) adopted a modified Frechet no. 2 distribution instead of polynomial to fit the numerous data extracted from three databases. Kinematic velocity and impact pressure could also be the disaster intensity factors when the numerical simulation of debris flow was conducted (Luna et al., 2011).

So far, vulnerability curves of debris flow were mostly established based on the historic data. However, few useful data could be applied for vulnerability assessment in China since the main attention of engineering and government paid on the disasters themselves and resettlement of disaster-affected elements. Therefore, physical experiments can be an alternative method. The primary damage of building occurred during the deposit process due to the debris flow impact. Hu et al. (2011) monitored the impact force caused by the real-scale debris flow at Jiangjia Ravine, China and exhibited the force distribution along the flow height. Bugnion et al. (2012) and Scheidl et al. (2012) analyzed the debris flow impact model based on field-scale flows and small-scale flows, respectively. They both provided the precious data for flow impact process. Canelli et al. (2012) took the material of impacted elements into account and discussed the different effects of the type of barrier on flow impact. Although some impact research of debris flow has been conducted, most of it was based on the purpose of optimizing the mitigation measure for debris flow. Therefore, they focused on the dynamics of debris flow impact rather than the response of damaged elements, and the force plate was always utilized as the substitute for impacted elements. Thus, it is hard to draw vulnerability curves from the prior flow-impact research. Borrowing the evaluation models of earthquake did not work well probably because the destruction mechanisms of the elements induced by these two disasters are different (HAZUS, 2006; Haugen and Kaynia, 2008).

Brick and concrete buildings are the typical civil architecture in the southwestern mountain area of China. Generally, the destruction of load-bearing walls mainly leads to the collapse of the building. Zhang (2005) studied the ultimate load-bearing capacity of brick and concrete wall in 1 : 2 scale impacted by iron spheres which were took as the substitute of debris flow. The iron spheres had the weights from 27.4 to 41.4 kg and the equivalent falling height was 2.9 m representing the potential energy that transformed to the horizontal kinetic energy. Therefore, the kinetic energy of the substitute impacting on the walls varied from 794.6 to 1200.6 $\mathrm{kg\,m^{-2}\,s^{-2}}$. There are two problems if the vulnerability curve is going to be drawn from this kind of experiment:

1. vulnerability curves contain various damage degrees of the building, while, in her research, only the destruction status was considered;
2. since the similarity law between load and geometry in structure impact experiments is unknown, the vulnerability evaluation cannot be applied in prototype.

2 Experiment description

2.1 Assumption

In this study, experiments were conducted in order to obtain the vulnerability curve for estimating the brick and concrete building with one storey. Since the collapse of building was mainly caused by the destruction of the load-bearing walls, the vulnerability assessment of the load-bearing wall of impacted building is rather representative and important.

Debris flow consists of slurry and particles. Slurry produces distributed load on elements while particles, of which the size varies in great range, produce concentrated load, and the caused damage resembles that of rock fall events (Mavrouli and Corominas, 2010). Simulating experimental or real-scale debris flow with a specific momentum is rather difficult, experimental boulders are feasible to be given certain momentum depending on their mass and falling heights or velocity. However, compared to debris flow, boulder impact produces more serious damage within the contact area because of stress concentration (e.g., a through hole). Iron board and rubber cushion were applied in Zhang's (2005) experiments to transmit the concentrated force induced by an iron sphere to distributed force which was finally exerted on the wall. According to her experiments, the wall was loaded a delta impulse which was similar to the impact loading in the field observation of the real-scale debris flow in Jiangjia

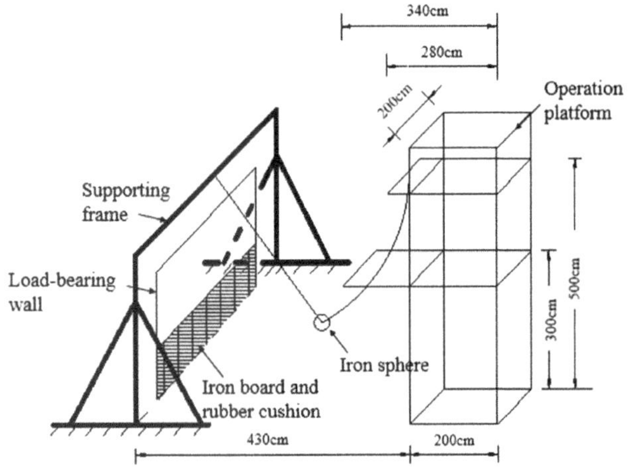

Figure 1. Experiment setup.

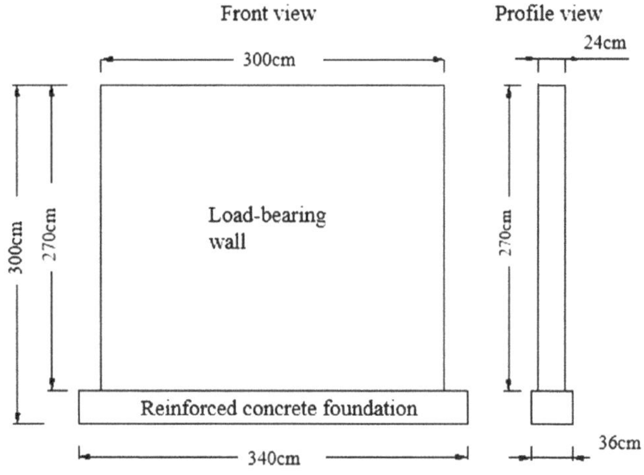

Figure 2. Sketch of load-bearing wall.

Ravine in China on 25 August 2004 (Hu et al., 2011): at first the impact load increased to the peak rapidly then decayed gradually in relatively long time with fluctuation. However, the impact process in field continues for more than 10 s, and the maximum load fluctuated in approximate 5 s since the flow impacted the obstacle constantly for a while. In experimental conditions, the impact process associated with an iron sphere was short and had a single peak value. If taking the real-scale debris flow as a stacking process consisted of every debris flow in unit time, then there should be several peak values that represent the fluctuation of maximum load. Furthermore, if a debris flow in unit time was viewed as an overall, an iron sphere with a certain motion energy can be applied to simulate it under laboratory condition. Additionally, a delta shape signal also observed in the impact process caused by debris flow model in other research (Scheidl et al., 2012; Bugnion et al., 2012). As a result, it is assumed that the impact process induced by an experimental boulder with an iron board and a rubber cushion can represent the impact process induced by debris flows.

Generally, the impact force is $F \cdot \cos\alpha$, in which α is the intersection angle of the impact force and wall surface. In this study, debris flow is assumed to impact the wall vertically ($\alpha = 90°$). Then width of the contact area of flow and elements is always equal to the width of the wall.

2.2 Experiment setup

Based on the assumptions mentioned above, the experiments used an iron sphere as an experimental boulder to create the concentrated load which was further distributed with an iron board and a rubber cushion. At the beginning of the experiments, the iron sphere, which was jointed to the top of supporting frame with a chain, was dragged by the dynamic system up to a certain height of operation platform. When the system power was off, the sphere would fall in a circle under gravity force and then hit the iron board in front of the wall. The impact point was at the geometrical center of the boards. Since the iron board was rigid, the concentrated load could spread onto the rest of the wall area covered by the board due to the displacement of the overall board. Hence, the concentrated load associated with iron sphere was transmitted to distributed load on walls. The boards represent the contact area of debris flow, and the wall since debris flow is probably wider than the wall. Therefore, the board height was the flow depth, and the board width was the average width of the contact area, which was equal to the wall width in all tests. A rubber cushion was set between the board and wall to delay the impact duration. The experiment setup is shown in Fig. 1. The operation platform (see Fig. 1) provided two falling heights (3 and 5 m) for spheres. The support frame mounted on the ground was welded with steel. The iron sphere releasing from a certain height rotated with the circle center fixed on the beam of the frame. The debris flow in various magnitudes was simulated by releasing different spheres from different heights. In the experiments, the length of the chain and the height of the iron board were both adjustable.

The load-bearing walls in the experiments were designed and constructed according to the Chinese Brick and Concrete Structural Design Specifications (Construction Industry of China, 2011) and in alignment with the structure of the buildings mostly used in the southwestern mountain area of China. The walls were 240 mm thick, 3.0 m high and 3.0 m wide. The net height of the walls was 2.7 m, and the foundation, which was 0.3 m deep, was made with reinforced concrete (see Fig. 2). The material of the walls was stated as follows: the brick was 240 mm × 115 mm × 53 mm and was bought from a factory. The compressive strength of the brick was 10 MPa. The mortar made of water, cement and sand was in standard curing for 28 days (temperature = $(20 \pm 2)°$, relative humidity ≥ 90 %). The test cubes in the same curing condition for 29 days had the average compressive strength

8.2 MPa. Make the surface of the walls as smooth as possible so that the board will attach perfectly to the wall.

2.3 Measurement device

An impact force gauge system including a sensor and a peak force instrument was applied to obtain the force suffered by the load-bearing wall during the impact process. At first, the sensor fixed onto the back of the wall changed the weight signal of force to electronic signal; then the electronic signal was sent to the digital display on the instrument; finally, the impact force graph during the whole process could be read and recorded by the computer. The sample frequency was 100 Hz. Maximum dynamic and static displacement were measured with a self-made displacement gauge. The inclination of the wall i is the ratio of the maximum dynamic displacement L_{dd} and the wall height H. The cracks with different widths on the wall could be identified by a crack-comparing ruler with accuracy 0.1 mm.

2.4 Conditions

According to the historic debris flow events in China, the parameters of debris flow vary in a wide range (Generally, the density ρ is 1.2–2.3 g cm^{-3}; the velocity v is 3–10 m s^{-1}; the flow depth h is 0–10 m) (Cui et al., 2005; Tang et al., 2011b). The destructive ability of debris flows depends not only on their velocity but also volume and density. Momentum involves all these parameters and would be a good index to represent the disaster intensity of flow. The momentum of debris flow in unit time can be calculated with $m\,v = \rho\,b\,h\,v^2$ in which m, ρ, b, h and v are the mass quality, density, width and velocity of debris flow, respectively. In view of the size of lab site and the sufferance of the supporting frame, the weights of iron spheres are 49 and 86 kg determined by back stepping with the platform height (assuming the density of viscous debris flow ρ is 2.2 g cm^{-3}). The spheres are made of iron with density 7.8 g cm^{-3}. The heights of the boards in front of the walls are 1, 1.5 and 2 m. There are nine total experiments numbered as A1, A2, A3, B1, B2, B3, C1, C2 and C3. Details on the experimental programme are summarized in Table 1.

3 Estimation method

3.1 Failure criterion

Since it has been proven that the failure mode of a load-bearing wall is out-of-plane bending failure, similar to the static load condition (Zhang, 2005), cracks and inclination are important indicators of the failure of load-bearing walls impacted by debris flow. Additionally, several damage classifications of brick and concrete building from specifications of different industries (coal mining and construction) are also considered (Qian, 2013). Then, the failure criterion for the load-bearing wall impacted by debris flow is established in Table 2. However, directly applying this criterion will lead the unreasonable results when the value around the critical number is judged. For example, there is no significant different between the damaged element with maximum crack length 750 mm and the element with 751 mm. However, the damage assessment of these two elements belongs to different classes according to Table 2. In this case, fuzzy mathematical theory is helpful to solve this problem.

3.2 Estimation method based on fuzzy mathematics

Based on fuzzy mathematics, the procedure of the vulnerability estimation of wall is stated as follows.

3.2.1 Single index evaluation

$U = \{a, b, c, d\}$ is the influence indicators aggregate and a, b, c, d denote the influence indicators listed in Table 2 ,respectively. $T = \{\mathrm{I}, \mathrm{II}, \mathrm{III}, \mathrm{IV}\}$ is the damage results aggregate. Then, the fuzzy relation between influence indicators and damage results can be represented with evaluation matrix $\mathbf{R}$:

$$\mathbf{R} = \begin{bmatrix} r_{11} & r_{12} & r_{13} & r_{14} \\ r_{21} & r_{22} & r_{23} & r_{24} \\ r_{31} & r_{32} & r_{33} & r_{34} \\ r_{41} & r_{42} & r_{43} & r_{44} \end{bmatrix}, \tag{1}$$

in which $r_{mn} = \mu_{T_n}(U_{\mathrm{m}})$ $(0 \le r_{mn} \le 1, 1 \le m \le 4, 1 \le n \le 4)$ denotes the membership degree of the result element T_n from the view of the indicator element U_m, and $\mathbf{R}_m = (r_{m1}, r_{m2}, r_{m3}, r_{m4})$ is the assessment aggregate of U_m and also the fuzzy subset of T. The membership function $\mu_{T_n}(U_m)$ has the formulas below.

$$\mu_{\mathrm{I}}(U_m) = \begin{cases} 1, & U_m \le k_{m1}; \\ (k_{m2} - U_m)/(k_{m2} - k_{m1}), & k_{m1} < U_m \le k_{m2}; \\ 0, & \text{otherwise} \end{cases} \tag{2}$$

$$\mu_{\mathrm{II}}(U_m) = \begin{cases} (U_m - k_{m1})/(k_{m2} - k_{m1}), & k_{m1} < U_m \le k_{m2}; \\ (k_{m3} - U_m)/(k_{m3} - k_{m2}), & k_{m2} < U_m \le k_{m3}; \\ 0, & \text{otherwise} \end{cases} \tag{3}$$

$$\mu_{\mathrm{III}}(U_m) = \begin{cases} (U_m - k_{m2})/(k_{m3} - k_{m2}), & k_{m2} < U_m \le k_{m3}; \\ (k_{m3} - U_m)/(k_{m3} - k_{m2}), & k_{m3} < U_m \le k_{m4}; \\ 0, & \text{otherwise} \end{cases} \tag{4}$$

$$\mu_{\mathrm{IV}}(U_m) = \begin{cases} (U_m - k_{m3})/(k_{m4} - k_{m3}), & k_{m3} < U_m \le k_{m4}; \\ 1, & U_m > k_{m4}; \\ 0, & \text{otherwise} \end{cases} \tag{5}$$

In these matrices, $\mathbf{K}_m = (k_{m1}, k_{m2}, k_{m3}, k_{m4})$ denotes the membership matrix of an influence indicator and k_{m1}–$k_{m4}$are the evaluation parameters of the four damage classes, respectively. The parameter adopts the mean value of the adjacent critical number. For example, the matrix of the maximum crack width $\mathbf{K}_1 = (0.1, 0.3, 0.55, 0.8)$. Thus, the evaluation matrix $\mathbf{R}$ can be obtained by Eqs. (2)–(5).

3.2.2 Weight determination

The weight determination mostly considers the over limit of indicators. Therefore, the weights of the indicators are ob-

Table 1. Conditions of the experiments.

No.	A1	A2	A3	B1	B2	B3	C1	C2	C3
Height of board (m)	1	1	1	1.5	1.5	1.5	2	2	2
Height of falling (m)	3	3	5	3	3	5	3	3	5
Weight of sphere (kg)	49	86	86	49	86	86	49	86	86

Table 2. Failure criterion for the load-bearing wall impacted by a single iron sphere.

Damage class	Maximum crack width (mm)	Maximum crack length (mm)	Total area of cracks (mm^2)	Inclination (10^3)	Damage description	Required repair
I	0–0.2	0–750	0–500	0–1	slight	simple
II	0.2–0.4	750–1500	500–1000	1–1.5	minor	minor
III	0.4–0.7	1500–2250	1000–2000	1.5–2	mediate	mediate
IV	> 0.7	2250–3000	2000–3500	2–2.5	serious	thorough repair or rebuild

tained by Eq. (6) and compose the weight matrix $\mathbf{A} = [W_a, W_b, W_c, W_d]$.

$$W_{U_\mathrm{m}} = \frac{P_{U_\mathrm{m}}/S_{U_\mathrm{m}}}{\sum_{m=1}^{4}\left(P_{U_\mathrm{m}}/S_{U_\mathrm{m}}\right)} \tag{6}$$

In this matrix, P_{U_m} is the measured value of influence indicators; S_{U_m} is the mean value of all critical numbers. For example, the critical numbers of the maximum crack width are 0.2, 0.4 and 0.7 mm, then $S_a = (0.2 + 0.4 + 0.7)/3 = 0.433$ mm.

4 Vulnerability assessment

Multiply the matrix **A** and **R**, then generate a new matrix $[x_1, x_2, x_3, x_4]$, in which $x_1 + x_2 + x_3 + x_4 = 1$. x_1, x_2, x_3, x_4 represent the membership degrees of indicators to the damage classes (I, II, III and IV), respectively. The loss percentage l_n for different types of damage condition is defined as below:

a. slight damage: 0–10 %, $l_1 = 10$ %;

b. minor damage: 10–30 %, $l_2 = 30$ %;

c. mediate damage: 30–60 %, $l_3 = 60$ %;

d. serious damage or collapse: 60–100 %, $l_4 = 100$ %.

As shown above, the upper limit of damage loss is defined as the loss percentage in every damage class so that the estimation will overestimate the element loss. Finally the vulnerability assessment is determined as the following:

$$V = \sum_{n=1}^{4} (l_n \cdot x_n), \tag{7}$$

in which V denotes the vulnerability of elements, namely the damage loss percentage of elements.

5 Results and analysis

5.1 Damage description

Three types of loads are discussed as follows to be the candidates of disaster intensity index for vulnerability curve.

1. Momentum – The velocity and flow depth are the basic physical descriptors of debris flow in unit width. Momentum includes both two descriptors and represents flow energy.

2. Maximum impact force – The overload of structure material is an important cause of structure failure. Therefore, maximum impact force should be taken into the consideration.

3. Maximum impact bending moment – As mentioned above, the out-of-plane bending failure was observed in previous studies as the failure mode of brick and concrete wall. The impact bending moment also contains the velocity and flow depth. Therefore, it is reasonable to consider the maximum impact bending moment as the candidate of disaster intensity index.

In order to choose the most suitable disaster intensity index from these candidates, analysis and comparison are conducted with the experimental results. The momentum, maximum impact force and maximum impact bending moment under each experimental condition are listed in Table 3.

Figures 3–5 show the cracks in series A, B and C experiments, respectively. The crack distribution is analyzed from two aspects – under the same board height and the same falling height. Directly impacted place denotes the wall area covered by board, while the rest area of the wall is indirectly impacted place.

Table 3. Loads in experiments.

Type of load	A1	A2	A3	B1	B2	B3	C1	C2	C3
Momentum ($\mathrm{kg\,m^{-1}\,s^{-1}}$)	329	577.4	789.5	310.6	545.1	766.2	291	510.8	742.2
Maximum impact force (kN)	39	45	102.3	20.6	23.2	45.8	29.4	37.5	80.7
Maximum impact bending moment ($\mathrm{kN\,m^{-1}}$)	18	22.5	51.15	15.45	17.4	34.35	29.4	37.5	80.7

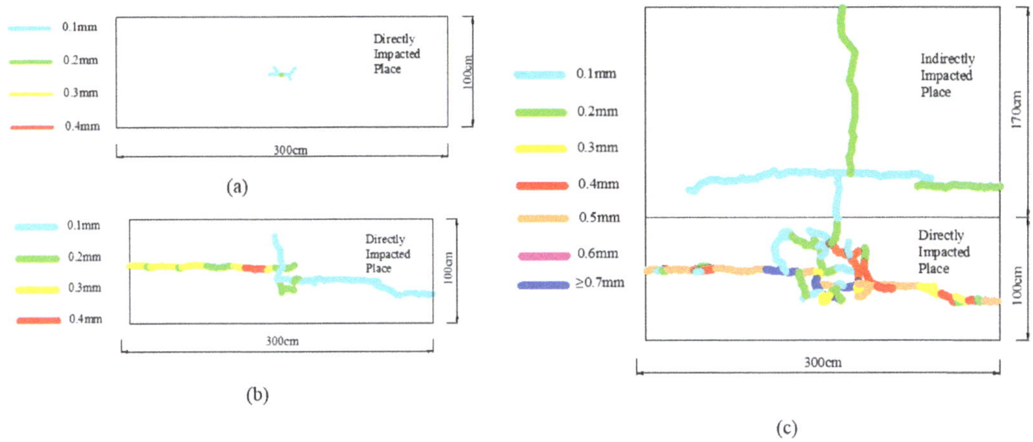

Figure 3. Distribution of the cracks on the element in series A (**a**, **b** and **c** represent the A1, A2 and A3, respectively).

5.1.1 The same height of the iron board

Taking the 1.0 m height as an example, the crack distributions of series A are shown in Fig. 3 (different motion energies or flow velocities). In A1, A2 and A3, the ratio of maximum impact force is 1 : 1.15 : 2.62, the ratio of maximum bending moment is 1 : 1.25 : 2.84, and the ratio of momentum is 1 : 1.76 : 2.4. Under the same flow depth, the crack width, length and quantity increase when the load acting on the wall increases. In the A1 experiment, the wall still maintains completeness except for several short and tiny cracks, among which the maximum width is 0.2 mm, and the maximum length is 273 mm. In the A2 experiment, the cracks spread wider in horizontal direction. A part of the wall dropped the surface layer due to the deformation and tremor. Most of the cracks are 0.1–0.2 mm width. The maximum crack width is 0.4 mm, and the maximum length is 2022 mm. In the A3 experiment, the cracks spread onto the area above the iron broad. Both the cracks in horizontal and vertical direction extend constantly to the edge of the wall. Thus, the wall has major visible deformation in profile, and the surface layer dropped significantly from the wall. The maximum crack width is 1 mm, and the maximum length is 3216 mm. For the experiments with iron board heights of 1.5 m (B1, B2 and B3) and 2.0 m (C1, C2 and C3) experiments, the results of the crack development in relation to impact load are similar.

5.1.2 The same falling height

Taking the iron sphere with a weight of 86 kg and falling from 3 m height for example (different board heights or flow depths), the crack distribution of A2, B2 and C2 experiment is shown in Figs. 3b, 4b and 5b, respectively. The measured impact force in these three experiments is 45, 23.2 and 28 kN, respectively. It is found that the measured data of series B are lower than the expected value. There are at least two factors that can influence the load impacting on the wall: first, since the experiments were conducted in an outdoor lab, the weather (for example wind) can accelerate or decelerate the velocity of the iron sphere depending on their relative movement direction; second, due to the device malfunction, the tractive force between the sphere and the dynamic system did not vanish completely until the sphere departed from the system for a certain distance. This residual tractive force decreased the kinetic energy which the sphere possessed before the impact. Therefore, the spheres did not get the expected impact energies in the experiments of series B. However, the relevant damage indicators and the vulnerability also were lower than the expected value and cohered with the actual impact loads. This deviation will not influence the data analysis if only the measured data are used to establish the vulnerability curve. Table 4 includes the measured data of the indicators and the final vulnerability results based on the proposed method. From the comparison of the three figures, the total area and maximum crack length in C2 experiment exceed those in the A2 experiment, even though the impact force in

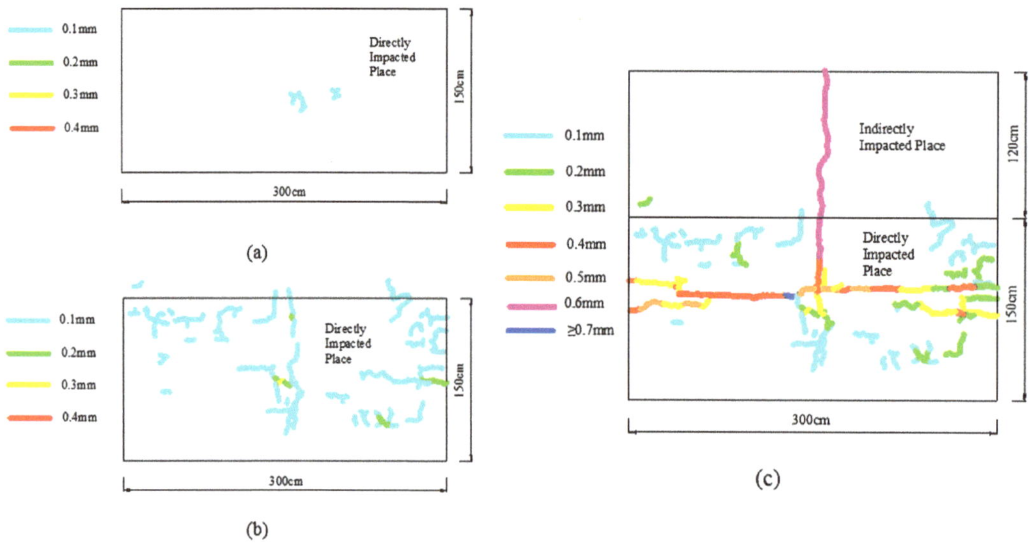

Figure 4. Distribution of the cracks on the element in series B (**a**, **b** and **c** represent the B1, B2 and B3, respectively).

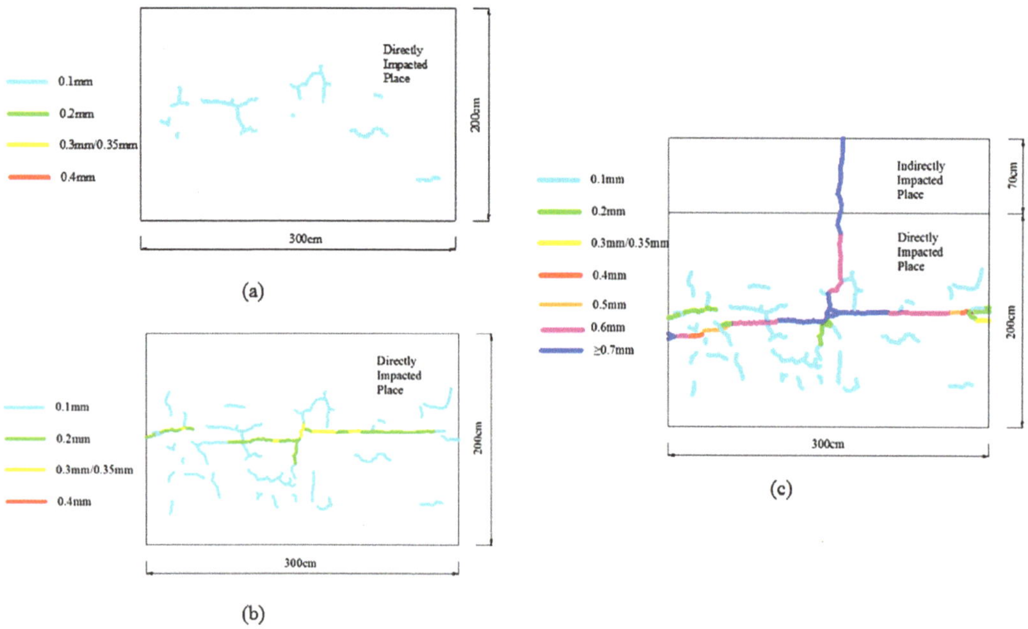

Figure 5. Distribution of the cracks on the element in series C (**a**, **b** and **c** represent the C1, C2 and C3, respectively).

A2 is 1.6 times larger than that in C2. The vulnerability in C2 is 1.5 times larger than that in A2, and the dynamic displacement in A2 and C2 is 3.8 and 5.5 mm, respectively. Therefore it might be deduced that the cracks are mainly caused by the dynamic displacement when the wall is swinging under the impact force.

5.2 Vulnerability curve

Based on the failure criterion in Table 2, the indicators of the experiments are collected in Table 4 below.

Assuming a piecewise linear function to represent vulnerability curve, therefore the curve consists of three lines. (1) At the first section of vulnerability, the value is a constant 10 %. (2) The vulnerability linearly varies with the disaster intensity factor when the loss percentage is between 10 and 100 %. (3) The vulnerability stays at 100 %, even if the loads increase. Arrange the vulnerability results in relation to momentum, maximum impact force and maximum bending moment, respectively, and plot them in coordinate system (see Figs. 6–8).

Table 4. Statistical table of the influence indicators.

No.	Cracks			Inclination	Maximum	Vulnerability	Crack description
	Maximum width (mm)	Maximum length (mm)	Total area (mm^2)	(10^3)	dynamic displacement (mm)		
A1	0.2	273	45.1	0.89	2.4	18.4 %	a few and tiny
A2	0.4	2022	611.2	1.41	3.8	48.2 %	more and wider
A3	1	3267	2944.3	5.00	13.5	100.0 %	constantly extend to the edge; spread out of the board
B1	0.1	204	48.9	0.78	2.1	14.0 %	a few and tiny
B2	0.3	984	961.2	1.22	3.3	31.4 %	more and wider
B3	0.7	3216	4196	1.00	2.7	90.6 %	constantly extend to the edge; spread out of the board
C1	0.1	591	303.1	1.22	3.3	21.0 %	separated and tiny
C2	0.3	2475	1296	2.04	5.5	70.9 %	more and wider
C3	1.1	3208	4192.7	1.96	5.3	96.8 %	constantly extend to the edge; spread out of the board

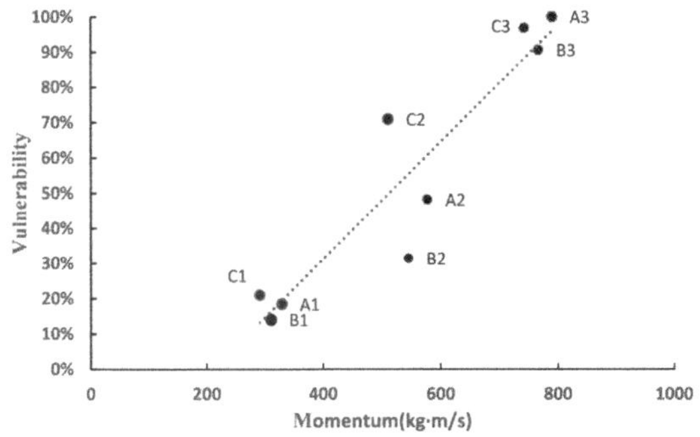

Figure 6. Vulnerability scatter chart with momentum.

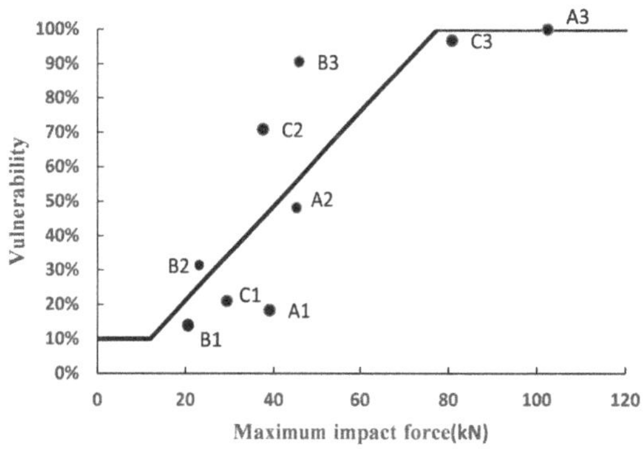

Figure 7. Vulnerability curve with maximum impact force.

5.2.1 Momentum

From Fig. 6, it can be found that vulnerability has a linear relation with momentum, and the data with the same momentum cluster together. However, the momentum here is just theoretical value and energy loss exists during the impact process. For example, the inelastic collision between the sphere and the board consumes the energy; the rubber cushion between the board and the wall also absorbs a part of the energy. Therefore, the actual momentum on the wall is less than the theoretical value. Unfortunately, no function can be utilized to calculate the actual momentum. As a result, the momentum in Fig. 6 is not reliable to establish the function of vulnerability.

5.2.2 Maximum impact force

The piecewise linear function with maximum impact force is written as Eq. (8) and the relation coefficient r is 0.78. The corresponding fitting curve is drawn in Fig. 7. From Fig. 7, it can be observed that the data are not closely clustered around the curve.

$$\begin{cases} V = 0.1 & F \leq 11.87\,\text{kN} \\ V = 0.014F - 0.063 & 11.87\,\text{kN} < F \leq 77.34\,\text{kN} \\ V = 1 & F > 77.34\,\text{kN} \end{cases} \tag{8}$$

5.2.3 Maximum bending moment

Maximum bending moment is defined as the multiplication of maximum impact force F and the arm of resultant force $L = h/2$. The piecewise linear function with maximum bending moment is written as Eq. (9) and the corresponding fitting curve is drawn in Fig. 8. The relation coefficient is 0.87 which is higher than the coefficient with maximum impact force. From the comparison of these two figures, it can be observed that the curve with bending moment fits better with the data. If the wall is not entirely destroyed because of the impact force, the wall swings producing the dynamic displacement due to bending moment since the reinforced concrete foundation is immobile. Deformation can be partly recovered. The cracks and the static displacement are the unrecovered deformation, and the cracks are caused by the tension stress during the swing. As a result, the maximum bending moment is more likely to be the disaster intensity factor in the vulnerability estimation of brick and concrete wall.

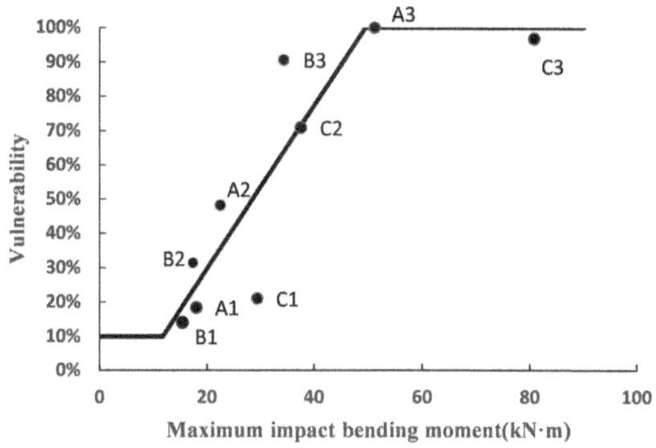

Figure 8. Vulnerability curve with maximum impact bending moment.

$$\begin{cases} V = 0.1 & F \cdot L \leq 11.75\,\text{kN}\,\text{m}^{-1} \\ V = 0.024(F \cdot L) - 0.18 & 11.75\,\text{kN} \cdot m < F \cdot L \leq 49.46\,\text{kN}\,\text{m}^{-1} \\ V = 1 & F \cdot L > 49.46\,\text{kN}\,\text{m}^{-1} \end{cases} \quad (9)$$

6 Discussion

The impact research of debris flow is mostly related to the characteristics of flow itself in the past. As for the elements potentially threatened by debris flow, little information can be applied to estimate or predict their vulnerability. The estimation method based on fuzzy mathematics allows the quantitatively estimation on the walls built with brick and concrete. The damage loss percentage of the walls can be calculated by collecting the indicators in Table 2. With the derived vulnerability curve and formula, it is convenient to predict the vulnerability of the walls if the dynamic characteristics of debris flow are known. The application of the results is benefit for the insurance to be a popular kind of mitigation measure for debris flow. However, these experimental results are limited to the assumptions of experiments. Therefore, modification is necessary when the results are used for the impact event beyond the scope of assumptions. Since the quantity of the data in this study is small, the derived vulnerability curve needs more field or experimental data to promote the precision. Additionally, several problems about the experiment setup have to be clarified as below.

According to the momentum theorem ($m\,v = f \cdot t$), collision duration which depends on the characteristics of material (such as elastic modulus and Poisson ratio) governs the value of impact force when the momentum remains the same. Therefore, the impact force of iron materials is 2 times larger than rock materials (or concrete materials) due to the difference of elastic modulus. In the experiments, both the experimental boulders and the boards are made of iron so the collision will create great impact force beyond the value induced by debris flow (including slurry and particles) and concrete. The composite medium consisted of rubber cushion and iron board has an elastic modulus between iron and rubber since the elastic modulus of rubber is rather small. With the increasing thickness of rubber cushion, the elastic modulus of composite medium and the force acting on the wall decrease. Theoretically, the elastic modulus of the medium should be adjusted to cohere with historic data through preliminary experiments. Unfortunately, rare data of impact force was recorded in prototype since debris flow always occurred abruptly.

Since the rigidity of the iron board is not infinite, the board does not move as an overall. Therefore, the concentrated load cannot be transmitted to homogenous distributed load. As for the particles of debris flow, their location is random at the cross section of flow. As for the slurry of debris flow, both hydrostatic and hydrodynamic characteristics should be taken into consideration when determining the pressure application point. Thus, the load induced by debris flow is the resultant of the concentrated load with particles and the distributed load with slurry and it also deviates from homogenous distributed load. From Figs. 3–5, it seems that the cracks distributed dispersedly under most experimental conditions (A1, A2, B1, B2, C1 and C2) and serious damage was observed to occur around the hard contact under relatively high load. In addition, the damage degree of the wall was expressed from slight to serious under different loads. On the whole, the experiments set up can simulate the impact of debris flow properly.

The influence of transversal wall and rooftop around the load-bearing wall was not taken into consideration and both of them can provide constraint force. Since the displacement and swing are restrained, the actual loss percentage will be lower than the vulnerability evaluated based on this approach. The metal frame mounted around the wall can be applied to simulate this constraint force which should be determined by numerical or experimental test.

7 Conclusions

It is assumed that the impact process induced by an experimental boulder with an iron board and a rubber cushion can represent the impact process induced by debris flows since they both exert delta impulse loads on elements. The experimental results demonstrate that the maximum impact bending moment is the principal cause of the wall damage and is more suitable to be the disaster intensity index in the vulnerability curve of the wall impacted by debris flow compared with the maximum impact force. The results also verify the conclusion of Zhang's (2005) research further, and in turn it also proves the reliability of the results. Since it takes a long time to cure the concrete, and the construction of the models is rather costly, the experimental data are limited but very precious to the vulnerability research of debris flow. The curve and formula proposed above need more field or experimental data to be more reliable. Vulnerability study is much concern on the type of the vulnerable element. The curve and formula need modification when they are applied on brick and concrete wall elements with different construction details.

Author contributions. J. Zhang, H. Qian and Z. X. Guo designed the experiments. H. Qian carried out the experiments and processed the data. J. Zhang prepared the manuscript with contributions from all co-authors.

Acknowledgements. The authors thank the support from the National Natural Science Foundation of China (No. 51409224) and the open fund program of SKLH, Sichuan University, China (No. SKHL1312). The research has also been supported by the Key Laboratory of Fluid and Power Machinery, Ministry of Education, Xihua University.

References

Bell, R. and Glade, T.: Quantitative risk analysis for landslides – Examples from Bíldudalur, NW-Iceland, Nat. Hazards Earth Syst. Sci., 4, 117–131, doi:10.5194/nhess-4-117-2004, 2004.

Bugnion, L., McArdell, B., Bartelt, P., and Wendeler, C.: Measurements of hillslope debris flow impact pressure on obstacles, Landslides, 9, 179–187, doi:10.1007/s10346-011-0294-4, 2012.

Canelli, L., Ferrero, A. M., Migliazza, M., and Segalini, A.: Debris flow risk mitigation by the means of rigid and flexible barriers – experimental tests and impact analysis, Nat. Hazards Earth Syst. Sci., 12, 1693–1699, doi:10.5194/nhess-12-1693-2012, 2012.

Construction Industry of China: Specification of the design of masonry structures, Beijing, 2011.

Cui, P.: Advances in debris flow prevention in China, Sci. Soil Water Conserv., 7, 7–13, 2009.

Cui, P., Chen, X., Wang, Y., Hu, K., and Li, Y.: Jiangjia Ravine debris flows in south-western China. Debris-flow Hazards and Related Phenomena, Springer Praxis Books, Springer, Berlin, Heidelberg, 565–594, 2005.

Douglas, J.: Physical vulnerability modelling in natural hazard risk assessment, Nat. Hazards Earth Syst. Sci., 7, 283–288, doi:10.5194/nhess-7-283-2007, 2007.

Fuchs, S., Heiss, K., and Hübl, J.: Towards an empirical vulnerability function for use in debris flow risk assessment, Nat. Hazards Earth Syst. Sci., 7, 495–506, doi:10.5194/nhess-7-495-2007, 2007.

Haugen, E. D. and Kaynia, A .M. (Eds.): Vulnerability of structures impacted by debris flow, Taylor & Francis Group, London, 2008.

HAZUS: Multi-hazard Loss Estimation Methodology, Earthquake Model, HAZUS MH-MR2, User Manual, FEMA and NIBS, USA, 2006.

Hu, K.,Wei, F., Li, Y.: Real-time measurement and preliminary analysis of debris flow impact force at Jiangjia Ravine, China, Earth Surf. Proc. Land., 36, 1268–1278, doi:10.1002/esp.2155, 2011.

Jakob, M., Stein, D., and Ulmi, M.: Vulnerability of buildings to debris flow impact, Nat. Hazards, 60, 241–261, doi:10.1007/s11069-011-0007-2, 2012.

Liu, K. F., Lee, H. C., and Wu, Y. H.: Assessment social impact of debris flow disaster by social vulnerability index, in: Kyoto Conference Proceeding, Kyoto, 24–31, 2012.

Liu, X. L., Mo, D. W., and Wang, X. D.: Regional vulnerability assessment of debris flows, Chinese J. Geol. Hazard Control, 12, 7–12, 2001.

Luna, Q. B., Blahut, J., Westen, J. V., Sterlacchini, S., Asch, T. W. J. V., and Akbas, S. O.: The application of numerical debris flow modelling for the generation of physical vulnerability curves, Nat. Hazards Earth Syst. Sci., 11, 2047–2060, doi:10.5194/nhess-11-2047-2011, 2011.

Mavrouli, O. and Corominas, J.: Vunerability of simple reinforced concrete buildings to damage by rockfalls, Landslides, 7, 169–180, doi:10.1007/s10346-010-0200-5, 2010.

Michael, L., Baynes, F., Scott, G., and Granger, K.: Regional landslide risk to the cairns community, Nat. Hazards, 30, 233–249, 2003.

Papathoma, K. M., Kappes, M., Keiler, M., and Glade, T.: Physical vulnerability assessment for alpine hazards: state of the art and future needs, Nat. Hazards, 58, 645–680, doi:10.1007/s11069-010-9632-4, 2011.

Papathoma, K. M., Keiler, M., Totschnig, R., and Glade, T.: Improvement of vulnerability curves using data from extreme events: debris flow event in South Tyrol, Nat. Hazards, 64, 2083–2105, 2012.

Qian, H.: Study of Vulnerability based on Debris Flows Dynamic Factors, Master Thesis, Sichuan Univ., Chengdu, 2013.

Rickenmann, D.: Empirical relationships for debris flows, Nat. Hazards, 19, 47–77, 1999.

Romang, H. (Ed.): Wirksamkeit and Kosten von Wildbach-Schutzmassnahmen, Verlag des Geographischen Instituts der Universität Bern, Bern, Switzerland, 2004.

Scheidl, C., Chiari, M., Kaitna, R., Müllegger, M., Krawtschuk, A., Zimmermann, T., and Proske, D.: Analysing Debris-Flow Impact Models, Based on a Small Scale Modelling Approach, Surv. Geophys., 34, 121–140, doi:10.1007/s10712-012-9199-6, 2012.

Tang, C., Zhang, J., and Zhou, C. H.: Vulnerability assessment of urban debris flow hazard, J. Catastrophol., 20, 11–17, 2005.

Tang, C., Rengers, N., van Asch, Th. W. J., Yang, Y. H., and Wang, G. F.: Triggering conditions and depositional characteristics of a disastrous debris flow event in Zhouqu city, Gansu Province, northwestern China, Nat. Hazards Earth Syst. Sci., 11, 2903–2912, doi:10.5194/nhess-11-2903-2011, 2011a.

Tang, C., Zhu, J., Ding, J., Cui, X. F., Chen, L., and Zhang, J. S.: Catastrophic debris flows triggered by a 14 August 2010 rainfall at the epicenter of the Wenchuan earthquake, Landslides, 8, 485–497, doi:10.1007/s10346-011-0269-5, 2011b.

Tie, Y. B.: The methodology and framework study of Urban Debris Flow Risk Assessment, PhD thesis, Chengdu University of Technology, Chengdu, 2009.

Totsching, R., Sedlacek, W., and Fuchs, S.: A quantitative vulnerability function for fluvial sediment transport, Nat. Hazards, 58, 681–703, 2011.

Zhang, J., Guo, Z. X., Cao, S. Y., and Singh, V. P.: Scale model for the confluent area of debris flow and main river: a case study of the Wenjia Gully, Nat. Hazards Earth Syst. Sci., 13, 3083–3093, doi:10.5194/nhess-13-3083-2013, 2013.

Zhang, Y.: Dynamic response of mountainous civil architecture in east China impacted by debris flows, PhD thesis, Institute of Mountain Hazards and Environment, CAS, Chengdu, 2005.

11

Risk of large-scale fires in boreal forests of Finland under changing climate

I. Lehtonen[1], A. Venäläinen[1], M. Kämäräinen[1], H. Peltola[2], and H. Gregow[1]

[1]Finnish Meteorological Institute, Helsinki, Finland
[2]University of Eastern Finland, School of Forest Sciences, Joensuu, Finland

Correspondence to: I. Lehtonen (ilari.lehtonen@fmi.fi)

Abstract. The target of this work was to assess the impact of projected climate change on forest-fire activity in Finland with special emphasis on large-scale fires. In addition, we were particularly interested to examine the inter-model variability of the projected change of fire danger. For this purpose, we utilized fire statistics covering the period 1996–2014 and consisting of almost 20 000 forest fires, as well as daily meteorological data from five global climate models under representative concentration pathway RCP4.5 and RCP8.5 scenarios. The model data were statistically downscaled onto a high-resolution grid using the quantile-mapping method before performing the analysis. In examining the relationship between weather and fire danger, we applied the Canadian fire weather index (FWI) system. Our results suggest that the number of large forest fires may double or even triple during the present century. This would increase the risk that some of the fires could develop into real conflagrations which have become almost extinct in Finland due to active and efficient fire suppression. However, the results reveal substantial inter-model variability in the rate of the projected increase of forest-fire danger, emphasizing the large uncertainty related to the climate change signal in fire activity. We moreover showed that the majority of large fires in Finland occur within a relatively short period in May and June due to human activities and that FWI correlates poorer with the fire activity during this time of year than later in summer when lightning is a more important cause of fires.

1 Introduction

Fire is one of the major natural disturbances affecting forest dynamics and biodiversity in boreal conditions (e.g. Granström, 2001; Kuuluvainen, 2002). Globally, over 10 million hectares of boreal forest burns during a typical year; mostly in Siberia, Canada, and Alaska (Flannigan et al., 2009). A small number of large-scale fires are responsible for a large part of the burned area. For example, in Canada, fires larger than 200 ha represent 3 % of the total number of fires but account for 97 % of the total area burned (Stocks et al., 2002). Since small fires are much easier to control than large fires, it is essential for fire management agencies to try to suppress forest fires before they escalate to large fires that pose a risk for devastating-scale conflagrations.

In Finland, suppression of forest fires has been effective during the recent decades. Although roughly about 1000 forest fires occur annually in Finland, the average size of fires is less than 1 hectare. Fire survey flights contribute to the early detection of ignited fires and the dense forest road network in Finland aids fire fighters to reach and suppress the fires. During the nineteenth century and early twentieth century, large forest fires were still not uncommon in Finland. Back then, the average size of forest fires was over 50 ha in many years and, for instance, in 1868, over 60 000 ha of state-owned forest was burned within a single year (Saari, 1923; Osara, 1949). The steep decline in forest fires across Fennoscandia in the late nineteenth century has been attributed to the cultural transition to modern agriculture and forestry (Wallenius, 2011). At the same time, no significant change in the climatological fire proneness of Finnish forests has been observed (Mäkelä et al., 2012), illustrating that the possibility of conflagrations under the current climatological conditions still exists. This was recently demonstrated in 2014, when a single fire in Västmanland in central Sweden burned 15 000 ha of forest in climatological and environmental conditions similar to Finland.

In determining the risk of forest fires, weather and climate play a key role, along with the fuel amount. High temperatures accompanied by low relative humidity and strong winds enhance the evaporation and drying of the soil and continue to make forest fuels easily flammable. Natural sources, i.e. lightning strikes, ignite less than 15 % of all forest fires in Finland (Larjavaara et al., 2005), and although human activities are responsible for most forest fires, weather makes the conditions favourable for the occurrence and spreading of fires. Studies of historical fire records (e.g. Power et al., 2008; Olsson et al., 2010) have moreover linked changes in fire activity to climatic variations before any human impact was present, illustrating the crucial role of climate on fire activity. Furthermore, increased large fire activity in Canada and Alaska during the late twentieth century has been attributed to increased drought in the area (Xiao and Zhuang, 2007).

In response to global warming, the forest-fire danger is generally projected to increase in the circumboreal region, which may hamper the effectiveness of fire management (e.g. Flannigan et al., 2009). Recent studies have indicated that the forest-fire danger will also most likely increase in Finland due to global warming (Kilpeläinen et al., 2010; Lehtonen et al., 2014b; H. M. Mäkelä et al., 2014), as well as in neighbouring Russia (Sherstyukov and Sherstyukov, 2014); whereas Yang et al. (2015) concluded that northern Sweden will have a lower risk of forest fire in the future. These studies were based either on multi-model mean response (Kilpeläinen et al., 2010; Lehtonen et al., 2014b; Sherstyukov and Sherstyukov, 2014) or basically only on one climate model (H. M. Mäkelä et al., 2014; Yang et al., 2015). Moreover, only Yang et al. (2015) used simulated climate data on a daily timescale, making it possible to take predicted changes in climate variability into account when assessing the changes in forest-fire danger.

To complement the above-mentioned studies, we estimate the impact of climate change on forest-fire danger in Finland by using daily input from five independent general circulation models (GCMs). This allows us to explore the uncertainty ranges related to the projected change in forest-fire danger, an aspect that is poorly covered in the previous studies. In spite of the continuous development of climate models, the range of model uncertainty has not considerably decreased since the 1990s (Räisänen and Ylhäisi, 2015) and with lead times of a few decades, this model uncertainty can account for more than half of the total uncertainty related to climate projections (Hawkins and Sutton, 2009). As the previous studies have, in addition, mainly focused on changes in mean conditions, we direct our special interest to large-scale fires which are more relevant with regard to fire management and ecological consequences. We are moreover motivated by the fact that in countries like Finland, forest-based bioeconomy has a key role in climate change mitigation and it is thus particularly important to understand the impact of climate change on the risks affecting forests and to take them into account in forest management. That is because efficient mitigation requires increasing carbon sequestration and use of forest biomass to substitute fossil-intensive fuels, materials, and products (Kilpeläinen et al., 2015).

In this work, we first use forest-fire statistics from Finland covering 19 fire seasons to study the relationship between weather and fire occurrence in the present climate. Then, we use daily data from five GCMs participating in the Coupled Model Intercomparison Project (CMIP) phase 5 (Taylor et al., 2012) under representative concentration pathway (RCP) scenarios RCP4.5 and RCP8.5 (van Vuuren et al., 2011) to study the impact of climate change on fire occurrence over the period 1980–2099. An intrinsically similar approach has previously been used in studying the impact of climate change on regional fire activity, e.g. by Pereira et al. (2013). In this work, modelled values of weather variables are downscaled onto a high-resolution grid covering Finland using the quantile-mapping approach. In assessing the forest-fire potential, we apply the widely used Canadian forest-fire weather index (FWI) system (Van Wagner, 1987), which provides a numerical rating of fire danger, as well as indices for the moisture content of forest fuels.

2 Materials and methods

2.1 Fire statistics

To study the spatial and temporal occurrence of forest fires in Finland, we used fire data that consisted of fire reports collected from the national Finnish Rescue Service database, available from 1996 onwards. The fire reports include information on date, time, location, burned area, and ignition source of a fire, as well as vegetation type (e.g. forest, clearing, peat land, grassland, park) of a fire site. In this study, we only consider those fires that reportedly burned forested area. The fires that occurred in the autonomous Åland Islands (the group of islands located in the south-westernmost part of Finland, consisting of about 0.5 % of the Finnish land area) were not included in the database.

In most cases, the locations of fires were only given at municipality level prior to 2005, but thereafter, the exact coordinates of the fire sites were usually provided. In this study, the fires were located on a $0.1° \times 0.2°$ (approximately $10\,km \times 10\,km$) latitude–longitude grid. In cases when the spatial coordinates were missing, the fires were assumed to be located in the centroid of the municipality where they had occurred.

According to the statistics, almost 20 000 forest fires occurred in Finland from 1996 to 2014. Indeed, 112 of these

Table 1. CMIP5 models used in this study with information on country of origin and resolution of the models (*L* refers to number of vertical levels, *T* to triangular truncation, and *C* to cubed sphere).

Model	Country of origin	Resolution (long × lat), level	Reference
CanESM2	Canada	T63 (1.875° × 1.875°), L35	von Salzen et al. (2013)
CNRM-CM5	France	T127 (1.4° × 1.4°), L31	Voldoire et al. (2013)
GFDL-CM3	United States	C48 (2.5° × 2.0°), L48	Donner et al. (2011)
HadGEM2-ES	United Kingdom	1.25° × 1.875°, L38	Collins et al. (2011)
MIROC5	Japan	T85 (1.4° × 1.4°), L40	Watanabe et al. (2010)

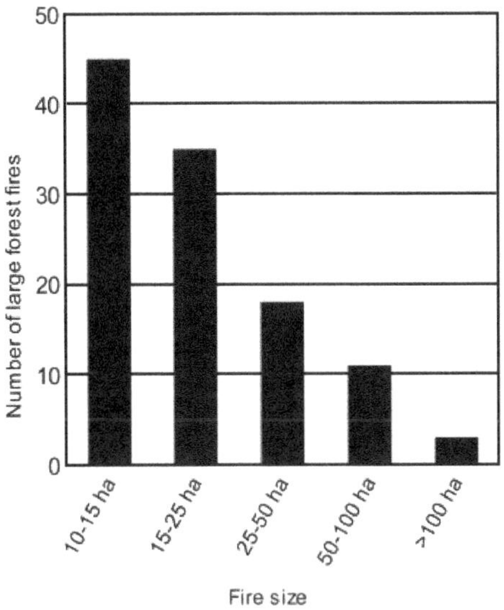

Figure 1. Large forest fires in Finland during 1996–2014 as classified on the basis of fire size.

fires (approximately 0.6 % of all forest fires) burned 10 ha or more forest. Hereafter, we refer to these fires as large forest fires. The largest forest fire in the database burned 200 ha of forest in Tammela in 1997 but most of the large forest fires burned only 10–25 ha (Fig. 1).

2.2 Climate data

In order to build a relationship between the fire data and prevailing weather conditions, we used gridded daily weather data covering Finland over the period 1996–2014 for which the fire data existed. Air temperature (daily mean, maximum and minimum) and daily mean relative humidity at a height of 2 m, as well as precipitation, observed by the Finnish Meteorological Institute weather observation network were interpolated onto the same high-resolution 0.1° × 0.2° grid with the fire data by applying kriging with external drift (Aalto et al., 2013). Because the quality of wind speed observations did not support the creation of a homogenous gridded daily data set for Finland, we used coarser daily wind speed data from the European Centre for Medium-Range Weather Forecasts ERA-Interim reanalysis (Dee et al., 2011). These data were provided on a regular 0.75° × 0.75° grid and bilinearly interpolated onto the same 0.1° × 0.2° grid with the abovementioned other variables.

To estimate the effects of changing climate on the forest-fire risk, we used daily data from five CMIP5 models (Table 1). The models were chosen on the basis of their skill to simulate the present-day average monthly temperature and precipitation climatology in northern Europe and the availability of all required variables on a daily timescale. Our study period consisted of the years 1980–2099 and historical simulations until 2005 were combined with simulations under RCP4.5 and RCP8.5 emission scenarios (van Vuuren et al., 2011) for the period 2006–2099. Because climate model outputs are often biased high or low in relation to the observed climate (e.g. Cattiaux et al., 2013), and in addition, presented on a relative coarse grid, we performed a combined statistical downscaling and bias correction on the modelled daily values before calculating the forest-fire-risk index. We performed the downscaling onto the Finnish 0.1° × 0.2° grid by applying a quantile-mapping technique using smoothing (Räisänen and Räty, 2013; Räty et al., 2014).

Figure 2 illustrates projected changes in climate variables in our data set. The mean daily maximum temperature of the forest-fire season is projected to increase in Finland by 1–3 °C for the period 2010–2039, 2–6 °C for the period 2040–2069, and 2–8 °C for the period 2070–2099, relative to the baseline period 1980–2009 depending on the scenario and model. The projected change is greater in RCP8.5 than in RCP4.5, although there is a considerable amount of variability in the rate of change among different models for temperature and other variables. As for temperature, the projected change is uniformly positive for precipitation. April–October precipitation is likely to increase in Finland by about 20 % by the end of the twenty-first century. For relative humidity, the projections indicate a decrease by 0–6 percentage points within the present century. For wind speed, the projected changes vary around 0 % with considerable inter-model variation for all periods under both scenarios. Regionally, both temperature and precipitation are projected to increase more in northern than in southern Finland (not shown). Moreover, the projected increases in temperature and precipitation are in general larger for early and late forest-fire season than for the midsummer months. Consequently, in the southern and

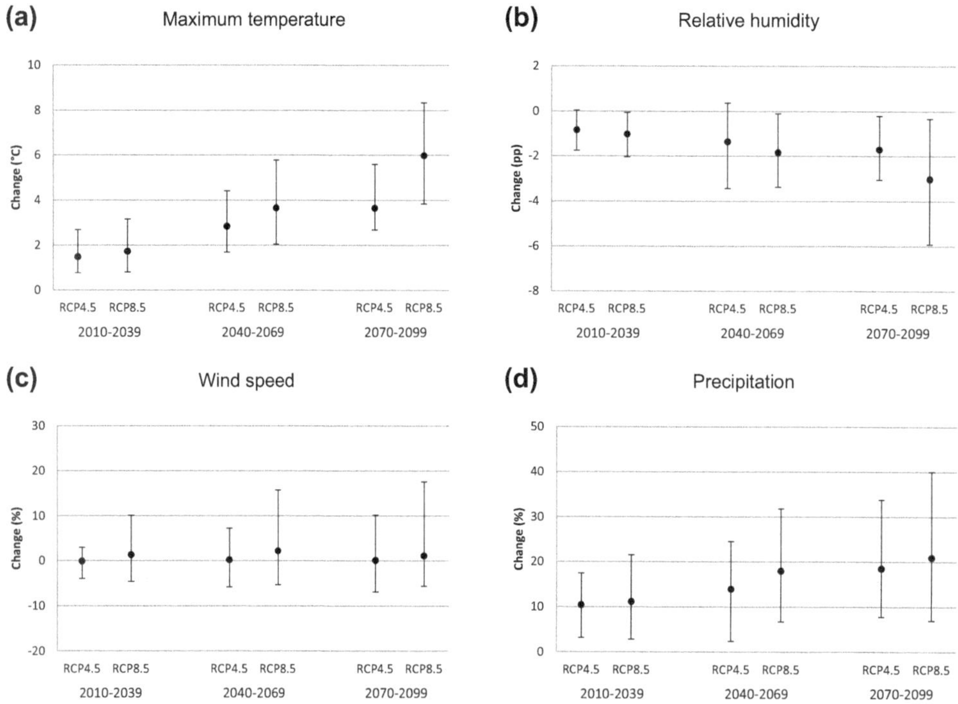

Figure 2. Projected changes in April–October mean daily maximum air temperature at 2 m **(a)**, daily average relative humidity at 2 m **(b)**, daily average wind speed at 10 m **(c)**, and total precipitation **(d)** compared to the period 1980–2009 and averaged over the whole of Finland. Dots indicate the multi-model mean change and whiskers extend to the maximum and minimum projections.

eastern parts of the country, precipitation may even decrease between June and August.

2.3 Forest fire risk assessment based on the fire weather index system

We assessed the forest-fire risk by applying the FWI system following Van Wagner and Pickett (1985). In the FWI system, three soil moisture codes are calculated on a daily basis based on air temperature, relative humidity, and wind speed observations at local noon and the total precipitation sum of the preceding 24 h. Affected by wind speed, these codes are then converted into three fire behaviour indices. The final FWI rating is a dimensionless quantity indicating the likely intensity of fire. The FWI rating can be further converted into daily severity rating (DSR) according to

$$\mathrm{DSR} = 0.0272 \times \mathrm{FWI}^{1.77}. \tag{1}$$

The DSR emphasizes higher FWI values through the power relation and reflects the expected efforts required for fire suppression more accurately than FWI. The DSR can be averaged over time to give the seasonal severity rating (SSR):

$$\mathrm{SSR} = \sum_{i=1}^{n} \mathrm{DSR}_i / n, \tag{2}$$

where DSR_i is the DSR value for the ith day, and n is the total number of days. The DSR averaged over a 1 month period is referred to as the monthly severity rating (MSR).

Because climate model data were provided as daily mean values, we converted the daily means of relative humidity and wind speed to correspond to the afternoon values better and took advantage of available daily maximum temperature data by using the bias-corrected daily maximum temperatures for calculating the FWI. The bias-corrected precipitation sums were used unaltered since 24 h precipitation sums are intended to be used in the FWI system. In converting the daily means of relative humidity into afternoon values, we used daily maximum temperatures and assumed specific humidity to stay constant throughout a day. In the case that this lead to night-time supersaturation according to the bias-corrected daily minimum temperatures, the moisture content of air at night was reduced to give a maximum relative humidity of 100 % at the time of the minimum temperature. The moisture content of air at the time of maximum temperature was correspondingly increased so that the daily mean specific humidity remained unaltered. To reflect the diurnal cycle in wind speed, we simply multiplied the bias-corrected daily mean wind speeds by 1.2, as on average, wind speed peaks in the early afternoon in phase with diurnal cycle of near-surface air temperature. This was based on 30 years (1980–2009) of meteorological observations from four locations (Vantaa, Jokioinen, Jyväskylä, and Sodankylä) across Finland, which

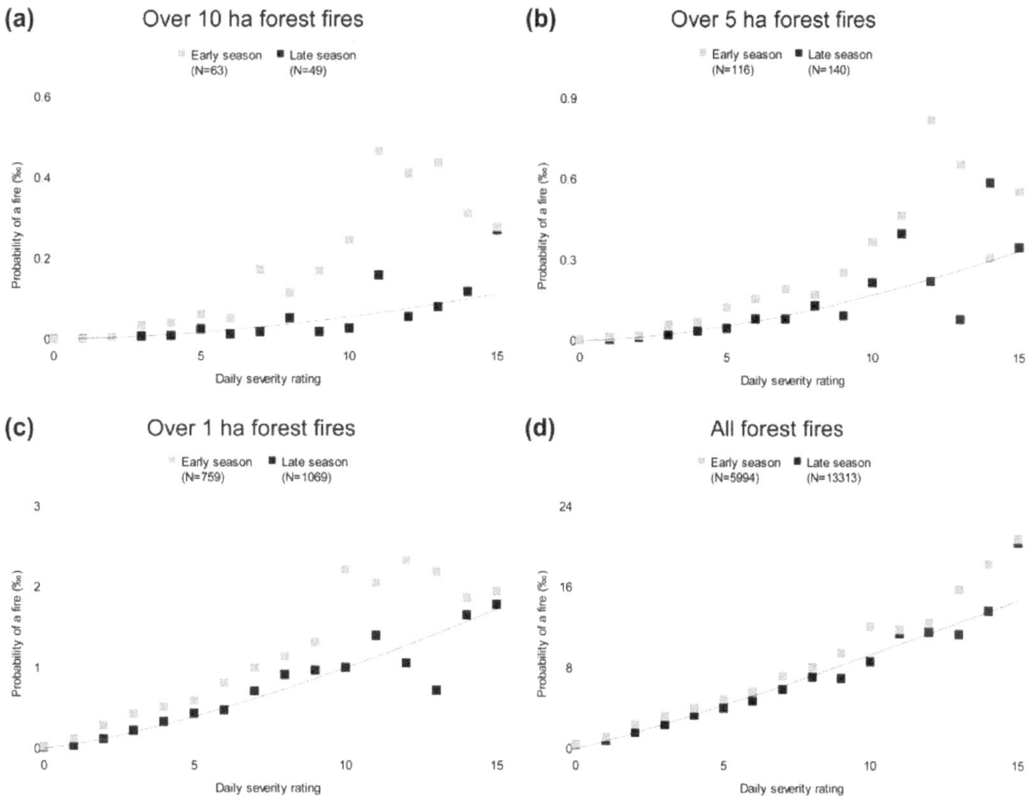

Figure 3. The relationship between daily severity rating (DSR) and occurrence of forest fires of different size in Finland during 1996–2014, performed separately for the early (effective temperature sum below 250 °C days; grey squares) and late season (effective temperature sum above 250 °C days; black squares). **(a)** Forest fires over 10 ha (i.e. large forest fires). **(b)** Forest fires over 5 ha. **(c)** Forest fires over 1 ha. **(d)** All forest fires. The numbers of fires in each class is shown as well.

showed that on average, wind speed exceeds the daily mean in the afternoon by about 20 %. The same set of observations also showed that the procedure of transforming daily mean relative humidities into afternoon values was valid and produced correct results, on average.

2.4 Regression models for fire-danger estimations

In order to estimate the impact of climate change on fire danger, we first compared the gridded DSR values with the information on locations of forest fires during 1996–2014. As the ignition probability with the same FWI value varies considerably between different stages of seasonal vegetation development (Tanskanen and Venäläinen, 2008), the inspection was performed separately for the early and late season (Fig. 3). We used the early season probability from the beginning of the growing season until the effective temperature sum reached 250 degree days when understorey vegetation is fully developed (Tanskanen and Venäläinen, 2008). In the current climate, this happens over most of Finland typically during the first half of June. Then, the late season probability was used until the end of October when the forest-fire season in Finland is virtually over (Tanskanen and Venäläinen, 2008). The commencement of the growing season was annually defined to occur on a date, which after daily mean temperature, remained above 5 °C on average. In general, the ignition probabilities were higher during the early season than the late season. Moreover, the larger the fires that were inspected, the larger this difference was. Among the tested relations (linear relation, exponential relation, and power-law relation), the ignition probabilities best followed the power-law relation as a function of DSR. For large forest fires (i.e. fires over 10 ha), the occurrence probability in a single grid cell in a given day as a function of DSR was estimated via the power relation:

$$\begin{aligned} &P(\mathrm{DSR}) = 0, \text{ when DSR} = 0 \\ &P(\mathrm{DSR}) = a \times \mathrm{DSR}^b, \text{ when } 0 < \mathrm{DSR} \leq 15 \\ &P(\mathrm{DSR}) = a \times 15^b, \text{ when DSR} > 15. \end{aligned} \tag{3}$$

For the early season, we used the coefficients $a = 0.002114452978079$ and $b = 2.02257786261162$ and for the late season $a = 0.000919759277827$ and $b = 1.77233673026624$. By summing the probabilities over the whole of Finland (excluding the Åland Islands) and the fire season, we modelled the annual number of large forest

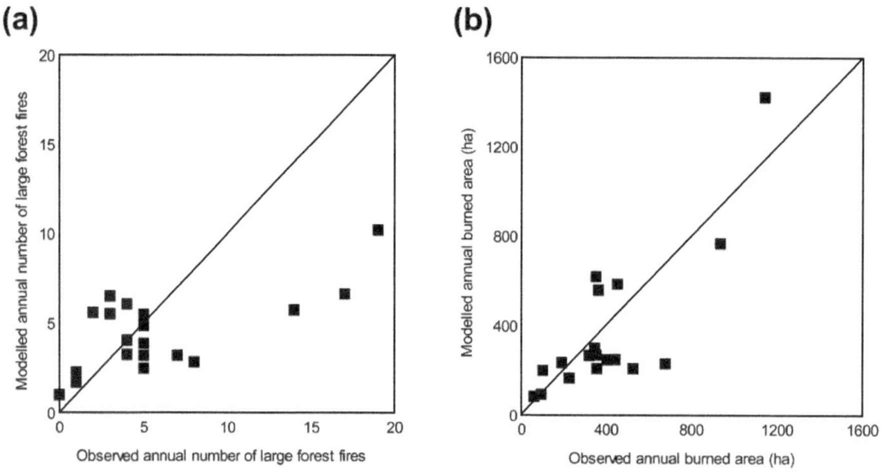

Figure 4. **(a)** Modelled annual number of large forest fires as a function of observed annual number of large forest fires in Finland during 1996–2014. **(b)** As in **(a)** but for annual burned area.

Table 2. Coefficients a and b used in Eq. (4) to estimate the total burned area by month as a function of MSR averaged over the whole of Finland. R^2 is the coefficient of determination.

Month	a	b	R^2
April	69.58	1.97	0.54
May	52.96	1.07	0.28
June	6.85	2.71	0.67
July	7.67	2.58	0.83
August	10.33	2.61	0.56
September	16.37	2.97	0.67
October	7.96	1.23	0.42

fires in Finland. As only 0.2 % of considered days (and 8 % of large forest fires) showed a DSR above 15, we assumed the fire probability to stay constant when the DSR was above 15, as it was hard to say whether the same power relation still applies with such high DSR values. Nevertheless, we repeated all of our calculations, assuming that the power relation would hold with DSRs above 15 and the estimated numbers of large forest fires were only limitedly increased because high DSRs occur relatively seldom.

A similar power relation was created to estimate the annual burned area in Finland based on MSRs averaged over the whole of Finland from April to October:

$$A(\mathrm{MSR}) = a \times \mathrm{MSR}^b, \qquad (4)$$

where A is the monthly burned area in hectares. We defined the coefficients a and b separately for each month (Table 2) and estimated the annual burned area by summing the estimated burned areas in each month.

Performance of the regression models is illustrated in Fig. 4 and the statistics for model validation are summarized in Table 3. In general, the regression model for burned area showed higher correlation with observations than the model for the number of large forest fires. In addition, the highest annual peaks in the number of large forest fires are underestimated, leading to a negative mean bias error. The nonparametric Spearman's rank correlations between the models and observations were weaker than the parametric Pearson's correlations. Nevertheless, the Spearman's correlation for burned area was still statistically significant at 1 % level.

2.5 Data analysis

First, we studied the distribution of large forest fires in Finland and the fire activity with regard to population density based on the fire statistics during 1996–2014. We assumed that no significant impact on the fire regime is caused due to differences in fuel amount or type because forest fuels are relatively similar throughout Finland with the exception of the tundra vegetation in northernmost Lapland above 68° N (Reinikainen et al., 2000). Then, we used Eqs. (3) and (4) to estimate the number of large forest fires and burned area in Finland until 2099 by utilizing the climate model data.

3 Results

3.1 Fire regime in Finland

The distribution of large forest fires in Finland during 1996–2014 is shown in Fig. 5a along with population density. The average size of forest fires in Finland steadily decreases with increasing population density (Fig. 5b), while the population density tends to strongly decrease towards the north. Hence, although the occurrence of forest fires has a strong positive correlation ($R^2 = 0.85$) with population density on a regional scale (Fig. 5c), this dependency is largely absent when con-

Table 3. Statistics for the model validation over the period 1996–2014.

	Pearson correlation	Spearman correlation	Mean bias error	Root-mean-square error
Large forest fires	0.67**	0.39*	−1.43	4.25
Burned area	0.81***	0.58**	−34.34	184.87

Asterisks denote statistical significance (* $p<0.1$; ** $p<0.01$; *** $p<0.001$).

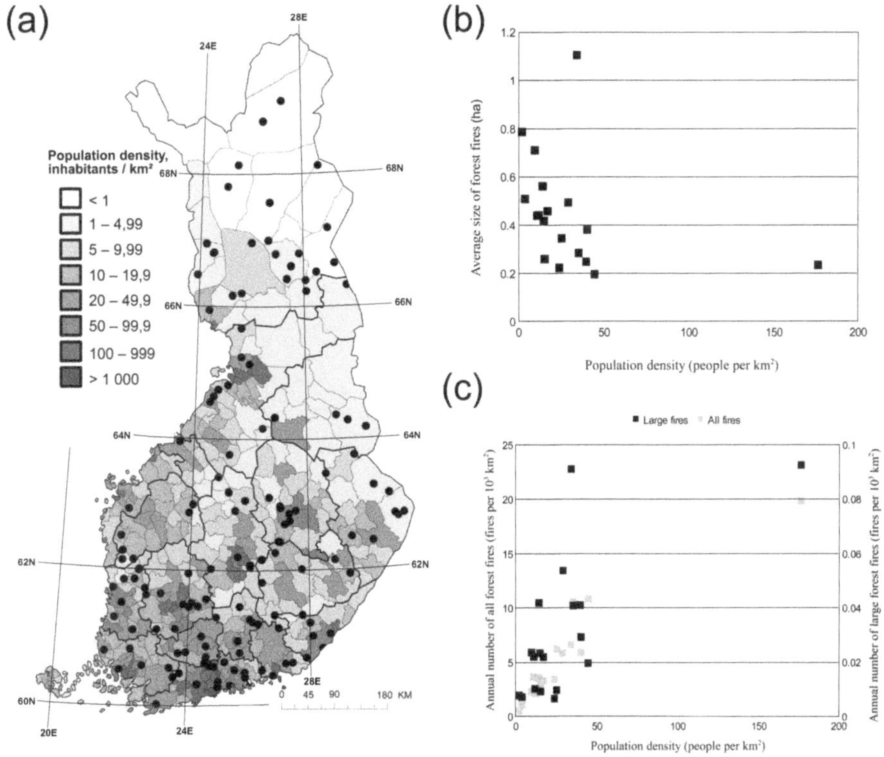

Figure 5. **(a)** Locations of large forest fires in Finland during 1996–2014 along with population density by municipality. **(b)** Average size of forest fires in Finland by region during 1996–2014 as a function of population density. **(c)** Annual mean number of all forest fires (grey squares) and large forest fires (black squares) per 10^3 km^2 in Finland by region during 1996–2014 as a function of population density.

sidering large forest fires. This is one reason why we decided to use 10 ha as a threshold for large forest fires.

Figure 6 shows MSRs based on observational weather data during 1996–2014 along with monthly burned forest areas. As seen in Table 2, the burned area correlates best with MSR in July and worst in May. In general, variations in the annual burned area reflect variations in SSR fairly well. The Pearson product-moment correlation coefficient between these two variables proved to be as high as 0.75 during 1996–2014. Nevertheless, when the annual burned area is estimated on the basis of MSRs by using Eq. (4), the correlation with the actual burned area is even higher (0.81; Table 3).

Annual modelled and observed numbers of large forest fires and burned area in Finland during 1996–2014 are displayed in Fig. 7. As can be expected based on the positive correlations in Table 3, the modelled fire activity follows the observed fire activity well enough to depict the main temporal variations. The highest modelled annual peaks in the number of large forest fires are underestimated largely due to weak correlation between the fire weather and occurrence of large forest fires in May. For instance, in 1997 and 2008, two years with relatively many large forest fires, all large forest fires occurred before mid-June and most of them in May. However, May and early June also displayed similarly dry fire weather conditions in 1999, 2000, and 2002; but only a few large forest fires occurred during these years.

Classification of large forest fires based on the reported ignition source reveals interestingly that early season fires are almost entirely human-induced; whereas in July, most large forest fires are ignited by a lightning strike (Fig. 7c). Outstandingly common human-caused large forest fires are in May and early June. The large majority of all human-caused large forest fires in Finland during 1996–2014 occurred during this relatively short period. At that time of year, the large

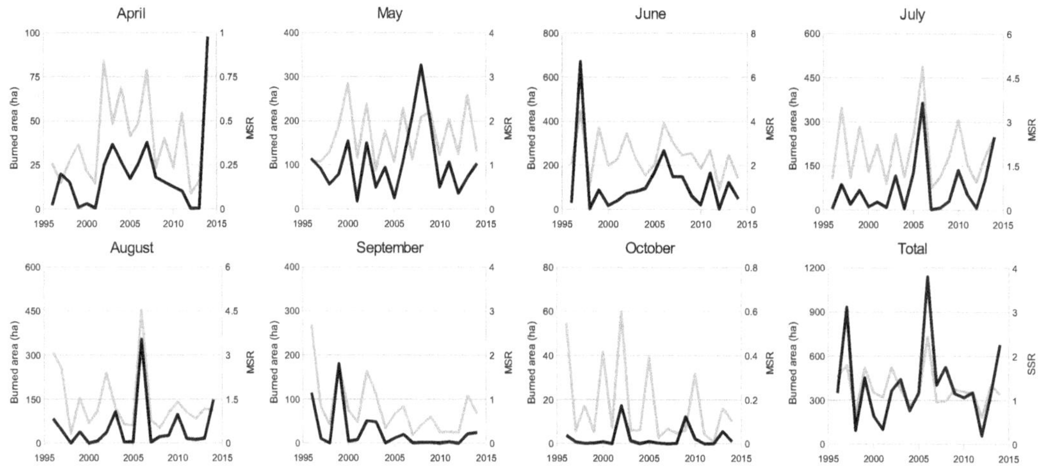

Figure 6. Burned forest area (black lines) in Finland by month and monthly severity rating (grey lines) averaged over the whole of Finland during 1996–2014.

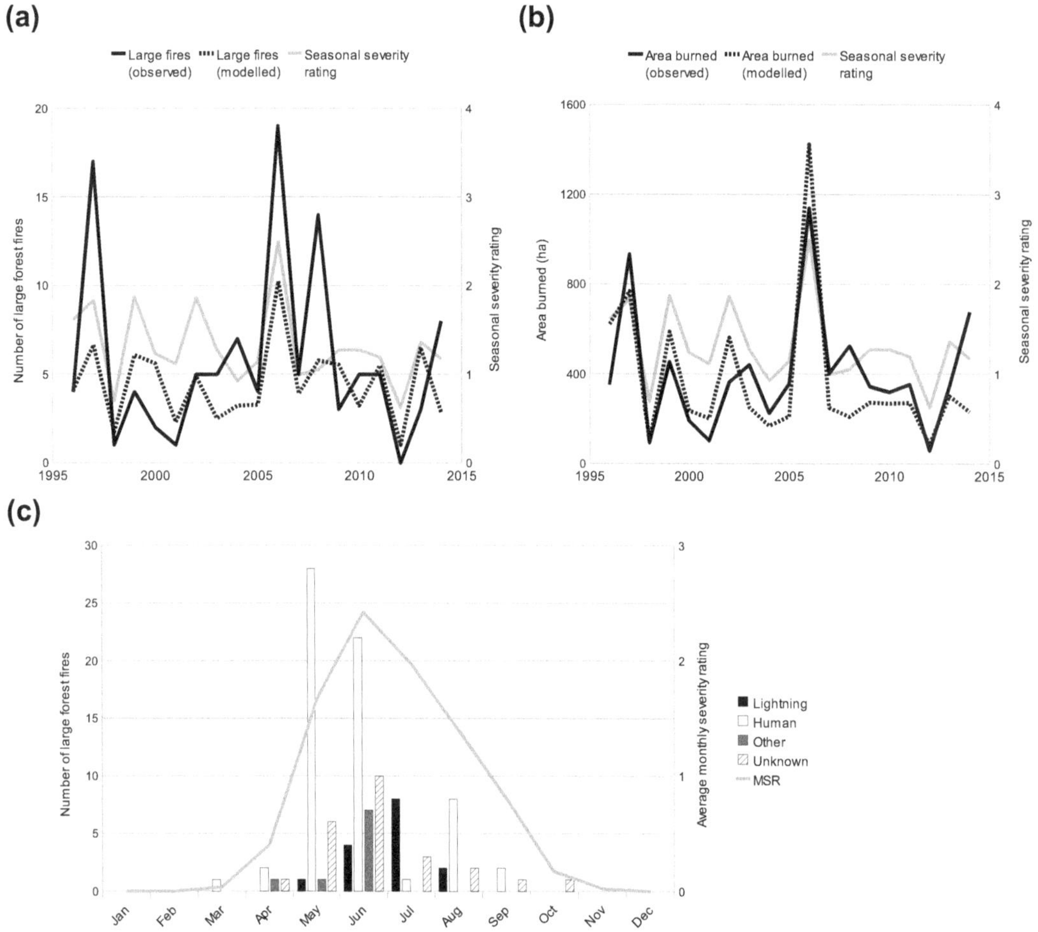

Figure 7. (a) Annual observed (solid black line) and modelled (dashed black line; based on Eq. 3) numbers of large forest fires in Finland during 1996–2014, as well as the annual April–October seasonal severity rating averaged over the whole of Finland (grey line). **(b)** Annual observed (solid black line) and modelled (dashed black line; based on Eq. 4) area burned in Finland during 1996–2014, as well as the annual April–October seasonal severity rating averaged over the whole of Finland (grey line). **(c)** Nationwide average of monthly severity rating (MSR) in Finland during 1996–2014 (grey line) and monthly distribution of large forest fires in Finland within the same period, divided by the source of ignition (bars).

Table 4. Proportions (in %) of forest fires of different sizes, divided according to the daily severity index (DSR) classes in Finland during 1996–2014.

DSR	< 1 ha	1–5 ha	5–10 ha	10–20 ha	> 20 ha
< 1	96.4	3.2	0.2	0.2	0.0
1–5	90.8	8.2	0.6	0.3	0.1
5–10	87.6	10.5	1.1	0.4	0.4
> 10	89.2	8.1	1.2	0.7	0.8

fires tend to be often caused by escalated prescribed burning or the burning of rubbish. These activities are not practised anymore later in summer.

The average size of forest fires in Finland increases with increasing severity of prevailing fire weather. The large majority of all forest fires still burn less than 1 ha of forest with high DSR values, but the share of large forest fires (we recall that with large forest fires, we refer to all forest fires larger than 10 ha) increases from 0.2 to 1.5 % when the DSR increases from below 1 to over 10 (Table 4).

3.2 Projected climate change impact on the forest-fire risk

The SSR averaged over the April–October period likely already increased during the early twenty-first century, and by the period 2070–2099, the nationwide multi-model mean change exceeds 100 % under the RCP8.5 scenario (Fig. 8a). However, among different model projections, the increase varies between 28 and 200 %. For the number of large forest fires, the projected change is slightly larger than for SSR (Fig. 8b). For instance, under the RCP8.5 scenario the range for the projected increase from 1980–2009 to 2070–2099 is from 54 to 238 %. For the burned area, future estimates have a huge variability among different model projections (Fig. 8c). Already by the period 2010–2039, the projected change varies approximately between 5 and 200 %. By the period 2070–2099, the burned area is projected to increase under the RCP8.5 scenario by 35–1271 %, depending on the model, and under the RCP4.5 scenario by 56–441 %. However, as the burned area has been small during recent years, even a single fire comparable in size to the Västmanland wildfire in Sweden in 2014 would burn about twice as much forest area than was burned in Finland during the years 1996–2014 in total. Hence, occurrence of only a couple of conflagrations could lead to an increase of hundreds of percent in the burned area. For all statistics, the projected multi-model mean change and the range among different model projections are smaller under the RCP4.5 than RCP8.5 scenario.

Regionally, the forest-fire danger is projected to increase rather similarly throughout Finland (Fig. 9). Under the RCP8.5 scenario, multi-model mean SSR averaged over the April–October period increases in the south from about 2–3

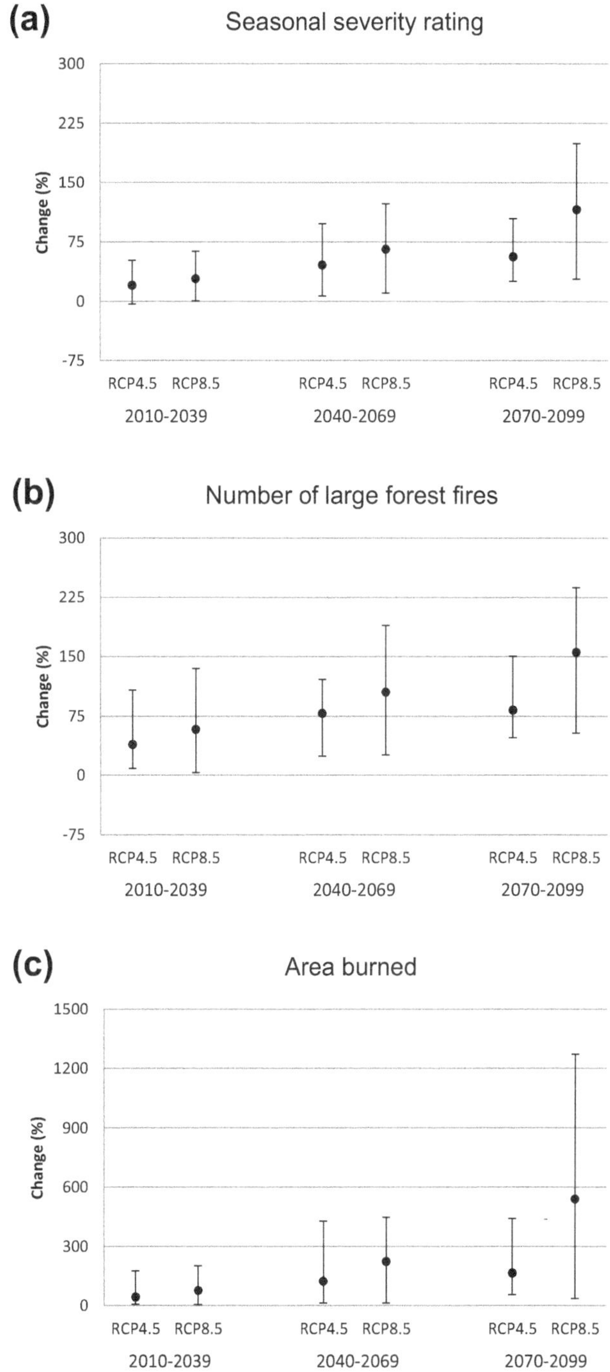

Figure 8. Projected changes in April–October seasonal severity rating averaged over the whole of Finland **(a)**, in the number of large forest fires in Finland **(b)**, and in the area burned **(c)** compared to the period 1980–2009. Dots indicate the multi-model mean change and whiskers extend to the maximum and minimum projections.

Table 5. The 90th, 50th, and 10th percentiles of multi-model mean annual number of large forest fires in Finland, excluding the Åland Islands. The range in the number of modelled large forest fires among the model projections is shown in parentheses.

	RCP4.5 1980–2009	RCP4.5 2010–2039	RCP4.5 2040–2069	RCP4.5 2070–2099
90th percentile	11 (6–18)	12 (10–15)	15 (13–18)	14 (12–16)
50th percentile	5 (4–6)	6 (5–7)	9 (5–10)	9 (8–9)
10th percentile	2 (1–2)	3 (2–4)	5 (2–6)	4 (3–6)
	RCP8.5 1980–2009	RCP8.5 2010–2039	RCP8.5 2040–2069	RCP8.5 2070–2099
90th percentile	9 (7–11)	15 (11–24)	16 (12–20)	18 (11–21)
50th percentile	4 (4–5)	7 (4–10)	10 (6–14)	12 (8–16)
10th percentile	2 (1–2)	3 (2–4)	5 (4–8)	7 (4–12)

to 4–6 and in the north from about 1 to 2 until the end of the twenty-first century; i.e. it approximately doubles.

Moreover, the fire danger is projected to increase both during the driest and wettest summers but in relative terms, the number of large forest fires is expected to increase most in the summers that express a relatively small number of large fires (Table 5). In spite of large inter-model variability, the number of large forest fires during a typical year in the late 21st century is expected to be close to what it was was during the recent years that experienced the highest number of large forest fires (e.g. 1997, 2006 and 2008). Similarly, the easiest future fire seasons would be comparable to the current average fire seasons.

4 Discussion and conclusions

4.1 Evaluation of methodology

In this study, we used statistically downscaled climate model simulations to evaluate the impact of climate change on the number of large fires and total burned area in the boreal forests of Finland. In assessing the fire risk, we applied the FWI system, and the statistical downscaling was performed with the quantile-mapping technique. Quantile mapping has proven to be among the best-performing empirical bias-correction methods for temperature (Räisänen and Räty, 2013) and precipitation (Räty et al., 2014) throughout the probability distribution and it has been suggested the most in recent studies (e.g. Teutschbein and Seibert, 2012). Quantile mapping has also been previously successfully applied for correcting relative humidity and wind speed simulations (Wilcke et al., 2013). Moreover, Yang et al. (2015) used a rather similar approach for correcting regional climate model output in order to assess forest-fire risk in Sweden. However, the method is still by no means perfect. Where the local differences between simulated and observed climates are fairly large, the downscaling technique is less likely to yield accurate results. In Finland, these areas include many coastal regions and, in addition, northernmost Lapland, where the relatively scarce station density is compounded with com-

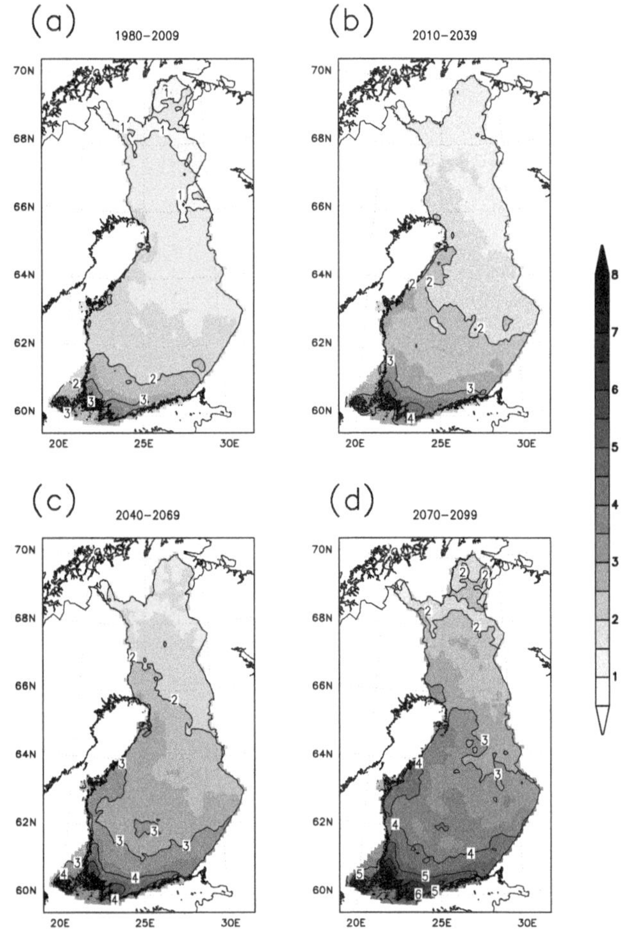

Figure 9. Projected multi-model mean for the April–October seasonal severity rating (SSR) in 1980–2009 **(a)**, 2010–2039 **(b)**, 2040–2069 **(c)**, and 2070–2099 **(d)** under the RCP8.5 scenario.

plex topography. One shortcoming of the quantile-mapping method is that averaging the downscaled time series back to the original resolution leads to overestimation of extreme values if the variable in question has much small-scale variability (Maraun, 2013). This holds particularly for precipitation. This effect is only visible for the area-averaged time series, and in the present study, it probably somewhat increased the inter-annual variability in the fire weather projections.

The FWI system applied in fire-risk estimation was initially developed empirically for Canadian boreal conditions, but it has become widely implemented in other countries around the world as well. Eventually, the FWI system has been suggested as the basis for a global early warning system for wildland fires (de Groot et al., 2006). Comparison of FWI to the forest-fire index used operationally in Finland revealed that the two indices perform similarly in Finnish conditions (Vajda et al., 2014).

The developed regression models for estimation of the number of large forest fires and burned area have marked uncertainties. Firstly, the period consisting of information on fire locations and used in developing the regression models is fairly short, only 19 years. Secondly, it is uncertain that a similar relationship between fire weather and fire activity would still hold in the future if the fire weather turns much more severe. However, as most of large forest fires occur when the fire danger is only moderately high, the change in the most extreme conditions has less relevance because those situations will, in any case, occur relatively rarely. Thirdly, weather explains only a part of the variability in fire activity. Our results suggest that roughly about half of the variability in the annual burned area can be explained by variations in MSRs, but in the long term, other factors may be more important.

The use of the DSR instead of FWI ameliorated our results: the correlation between annual burned area and SSR ($\sim$ 0.75) was larger than reported by Venäläinen et al. (2014) between annual burned area and seasonal mean FWI ($\sim$ 0.60) in Finland. By taking into account the seasonal variations in the correlations between fire activity and fire-danger indices, limited improvements were achieved in the performance of our regression models; though this is also one source of uncertainty. Currently, most large forest fires in Finland occur within a relative short period in May and early June as a result of human activities, often including prescribed burning and the burning of rubbish. It may have an impact on the fire activity whether these activities are still conducted during the same time of year in the future or whether they will be preponed as the commencement of the growing season is projected to take place earlier in a warmer climate (Ruosteenoja et al., 2011). In addition, the correlation between fire activity and fire danger was poorest during this time of year, indicating that the use of fire by humans is probably reduced while the fire danger is high. Later in summer, when lightning is more important cause of large forest fires, fire-danger indices correlate much better with the observed fire activity. Consequently, the projected increase in the burned area is, by a large part, caused by the projected increase in fire danger during midsummer and late summer.

4.2 Evaluation of main results

In accordance with previous studies (Kilpeläinen et al., 2010; Lehtonen et al., 2014b; H. M. Mäkelä et al., 2014), we found that in response to climate change, the forest-fire risk in Finland will increase with a high probability. In these previous studies, the projected change in fire danger was converted into the change in the number of days expressing a high forest-fire danger. Because extreme conditions are more relevant with regard to fire management efficiency, we estimated the climate change impact on potential large-scale forest fires and burned area. Our results suggest that the number of large forest fires could easily double by 2100, but there is large variability in the projected change among different models and also between the two emission scenarios considered here. Hence, the change can be, in the worst case, even larger. This large inter-model variability is already evident with the subset of five different GCMs used in this study. Within a larger model set, this variability would probably be even larger. To exemplify this, based on results of a single climate model, Yang et al. (2015) estimated that northern Sweden, which is in close proximity to Finland, would face lower fire risk in the future than today. It was mainly because in their simulation, climate was projected to become more humid, while our projections indicated either drier future conditions or little change in relative humidity. In general, the large uncertainty ranges related to the fire-danger projections reflect that uncertainties related to changes in temperature, precipitation, wind, and humidity climates all add uncertainty to the estimation of forest-fire danger. Possible changes in wind climate are particularly important because the FWI rating has been found to be most sensitive to wind speed (Dowdy et al., 2010) and as the multi-model mean change for wind speed is close to zero, it is uncertain whether the actual change will be positive or negative.

The estimates given for burned area are highly uncertain, mostly because the occurrence of only a few conflagrations would increase the burned area from the present level by hundreds of percent. Nevertheless, the likely increase in the number of large fires driven by general increase in the fire danger increases the probability that some of these fires would escalate to conflagrations. It is thus of the utmost importance to suppress the fires as quickly as possible, which may prove to be problematic if multiple fires are ignited within a short time in isolated locations.

Considering the multi-model mean, the present projections for the number of large forest fires and burned area clearly show larger increases than previously estimated for the increase in the number of fire-danger days. This is partly because a larger portion of all fires spread into large fires when fire weather becomes more severe. An additional explanation is that the RCP8.5 scenario is a more extreme climate change scenario than any of the scenarios used in the previous studies which applied the Special Report on Emission Scenarios (SRES) (Nakićenović et al., 2000). While based on the multi-model mean under the high-emission SRES A2 scenario, summer temperatures in Finland were projected to increase by about 3 °C by the end of the present century (Giorgi and Coppola, 2009); this increase is almost 5 °C in the RCP8.5 scenario (Cattiaux et al., 2013). Moreover, among the models involved in this study, the warming is, on average, slightly larger. Actually, the projected summertime warming in Finland under the RCP4.5 scenario corresponds closely to that under the SRES A2 scenario. For wind speed, relative humidity, and precipitation, the projected changes among the models involved were, on average, rather similar to projected changes from the multi-model means under the SRES scenarios (Gregow et al., 2012; Ruosteenoja and Räisänen, 2013; Lehtonen et al., 2014a).

The impact of climate change on the annual burned area has been previously estimated with the FWI system in North America (Flannigan et al., 2005; Balshi et al., 2009) and in the Mediterranean region (Amatulli et al., 2013). Flannigan et al. (2005) suggested that in Canada, the annual burned area could approximately double by the end of this century, and an even greater increase was projected by Balshi et al. (2009). Recently, Migliavacca et al. (2013) estimated the future burned area in Europe by using a land–atmosphere model that computes the probability of fire occurrence as the product of three terms: the probability related to biomass availability, the probability conditioned on the moisture, and the probability of ignition. They demonstrated that a reduction in productivity reduces the increase in fire activity over semiarid regions, but this is unlikely to happen in northern Europe where forest productivity and biomass stock are projected to increase under a warming climate (Kellomäki et al., 2008; Dury et al., 2011), increasing the forest-fuel load. In northern Europe, Migliavacca et al. (2013) found temperature to be the most important driver of fire activity. For burned area, their results curiously showed an abrupt doubling of the annual burned area in northern Europe around 2010 and no coherent change after that under the modest SRES A1B emission scenario.

Our results indicating substantial increase in the number of large forest fires and burned area in Finland due to a warming climate are generally quantitatively similar with the findings of the above-mentioned studies. The projected increase in fire danger is essentially due to the reduction in forest-fuel moisture content. Previously, Dai (2013) has shown that CMIP5 models consistently project soil moisture to decrease over all of Europe. In Finland, the drying of soil is mostly a result of the increase in evaporative demand exceeding the increase in precipitation. In the future, the fire season is also expected to start earlier because of earlier snowmelt (Räisänen and Eklund, 2012) and earlier commencement of the growing season (Ruosteenoja et al., 2011). In autumn, a considerable lengthening of the fire season is not probable because air humidity increases towards winter due to a shortening of the day length.

Tanskanen and Venäläinen (2008) had previously demonstrated that there are three peaks in annual fire activity in Finland: the first in late May and early June, the second after mid-July, and the third in September. They did not directly inspect the ignition sources of fires but hypothesized that the second peak may be associated with lightning, and the last peak, mainly consisting of small-scale fires, would occur because as a result of the open season for elk hunting, as well as various gathering activities, people fill the forests and light campfires. Consistent with their hypothesis, we showed that most large fires in July are ignited by lightning strikes. Moreover, the annual course of lightning-ignited large forest fires follows the annual lightning activity closely, with a peak in July (A. Mäkelä et al., 2014). The first and most prominent peak in fire activity in late May was considered surprising by Tanskanen and Venäläinen (2008) because May had previously been considered a marginal part of the fire season. They assumed that the majority of fires originating from silvicultural slash burning of cured vegetation and rubbish are likely to occur during this time of year. Again, our results confirm this assumption: the large fires in May and early June are almost entirely human-caused, and mainly because of the above-mentioned activities. Moreover, because humans ignite many more large fires before mid-June than later in summer, the seasonal vegetation development might not be the main reason for the higher ignition probabilities in the early season found in fire statistics.

4.3 Conclusions

The impact of climate change on forest-fire danger in Finland with emphasis on large-scale fires and model-based uncertainty was studied using the statistically downscaled and bias-corrected daily output of five CMIP5 models. The regression models for estimating the number of large forest fires and burned area were constructed based on the fire statistics covering the years 1996–2014. Our results show that the number of large forest fires may double or even triple by the end of this century but above all, the projections show large inter-model variability. Because of several uncertainties related to this study, the results should be considered to be only approximate; though they highlight the large uncertainty in the rate of the projected increase of forest-fire danger, which is moreover only partly covered by climatological factors considered in this study.

Our results largely confirmed the previous presumptions of Tanskanen and Venäläinen (2008) about the ignition sources of fires at different times of the year. Human-caused large fires are greatly over-represented in late May and early June; whereas in July, lightning ignites the majority of large fires. We also showed that the correlation between fire activity and fire weather indices is poorest in May when humans ignite more large fires than during any other month. However, our results did not indicate that population density is a key driver in the occurrence of large forest fires in Finland. That is because although the number of forest fires steadily increases with increasing population density, the average size of fires simultaneously decreases.

Climatological conditions do not prevent conflagrations from occurring in Finland. An increase in fire danger increases the proportion of large-scale fires because the fire managers have less time to suppress the fires if the conditions for vigorous spread of fire are favourable. Even a single conflagration could burn more forest area than has been typically burned within 1 decade in Finland during the last half a century. Our results suggest that the probability of such an event occurring will increase. For the highest projections of burned area to become realized, some fires comparable to the Västmanland wildfire would virtually be required to take place during the present century.

Acknowledgements. This research has been supported by the Consortium project ADAPT (Adaptation of forest management to climate change: uncertainties, impacts, and risks to forests and forestry in Finland), which is a collaboration project between the University of Eastern Finland (UEF proj. 14907) and the Finnish Meteorological Institute (FMI proj. 260785), and is jointly funded by the Academy of Finland, UEF, and FMI. Support was also received from the Strategic Research Council at the Academy of Finland through the FORBIO (Sustainable, climate-neutral, and resource-efficient forest-based bioeconomy) project. We acknowledge the World Climate Research Programme's Working Group on Coupled Modelling, which is responsible for CMIP, and we thank the climate modelling groups (listed in Table 1 of this paper) for producing their model output and making it available. For CMIP, the US Department of Energy's Program for Climate Model Diagnosis and Intercomparison provided coordinating support and led the development of software infrastructure in partnership with the Global Organization for Earth System Science Portals. The ERA Interim data were downloaded from the European Centre for Medium-Range Weather Forecasts' data server. Kimmo Ruosteenoja is thanked for downloading and preprocessing the model data. We appreciate the help of Pentti Pirinen in interpolating the observational data onto the Finnish grid. Olle Räty and Jouni Räisänen from the Department of Physics, University of Helsinki, are acknowledged for developing the applied bias correction software. Finally, we are grateful to Mário Pereira and one anonymous referee for their assistance in improving the paper.

References

Aalto, J., Pirinen, P., Heikkinen, J., and Venäläinen, A.: Spatial interpolation of monthly climate data for Finland: comparing the performance of kriging and generalized additive models, Theor. Appl. Climatol., 112, 99–111, 2013.

Amatulli, G., Camia, A., and San-Miguel-Ayanz, J.: Estimating future burned areas under changing climate in the EU-Mediterranean countries, Sci. Total Environ., 450–451, 209–222, 2013.

Balshi, M. S., McGuire, A. D., Duffy, P., Flannigan, M., Walsh, J., and Melillo, J.: Assessing the response of area burned to changing climate in western boreal North America using a multivariate adaptive regression splines (MARS) approach, Global Change Biol., 15, 578–600, 2009.

Cattiaux, J., Douville, H., and Peings, Y.: European temperatures in CMIP5: origins of present-day biases and future uncertainties, Clim. Dynam., 41, 2889–2907, 2013.

Collins, W. J., Bellouin, N., Doutriaux-Boucher, M., Gedney, N., Halloran, P., Hinton, T., Hughes, J., Jones, C. D., Joshi, M., Liddicoat, S., Martin, G., O'Connor, F., Rae, J., Senior, C., Sitch, S., Totterdell, I., Wiltshire, A., and Woodward, S.: Development and evaluation of an Earth-System model – HadGEM2, Geosci. Model Dev., 4, 1051–1075, doi:10.5194/gmd-4-1051-2011, 2011.

Dai, A.: Increasing drought under global warming in observations and models, Nat. Clim. Change, 3, 52–58, 2013.

Dee, D. P., Uppala, S. M., Simmons, A. J., Berrisford, P., Poli, P., Kobayashi, S., Andrae, U., Balmaseda, M. A., Balsamo, G., Bauer, P., Bechtold, P., Beljaars, A. C. M., van de Berg, L., Bidlot, J., Bormann, N., Delsol, C., Dragani, R., Fuentes, M., Geer, A. J., Haimberger, L., Healy, S. B., Hersbach, H., Hólm, E. V., Isaksen, L., Kållberg, P., Köhler, M., Matricardi, M., McNally, A. P., Monge-Sanz, B. M., Morcrette, J.-J., Park, B.-K., Peubey, C., de Rosnay, P., Tavolato, C., Thépaut, J.-N., and Vitart, F.: The ERA-Interim reanalysis: configuration and performance of the data assimilation system, Q. J. Roy. Meteorol. Soc., 137, 553–597, 2011.

de Groot, W. J., Goldhammer, J. G., Keenan, T., Brady, M. A., Lynham, T. J., Justice, C. O., Csiszar, I. A., and O'Loughlin, K.: Developing a global early warning system for wildland fire, in: Proceedings of the V International Conference on Forest Fire Research, 27–30 November 2006, Coimbra, edited by: Viegas, D. X., Elsevier, Amsterdam, 12 pp., 2006.

Donner, L. J., Wyman, B. L., Hemler, R. S., Horowitz, L. W., Ming, Y., Zhao, M., Golaz, J.-C., Ginoux, P., Lin, S.-J., Schwarzkopf, M. D., Austin, J., Alaka, G., Cooke, W. F., Delworth, T. L., Freidenreich, S. M., Gordon, C. T., Griffies, S. M., Held, I. M., Hurlin, W. J., Klein, S. A., Knutson, T. R., Langenhorst, A. R., Lee, H.-C., Lin, Y., Magi, B. I., Malyshev, S. L., Milly, P. C. D., Naik, V., Nath, M. J., Pincus, R., Ploshay, J. J., Ramaswamy, V., Seman, C. J., Shevliakova, E., Sirutis, J. J., Stern, W. F., Stouffer, R. J., Wilson, R. J., Winton, M., Wittenberg, A. T., and Zeng, F.: The dynamical core, physical parameterizations, and basic simulation characteristics of the atmospheric component AM3 of the GFDL global coupled model CM3, J. Climate, 24, 3484–3519, 2011.

Dowdy, A. J., Mills, G. A., Finkele, K., and de Groot, W.: Index sensitivity analysis applied to the Canadian forest fire weather index and the McArthur forest fire danger index, Meteorol. Appl., 17, 298–312, 2010.

Dury, M., Hambuckers, A., Warnant, P., Henrot, A., Favre, E., Ouberdous, M., and François, L.: Responses of European forest ecosystems to 21st century climate: assessing changes in interannual variability and fire intensity, iForest, 4, 82–99, 2011.

Flannigan, M., Logan, K. A., Amiro, B. D., and Skinner, W. J.: Future area burned in Canada, Clim. Change, 72, 1–16, 2005.

Flannigan, M., Stocks, B., Turetsky, M., and Wotton, M.: Impacts of climate change on fire activity and fire management in the circumboreal forest, Global Change Biol., 15, 549–560, 2009.

Giorgi, F. and Coppola, E.: Projections of twenty-first century climate over Europe, Eur. Phys. J. Conf., 1, 29–46, 2009.

Granström, A.: Fire management for biodiversity in the European boreal forest, Scand. J. Forest Res., 16, 62–69, 2001.

Gregow, H., Ruosteenoja, K., Pimenoff, N., and Jylhä, K.: Changes in mean and extreme geostrophic wind speeds in northern Europe until 2100 based on nine global climate models, Int. J. Climatol., 32, 1834–1846, 2012.

Hawkins, E. and Sutton, R.: The potential to narrow uncertainty in regional climate predictions, B. Am. Meteorol. Soc., 90, 1095–1107, 2009.

Kellomäki, S., Peltola, H., Nuutinen, T., Korhonen, K. T., and Strandman, H.: Sensitivity of managed boreal forests in Finland to climate change, with implications for adaptive management, Philos. T. Roy. Soc. B, 363, 2341–2351, 2008.

Kilpeläinen, A., Kellomäki, S., Strandman, H., and Venäläinen, A.: Climate change impacts on forest fire potential in boreal conditions in Finland, Clim. Change, 103, 383–398, 2010.

Kilpeläinen, A., Torssonen, P., Strandman, H., Kellomäki, S., Asikainen, A., and Peltola, H.: Net climate impacts of forest biomass production and utilization in managed boreal forests, Global Change Biol. Bioenergy, doi:10.1111/gcbb.12243, online first, 2015.

Kuuluvainen, T.: Disturbance dynamics in boreal forests: defining the ecological basis of restoration and management of biodiversity, Silva Fenn., 36, 5–11, 2002.

Larjavaara, M., Kuuluvainen, T., and Rita, H.: Spatial distribution of lightning-ignited forest fires in Finland, Forest Ecol. Manage., 208, 177–188, 2005.

Lehtonen, I., Ruosteenoja, K., and Jylhä, K.: Projected changes in European extreme precipitation indices on the basis of global and regional climate model ensembles, Int. J. Climatol., 34, 1208–1222, 2014a.

Lehtonen, I., Ruosteenoja, K., Venäläinen, A., and Gregow, H.: The projected 21st century forest-fire risk in Finland under different greenhouse gas scenarios, Boreal Environ. Res., 19, 127–139, 2014b.

Mäkelä, A., Enno, S.-V., and Haapalainen, J.: Nordic lightning information system: thunderstorm climate of Northern Europe for the period 2002–2011, Atmos. Res., 139, 46–61, 2014.

Mäkelä, H. M., Laapas, M., and Venäläinen, A.: Long-term temporal changes in the occurrence of a high forest fire danger in Finland, Nat. Hazards Earth Syst. Sci., 12, 2591–2601, doi:10.5194/nhess-12-2591-2012, 2012.

Mäkelä, H. M., Venäläinen, A., Jylhä, K., Lehtonen, I., and Gregow, H.: Probabilistic projections of climatological forest fire danger in Finland, Clim. Res., 60, 73–85, 2014.

Maraun, D.: Bias correction, quantile mapping, and downscaling: revisiting the inflation issues, J. Climate, 26, 2137–2143, 2013.

Migliavacca, M., Dosio, A., Camia, A., Hobourg, R., Houston-Durrant, T., Kaiser, J. W., Khabarov, N., Krasovskii, A. A., Marcolla, B., San Miguel-Ayanz, J., Ward, D. S., and Cescatti, A.: Modeling biomass burning and related carbon emissions during the 21st century in Europe, J. Geophys. Res.-Biogeo., 118, 1732–1747, 2013.

Nakićenović, N., Joseph, A., Gerald, D., de Vries, B., Fenhann, J., Graffin, S., Gregory, K., Grubler, A., Jung, T. J., Kram, T., La Rovere, E. L., Michaelis, L., Mori, S., Morita, T., Pepper, W., Pitcher, H. M., Price, L., Riahi, K., Roehrl, A., Rogner, H.-H., Sankovski, A., Schlesinger, M., Shukla, P., Smith, S. J., Swart, R., van Rooijen, S., Victor, N., and Dadi, Z.: Special Report on Emission Scenarios, A Special Report of Working Group III of the Intergovernmental Panel on Climate Change, Cambridge University Press, Cambridge, 2000.

Olsson, F., Gaillard, M.-J., Lemdahl, G., Greisman, A., Lanos, P., Marguerie, D., Marcoux, N., Skoglund, P., and Wäglind, J.: A continuous record of fire covering the last 10,500 calendar years from southern Sweden – The role of climate and human activities, Palaeogeogr. Palaeocl., 291, 128–141, 2010.

Osara, N. A.: Kulovalkeat, in: Suuri Metsäkirja, 1 osa, edited by: Kaleva, E. K., Werner Söderström osakeyhtiö, Porvoo, 455–486, 1949.

Pereira, M. G., Calado, T. J., DaCamara, C. C., and Calheiros, T.: Effects of regional climate change on rural fires in Portugal, Clim. Res., 57, 187–200, 2013.

Power, M. J., Marlon, J., Ortiz, N., Bartlein, P. J., Harrison, S. P., Mayle, F. E., Ballouche, A., Bradshaw, R. H. W., Carcaillet, C., Cordova, C., Mooney, S., Moreno, P. J., Prentice, I. C., Thonicke, K., Tinner, W., Whitlock, C., Zhang, Y., Zhao, Y., Ali, A. A., Anderson, R. S., Beer, R., Behling, H., Briles, C., Brown, K. J., Brunelle, A., Bush, M., Camill, P., Chu, G. Q., Clark, J., Colombaroli, D., Connor, S., Daniau, A.-L., Daniels, M., Dodson, J., Doughty, E., Edwards, M. E., Finsinger, W., Foster, D., Frechette, J., Gaillard, M. J., Gavin, D. G., Gobert, E., Haberle, S., Hallett, D. J., Higuera, P., Hope, G., Horn, S., Inoue, J., Kaltenrieder, P., Kennedy, L., Kong, Z. C., Larsen, C., Long, C. J., Lynch, J., Lynch, E. A., McGlone, M., Meeks, S., Mensing, S., Meyer, G., Minckley, T., Mohr, J., Nelson, D. M., New, J., Newnham, R., Noti, R., Oswald, W., Pierce, J., Richard, P. J. H., Rowe, C., Sanchez Goñi, M. F., Shuman, B. N., Takahara, H., Toney, J., Turney, C., Urrego-Sanchez, D. H., Umbanhowar, C., Vandergoes, M., Vanniere, B., Vescovi, E., Walsh, M., Wang, X., Williams, N., Wilmshurst, J., and Zhang, J. H.: Changes in fire regimes since the Last Glacial Maximum: an assessment based on a global synthesis and analysis of charcoal data, Clim. Dynam., 30, 887–907, 2008.

Räisänen, J. and Eklund, J.: 21st century changes in snow climate in northern Europe: a high-resolution view from ENSEMBLES regional climate models, Clim. Dynam., 38, 2575–2591, 2012.

Räisänen, J. and Räty, O.: Projections of daily mean temperature variability in the future: cross-validation tests with ENSEMBLES regional climate simulations, Clim. Dynam., 41, 1553–1568, 2013.

Räisänen, J. and Ylhäisi, J. S.: CO_2-induced climate change in northern Europe: CMIP2 versus CMIP3 versus CMIP5, Clim. Dynam., 45, 1877–1897, 2015.

Räty, O., Räisänen, J., and Ylhäisi, J. S.: Evaluation of delta change and bias correction methods for future daily precipitation: intermodel cross-validation using ENSEMBLES simulations, Clim. Dynam., 42, 2287–2303, 2014.

Reinikainen, A., Mäkipää, R., Vanha-Majamaa, I., and Hotanen, J.-P. (Eds): Kasvit muuttuvassa metsäluonnossa, Tammi, Helsinki, 2000.

Ruosteenoja, K. and Räisänen, P.: Seasonal changes in solar radiation and relative humidity in Europe in response to global warming, J. Climate, 26, 2467–2481, 2013.

Ruosteenoja, K., Räisänen, J., and Pirinen, P.: Projected changes in thermal seasons and the growing season in Finland, Int. J. Climatol., 31, 1473–1487, 2011.

Saari, E.: Kuloista etupäässä Suomen valtionmetsiä silmällä pitäen, Acta Forestalia Fenn., 26, 1–142, 1923.

Sherstyukov, B. G. and Sherstyukov, A. B.: Assessment of increase in forest fire risk in Russia till the late 21st century based on scenario experiments with fifth-generation climate models, Russ. Meteorol. Hydrol., 39, 292–301, 2014.

Stocks, B. J., Mason, J. A., Todd, J. B., Bosch, E. M., Wotton, B. M., Amiro, B. D., Flannigan, M. D., Hirsch, K. G., Logan, K. A., Martell, D. L., and Skinner, W. R.: Large forest fires in Canada, 1959–1997, J. Geophys. Res., 108, D8149, doi:10.1029/2001JD000484, 2002.

Tanskanen, H. and Venäläinen, A.: The relationship between fire activity and fire weather indices at different stages of the growing season in Finland, Boreal Environ. Res., 13, 285–302, 2008.

Taylor, K. E., Stouffer, R. J., and Meehl, G. A.: An overview of CMIP5 and the experimental design, B. Am. Meteorol. Soc., 93, 485–498, 2012.

Teutschbein, C. and Seibert, J.: Bias correction of regional climate model simulations for hydrological climate-change impact studies: review and evaluation of different methods, J. Hydrol., 456–457, 12–29, 2012.

Vajda, A., Venäläinen, A., Suomi, I., Junila, P., and Mäkelä, H. M.: Assessment of forest fire danger in a boreal forest environment: description and evaluation of the operational system applied in Finland, Meteorol. Appl., 21, 879–887, 2014.

van Vuuren, D. P., Edmonds, J., Kainuma, M., Riahi, K., Thomson, A., Hibbard, K., Hurtt, G. C., Kram, T., Krey, V., Lamarque, J.-F., Masui, T., Meinshausen, M., Nakicenovic, N., Smith, S. J., and Rose, S. K.: The representative concentration pathways: an overview, Clim. Change, 109, 5–31, 2011.

Van Wagner, C. E.: Development and structure of the Canadian Forest Fire Weather Index System, Forestry Technical Report 35, Canadian Forestry Service, Ottawa, ON, 1987

Van Wagner, C. E. and Pickett, T. L.: Equations and FORTRAN Program For the Canadian Forest Fire Weather Index System, Forestry Technical Report 33, Canadian Forestry Service, Ottawa, ON, 1985.

Venäläinen, A., Korhonen, N., Hyvärinen, O., Koutsias, N., Xystrakis, F., Urbieta, I. R., and Moreno, J. M.: Temporal variations and change in forest fire danger in Europe for 1960–2012, Nat. Hazards Earth Syst. Sci., 14, 1477–1490, doi:10.5194/nhess-14-1477-2014, 2014.

Voldoire, A., Sanchez-Gomez, E., Salas y Mélia, D., Decharme, B., Cassou, C., Sénési, S., Valcke, S., Beau, I., Alias, A., Chevallier, M., Déqué, M., Deshayes, J., Douville, H., Fernandez, E., Madec, G., Maisonnave, E., Moine, M.-P., Planton, S., Saint-Martin, D., Szopa, S., Tyteca, S., Alkama, R., Belamari, S., Braun, A., Coquart, L., and Chauvin, F.: The CNRM-CM5.1 global climate model: description and basic evaluation, Clim. Dynam., 40, 2091–2121, 2013.

von Salzen, K., Scinocca, J. F., McFarlane, N. A., Li, J., Cole, J. N. S., Plummer, D., Verseghy, D., Reader, M. C., Ma, X., Lazare, M., and Solheim, L.: The Canadian fourth generation atmospheric global climate model (CanAM4) – Part I: representation of physical processes, Atmos. Ocean, 51, 104–125, 2013.

Wallenius, T.: Major decline in fires in coniferous forests – reconstructing the phenomenon and seeking for the cause, Silva Fenn., 45, 139–155, 2011.

Watanabe, M., Suzuki, T., O'ishi, R., Komuro, Y., Watanabe, S., Emori, S., Takemura, T., Chikira, M., Ogura, T., Sekiguchi, M., Takata, K., Yamazaki, D., Yokohata, T., Nozawa, T., Hasumi, H., Tatebe, H., and Kimoto, M.: Improved climate simulation by MIROC5: mean states, variability, and climate sensitivity, J. Climate, 23, 6312–6335, 2010.

Wilcke, R. A. I., Mendlik, T., and Gobiet, A.: Multi-variable error correction of regional climate models, Clim. Change, 120, 871–887, 2013.

Xiao, J. and Zhuang, Q.: Drought effects on large fire activity in Canadian and Alaskan forests, Environ. Res. Lett., 2, 044003, doi:10.1088/1748-9326/2/4/044003, 2007.

Yang, W., Gardelin, M., Olsson, J., and Bosshard, T.: Multi-variable bias correction: application of forest fire risk in present and future climate in Sweden, Nat. Hazards Earth Syst. Sci., 15, 2037–2057, doi:10.5194/nhess-15-2037-2015, 2015.

12

Variations in water storage in China over recent decades from GRACE observations and GLDAS

X. Mo, J. J. Wu, Q. Wang, and H. Zhou

Academy of Disaster Reduction and Emergency Management, MOCA/MOE, Beijing Normal University, Beijing 100875, China

Correspondence to: J. J. Wu (jjwu@bnu.edu.cn)

Abstract. We applied Gravity Recovery and Climate Experiment (GRACE) Tellus products in combination with Global Land Data Assimilation System (GLDAS) simulations and data from reports, to analyze variations in terrestrial water storage (TWS) in China as a whole and eight of its basins from 2003 to 2013. Amplitudes of TWS were well restored after scaling, and showed good correlations with those estimated from models at the basin scale. TWS generally followed variations in annual precipitation; it decreased linearly in the Huai River basin ($-0.56\,\mathrm{cm\,yr^{-1}}$) and increased with fluctuations in the Changjiang River basin ($0.35\,\mathrm{cm\,yr^{-1}}$), Zhujiang basin ($0.55\,\mathrm{cm\,yr^{-1}}$) and southeast rivers basin ($0.70\,\mathrm{cm\,yr^{-1}}$). In the Hai River basin and Yellow River basin, groundwater exploitation may have altered TWS's response to climate, and TWS kept decreasing until 2012. Changes in soil moisture storage contributed over 50 % of variance in TWS in most basins. Precipitation and runoff showed a large impact on TWS, with more explained TWS in the south than in the north. North China and southwest rivers region exhibited long-term TWS depletions. TWS has increased significantly over recent decades in the middle and lower reaches of Changjiang River, southeastern coastal areas, as well as the Hoh Xil, and the headstream region of the Yellow River in the Tibetan Plateau. The findings in this study could be helpful to climate change impact research and disaster mitigation planning.

1 Introduction

Terrestrial water storage (TWS) is a key component of the global hydrological cycle and plays a critical role in Earth's climate system (Famiglietti, 2004). Despite its importance, there are still many gaps in the existing water storage observation networks at both the global and regional scale (Lettenmaier and Famiglietti, 2006). Although recent advances in satellite imaging and altimetry have strengthened our monitoring capability over a vast area, these technologies primarily provide only variation information for single factors related to TWS, such as precipitation estimates, surface soil moisture, snow cover, and river/lake level. With the progress in satellite gravimetric techniques, direct observation of TWS has become available. The Gravity Recovery and Climate Experiment (GRACE) twin satellites were launched in 2002 as a joint space mission between NASA (US) and DLR (German) to observe variations in Earth's gravity field. Over land, these observations provide information of integrated water storage changes in the vertical profile, including surface water reservoirs, upper layers of soil, and underground water reservoirs.

At global, regional, and basin scale, GRACE data have been applied to analyze seasonal cycle characteristics of TWS (Schmidt et al., 2006; Syed et al., 2008; Strassberg et al., 2007). Because of their sensitivity to water amount over large areas, GRACE data can also be a useful tool for identifying impact caused by extreme climate events like droughts and floods, or for tracking climate change's influence on local water resources (Andersen et al., 2005; Long et al., 2013; Phillips et al., 2012). Scientists found that inclusion of GRACE-based total water storage information allows the predisposition of a river basin to flooding to be assessed as much as 5–11 months in advance (Reager et al., 2014). Chen et al. (2009, 2010) quantified an extreme drought in 2005 and an extreme flood in 2009 in the Amazon river, and found that local interannual TWS changes are closely connected to ENSO events in the tropical Pacific. Because of the lack of direct observations independent of GRACE TWS, TWS estimated from the atmospheric water balance, land water balance, and model simulations was used to compare with and verify the GRACE TWS (Yirdaw et al., 2008; Zeng et al., 2008; Syed et al., 2005; Schmidt et al., 2006). These papers have demonstrated that GRACE data are capable of identifying seasonal and long-term variations in TWS and have also made contributions to the development of climate and hydrological models.

GRACE TWS combined with hydrological information from other observations or models could help us further understand and manage variables in the hydrological cycle. Land surface model simulations were used to infer the roles of water components (snow water, canopy water, and soil water) in GRACE terrestrial water storage change (TWSC) and to understand the effect of hydrologic fluxes fluctuation on water storage (Syed et al., 2008; Kim et al., 2009). With the help of TRMM data and NOAA's Climate Prediction Center (CPC) model simulations, Crowley et al. (2008) found that the source (precipitation) is more important than sink (evapotranspiration and runoff) to the water balance in the Amazon basin. Other papers tried to separate variations in groundwater storage from GRACE TWS, and their results showed agreement with in situ observations (Rodell et al., 2009; Leblanc et al., 2009; Famiglietti et al., 2011; Jin and Feng, 2013).

In China, GRACE data have been compared with several model simulations and used to extract TWS's spatial and temporal variation characteristics as well as its responses to droughts (Duan et al., 2007; Zhong et al., 2009; Wang and Yang, 2013; Hu et al., 2006; Xu et al., 2013; Tang et al., 2013, 2014). The serious TWS depletion in north China has gained much attention in recent years (Su et al., 2011; Moiwo et al., 2009; Feng et al., 2013). Previous research has mostly focused on characteristics of the seasonal cycle in TWS in certain regions, and there has been less further analysis of long-term variations over China. Moreover, early research was typically limited by the short period of data availability and the obsolete version, and leakage errors in TWS from processed GRACE data could also misguide analyses at the regional scale.

China is one of the countries that is confronted with problems of water scarcity and has suffered several regional extreme climate events in recent decades. The knowledge of TWS variations over recent decades is necessary for understanding the large-scale water storage variation process. In this study, the GRACE Tellus land products and Global Land Data Assimilation System (GLDAS) products, combined with data records from national water resources bulletins, were used to analyze long-term TWS variations in China as a whole, as well as in its eight major basins. This study could give guidance to water resource management and future research on areas with critical water storage changes in China. For better understanding, main acronyms and variables used in this paper are listed in Table 1.

2 Data and methods

2.1 Data

The monthly grids from GRACE Tellus land data are applied to analyze TWS variations. The product is derived from the latest Release-05 spherical harmonics, which is an improvement over the previous 04 version. Several institutions provide gravity solutions, such as the University of Texas Center for Space Research (CSR), NASA's Jet Propulsion Laboratory (JPL), and Deutsches GeoForschungsZentrum (GFZ). A recent comparison suggested that TWS estimates from GFZ, CSR, and JPL solutions were highly correlated with one other, and tiny differences among them were within the margin of solution's error (Sakumura et al., 2014). Among these three products, the one from CSR had the smallest root-mean-square (RMS) of deviations between the ensemble mean and itself in 156 basins around the world. In this study, we chose products derived from CSR's solution for the following analyses. The paper by Swenson and Wahr (2006) describes the details of the post-processing for the spherical harmonics. The final grid (1° in both latitude and longitude) values are presented in the form of changes in equivalent water thickness (unit: cm) relative to a time–mean baseline. The data period is from January 2003 to December 2013 and months of absent data are as follows: June 2003, January and June 2011, May and October 2012, and March, August, and September 2013. Grid-scale factors, which correspond to the gridded product, were used to partially correct leakage errors and restore the amplitude-damping caused by the filtering process. Errors and uncertainties in mass variation can be computed from the scaled gridded data (Launderer and Swenson, 2012). In this paper, gridded fields of scale factors and error estimates provided along with the GRACE Tellus products were calculated following Landerer's method based on NCAR's CLM4 model (Oleson et al., 2008). Recently, Long et al. (2015) conducted comprehensive comparisons to assess skills and uncertainties of different approaches for processing GRACE data to restore signal losses caused by spatial filtering.

Monthly flux/state variables (Table 2) from GLDAS (Rodell et al., 2004) were applied to estimate water storage variations, and these variables were also used to address variations in the components of water storage. GLDAS drives four land surface models: Mosaic (Koster and Suarez, 1996), Noah (Chen et al., 1996; Koren et al., 1999; Betts et al., 1997; Ek et al., 2003), Community Land Model (CLM) (Bonan et al., 1998; Dickinson et al., 1993; Dai and Zeng, 1997), and Variable Infiltration Capacity (VIC) (Liang et al., 1994, 1996). Satellite-based and ground-based observations are integrated into these models to generate optimal fields of land

Table 1. List of main acronyms and variables used in the paper.

Acronym/ variable	Full name
GRACE	Gravity Recovery and Climate Experiment
NASA	National Aeronautics and Space Administration
DLR	Deutsches Zentrum für Luft- und Raumfahrt
CSR	University of Texas, Center for Space Research
JPL	Jet Propulsion Laboratory
GFZ	Deutsches GeoForschungsZentrum
CMAP	Climate Prediction Center Merged Analysis of Precipitation
CPC	Climate Prediction Center
CMA	China Meteorological Administration
GLDAS	Global Land Data Assimilation System
CLM	Community Land Model
VIC	Variable Infiltration Capacity model
PDF	Probability density function
RMS	Root-mean-square
TWS	Terrestrial water storage
TWSC	Terrestrial water storage change
SWE	Snow water equivalent
CWS	Canopy water storage
SM	Soil moisture
P	Precipitation
E	Evapotranspiration
Q	Runoff

Table 2. Variables used from the GLDAS simulations.

GLDAS variables	Unit	Temporal resolution	Spatial resolution
Soil moisture (SM)	kg m^2	Monthly	$1° \times 1°$
Snow water equivalent (SWE)	kg m^2	Monthly	$1° \times 1°$
Canopy water storage (CWS)	kg m^2	Monthly	$1° \times 1°$
Precipitation (P)	$\text{kg m}^{-2}\text{s}^{-1}$	Monthly	$1° \times 1°$
Evapotranspiration (E)	$\text{kg m}^{-2}\text{s}^{-1}$	Monthly	$1° \times 1°$
Runoff (Q)	$\text{kg m}^{-2}\text{s}^{-1}$	Monthly	$1° \times 1°$

surface states and fluxes. The forcings for these models from 2001 to present are a combination of NOAA/GDAS atmospheric analysis fields, spatially and temporally disaggregated NOAA Climate Prediction Center Merged Analysis of Precipitation (CMAP) fields (Xie and Arkin, 1996) and observation-based downward shortwave and long-wave radiation fields from the Air Force Weather Agency (AFWA).

Drainage networks are mostly distributed in the monsoon-dominated middle and east China (Fig. 1a), which are also highly populated areas with high levels of water consumption. In this study, we specifically focus on eight large basins: Heilongjiang River, Liao River, Hai River, Huai River, Yellow River, Changjiang River, Zhujiang, southeast rivers (with abbreviations of HLJ, LR, HaiR, HuaiR, YR, CJ, ZJ, and SERs, respectively, in the following tables). Desert is the dominant land cover in northwestern China, while glacier, snow cover, and frozen soil are widely distributed across the Tibetan Plateau (Fig. 1b). Vector data for the desert are acquired from the Data Sharing Infrastructure of Earth System Science (http://www.geodata.cn). The Second Glacier Inventory Dataset of China (Version 1.0) is acquired from Science Data Center for Cold and Arid Regions (http://westdc.westgis.ac.cn). Annual Chinese water resources bulletins from 2003 to 2012 are acquired from the Ministry of Water Resources to assist with the analysis. Values for surface water resources, groundwater resources and gross water resources provided in the water resources bulletins are the results of existing monitoring and statistical analyses (Fig. 5). Surface water resources refer to water storage in rivers, lakes, and glaciers, and groundwater resources mainly refer to water storage in underground shallow aquifers.

2.2 Methods

2.2.1 Data preprocessing

Unlike scale factor applied for region-averaged TWS time series in previous research (Chen et al., 2007; Landerer et al., 2010; Feng et al., 2013), monthly products from GRACE Tellus land data were multiplied by grid-scale factors to restore signal attenuation. Next, the average value for each grid from Janunary in 2003 to December 2013 was subtracted from all other scaled monthly grids. The deviations to time-averaged TWS were used for the following analyses. At the regional scale, all grids in a basin were averaged with the cosine of latitude as the weight, and missing values for absent months were interpolated from adjacent available months. For regionally averaged TWS, total errors were calculated from error fields provided along with the GRACE products (Eqs. 1, 2, Table 5). Because of spatial correlation among neighboring grids, covariance was considered in the calculation of regional-scale error $\text{Error}_{\text{region}}$ (Landerer and Swenson, 2012; Eq. 1). The dist in Eq. (1) is the geometric distance between any two grids in the basin (unit: km); n is the number of valid grids in a specific basin; β is the de-correlation length, which is set to 300 for measurement error and 100 for leakage error; i and j indicate the value in the ith column and jth row of the grid data. The regional-scale total error $\text{Error}_{\text{total}}$ included both regional-scale measurement error $\text{Error}_{\text{measure}}$ and regional-scale leakage error $\text{Error}_{\text{leakage}}$ (Eq. 2). Early analysis suggested that the TWS variations could be distinguished from GRACE monthly data over regions larger than 200 000 km^2, with an accuracy of 1.5 cm equivalent water thickness (Rodell and Famiglietti, 1999), and the larger the spatial scale of the research area was, the better the accuracy the results could acquire (Swenson and Wahr, 2003; Wahr et al., 2004).

$$\text{Error}_{\text{region}} = \sqrt{\sum_{i=1}^{n}\sum_{j=1}^{n} \text{Error}_i \cdot \text{Error}_j \cdot e^{-\left(\text{dist}_{i,j}^2\right)/\left(2\cdot\beta^2\right)}} \Big/ n \quad (1)$$

$$\text{Error}_{\text{total}} = \sqrt{(\text{Error}_{\text{measure}})^2 + \left(\text{Error}_{\text{leakage}}\right)^2} \quad (2)$$

State variables (snow water equivalent, SWE; canopy water storage, CWS; total soil moisture storage in all layers, SM)

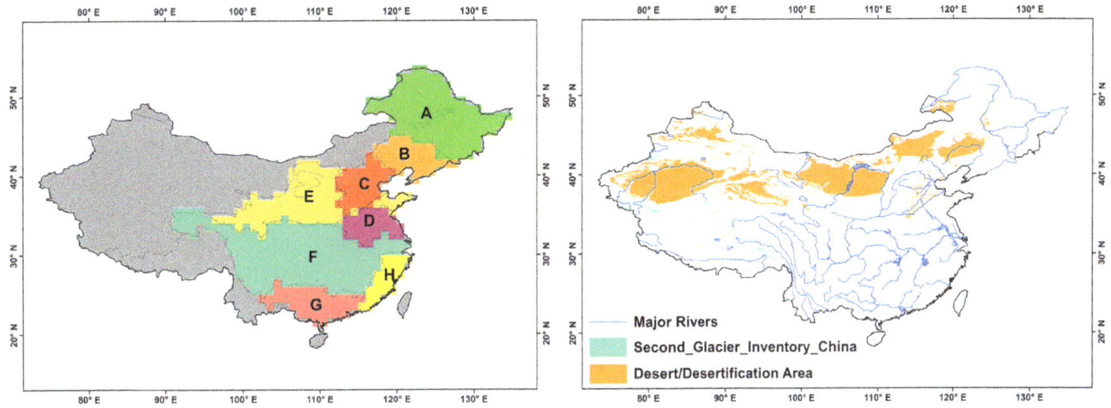

Figure 1. Schematic diagrams of research area (A: Heilongjiang River, B: Liao River, C: Hai River, D: Huai River, E: Yellow River, F: Changjiang River, G: Zhujiang, H: southeast rivers).

and flux variables (precipitation, P; evapotranspiration, E; runoff, Q) from the four models in GLDAS were presented in the form of equivalent water thickness (unit: cm). The time averages were removed from these variables following the process used for the GRACE data to keep the same time base for comparison. The ensemble mean (arithmetic average) of the four models' simulations was also calculated to be used as the representative of model results.

2.2.2 TWS estimates from model simulations

In despite of deficiencies in model simulations, state variables SWE, CWS, and SM from GLDAS outputs can be combined to estimate TWS (Eq. 3). Although the estimates are not able to fully reflect the information in the actual TWS variations, they can still capture the fluctuation and magnitude in land hydrology, which is necessary for assessing and understanding the TWS observation from GRACE (Syed et al., 2004).

$$\text{TWS} = \text{SWE} + \text{CWS} + \text{SM} \quad (3)$$

Pearson correlation coefficients R between TWS from scaled GRACE and model simulations are listed in Table 3. TWS estimates from CLM and VIC have relatively poor correlations with GRACE observations at the national scale. However, all model estimates generally have high correlation coefficients at the regional scale, except in the Heilongjiang River basin. The differences between models and regions showed that model simulations have a high degree of uncertainties, and TWS estimates from NOAH and the GLDAS ensemble mean have a good agreement with TWS from scaled GRACE at both the national and regional scale. The differences between the GLDAS simulations and the GRACE observations are mainly the result of missing information on components of land hydrology, such as groundwater and reservoirs, and poor parameterization (snow cover, frost soil, etc.) in the model mechanism (Syed et al., 2008). These components or processes could be critical to TWS in some parts of the world (Rodell and Famiglietti, 2001). The RMS of deviations from the ensemble mean was calculated as the bias of TWS estimates from the GLDAS simulations (Table 4).

Table 3. Pearson correlation coefficients r between regionally averaged TWS from the scaled GRACE data and model simulations in China as a whole and eight of its basins.

Region	CLM	VIC	MOSAIC	NOAH	Ensemble mean
HLJ	0.83	0.84	0.74	0.87	0.86
LR	0.71	0.65	0.54	0.64	0.64
HaiR	0.43	0.54	0.66	0.61	0.61
HuaiR	0.68	0.54	0.68	0.79	0.72
YR	0.77	0.62	0.62	0.70	0.69
CJ	0.61	0.51	0.47	0.77	0.60
ZJ	0.70	0.77	0.79	0.82	0.81
SERs	0.70	0.69	0.83	0.76	0.81
CHN	0.25	0.31	0.53	0.79	0.55

Table 4. Error statistics of regionally averaged TWS (unit: cm).

Region	Area (km^2)	Measurement error	Leakage error	Total error	Bias for GLDAS
China	9 510 610	0.38	0.31	0.54	0.63
Changjiang River	1 815 855	0.90	0.79	1.27	1.47
Heilongjiang River	956 832	0.98	0.72	1.39	1.10
Yellow River	860 883	0.78	0.73	1.10	0.72
Zhujiang	463 050	1.86	1.62	2.63	3.35
Hai River	327 096	1.30	1.36	1.84	1.59
Liao River	310 881	1.13	0.98	1.60	1.61
Huai River	288 152	1.74	1.54	2.46	3.51
Southeast rivers	242 524	1.44	1.29	2.04	3.69

2.2.3 Scaling effect assessment

To understand how many changes scale factors could make to GRACE TWS in China, we compared the spatial distributions of the amplitude of TWS variations from observations and simulations. The RMS of the time series at each grid was taken as a proxy for local TWS amplitude, and then the empirical probability density functions (PDFs) for RMS values over China were calculated for TWS derived from scaled and unscaled GRACE data and the model simulations. To avoid abnormal values, only RMS values between the 5th and 95th quantile over China were considered. For regionally averaged TWS, the slopes and coefficients of determination R^2 were calculated with a linear least squares fit to assess the damping influence of leakage errors (Eq. 4).

$$\mathrm{TWS}_{\mathrm{scaled}} = a \cdot \mathrm{TWS}_{\mathrm{unscaled}} \quad (4)$$

2.2.4 Analysis of TWS long-term variations

TWSC, the integrated changes in the vertical components of TWS, is the difference between current and previous months' TWS (Eq. 5). This value can also be inferred using the water balance with precipitation, evapotranspiration and runoff data in a specific basin (Hirschi et al., 2006, Eq. 6). With GLDAS and GRACE products, we applied correlation analysis to find out how much state/flux variables can contribute to TWSC's variance in different basins. After applying a 13-point moving average to remove intra-annual variations in times series, we analyzed annual variations based on the regionally averaged TWS from scaled GRACE and the GLDAS ensemble mean, in combination with annual Chinese water resources bulletins. Annual data from the water resources bulletins were converted from volume (unit: million m^3) to equivalent water thickness (unit: cm), and the multi-year average was removed. To identify a major area with significant TWS increase or depletion in the recent decade, a linear trend of scaled GRACE TWS for each grid was calculated based on linear regression, and the long-term trends of seasonal average TWS were also analyzed. Grids with trends which passed the F test (significant at 95 % confidence level) are marked with black dots in Figs. 7 and 8.

$$\mathrm{TWSC}_N = \mathrm{TWS}_N - \mathrm{TWS}_{N-1} \quad (5)$$

$$\mathrm{TWSC} = P - E - Q \quad (6)$$

3 Results and discussion

3.1 Effect of the scaled factor in China

The effect of the truncation ($\mathrm{Order}_{\max} = 60$) and filtering processes (300 km Gaussian filtering) on the GRACE spherical harmonics is equivalent to a low-pass filter; thus the effective resolution of the GRACE TWS product is several hundred kilometers (Tapley et al., 2004). The TWS time series in a 1° grid was mixed with TWS signals from the surrounding area, leading to leakage errors. When the outside TWS signal was stronger than the inside, the grid value was exaggerated by leakage errors and vice versa. In addition, the sign of gridded TWS could even be changed in cases where inside and outside TWS signals had opposite phases caused by extreme changes in topography, such as in the Turpan basin in northwestern China. As relationships between TWS series at different spatial scales were inferred from land hydrology model simulations, grid-scale factors calculated based on this information could partially correct GRACE TWS and to some extent recover small-scale information (Landerer and Swenson, 2012); thus, these scale factors can be quite helpful for extracting TWS over arbitrary shaped region.

The RMS value of TWS time series in a specific grid is an indicator for the amplitude of local TWS. And the empirical probability density distribution (empirical PDF) curve for RMS values in the research region described the statistical distribution of TWS amplitude within the area. In Fig. 2, empirical PDF curves based on TWS data from modeled TWS data (MOSAIC, VIC, CLM, NOAH, and GLDAS ensemble mean) and observation TWS data (scaled and unscaled GRACE data) were compared. Empirical PDF curves based on scaled GRACE data and modeled data (except CLM) all showed a larger RMS value range on the x axis than that based on unscaled GRACE data. This means the range of TWS amplitude within the research area has been stretched after scaling. In addition, empirical PDF curves based on scaled GRACE data and most modeled data showed that RMS values concentrated in the relatively low numerical zone, with lowest RMS values close to 0 cm. Spatially, areas with low RMS values correspond to northwest China, which is an arid climate zone with vast deserts (Figs. 1, 3b and c). From the comparison in Fig. 3, we can also see that scaled GRACE TWS has a similar distribution of amplitude to that from the GLDAS ensemble mean over China, particularly the boundary with RMS of 3cm, separating arid and humid climate zones. TWS is quite stable over some part of the oceans and major deserts around the world; thus a small RMS for TWS in these areas indicates small data noise in GRACE TWS (Sakumura et al., 2014). Both the empirical PDFs and the spatial distribution of the RMS of the TWS suggested that grid-scale factors could correct the amplitude of TWS in space. Previous research has demonstrated that correction for leakage is critical to regional TWS analysis (Chen et al., 2014).

Regionally averaged TWS time series from scaled and unscaled GRACE data are highly correlated, and this means that the fluctuation process in TWS has not been heavily influenced by the scale factors. At the same time, the values of TWS were all amplified to different degrees (Table 5), with the amplitudes in the Huai River basin and Zhujiang basin increasing over 50 %. However, in the Liao River basin and Yellow River basin, only small changes occurred (< 10 %). The slopes in Table 5 can be regarded as the basin-specific scale factors for GRACE TWS. Generally, basins with large area are less affected by leakage errors and have slopes close to 1.

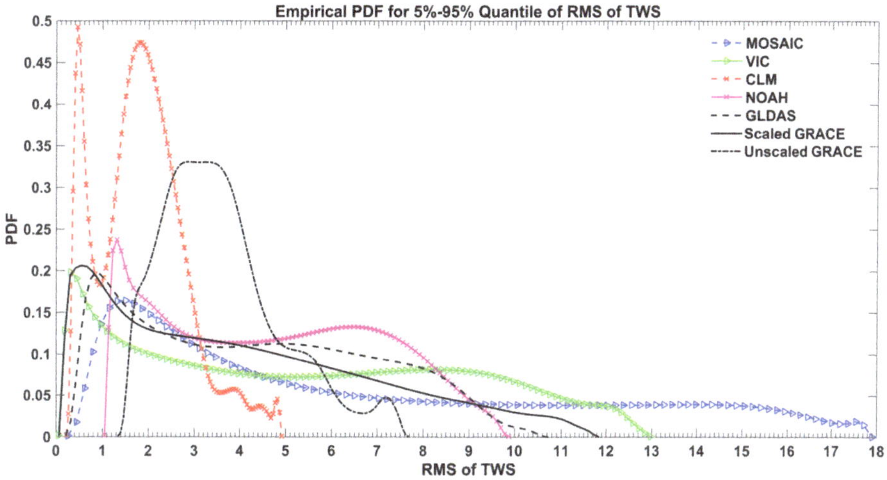

Figure 2. Empirical probability density distributions of root-mean-square of TWS from the scaled GRACE data, the unscaled GRACE data, and model simulations (only TWS values between the 5th and 95th quantiles are considered); unit of RMS is centimeters.

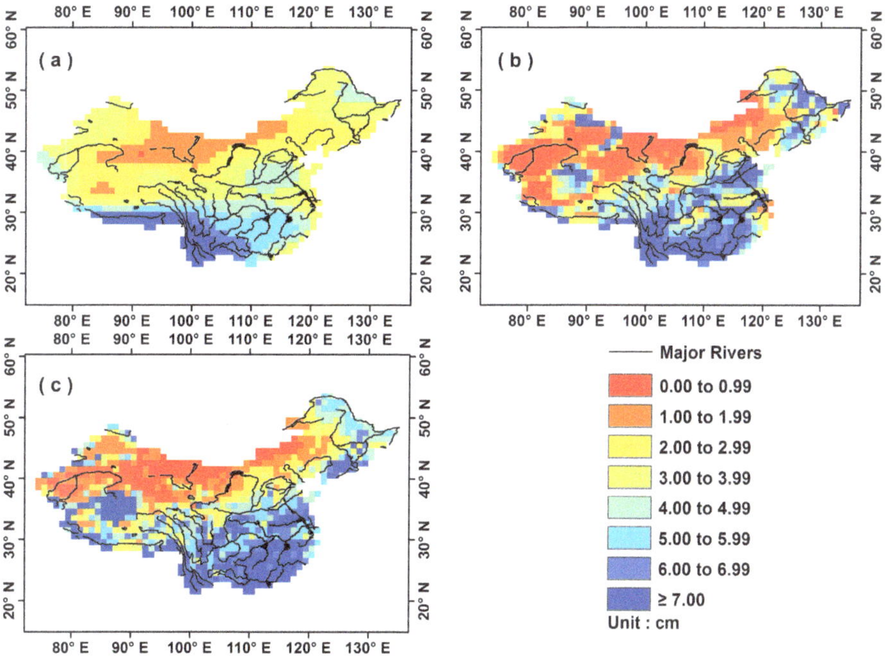

Figure 3. **(a)** Root-mean-square of TWS from the unscaled GRACE data, **(b)** the scaled GRACE data, and **(c)** the GLDAS ensemble mean.

Table 5. Slopes of linear least squares fittings for basin-averaged TWS from the scaled and the unscaled GRACE data calculated in Eq. (4).

Region	HLJ	LR	HaiR	HuaiR	YR	CJ	ZJ	SERs	CHN
Factor	1.26	1.10	1.32	1.57	1.08	1.34	1.54	1.11	1.19
R^2	0.996	0.989	0.988	0.994	0.982	0.992	0.996	0.991	0.982

3.2 Annual variations in regional TWS time series

In general, fluctuations in annual precipitation could appropriately characterize the interannual variability in regionally averaged TWS, but distinct processes also exist in certain basins or over certain periods because of the influence of other factors (Figs. 4, 6). TWS in China was at a relatively high level before 2006, but then stayed at a continuously low level with a high variability from 2006 to 2012. During this

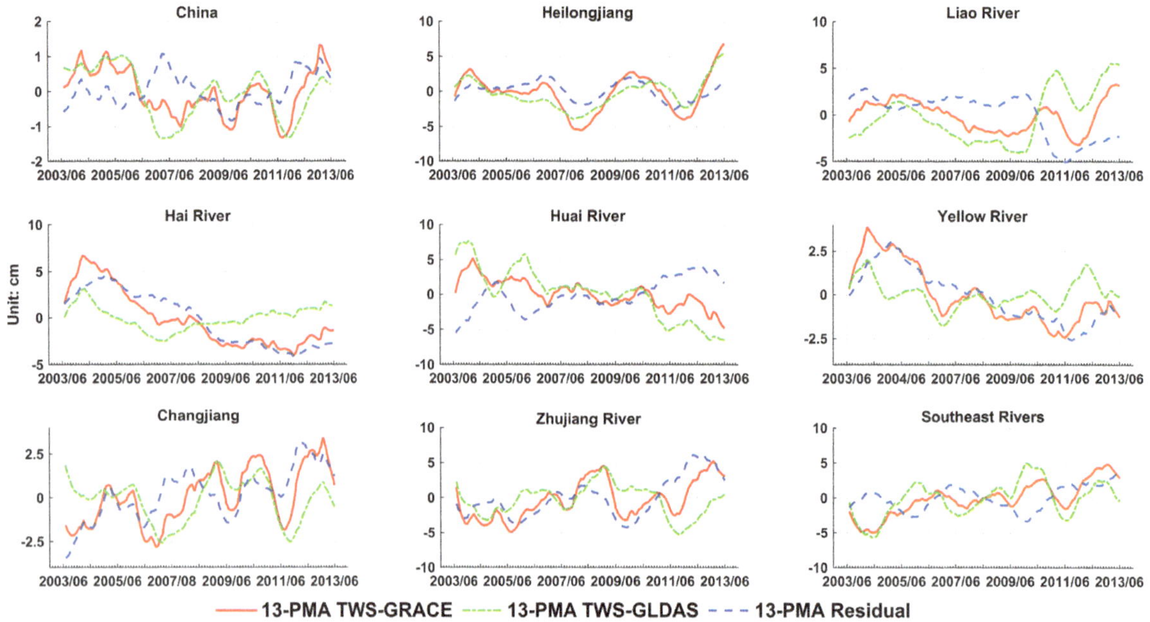

Figure 4. Regionally averaged monthly TWS series (2003–2013) from the scaled GRACE data and the GLDAS ensemble mean, and their residual time series for China and eight of its basins (all time series have been processed with 13-point moving average; unit: cm).

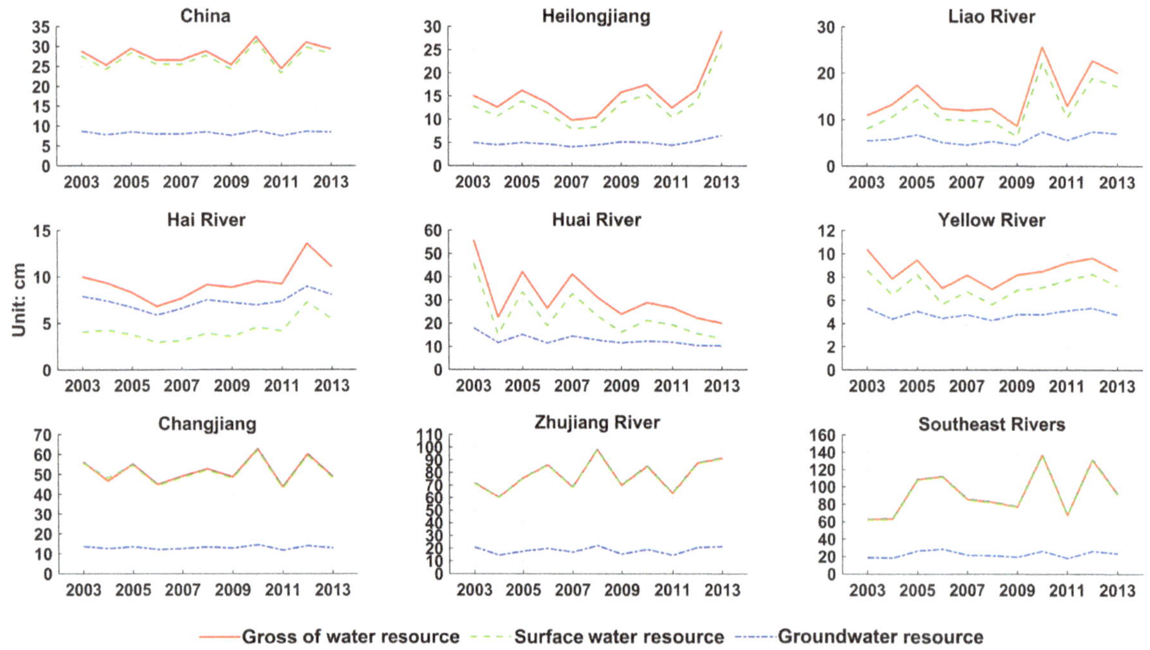

Figure 5. Annual variations of water resources in China as a whole and eight of its basins from 2003 to 2013 (unit: cm).

period, severe droughts occurred frequently and caused particularly sharp declines in TWS in 2007, 2009, and 2011. TWS in China did not recover to the same level as in 2005 until 2013. This periodic process was partially reflected in the TWS estimates from the GLDAS ensemble mean (mostly soil moisture storage) but not in the residual series or water resources records (Figs. 4, 5).

In northeast China, TWS observations and simulation estimates in the Heilongjiang River basin were consistent and showed no long-term trend; the region mainly suffered from two severe regional droughts in 2007 and 2011 and a significant basin flood event in 2013. Annual precipitation in the Liao River basin continuously declined from 2005 to 2009 and then increased rapidly in following years; TWS estimates

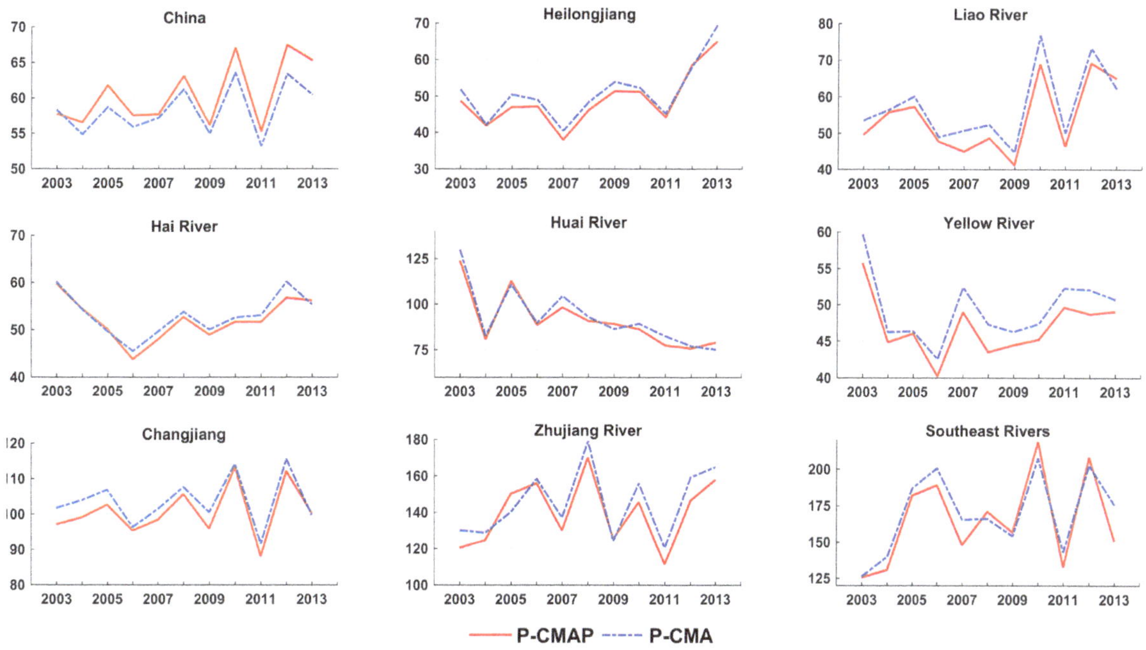

Figure 6. Regionally averaged annual precipitation for China and eight of its basins from 2003 to 2013 (unit: cm). *P* from CMAP refers to precipitation grid data from GLDAS forcing, and *P* from CMA refers to precipitation grid data based on station observations provided by China Meteorological Administration.

and water resources records both captured this process precisely. Nevertheless, it seemed that the TWS observations failed to respond to heavy precipitation in 2010, with a 9 cm increase in TWS estimates and only a 3 cm increase in TWS observations.

In north China, annual precipitation in Hai River basin changed following a V-shaped process, with year 2006 as the turning point, which can also be recognized from the TWS estimates and water resources records. After a rapid decline from 2004 to 2006 ($-2.48\,\mathrm{cm\,yr^{-1}}$), the TWS from the scaled GRACE data in Hai River basin became stable around 2007. Contrary to increasing precipitation, TWS dropped 3 cm in 2008 and continued to decline slowly at a rate of $-0.22\,\mathrm{cm\,yr^{-1}}$ until it started to recover in 2012. The TWS in Hai River basin generally showed a linear decreasing trend ($-1.27\,\mathrm{cm\,yr^{-1}}$) during 2004–2011. The residual between TWS from scaled GRACE data and the GLDAS ensemble mean could be treated approximately as the sum of surface reservoir and groundwater storage. Moreover, detection depth of GRACE is much deeper than the layers considered in models (1.90 m in VIC, 2.00 m in NOAH, 3.50 m in Mosaic, and 3.43 m in CLM) and in field monitoring (shallow aquifer). Although the increasing precipitation seemed to have alleviated the depletion trend in these areas, we should not ignore the large gap ($-1.80\,\mathrm{cm\,yr^{-1}}$) between the trends of the time series of the residuals and summed water resources records (-1.28 and $0.50\,\mathrm{cm\,yr^{-1}}$) during 2006–2011. The gap between these time series probably suggests that the long-term effect of over-exploitation of groundwater still remained, even though water-saving management practices had already been carried out in this basin, and water storage would suffer even worse depletion in the future drought years. Similar to the Hai River basin, the TWS from scaled GRACE data in the Yellow River basin followed a nearly linear decreasing trend ($-0.73\,\mathrm{cm\,yr^{-1}}$) during 2004–2011, and it changed more slowly ($0.13\,\mathrm{cm\,yr^{-1}}$) after 2007. The basin-averaged TWS, gross water resource, and precipitation also showed different processes in the latter half of research period. However, Fig. 7 reveals that the area with large long-term decreasing trends is mainly located midstream of the Yellow River basin (Shanxi and Shaanxi provinces), which is famous for coal mining. To identify the exact causes for decreasing TWS, more local statistical data and groundwater level records should be collected. Over recent decades, the TWS in the Huai River basin has shown a long-term descending trend ($-0.56\,\mathrm{cm\,yr^{-1}}$), which is similar to annual precipitation over this basin. The TWS from GLDAS ensemble mean also showed good agreement.

Annual variations in TWS from scaled GRACE data, the GLDAS ensemble mean, and water resources records are more similar across basins in south China than they are in north China. The TWS in the Changjiang River basin, Zhujiang basin, and southeast rivers basin all followed an increasing trend from 2003 to 2013, at 0.35, 0.55, and $0.70\,\mathrm{cm\,yr^{-1}}$, respectively. As a result of typhoons and tropical storms, the TWS in these basins also had much stronger fluctuations than in north China.

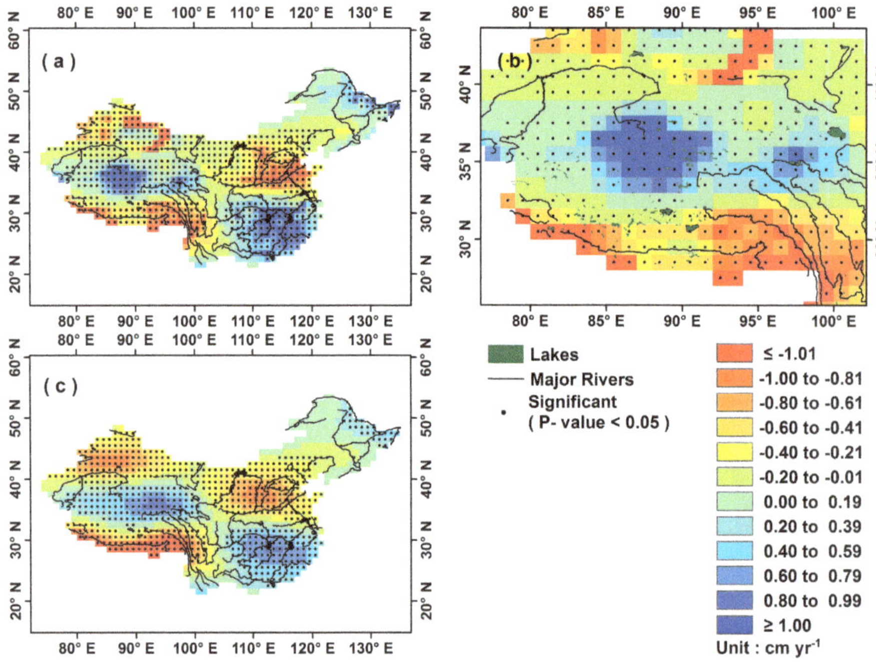

Figure 7. Spatial distribution of trends derived from linear least squares regression for TWS in 2003–2013 (unit: cm yr^{-1}); **(a)** and **(b)** are linear trends from the scaled GRACE data and a detailed diagram of the data for west part of China, **(c)** is the linear trend from the unscaled GRACE data. Grids with trends significant (F test) at 95 % confidence level are marked with black dots.

The TWS variations observed by GRACE are integrated information from different components in the vertical profile. Compared to water storage in surface soil layers, snow cover and canopy water storage are almost negligible in most regions, and analysis results in China suggested that changes in soil moisture contributed significantly to TWSC (Table 6). The percentage of TWSC variance explained by SMC could be as high as 62 % at the national scale. In most basins, over half of the TWSC variance could be attributed to SMC, with high percentages in the Changjiang River basin and Zhujiang basin (64 and 67 %). In the Heilongjiang River basin, SMC played a less important role in TWSC (38 %). Fluctuations in hydrologic fluxes over the basin jointly affected water storage. According to correlation analysis based on GRACE TWS and GLDAS fluxes, precipitation, evapotranspiration, and runoff each contributed 46, 41, and 32 %, respectively, to the TWSC variance in China (Table 6). As most basins we focused on are under control of the monsoon climate, precipitation and runoff generally showed higher contributions to the TWSC variance than evapotranspiration did. Overall, precipitation was found to have a much higher impact on TWSC in the south than in the north, with the highest explained variance in the Zhujiang basin (60 %), followed by that in the Changjiang River basin (44 %).

3.3 Spatial pattern of linear trend analysis

When focusing on differences between large regions, spatial patterns of linear trends calculated from scaled and unscaled

Table 6. Coefficient of determination R^2, based on Pearson correlation, between precipitation (P), evapotranspiration (E), runoff (Q), soil moisture change (SMC) from GLDAS ensemble mean, and TWSC from the scaled GRACE data in 2003–2013.

	HLJ	LR	HaiR	HuaiR	YR	CJ	ZJ	SERs	CHN
P	0.04	0.20	0.18	0.35	0.41	0.44	0.60	0.23	0.46
E	0.00	0.08	0.10	0.09	0.28	0.25	0.30	0.00	0.32
Q	0.08	0.26	0.291	0.31	0.38	0.47	0.50	0.18	0.41
SMC	0.38	0.52	0.527	0.57	0.46	0.64	0.67	0.51	0.67

GRACE TWS are consistent (Fig. 7a, c). But at a local scale, results from scaled GRACE TWS better correspond to natural features of the TWS intensity distribution. Areas around river networks usually have a high level of TWS, thus presenting large absolute values of trends. From 2003 to 2013, four main regions were identified with intensive and significant long-term trends in TWS. Results also revealed that seasons in a year made different contributions to these trends (Fig. 8).

According to the analysis in the previous section, we inferred that human activities rather than climate parameters could be responsible for the significant TWS depletion in north China, as withdrawals usually surpass net recharge in arid and semiarid regions. Severe areas are mainly located in the province of Shanxi and the southern part of the Hebei province, with decreasing trends less than $-0.80\,\mathrm{cm\,yr^{-1}}$. And offset to loss rate caused by mass gains from reser-

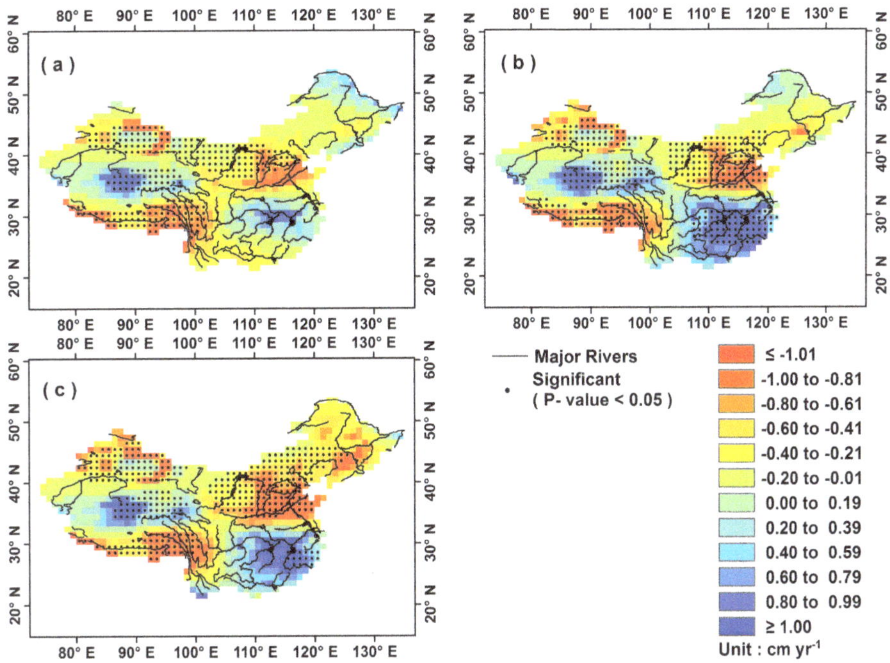

Figure 8. Spatial distribution of trends derived from linear least squares regression for seasonally averaged TWS in 2003–2013 from the scaled GRACE data (unit: $cm\,yr^{-1}$); **(a)** spring (March–May); **(b)** summer (June–August); **(c)** autumn (September–November). Grids with trends significant (F test) at 95 % confidence level are marked by black dots.

voir regulation, water diversion and coal transport in this region was estimated to $0.76\,cm\,yr^{-1}$ (Tang et al., 2013). In the Shanxi province, the east to mid-section of Yellow River, coal mining not only has disturbed normal recharge to the nearby aquifer but has also caused over-exploitation of groundwater. The groundwater is a major source of water consumption in the Huang–Huai–Hai Plain; agricultural irrigation consumes large amounts of freshwater pumped from deep wells every year (Foster et al., 2004; Kendy et al., 2004). This poor condition deteriorates with seasons and the depletion becomes most severe, impacting the largest area in autumn. Considering irrigation demand concentrated mostly in MMA and high social water consumption comes with JJA, TWS probably needs time to show all this influence in SON. The southwest rivers region (Yarlung Zangbo River, Nu River, and Lancang River) also showed significant TWS depletion, particularly in the upstream and downstream portions of the Yarlung Zangbo River. The area impacted by significant depletion was the largest in spring, while the depletion also became the most severe in autumn. Climate observation across this region proved that annual precipitation was decreasing over 11 years, with significant droughts in 2006, 2009, and 2012. Moreover, previous research also found ice loss in Himalaya from 2003 to 2007 (Matsuo and Heki, 2010).

Along with increasing precipitation in southeast China, there were significant increasing trends in TWS over the middle and lower reaches of the Changjiang River and southeastern coastal area. The main contribution to this significant increase occurred during the summer, when precipitation was the most concentrated of the year. There are two places that showed significant TWS increases in the Tibetan Plateau. One of these areas is around the Hoh Xil Mountains and the other is the headstream region of the Yellow River; they have maximum increasing trends of 2.59 and $1.77\,cm\,yr^{-1}$, respectively. The Hoh Xil Mountains are located at the intersection of the inland lakes in the Qangtang Plateau and the north source of the Changjiang River. The headstream region of the Yellow River lies at the plateau's two largest freshwater lakes, Eling Lake and Zaling Lake. Previous research applied multi-source satellite data to reconstruct volume changes in the Tibetan Plateau's major lakes, and found that they showed similar spatial distribution to mass variations in GRACE during 2003–2010 (Song et al., 2013). According to satellite images, lakes in Hoh Xil overall showed a trend of expansion during 2000–2011. Further analysis suggested that increasing precipitation and decreasing evaporation were major factors contributing to this trend, and additional water recharge from melting glacier and frozen soil caused by climate warming was a minor factor (Yao et al., 2014; Duan et al., 2007). In the headstream region of the Yellow River, precipitation is the main recharge source to runoff, with a ratio of 63 %. Local observations revealed that there was an increasing trend in runoff as this region was becoming warmer and wetter during the period 2000–2012 (Lan et al., 2010, 2013; Wang et al., 2014). The Chinese government has launched an ecological protection and construction project in the Three-River Source region that started in 2005. According to monitoring data from the Qinghai Provincial Meteorological Bureau, average lake extents during 2005–2012 showed an increase of 34.7 and 64.4 km^2, compared to those during 2003–2004 for Eling Lake and Zaling Lake, respectively.

In addition to the above large areas, there are also some other small regions showing strong TWS changes from 2003 to 2013. Along the northeast country border, there is significant TWS increase ($0.34 \sim 1.19\,cm\,yr^{-1}$); this is mostly contributed by winter. The central part of the province of Jilin in northeast China shows severe TWS depletion in autumn ($-1.10 \sim -2.28\,cm\,yr^{-1}$), but these linear trends have not passed the significance test (P value < 0.05). In northwest China, the unscaled GRACE data only show significant water depletion mainly around Tianshan Mountains, which is also identified with ice loss in previous research (Matsuo and Heki, 2010). However, trends from the scaled GRACE TWS also illustrate a significant TWS increase in the Turpan basin, while there is a depletion in its surrounding mountains. The Turban basin is the lowest basin in China, and Fig. 3b shows that the basin has a much smaller TWS amplitude than that in surrounding mountains. An extreme arid climate and local topography features in this region could make TWS more sensitive to climate change. Complex terrain in this region leads to a more complicated GRACE TWS signal mixture at a large spatial scale. Even though scale factors might have separated mixed TWS signals, limitations in factors' production should also be taken care of (Launderer and Swenson, 2012).

4 Summary and conclusions

In this study we analyzed annual variations in TWS over 11 years in China as a whole and eight of its basins, based on scaled GRACE data in combination with GLDAS simulations and water resources records. Areas with significant long-term trends were also identified and discussed. The major points are summarized as follows.

1. Grid-scale factors could adequately correct leakage errors in the GRACE products, and the scaled data gained more spatial details of the TWS intensity distribution. The values of the regionally averaged TWS were amplified after scale factors were applied. These increased percentages reached up to over 50 % in the Huai River basin (57 %) and Zhujiang basin (54 %), but were tiny for basins with larger sizes, such as the Liao River basin (10 %) and Yellow River basin (8 %).

2. The TWS at the national scale stayed at a relatively low level. These values exhibited high-intensity variations from 2006 to 2012, before recovering to their 2003–2005 condition. The TWS in the Hai River basin, Huai River basin, and Yellow River basin almost decreased linearly, while it increased in fluctuations in the Changjiang River basin, Zhujiang basin, and southeast rivers basin. The TWS variations generally followed the variations in annual precipitation at the basin scale, but they showed inverse changes in 2007–2013 in both the Hai River basin and Yellow River basin.

3. Changes in soil moisture storage contributed 62 % of the TWSC variance at the national scale, and the percentages were generally beyond 50 % in all basins with exceptions in the Heilongjiang River basin (38 %) and Yellow River basin (46 %). Under the control of the monsoon climate, precipitation and runoff explained more variance in TWSC than evapotranspiration did, and the precipitation's ability to explain TWSC variations was stronger in the south basins than in the north, reaching up to 60 % in the Zhujiang basin.

4. From 2003 to 2013, the southwest rivers region and north China showed significant water storage depletions, and the area of depletion was largest in spring and summer, respectively. The middle and lower reaches of the Changjiang River and southeastern coastal area, as well as the Hoh Xil Mountains, and the headstream region of the Yellow River in the Tibetan Plateau, all exhibited significant increases in TWS. These identified trends reflected TWS's responses to regional climate changes and human activities.

The current data period of GRACE products is shorter than some existing remote sensing data sets or site records, and the resolution and accuracy of GRACE data also need to be improved. However, TWS from GRACE has proved to be valuable in understanding large-scale hydrological processes over land. The results in this study would be helpful for water resources management and climate change impact research. More sources of data will be added to further analyze regions or phenomena addressed in this study. The GRACE Follow-On mission has already been scheduled, and will continue to support monitoring and research on TWS in the future.

Acknowledgements. This research received financial support from the International Science & Technology Cooperation Program of China (grant number: 2013DFG21010). The GLDAS simulations were provided by Goddard Earth Sciences (GES) Data and Information Services Center (DISC) (http://disc.sci.gsfc.nasa.gov/services/disc/services/grads-gds/gldas). GRACE land data (available at http://grace.jpl.nasa.gov) processing algorithms were provided by Sean Swenson, and supported by the NASA MEaSUREs Program; the authors would also like to thank the anonymous reviewers for the valuable comments that helped to improve the manuscript.

References

Andersen, O. B., Seneviratne, S. I., Hinderer, J., and Viterbo, P.: GRACE-derived terrestrial water storage depletion associated with the 2003 European heat wave, Geophys. Res. Lett., 32, L18405, doi:10.1029/2005GL023574, 2005.

Betts, A. K., Chen, F., Mitchell, K., and Janjic, Z. I.: Assessment of the land surface and boundary layer models in two operational versions of the NCEP Eta model using FIFE data, Mon. Weather Rev., 125, 2896–2916, 1997.

Bonan, G. B., Oleson, K. W., Vertenstein, M., Levis, S., Zeng, X. B., Dai, Y. J., Dickinson, R. E., and Yang, Z. L.: The land surface climatology of the NCAR Land Surface Model coupled to the NCAR Community Climate Model, J. Climate, 11, 1307–1326, 1998.

Chen, F., Mitchell, K., Schaake, J., Xue, Y. K., Pan, H. L., Koren, V., Duan, Q. Y., Ek, M., and Betts, A.: Modeling of land-surface evaporation by four schemes and comparison with FIFE observations, J. Geophys. Res., 101, 7251–7268, 1996.

Chen, J. L., Wilson, C. R., Famiglietti, J. S., and Rodell, M.: Attenuation effect on seasonal basin-scale water storage changes from GRACE time-variable gravity, J. Geodesy., 81, 237–245, 2007.

Chen, J. L., Wilson, C. R., Tapley, B. D., Yang, Z. L., and Niu, G. Y.: The 2005 drought event in the Amazon River basin as measured by GRACE and estimated by climate models, J. Geophys. Res., 114, B05404, doi:10.1029/2008JB006056, 2009.

Chen, J. L., Wilson, C. R., and Tapley, B. D.: The 2009 exceptional Amazon flood and interannual terrestrial water storage change observed by GRACE, Water Resour. Res, 46, W12526, doi:10.1029/2010WR009383, 2010.

Chen, J. L., Li, J., Zhang, Z. Z., and Ni, S. N.: Long-term groundwater variations in Northwest India from satellite gravity measurements, Global Planet. Change, 116, 130–138, 2014.

Crowley, J. W., Mitrovica, J. X., Bailey, R. C., Tamisiea, M. E., and Davis, J. L.: Annual variations in water storage and precipitation in the Amazon Basin, J. Geodesy., 82, 9–13, 2008.

Dai, Y. J. and Zeng, Q. C.: A land surface model (IAP94) for climate studies, Part I: Formulation and validation in off-line experiments, Adv. Atmos. Sci., 14, 443–460, 1997.

Dickinson, R. E., Henderson-Sellers, A., and Kennedy, P. J.: Biosphere–atmosphere Transfer Scheme (BATS) Version 1e as Coupled to the NCAR Community Climate Model, NCAR Technical Note NCAR/TN-387+STR, NCAR, doi:10.5065/D67W6959, 1993.

Duan, J. B., Zhong, M., Yan, W. M., and Jiang, M.: Recovery of land water storage variations in Chinese mainland by use of grace data, J. Geodesy., 27, 68–71, 2007 (in Chinese).

Ek, M. B., Mitchell, K. E., Lin, Y., Rogers, E., Grunmann, P., Koren, V., Gayno, G., and Tarpley, J. D.: Implementation of Noah land surface model advances in the National Centers for Environmental Prediction operational mesoscale Eta model, J. Geophys. Res., 108, 8851, doi:10.1029/2002JD003296, 2003.

Famiglietti, J. S.: Remote sensing of terrestrial water storage, soil moisture and surface waters, in: The State of the Planet: Frontiers and Challenges in Geophysics, edited by: Sparks, R. S. J. and Hawkesworth, C. J., American Geophysical Union, Washington, D.C., 197–207, 2004.

Famiglietti, J. S., Lo, M., Ho, S. L., Bethune, J., Anderson, K. J., Syed, T. H., Swenson, S. C., de Linage, C. R., and Rodell, M.: Satellites measure recent rates of groundwater depletion in California's Central Valley, Geophys. Res. Lett., 38, L03403, doi:10.1029/2010GL046442, 2011.

Feng, W., Zhong, M., Lemoine, J. M., Biancale, R., Hsu, H. T., and Xia, J.: Evaluation of groundwater depletion in North China using the Gravity Recovery and Climate Experiment (GRACE) data and ground-based measurements, Water Resour. Res., 49, 2110–2118, 2013.

Foster, S., Garduno, H., Evans, R., Olson, D., Tian, Y., Zhang, W. Z., and Han, Z. S.: Quaternary aquifer of the North China Plain – assessing and achieving groundwater resource sustainability, Hydrogeol. J., 12, 81–93, 2004.

Hirschi, M., Seneviratne, S. I., and Schar, C.: Seasonal variations in terrestrial water storage for major mid-latitude river basins, J. Hydrometeorol., 7, 39–60, 2006.

Hu, X. G., Chen, J. L., Zhou, Y. H., Cheng, H., and Liao, X. H.: Seasonal water storage change of the Yangtze River Basin detected by GRACE, Sci. China Ser. D, 49, 483–491, 2006.

Jin, S. G. and Feng, G. P.: Large-scale variations of global groundwater from satellite gravimetry and hydrological models, 2002–2012, Global Planet. Change, 106, 20–30, 2013.

Kendy, E., Zhang, Y. Q., Liu, C. M., Wang, J. X., and Steenhuis, T.: Groundwater recharge from irrigated cropland in the North China Plain: case study of Luancheng County, Hebei Province, 1949–2000, Hydrol. Process., 18, 2289–2302, 2004.

Kim, H., Pat Yeh, J. F., Oki, T., and Kanae, S.: Role of rivers in the seasonal variations of terrestrial water storage over global basins, Geophys. Res. Lett., 36, L17402, doi:10.1029/2009GL039006, 2009.

Koren, V., Schaake, J., Mitchell, K., Duan, Q. Y., Chen, F., and Baker, J. M.: A parameterization of snowpack and frozen ground intended for NCEP weather and climate models, J. Geophys. Res., 104, 19569–19585, 1999.

Koster, R. D. and Suarez, M. J.: Energy and Water Balance Calculations in the Mosaic LSM, NASA Technical Memorandum 104606, NASA, 9, 76 pp., 1996.

Lan, Y. C., Zhao, G. H., Zhang, Y. N., Wen, J., Liu, J. Q., and Hu, X. L.: Response of runoff in the source of the Yellow River to climate warming, Quatern. Int., 226, 60–65, 2010.

Lan, Y. C., Lu, C. Y., La, C. F., and Sheng, Y. P.: The factor of climate shift to warm-humid in the source regions of the Yellow River and its hydrologic response, J. Glaciol. Geocryl., 35, 920–928, 2013 (in Chinese).

Landerer, F. W. and Swenson, S. C.: Accuracy of scaled GRACE terrestrial water storage estimates, Water Resour. Res., 48, W04531, doi:10.1029/2011WR011453, 2012.

Landerer, F. W., Dickey, J. O., and Gntner, A.: Terrestrial water budget of the Eurasian pan-Arctic from GRACE satellite measurements during 2003–2009, J. Geophys. Res., 115, D23115, doi:10.1029/2010JD014584, 2010.

Leblanc, M. J., Tregoning, P., Ramillien, G., Tweed, S. O., and Fakes, A.: Basin-scale, integrated observations of the early 21st century multiyear drought in southeast Australia, Water Resour. Res., 45, W04408, doi:10.1029/2008WR007333, 2009.

Lettenmaier, D. P. and Famiglietti, J. S.: Water from on high, Nature, 444, 562–563, 2006.

Liang, X., Lettenmaier, D. P., Wood, E. F., and Burges, S. J.: A simple hydrologically based model of land surface water and energy fluxes for GSMs, J. Geophys. Res., 99, 14415–14428, 1994.
Liang, X., Lettenmaier, D. P., and Wood, E. F.: One-dimensional statistical dynamic representation of subgrid spatial variability of precipitation in the two-layer variable infiltration capacity model, J. Geophys. Res., 101, 21403–21422, 1996.
Long, D., Scanlon, B. R., Longuevergne, L., Sun, A. Y., Fernando, D. N., and Save, H.: GRACE satellites monitor large depletion in water storage in response to the 2011 drought in Texas, Geophys. Res. Lett., 40, 3395–3401, 2013.
Long, D., Longuevergne, L., and Scanlon, B. R.: Global analysis of approaches for deriving total water storage changes from GRACE satellites, Water Resour. Res., 51, 2574–2594, 2015.
Matsuo, K. and Heki, K.: Time-variable ice loss in Asian high mountains from satellite gravimetry, Earth Planet. Sc. Lett., 290, 30–36, 2010.
Moiwo, J. P., Yang, Y., Li, H., Han, S., and Hu, Y.: Comparison of GRACE with in situ hydrological measurement data shows storage depletion in Hai River Basin, northern China, Water SA, 35, 663–670, 2009.
Oleson, K. W., Niu, G. Y., Yang, Z. L., Lawrence, D. M., Thornton, P. E., Lawrence, P. J., Stöckli, R., Dickinson, R. E., Bonan, G. B., Levis, S., Dai, A., and Qian, T.: Improvements to the community land model and their impact on the hydrological cycle, J. Geophys. Res., 113, G01021, doi:10.1029/2007JG000563, 2008.
Phillips, T., Nerem, R. S., Fox-Kemper, B., Famiglietti, J. S., and Rajagopalan, B.: The influence of ENSO on global terrestrial water storage using GRACE, Geophys. Res. Lett., 39, L16705, doi:10.1029/2012GL052495, 2012.
Reager, J. T., Thomas, B. F., Famiglietti, J. S.: River basin flood potential inferred using GRACE gravity observations at several months lead time, Nat. Geosci., 7, 588–592, 2014.
Rodell, M. and Famiglietti, J. S.: Detectability of variations in continental water storage from satellite observations of the time dependent gravity field, Water Resour. Res., 35, 2705–2723, 1999.
Rodell, M. and Famiglietti, J. S.: An analysis of terrestrial water storage variations in Illinois with implications for the Gravity Recovery and Climate Experiment (GRACE), Water Resour. Res., 37, 1327–1339, 2001.
Rodell, M., Houser, P. R., Jambor, U., Gottschalck, J., Mitchell, K., Meng, C. J., Arsenault, K., Cosgrove, B., Radakovich, J., Bosilovich, M., Entin, J. K., Walker, J. P., Lohmann, D., and Toll, D.: The Global Land Data Assimilation System, B. Am. Meteorol. Soc., 85, 381–394, 2004.
Rodell, M., Velicogna, I., and Famiglietti, J. S.: Satellite-based estimates of groundwater depletion in India, Nature, 460, 999–1002, 2009.
Sakumura, C., Bettadpur, S., and Bruinsma, S.: Ensemble prediction and intercomparison analysis of GRACE time-variable gravity field models, Geophys. Res. Lett., 41, 1389–1397, 2014.
Schmidt, R., Schwintzed, P., Flechtner, F., Reigbera, C., Güntnera, A., Döllb, P., Ramillienc, G., Cazenavec, A., Petrovica, S., Jochmanna, H., and Wünsch, J.: GRACE observations of changes in continental water storage, Global Planet. Change, 50, 112–126, 2006.
Song, C. Q., Huang, B., and Ke, L. H.: Modeling and analysis of lake water storage changes on the Tibetan Plateau using multimission satellite data, Remote Sens. Environ., 135, 25–35, 2013.
Strassberg, G., Scanlon, B. R., and Rodell, M.: Comparison of seasonal terrestrial water storage variations from GRACE with groundwater-level measurements from the High Plains Aquifer (USA), Geophys. Res. Lett., 34, L14402, doi:10.1029/2007GL030139, 2007.
Su, X. L., Ping, J. S., and Ye, Q. X.: Terrestrial water variations in the North China Plain revealed by the GRACE mission, Sci. China Earth Sci., 54, 1965–1970, 2011.
Swenson, S. C. and Wahr, J.: Monitoring changes in continental water storage with GRACE, Space Sci. Rev., 108, 345–354, 2003.
Swenson, S. C. and Wahr, J.: Post-processing removal of correlated errors in GRACE data, Geophys. Res. Lett., 33, L08402, doi:10.1029/2005GL025285, 2006.
Syed, T. H., Lakshmi, V., Paleologos, E., Lohmann, D., Mitchell, K., and Famiglietti, J. S.: Analysis of process controls in land surface hydrological cycle over the continental United States, J. Geophys. Res., 109, 1–11, 2004.
Syed, T. H., Famiglietti, J. S., Chen, J. L., Rodell, M., Seneviratne, S. I., Viterbo, P., and Wilson, C. R.: Total basin discharge for the Amazon and Mississippi river basins from GRACE and a land–atmosphere water balance, Geophys. Res. Lett., 32, L24404, doi:10.1029/2005GL024851, 2005.
Syed, T. H., Famiglietti, J. S., Rodell, M., Chen, J. L., and Wilson, C. R.: Analysis of terrestrial water storage changes from GRACE and GLDAS, Water Resour. Res., 44, W02433, doi:10.1029/2006WR005779, 2008.
Tang, J. S., Cheng, H. W., and Liu, L.: Assessing the recent droughts in Southwestern China using satellite gravimetry, Water Resour. Res., 50, 3030–3038, 2014.
Tang, Q. H., Zhang, X. J., and Tang, Y.: Anthropogenic impacts on mass change in North China, Geophys. Res. Lett., 40, 3924–3928, 2013.
Tapley, B., Bettadpur, S., Ries, J., Thompson, P., and Watkins, M.: GRACE measurements of mass variability in the Earth system, Science, 305, 503–505, 2004.
Wahr, J., Swenson, S. C., Zlotnicki, V., and Velicogna, I.: Time-variable gravity from GRACE: first results, Geophys. Res. Lett., 31, L11501, doi:10.1029/2004GL019779, 2004.
Wang, C. and Yang, T.: Spatial–temporal characteristics of terrestrial water storage variation in China based on GRACE satellite data, Water Resour. Power, 31, 20–23, 2013 (in Chinese).
Wang, H., Li, D. L., and Jiang, Y. C.: Characteristics and reasons of the runoff variation in source regions of the Yellow River during 1956–2012, J. Glaciol. Geocryol., 36, 403–412, 2014 (in Chinese).
Xie, P. and Arkin, P. A.: Global precipitation: a 17-year monthly analysis based on gaugeobservations, satellite estimates, and numerical model outputs, B. Am. Meteorol. Soc., 78, 2539–2558, 1996.
Xu, M., Ye, B. S., and Zhao, Q. D.: Temporal and spatial pattern of water storage changes over the Yangtz river basin during 2002–2010 based on GRACE satellite data, Progress in Geography, 32, 68–77, 2013 (in Chinese).
Yao, X. J., Liu, S. Y., Li, L., Sun, M. P., and Luo, J.: Spatial–temporal characteristics of lake area variations in Hoh Xil region from 1970 to 2011, J. Geogr. Sci., 24, 689–702, 2014.
Yirdaw, S. Z., Snelgrove, K. R., and Agboma, C. O.: GRACE satellite observations ofterrestrial moisture changes for drought

characterization in the Canadian Prairie, J. Hydrol., 356, 84–92, 2008.

Zeng, N., Yoon, J. H., Mariotti, A., and Swenson, S.: Variability of basin-scale terrestrial water storage from a P-E-R water budget method: the Amazon and the Mississippi, J. Climate, 21, 248–265, 2008.

Zhong, M., Duan, J. B., Xu, H. Z., Peng, P., Yan, H. M., and Zhu, Y. Z.: Trend of China land water storage redistribution at medi- and large-spatial scales in recent five years by satellite gravity observations, Chinese Sci. Bull., 54, 816–821, 2009.

13

Assessment of physical vulnerability of buildings and analysis of landslide risk at the municipal scale: application to the Loures municipality, Portugal

C. Guillard-Gonçalves, J. L. Zêzere, S. Pereira, and R. A. C. Garcia

RISKam, CEG, Institute of Geography and Spatial Planning, Universidade de Lisboa, Lisbon, Portugal

Correspondence to: C. Guillard-Gonçalves (cguillard@campus.ul.pt)

Abstract. This study offers a semi-quantitative assessment of the physical vulnerability of buildings to landslides in a Portuguese municipality (Loures), as well as the quantitative landslide risk analysis computed as the product of the landslide hazard by the vulnerability and the economic value of the buildings. The hazard was assessed by combining the spatiotemporal probability and the frequency–magnitude relationship of the landslides. The physical vulnerability assessment was based on an inquiry of a pool of European landslide experts and a sub-pool of landslide experts who know the study area, and the answers' variability was assessed with standard deviation. The average vulnerability of the basic geographic entities was compared by changing the map unit and applying the vulnerability to all the buildings of a test site, the inventory of which was listed on the field. The economic value was calculated using an adaptation of the Portuguese Tax Services approach, and the risk was computed for different landslide magnitudes and different spatiotemporal probabilities. As a rule, the vulnerability values given by the sub-pool of experts who know the study area are higher than those given by the European experts, namely for the high-magnitude landslides. The obtained vulnerabilities vary from 0.2 to 1 as a function of the structural building types and the landslide magnitude, and are maximal for 10 and 20 m landslide depths. However, the highest risk was found for the landslides that are 3 m deep, because these landslides combine a relatively high frequency in the Loures municipality with a substantial potential damage.

1 Introduction

Landslides are natural phenomena that can cause costly damage when occurring in or impacting constructed areas. Landslide risk analysis is used to estimate the risk of landslide hazard to individuals, populations, properties, or the environment (Fell et al., 2008; Corominas et al., 2014, 2015) and generally contains five main steps: (i) hazard identification, (ii) hazard assessment, (iii) inventory of elements at risk and exposure, (iv) vulnerability assessment, and (v) risk estimation. Landslide risk analysis is useful to locate the zones where the risk is highest, but it is a complex and time-consuming task, especially when the study is conducted at the municipal scale.

During the last three decades, the landslide risk (R) has been considered as the product of the landslide hazard (H), the vulnerability (V), and the value of the elements at risk (EV) (Varnes and the IAEG Commission on Landslides and other Mass-Movements, 1984; Michael-Leiba et al., 1999; Cardinali et al., 2002; Remondo et al., 2005; Uzielli et al., 2008; van Westen et al., 2008; Zêzere et al., 2008): $R = H \times V \times EV$, where R is the risk (annual loss of property value). Landslide hazard (H) is the probability of occurrence within a specified period of time and within a given area of a potentially damaging phenomenon (Varnes and the IAEG Commission on Landslides and other Mass-Movements, 1984) having a given magnitude (Jaiswal et al., 2011a), which is typically measured with the landslide area or the landslide volume (Lee and Jones, 2004; Li et al., 2010). The vulnerability (V) concept is defined in physical terms as the "degree of loss" of a given element or set of elements at risk exposed to the occurrence of a landslide of a given magnitude, expressed in a scale ranging from 0 (no loss) to 1 (total loss) (e.g. Varnes and the IAEG Commission on Landslides and other Mass-Movements, 1984; Remondo et al., 2008). The value of the elements at risk is the economic value (EV) of the elements at risk, which in this study correspond to the built environment.

Whereas the landslide susceptibility and the landslide hazard have been extensively studied in the last two decades, whether with heuristic, statistic-probabilistic, or deterministic methods (e.g. Fell et al., 2008; Corominas et al., 2014), less work has been done, for various reasons, on the spatial assessment of landslide vulnerability and on the assessment of the value of the elements at risk (e.g. Zêzere et al., 2007, 2008; Papathoma-Köhle et al., 2012a; Silva and Pereira, 2014).

First, for most types of landslide, very limited damage data are available (van Westen et al., 2005; Papathoma-Köhle et al., 2012a), which hamper the creation and validation of any reliable vulnerability model. Second, different physical mechanisms are associated with different types of landslides, which mean that the same elements at risk have different vulnerability to different types of landslides. Therefore, the method used for assessing rockfall vulnerability would not be directly transferable to the slow slide vulnerability assessment (Alexander, 2005; Papathoma-Köhle et al., 2011; Ciurean et al., 2013). Third, the vulnerability of the elements at risk depends on the landslide intensity, which is usually associated with the landslide velocity (Hungr, 1997; Lateltin et al., 2005) that may range from some millimetres per year to several metres per second (Cruden and Varnes, 1996).

Moreover, methods used to assess vulnerability should be selected according to the scope and the scale of the study, which influences the level of spatial detail requested (Papathoma-Köhle et al., 2011). A vulnerability study conducted at the municipal level typically implies the existence of a large number of elements at risk (e.g. buildings) and details about building characteristics and landslide damage. Due to this reason, landslide vulnerability assessment is usually performed in small study areas with a reduced number of exposed elements in order to ease the methodology demonstration (e.g. Uzielli et al., 2015).

Previous studies have attempted to assess the landslide vulnerability and to analyse the landslide risk. Some of them are qualitative, focusing on human lives (e.g. Santos, 2003) and in both buildings and human lives (Macquarie et al., 2004). Other physical vulnerability studies are semi-quantitative, assigning empirical weighting of a set of building resistance parameters to buildings exposed to landslides (e.g. Silva and Pereira, 2014), or applying vulnerability curves to buildings exposed to hydrometeorological hazards (e.g. Godfrey et al., 2015).

Quantitative vulnerability studies usually aim to estimate the physical vulnerability of buildings based on landslide intensity parameters (e.g. impact energy, average velocity) and resistance or susceptibility of the exposed elements (e.g. structure type, construction material, maintenance state) (e.g. Uzielli et al., 2008, 2015; Li et al., 2010; Du et al., 2013; Peng et al., 2015). Most of the time, landslide intensity parameters can be quantified (e.g. landslide velocity), while proposed values for resistance or susceptibility of the exposed buildings are usually assigned based on expert opinion (Peng et al., 2015; Uzielli et al., 2015), which may increase the subjectivity and uncertainty of the vulnerability estimation. In addition, expert surveys can be used to estimate physical vulnerability using the standard deviation of the expert answers to measure the variability of the average vulnerability (Winter et al., 2014).

Physical vulnerability assessment has several sources of uncertainty that can be either epistemic or aleatory (Ciurean et al., 2013). Epistemic uncertainties can come from the use of proxies for the landslide intensity assessment (e.g. velocity, depth of affected material, volume), or from the characterization of elements at risk (e.g. structural-morphological characteristics, state of maintenance, strategic relevance), from the vulnerability model (e.g. selection of parameters, mathematic model, calculation limitations), or from expert judgement about building resistance parameters and landslide damaging potential (Ciurean et al., 2013). Aleatory uncertainties come from the spatial variability of parameters (e.g. landslide intensities, population density) (Ciurean et al., 2013). For instance, the position of the element at risk (e.g. a building) on the track of a landslide is a source of aleatory uncertainty as the damage would not be the same if it is located on the crown of the landslide or on its run-out zone (van Westen et al., 2005).

Some examples of non-site-specific studies on landslide risk to buildings are available in the technical literature (e.g. Michael-Leiba et al., 1999; Cardinali et al., 2002; Remondo et al., 2008; Uzielli et al., 2008; Zêzere et al., 2008; Jaiswal et al., 2010, 2011b; Uzielli et al., 2015). Despite the progress already made, major limitations persist on the reliable assessment of landslide frequency and magnitude (which are both critical for the hazard assessment), and on the quantification of the buildings' vulnerability, which is frequently based on expert opinion. This work aims to contribute to the fulfilling of a research gap on the physical vulnerability assessment based on expert opinion. The main purposes of the study are to develop and apply a method for building vulnerability assessment in a Portuguese municipality (Loures), and to analyse the landslide risk to buildings in this study area.

Following the previous work of Guillard and Zêzere (2012), the susceptibility of the slopes was modelled for deep-seated and shallow slides, and the hazard was assessed, considering the magnitude probability of the landslide area and the annual and multiannual spatiotemporal probability of landslides.

In this study, there are few records on building damage caused by landslides, which constitutes a drawback in the construction and validation of the vulnerability model. Due to this reason, buildings' physical vulnerability assessment was based on expert judgment of a pool of European landslide experts. In addition, from this pool, we extracted a sub-pool constituted by experts that have been working in the study area, i.e. who have a deep knowledge of both the landslides and the built environment of the study area. With this methodology, we aimed to evaluate the variability of the expert judgments, comparing the answers from the pool of landslide European experts with the answers from the sub-pool of landslide experts who know the study area, assessing thus the epistemic uncertainty in buildings' vulnerability assessment and evaluating how vulnerability controls risk results.

The market economic value of the buildings was assessed per pixel and the buildings' landslide risk of the municipality of Loures was assessed for different spatiotemporal probabilities using pixel units in a GIS environment.

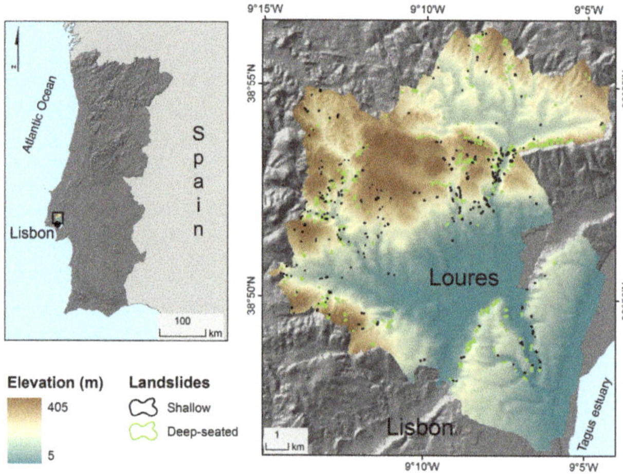

Figure 1. Loures municipality location, elevation, and location of the 686 inventoried landslides.

2 Study area

For various reasons we chose to analyse the risk of slides triggered by rainfall in the Loures municipality, near Lisbon. First, this municipality is prone to different natural hazards and in particular to landslides. Most of the landslides in the Loures municipality are rotational or translational and are triggered by rainfall (Zêzere et al., 2004, 2008). Landslides were classified according to the depth of slip surface in two groups: shallow slides (slip surface depth ≤ 1.5 m) and deep-seated slides (slip surface depth > 1.5 m). The landslide inventory includes 333 shallow slides (average area 961 m^2) and 353 deep-seated slides (average area 3806 m^2). Velocity of landslides is typically slow for shallow slides and very slow or extremely slow for deep-seated slides, according to Cruden and Varnes' (1996) classification. These landslides often affect buildings and roads with significant direct and indirect consequences. Out of 686 landslides (Fig. 1) inventoried by Guillard and Zêzere (2012), 462 occurred within 50 m of buildings and roads, and some of them had caused damage to a built environment in the past (Zêzere et al., 2008).

Second, Loures is adjacent to the city of Lisbon (Fig. 1), hence a large number of inhabitants, buildings, and infrastructures are exposed to landslide hazard; indeed, about 205 000 persons currently live in the Loures municipality (density around 1220 inhabitants per km^2), which is 6 % higher than in 2001 according to the National Institute of Statistics (INE, 2002, 2011). The mean age of the buildings is 37.5 years, 66.9 % of them with a structure of reinforced concrete, 30.6 % of masonry, 1.8 % of adobe, rammed earth, or loose stone, and 0.7 % of other materials (INE, 2011). The 32 495 buildings of the Loures municipality represent a total built-up area of 9.25 km^2 and the number of buildings, most of which were erected without taking into account the possibility of future landslide occurrence, increase every year. Indeed, according to the results obtained in the framework of the new Master Plan for the Lisbon Metropolitan Area, the construction on potentially unstable slopes within the Loures municipality increased by 64 % between 1995 and 2007.

Third, a study on the susceptibility of slopes to landslides was previously conducted in this municipality (Guillard and Zêzere, 2012). Therefore, we intend to complete the risk analysis for buildings in this study area.

Finally, a social vulnerability assessment was conducted for the greater Lisbon area (Guillard-Gonçalves et al., 2015), which opens up an avenue for a future study that combines these two dimensions of the vulnerability.

Additional information about the study area can be found in Guillard and Zêzere (2012).

3 Data and methods

The frequency–magnitude relationship of the inventoried landslides was established, plotting the probability of a landslide area. The susceptibility of deep-seated and shallow landslides was assessed by a bivariate statistical method and has been mapped. The annual and multiannual spatiotemporal probabilities were estimated, providing a landslide hazard model. Then, the physical vulnerability was assessed by analysing the answers to a questionnaire that had been sent to a pool and a sub-pool of landslide experts. The vulnerability map was based on statistical mapping units for the whole study area, and based on fieldwork building inventory for a test site included in the study area. Next, the market economic value of the buildings was calculated. Finally, the landslide risk (R) was computed by multiplying the potential loss (V × EV) by the hazard probability (H).

3.1 Frequency–magnitude of the landslides, susceptibility and hazard

3.1.1 Frequency–magnitude relationship

In order to complete the assessment of the landslide hazard and risk, we needed to establish a relationship between the magnitude of the landslides and their frequency. Ideally a landslide hazard model should incorporate not only the spatiotemporal probability of occurrence of the landslides, but also the landslide magnitude (Guzzetti et al., 1999; Cardinali et al., 2002). A landslide with a depth of 20 m can cause severe damage, but its frequency in the study area is much lower than a 1 m deep landslide. Which magnitude of landslide would present the highest risk for the Loures municipality?

Assuming that future landslides would have similar characteristics to the past ones, we considered the 686 landslides inventoried inside the Loures municipality. A curve representing the probability of a landslide versus its area was computed in the same way as Malamud et al. (2004) and Guillard and Zêzere (2012) for the deep-seated and shallow landslides of the Loures municipality. In this study, the landslides were considered all together (deep-seated and shallow rotational and translational slides) in order to know the probability associated with each scenario.

In addition, we linked the depth of the slide slip surface to the slide area and the height of accumulated material to the slide area. The relationship between the depth (d) and the area (A_L) of landslide used in this study is statistically based, and was established by Garcia (2012) ($A_L = 706 \times d$). The proximity of Garcia's study area from the Loures municipality and similarities in terms of landslide types and volumes were the main reasons for the choice of this relationship. As there is no established relationship between the height of accumulated material and the slide area, or between the height of accumulated material and the depth of the slide, we considered that the height-to-depth ratio is 0.5. This is an assumed relationship with significant uncertainty that can be an important source of bias, but which is based on landslides studied in the field whose depth is known (Zêzere et al., 1999).

3.1.2 Annual and multiannual spatiotemporal probabilities

The temporal probability has to be associated with the spatial probability in order to determine the spatiotemporal probability, which is part of the landslide hazard. First of all, the spatial probability of a shallow and a deep landslide occurrence was assessed by constructing two susceptibility maps. The susceptibility was mapped using a bivariate statistical method called the information value method (Yin and Yan, 1988). The first model represents the susceptibility of the slopes to shallow landslide occurrence, published in a previous study (Guillard and Zêzere, 2012). The total area of the shallow landslides is 319 975 m^2. The second model represents the susceptibility of the slopes to deep-seated landslide occurrence, and was built and validated by the joining of the 292 deep-seated rotational slides and the 61 deep-seated translational slides inventoried in the Loures municipality (Guillard and Zêzere, 2012). The total area of the deep-seated slides is 1 343 525 m^2. These two models provided two landslide susceptibility maps in a raster format with a pixel size of 5×5 m. Each map contains four landslide susceptibility classes that were defined by taking the predictive capacity of the model into account. Additional details on the landslide susceptibility assessment in the study area can be found in Guillard and Zêzere (2012).

The spatiotemporal values for shallow and deep-seated landslides were then calculated for each susceptibility class by dividing the product of the total affected area and the predictive capacity by the area of the class (Zêzere et al., 2004). As the inventoried landslides occurred from 1967 to 2004, we managed to calculate the hazard values for the next 38 years, and to deduce the 1-, 10-, 25-, and 50-year probability values.

3.2 Physical vulnerability of the buildings

Most of the landslides in the study area are slow (shallow slides), very slow, or extremely slow (deep-seated slides); therefore inhabitants' lives are unlikely to be endangered. However, buildings, roads, and infrastructures may suffer damage, thus generating relevant costs, both direct and indirect. That is why the vulnerability assessment is focused on the study of buildings, for which some data are available. Buildings were classified by structural elements and construction material (Table 1). Nevertheless, only direct costs are considered in the current study, due to scarcity of data.

3.2.1 Vulnerability matrix

In order to predict damage caused by landslides it is important to know the building resistance capacity. As the data related to the foundation properties of each building are not available for a large study area, such as a region or a municipality, mainly because of the huge number of elements at risk, other elements of buildings like age, structure type, and number of floors are generally used to assess the building resistance capacity (Douglas, 2007).

In contrast to social vulnerability, which is a measure of the sensitivity of a population to hazards and its ability to respond to and to recover from the hazards' impacts (Cutter and Finch, 2008), physical vulnerability is related to a specific scenario (Uzielli et al., 2008; Papathoma-Köhle et al., 2011). That is why we focused on rotational slides for which we considered nine magnitude scenarios: five scenarios in which the building location is on the body of the slide, assuming different depths of the slip surface (1, 3, 5, 10, and 20 m); and four scenarios in which the building loca-

Table 1. Structural building types in the Loures municipality (National Institute of Statistics, Census 2011, INE, 2011).

Structural building type	Structural elements and construction material	Number of buildings	%
SBT1	Wood or metal (light structures)	221	0.7
SBT2	Adobe, rammed earth, or loose stone walls	577	1.8
SBT3	Brick or stone masonry walls	9947	30.6
SBT4	Masonry walls confined with reinforced concrete	21 750	66.9
Total		32 495	100.0

tion is on the foot of the slide, assuming different heights of affected material (0.5, 1, 3, and 5 m) (Fig. 2). The maximum values considered for both the depth of the slip surface and the height of affected material were defined, taking the largest landslides inventoried in the study area into consideration (Zêzere, 2002; Zêzere et al., 2008). The remaining scenarios use standard values considered in landslide classifications (e.g. Záruba and Mencl, 1982). A building situated on the landslide body may suffer vertical and lateral displacements, whereas a building situated on the landslide foot may support dynamic pressures against the walls, and may be buried (Glade et al., 2005; van Westen et al., 2005; Léone, 2007).

Existing relationships between building damage patterns and height of affected material for debris flows (e.g. Papathoma-Köhle et al., 2012b) cannot be applied to the study area, as landslide types and velocities are not comparable. In this study, the landslide slip surface depth and the accumulated material height were used as proxies for landslide destructive capacity because of the following reasons. Landslides affecting the study area have generally slow, very slow, or extremely slow velocities, and in these circumstances, the landslide velocity is not the most appropriate parameter to assess the landslide destructive capacity. Moreover, there is no instrumental data about the velocity of each landslide. On the other hand, without relevant differences regarding landslide velocity, the depth of the slip surface is significant as a proxy for landslide destructiveness, namely through the comparison with the depth of the building foundation. In addition, it was possible to find a statistic relationship between the landslide slip surface depth and the landslide area, which is an accurate landslide morphometric parameter that is available in the landslide inventory.

A study realized at a local scale enables the landslide vulnerability to be assessed with a quantitative method, relying on expert judgment, damage records, or statistical analysis (Ciurean et al., 2013). Nevertheless, for a study at a municipal or regional scale, the physical vulnerability assessment is usually done by a semi-quantitative or a qualitative method, and is often based on historical records (Dai et al., 2002) and on expert judgments (Sterlacchini et al., 2007), and is largely subjective (Léone et al., 1996; Uzielli et al., 2008; Silva and

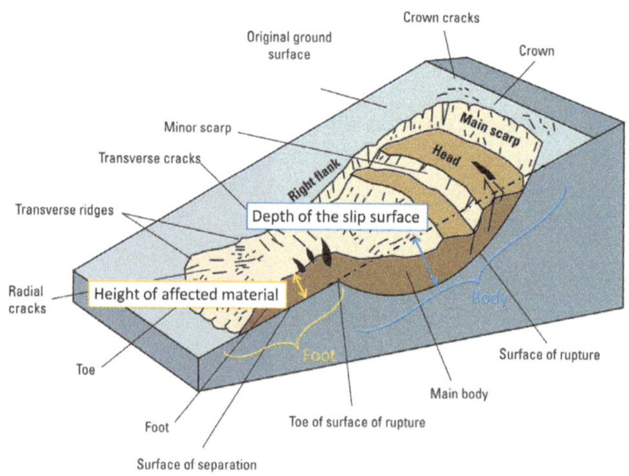

Figure 2. Rotational slide body and foot (adapted from Highland and Bobrowsky, 2008).

Pereira, 2014). In this work, we decided to ask the opinion of a pool of experts. A questionnaire was formulated and sent to more than 300 international experts on landslides and other natural risks who have worked with landslides in the past.

The experts were asked to fill in the questionnaire in which they attributed, on four structural types of buildings (Table 1), the corresponding potential damage caused by landslides of different magnitudes (Table 2); the magnitudes of the landslides were associated with the depth of the slip surface and with the height of the affected material. The experts provided 36 answers, corresponding to each situation (Supplement 1).

Fifty-two experts completed the questionnaire and their answers were used to obtain an average value of physical vulnerability for each type of building, for location within the landslide body and the landslide foot, and for each landslide magnitude. Each damage class was associated with the corresponding upper bound of its corresponding physical vulnerability, thus adopting a conservative approach (Table 2). We were also able to assess the variability of the obtained results by calculating and mapping the standard deviation of the answers. This vulnerability assessment exercise was repeated, keeping only a sub-pool with the answers of the 14 landslide experts who know the landslides and the buildings

Table 2. Damage level of buildings.

Damage class		Physical vulnerability	Damage level of buildings (based on Alexander, 1986; AGS, 2000; Tinti et al., 2011; Garcia, 2012)
1	Negligible damage	[0; 0.2]	No significant damage – slight accumulation of material causing aesthetic damage (dirt, chipping paint, etc.)
2	Slight damage	]0.2; 0.4]	No structural damage – minor repairable damage: chipping of plaster, slight cracks, damage to doors and windows
3	Significant damage	]0.4; 0.6]	No structural damage – major damage requiring complex repair: displacement or partial collapse of walls or panels without compromising structural integrity, highly developed cracks. Evacuation required.
4	Severe damage	]0.6; 0.8]	Structural damage that can affect the stability of the building: out-of-plane failure or collapse of masonry, partial collapse of floors, severe cracking or collapse of sections of structure due to settlement. Immediate evacuation; demolition of the element may be required.
5	Very severe damage	]0.8; 1]	Heavy damage seriously compromising the structural integrity: partial or total collapse of the building. Imperative and immediate evacuation and complete demolition.

of the study area, and the results obtained by the two different groups of experts were compared.

3.2.2 Vulnerability based on statistical mapping units

A geodatabase containing information about elements at risk was provided by the Loures municipality. Buildings of the municipality were compared with the most recent high-resolution images of the Loures municipality provided by the World Imagery File ESRI (2014) and buildings in ruins were excluded. However, the only data provided and used by this geodatabase are the geographical location of the buildings. In order to obtain more information about the buildings, like their structure, age, or functionality, we used data from the census of the INE. We chose, as a mapping unit, the smallest statistical unit, which is the Geographic Basis for Information Reference subsection (BGRI). The BGRI units are the basic geographic entities used for the 2011 census operations, which divide each basic administrative unit (which is the civil parish) into sections and subsections. The BGRI subsections are territorial units, whether built-up or not, which represent a block in urban areas, a locality or part of a locality in rural areas, or residual areas which may or may not have dwellings (INE, 2011). Their boundaries were defined by the INE, and the statistical information was also collected by the INE. The 3061 BGRI subsections of the Loures municipality used for the 2011 census were used in this study.

The buildings of the study area were classified into four structural types, corresponding to the data which are available for the whole area at the BGRI subsection scale, considering their structural elements and construction materials (Table 1). It should be noted that although the information provided at the BGRI subsection scale includes the number of structural types of buildings, no information was provided on the structural type of each individual building.

Therefore, the number of buildings pertaining to each structural building type class (from SBT1 to SBT4, see Table 1) is known for each BGRI, although the association of this information with each building polygon cannot be made directly. As the physical vulnerability of buildings was established for each structural building type, the vulnerability of the buildings was assessed for each BGRI subsection by calculating a weighting average, which takes into account the number of buildings of each structural building type within the BGRI (Eq. 1):

$$V_i = \frac{\sum_j V\left(\mathrm{SBT}_j\right) \times N\left(\mathrm{SBT}_j\right)}{\sum N(\mathrm{SBT})}, \quad (1)$$

where V_i is the vulnerability of the BGRI subsection to a landslide magnitude i, $V(\mathrm{SBT}_j)$ is the vulnerability of the structural building type j and $N(\mathrm{SBT}_j)$ is the number of buildings with a structural building type j.

Then, the average vulnerability was assigned to all the buildings of the BGRI subsection. This limitation of the study in which the value of vulnerability is the same for all the buildings of a BGRI comes from limited data. However, the average number of buildings per BGRI in the Loures municipality is 11, and most of the BGRI units have a large

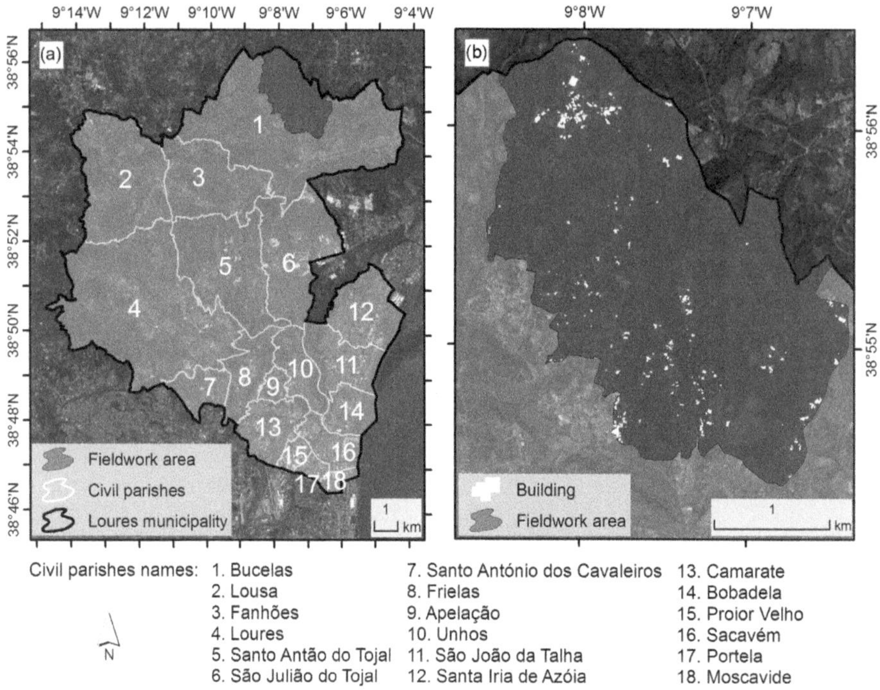

Figure 3. (**a**) Civil parishes of the Loures municipality and location of the fieldwork area; (**b**) buildings of the fieldwork area.

number of buildings belonging to the same structural building type (56 % of the BGRI have only one structural building type and 30 % have two structural building types). This means that the generalized vulnerability attributed to the BGRI buildings is in most cases quite close to what it would be for a vulnerability assessment made building by building.

The standard deviations of the answers given by the experts represent the variability of the vulnerability values and were calculated and mapped for each scenario and for each structural building type.

3.2.3 Vulnerability based on fieldwork building inventory

The above-mentioned vulnerability assessment approach based on statistical mapping units has the advantage of being time-saving, in contrast to a study that considers each building of the study area, as Silva and Pereira (2014) did for the Santa Marta de Penaguião municipality. In order to assess the accuracy of this approach, we selected a test site inside the Loures municipality to develop fieldwork, where the structural building type was inventoried for each individual building. The choice of the test site was made because of its proneness to landslides. The test site is located in the northern part of the Bucelas civil parish, has an area of $6.71\,\mathrm{km}^2$, and has 782 buildings (Fig. 3). Physical vulnerability of the test site was assessed using the same vulnerability matrix referred to in Sect. 3.2.1, but the vulnerability was attributed to each single building instead of being calculated per BGRI. With this approach, we evaluated the influence of the mapping unit in the final results of buildings' physical vulnerability.

3.3 Economic value of the buildings

The economic value (EV) of the buildings has been calculated using the same equation as Silva and Pereira (2014) (Eq. 2):

$$\mathrm{EV} = \mathrm{ACC} \times \mathrm{TA} \times \mathrm{FC} \times \mathrm{LC} \times \mathrm{AC}, \qquad (2)$$

where EV is the market economic value, ACC is the average cost of construction, TA is the total area, FC is the functionality coefficient, LC is the location coefficient, and AC is the age coefficient. The ACC is established by the Portuguese government (Decree Number 1456/2009) and expresses the costs associated with the construction of buildings. It was fixed at $603\,\mathrm{EUR\,m}^{-2}$ for the year 2011. As ACC is expressed per square metre, it had to be multiplied by the TA, which was calculated by multiplying the buildings area, provided by the Loures municipality geodatabase, by the average number of storeys in each BGRI subsection. The FC is related to the function of the buildings (residential, store or storages are the main functions of the Loures municipality buildings), also provided by the BGRI subsection data, and the coefficients were defined by the Portuguese Tax Services (Dec.-Law Number 287/2003 of 12 November), ranging from 0.35 (storage buildings) to 1.2 (buildings that have a commercial use). The AC values are also classified by Portuguese Tax Services (Law Number 64-A/2008 of 31 December), ranging from 0.40 (buildings older than 60 years) to 1

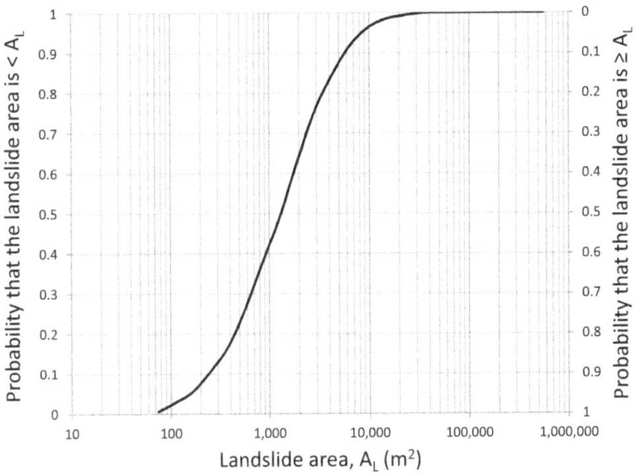

Figure 4. Probability of landslide area in the Loures municipality (based on the work done by Guillard and Zêzere, 2012).

(building less than 2 years old). The information about number of buildings per function and building age was obtained from BGRI data. The weighted average values were calculated for each BGRI for both coefficients and assigned to the buildings. LC is determined by the Portuguese Tax Services according to property market and accessibility (Law Number 64-B/2011 of 30 December). At the national level, the LC values range from 0.4 to 3.5; in the Loures municipality, the LC values vary between 0.85 for the rural areas and 2.25 for the zones of the Moscavide and Sacavém civil parishes (Fig. 3), which are located near Lisbon and have a better accessibility and proximity to social facilities and public transport.

The economic value per pixel (EVpix) was calculated from the EV value obtained for each building. Indeed, as the landslide hazard was calculated at a pixel base, we needed to obtain an economic value per pixel to calculate the risk. The EVpix value was obtained by dividing the EV value by the area of the building and multiplying it by 25, which is the pixel area in square metres.

3.4 Landslide risk

The buildings shape files were converted into raster files with a pixel size of 5 × 5 m. Then, the risk was computed according to Eq. 3, based on Varnes and the IAEG Commission on Landslides and other Mass-Movements (1984):

$$R_{ij} = H_i \times P_j \times \mathrm{PV}_j \times \mathrm{EVpix}, \tag{3}$$

where R is the risk, H is the spatiotemporal probability, P is the magnitude probability, PV is the physical vulnerability, and EVpix is the economic value per pixel. The index i takes the values of 1, 10, 25, and 50 years; the index j takes the values of 1, 3, 5, 10, and 20 m for the slip surface depth, and 0.5, 1, 3, and 5 m for the accumulated material height. The multiplication of the last two terms (the physical vulnerability and the economic value) represents the potential loss for the buildings.

Table 3. Magnitude probability of slides according to their slip surface depth in the Loures municipality.

Slip surface depth (m)	Landslide area (m^2)	Probability
1	706	0.57
3	2119	0.34
5	3532	0.19
10	7064	0.07
20	14 127	0.02

Table 4. Magnitude probability of slides according to the height of their accumulated material in the Loures municipality.

Accumulated material height (m)	Corresponding slip surface depth (m)	Landslide area (m^2)	Probability
0.5	1	706	0.57
1	2	1413	0.48
3	6	4238	0.16
5	10	7064	0.07

Annual spatiotemporal probability was considered (i.e. index $i = 1$ year) to calculate the landslide risk values for a year with different probabilities of occurrence according to the different landslide magnitude values. Box plots were computed to compare the effect of the landslide magnitude on the landslide risk. Then, the probability of occurrence was fixed (index $j = 10$ m deep) and the risk was calculated for different spatiotemporal probabilities.

4 Results

4.1 Frequency–magnitude of the landslides, susceptibility and hazard

The probability of the different landslide magnitudes was assessed using the curve shown in Fig. 4. The landslide area was used as a proxy for both the depth of landslide slip surface and the height of affected material in the landslide foot; the results are summarized in Tables 3 and 4. The corresponding slide areas range from 706 to 14 127 m^2. When a landslide occurs in the Loures municipality, the probability that this landslide has a slip surface depth higher than 1 m is 0.57; the probability that this landslide has a slip surface depth higher than 20 m is 0.02. In general terms, the probability of landslides decreases when their magnitude increases, which obeys the universal rule governing natural processes, and which is consistent with the results previously obtained by Guillard and Zêzere (2012) for this study area.

Table 5. Probability of occurrence of deep-seated landslides in 1, 10, 25, and 50 years in the Loures municipality.

Susceptibility class	Area (no. of pixels)	Predictive capacity	1-year probability	10-year probability	25-year probability	50-year probability
Very high	468 814	0.5	1.51×10^{-3}	1.51×10^{-2}	3.77×10^{-2}	7.54×10^{-2}
High	647 436	0.25	5.46×10^{-4}	5.46×10^{-3}	1.37×10^{-2}	2.73×10^{-2}
Low	1 246 342	0.15	1.70×10^{-4}	1.70×10^{-3}	4.26×10^{-3}	8.51×10^{-3}
Very low	4 362 465	0.1	3.24×10^{-5}	3.24×10^{-4}	8.10×10^{-4}	1.62×10^{-3}

Table 6. Probability of occurrence of superficial landslides in 1, 10, 25, and 50 years in the Loures municipality.

Susceptibility class	Area (no. of pixels)	Predictive capacity	1-year probability	10-year probability	25-year probability	50-year probability
Very high	400 890	0.5	4.20×10^{-4}	4.20×10^{-3}	1.05×10^{-2}	2.10×10^{-2}
High	810 140	0.25	1.04×10^{-4}	1.04×10^{-3}	2.60×10^{-3}	5.20×10^{-3}
Low	1 176 564	0.15	4.29×10^{-5}	4.29×10^{-4}	1.07×10^{-3}	2.15×10^{-3}
Very low	4 337 463	0.1	7.77×10^{-6}	7.77×10^{-5}	1.94×10^{-4}	3.88×10^{-4}

The deep-seated and shallow landslide susceptibility models were validated based on the random partition of the landslide inventories in two groups: the modelling group and the validation group. The modelling group was used to weight the classes of each landslide-predisposing factor and to build the landslide susceptibility models, whereas the validation group was crossed with the susceptibility results for their independent validation. The prediction-rate curves show the robustness of the models (Fig. 5): the area under curve (AUC) value is 0.87 for both models, which attests to the robustness of the models.

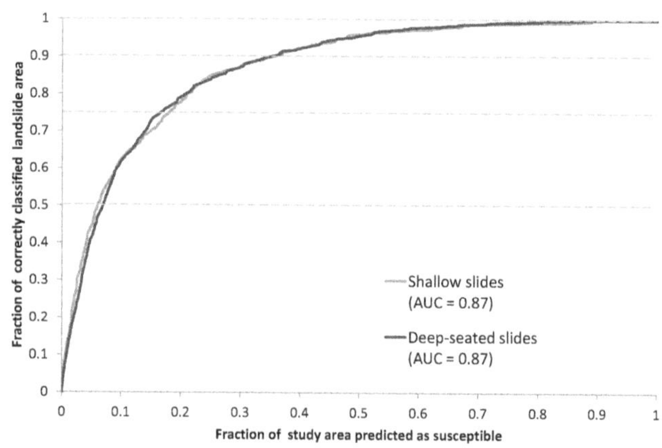

Figure 5. Prediction-rate curves and area under the curve (AUC) of landslide susceptibility models in the Loures municipality (based on the work done by Guillard and Zêzere, 2012).

The landslide susceptibility maps are shown in Fig. 6, with the landslides used for computing and for validating the models. In a previous work (Guillard and Zêzere, 2012), the conditional probability of both the landslide depletion areas and the landslide total areas were calculated for each class of each landslide predisposing factor, for shallow slides and deep-seated slides in the study area. The obtained results are very similar and we chose to model landslide susceptibility with the landslide total areas. Therefore, landslide susceptibility maps express the likelihood of an area to be involved in the rupture zone or the accumulation zone of a landslide (Guillard and Zêzere, 2012). The separation of the classes was done using the fraction of correctly classified landslide area (Fig. 5, and "predictive capacity" in Tables 5 and 6). Therefore, 50 % of future landslides should occur in the "very high" susceptibility classes, which represent only 7 and 6 % of the total area for the deep-seated and shallow landslides, respectively. Moreover, 25 % of future landslides should occur in the "high" susceptibility classes, which represent only 10 and 12 % of the total area for the deep-seated and shallow landslides, respectively.

Tables 5 and 6 show the probabilities of a pixel within a susceptibility class to be affected by a deep-seated (Table 5) or shallow (Table 6) slide, for different time periods (1, 10, 25, and 50 years). Probabilities of the total area to be affected by landslides in the future were calculated, as well as the area of the class and the class predictive capacity, as explained in Sect. 3.1.2. They can be calculated for any time period from the 1-year probabilities, but we selected 10, 25, and 50 years, which are significant time periods considering that stakeholders have to make choices that will have repercussions for decades. Indeed, even if a pixel within the high susceptibility class only has a probability of 5.46×10^{-4} (that is, a 1 in 1832 chance) of being affected by a deep-seated slide during the next year, it has a probability of 2.73×10^{-2} (that is, a 1 in 37 chance) of being affected by a deep-seated slide during the next 50 years (Table 5). Moreover, each pixel within the very high susceptibility class has a probability of

Table 7. Average vulnerability (Avg. vuln.) and standard deviation (SD) for each structural building type located on a landslide body (cf. Table 1 for building type).

		Landslide body: depth of slip surface									
		1 m		3 m		5 m		10 m		20 m	
		Avg. vuln.	SD	Avg. vuln.	SD	Avg. vuln.	SD	Avg. vuln.	SD	Avg. vuln.	SD
Pool of European experts (52)	SBT1	0.60	0.24	0.73	0.21	0.84	0.18	0.90	0.19	0.90	0.20
	SBT2	0.57	0.23	0.72	0.20	0.85	0.17	0.92	0.14	0.91	0.17
	SBT3	0.46	0.22	0.60	0.22	0.76	0.18	0.88	0.18	0.91	0.18
	SBT4	0.35	0.20	0.48	0.18	0.66	0.19	0.80	0.18	0.86	0.19
Sub-pool of study area experts (14)	SBT1	0.64	0.19	0.84	0.14	0.96	0.09	1.00	0.00	1.00	0.00
	SBT2	0.59	0.15	0.77	0.15	0.96	0.09	1.00	0.00	1.00	0.00
	SBT3	0.43	0.15	0.66	0.15	0.86	0.12	0.99	0.05	1.00	0.00
	SBT4	0.30	0.10	0.50	0.13	0.71	0.15	0.91	0.13	0.99	0.05

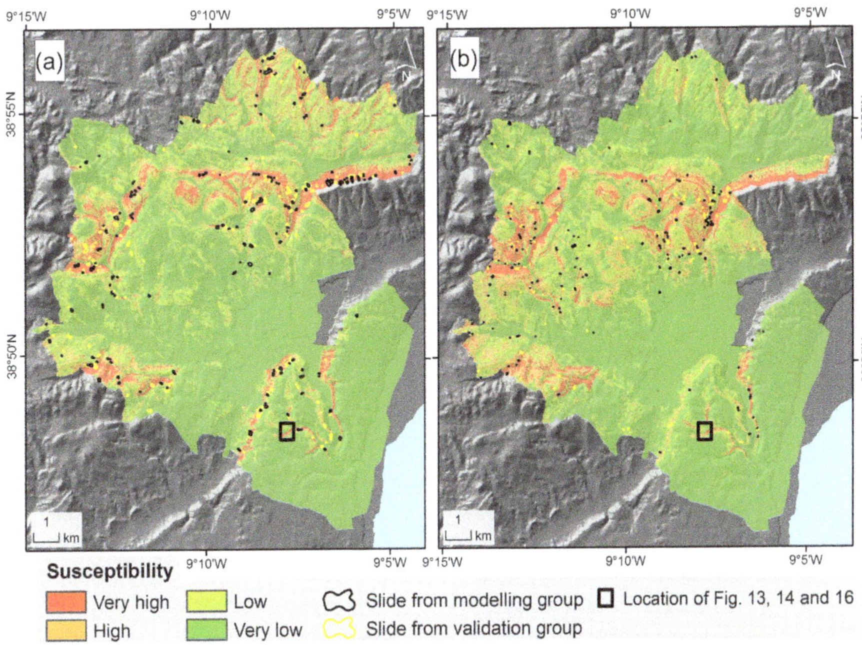

Figure 6. Landslide susceptibility maps in the Loures municipality for **(a)** deep-seated slides, **(b)** shallow slides (based on the work done by Guillard and Zêzere, 2012).

7.54×10^{-2} (that is, a 1 in 13 chance) of being affected by a deep-seated slide during the next 50 years.

4.2 Physical vulnerability of the buildings

Out of 52 questionnaires completed by the experts who have a research background or some experience in the landslide field, 30 came from Portuguese experts, 14 of whom have been doing research on landslides in the area north of Lisbon. As the damage level asked about in the questionnaire is a proxy for the physical vulnerability, the damage values provided by the experts, comprised between 1 and 5, were converted into vulnerability values, comprised between 0 and 1 (see Table 2).

The physical vulnerability of buildings was assessed twice, first with the total landslide expert answers and second with the sub-pool of landslide experts who have been working in the study area. The vulnerability averages of the two groups of experts are presented in Tables 7 and 8, along with the standard deviation for each scenario, which was calculated in order to evaluate the variability of the answers through the differences between the experts' answers. The vulnerability averages were used to calculate the vulnerability of each BGRI subsection. These averages range from 0.25

Table 8. Average vulnerability and standard deviation for each structural building type located on a landslide foot (cf. Table 1 for building type).

		Landslide foot: height of accumulated material							
		0.5 m		1 m		3 m		5 m	
		Avg. vuln.	SD	Avg. vuln.	SD	Avg. vuln.	SD	Avg. vuln.	SD
Pool of European experts (52)	SBT1	0.45	0.22	0.61	0.20	0.85	0.17	0.94	0.12
	SBT2	0.38	0.23	0.53	0.21	0.78	0.18	0.93	0.12
	SBT3	0.30	0.18	0.40	0.22	0.66	0.17	0.83	0.17
	SBT4	0.25	0.16	0.31	0.19	0.54	0.19	0.72	0.20
Sub-pool of study area experts (14)	SBT1	0.39	0.18	0.56	0.22	0.86	0.15	0.97	0.07
	SBT2	0.29	0.15	0.49	0.17	0.81	0.12	0.97	0.07
	SBT3	0.24	0.09	0.39	0.15	0.71	0.15	0.90	0.13
	SBT4	0.20	0.00	0.27	0.10	0.53	0.10	0.79	0.15

(for a SBT4 building on a 0.5 m high landslide foot) to 0.94 (for a SBT1 building on a 5 m high landslide foot) regarding the European expert answers, and from 0.20 (for a SBT4 building on a 0.5 m high landslide foot) to 1 (for a SBT1 building on a 5 m high landslide foot) regarding the answers of the sub-pool of experts. As expected, the vulnerability of the buildings increases with the landslide magnitude, and is lowest for SBT4 and SBT3. The standard deviation ranges from 0.12 (for SBT1 and SBT2 buildings located on a 5 m high landslide foot) to 0.24 (for a SBT1 building located on a 1 m deep landslide body) regarding the European expert answers, and from 0 (several times) to 0.22 (for a SBT1 building on a 1 m high landslide foot) regarding the answers of the sub-pool of experts.

The vulnerability assessment provided by the sub-pool of experts who know the study area has a larger scope than the European landslide experts. Indeed, according to the study area experts, the low-magnitude landslides (landslides that are 1 m deep for the SBT3 and SBT4 buildings, and 0.5 and 1 m high of accumulated material landslides for all the structural building types) cause less damage than according to the European experts, and the high-magnitude landslides cause more damage than according to the European experts (Tables 7 and 8). Moreover, the standard deviation values of the study area experts' answers are typically lower than the standard deviation values of the European experts' answers (Tables 7 and 8), which indicates the consistency of the answers given by the study area experts.

In each BGRI subsection, the average vulnerability was calculated, taking into account the number of buildings belonging to each structural building type. Then, the average vulnerability given by the sub-pool of study area experts was attributed to each building included into the BGRI subsection in order to obtain more explicit maps (Figs. 7 and 8). The average vulnerabilities of the Loures municipality buildings associated with the landslides that are 1, 3, 5, 10, and 20 m deep are 0.34, 0.55, 0.75, 0.92, and 0.97, respectively; the average vulnerabilities of the Loures municipality buildings associated with the landslides which have a height of accumulated material of 0.5, 1, 3, and 5 m are 0.21, 0.31, 0.58, and 0.81, respectively. The standard deviation of the BGRI subsection vulnerability was also represented in shades of blue in Figs. 7 and 8. As a rule, the standard deviation decreases as the landslide magnitude increases.

As expected, the average vulnerability depends on the structural building type, and increases with the landslide magnitude. However, when the magnitude is maximum – which is for a landslide 10 m or 20 m deep – all the buildings have maximum vulnerability (PV > 0.8 see Fig. 7d and e, and Table 7), independently of their structural building type. This means that the structure type may play a role when the landslide magnitude is low, but all the buildings have the same (maximum) vulnerability when the landslide magnitude reaches a certain level of potential damage. The variability in the expected damage to buildings among the study area experts is higher for damage generated by low-magnitude landslides (e.g. landslides 1 m deep, and landslides with a 0.5 to 1 m high of accumulated material) on SBT1, SBT2, and SBT3. This can be explained by the fact that the landslide experts have more facilities to assess the vulnerability to the high-magnitude landslides, which have a high potential for damage, than to the low-magnitude landslides, for which the potential for damage is more difficult to determine. The maps shown in Figs. 7 and 8 enable the location of the buildings and their vulnerabilities to be identified according to different landslide magnitudes, but they also highlight the uncertainty associated with the attributed vulnerabilities.

The vulnerability of the test site buildings inventoried during fieldwork (Fig. 3) is presented in Figs. 9 and 10 for locations in the landslide body and the landslide foot, respectively. As each building has its own vulnerability, the results are more accurate than when an average value is calculated

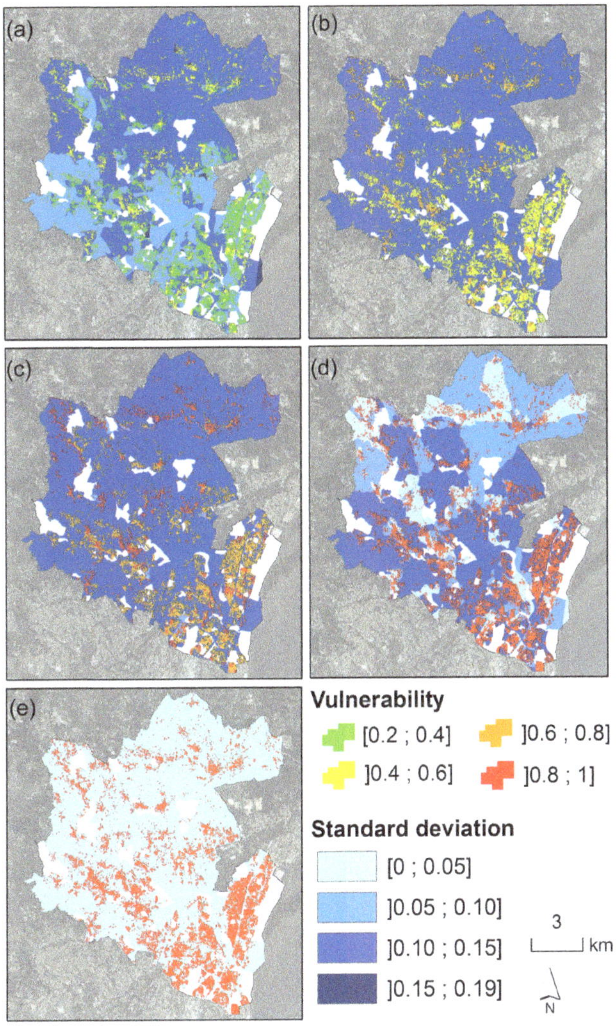

Figure 7. Average building vulnerability and standard deviation per BGRI subsection for buildings located on a landslide body, for a slip surface depth of **(a)** 1 m, **(b)** 3 m, **(c)** 5 m, **(d)** 10 m, and **(e)** 20 m. White polygons are BGRI subsections without buildings.

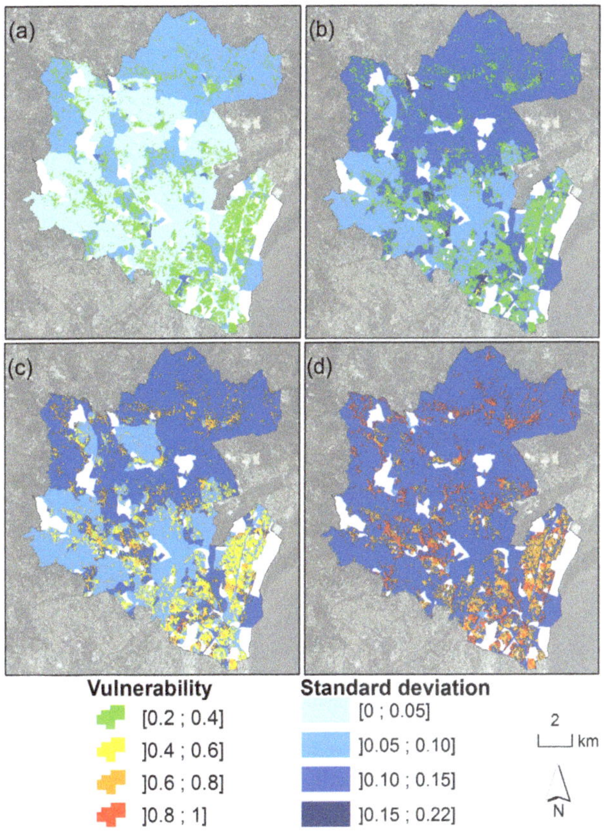

Figure 8. Average building vulnerability and standard deviation per BGRI subsection, for buildings located on a landslide foot with an affected material height of **(a)** 0.5 m, **(b)** 1 m, **(c)** 3 m, and **(d)** 5 m. White polygons are BGRI subsections without buildings.

for all the buildings of the BGRI subsection. However, the comparison of building vulnerability expressed in Figs. 9 and 10 with the corresponding area at the BGRI subsection level shows that global results are similar. In order to obtain a more accurate comparison, the box plots of the vulnerability values obtained by both vulnerability approaches for the test site are shown in Fig. 11. Indeed, Fig. 11 enables the comparison of vulnerability values of the test site buildings inventoried by fieldwork (in grey) with the vulnerability values of the buildings of the BGRI subsections (in black). In each case, the range of the vulnerability values obtained by fieldwork is wider than the one obtained by the BGRI subsections calculations. This can be explained by the fact that the data obtained by fieldwork are much more detailed because the buildings were considered one by one; therefore the results are less generalized. Moreover, for each scenario, the median of the fieldwork data is the same (or almost the same in the case of the landslides that are 10 m deep) as the one calculated from BGRI subsections data, which validates the accuracy of the vulnerability values obtained by calculations in the BGRI subsections. The vulnerability assessment procedure based on BGRI subsection mapping units is much less time-consuming than the fieldwork procedure and can easily be applied to other areas, because the data are available in the census. As the obtained results are satisfactory, we recommend the application of the first approach at the municipal level.

4.3 Economic value of the buildings

The economic value of the buildings was calculated using Eq. (2). We found that 3417 buildings have an economic value above EUR 100 000 per pixel (which corresponds to 4000 EUR m^{-2}), that is 3 % of the buildings of the whole municipality. Most of them are located in the southern half of the Loures municipality (near Lisbon), which is more urbanized than its northern half, and presents the highest concentration in the civil parishes of Portela, Moscavide, and

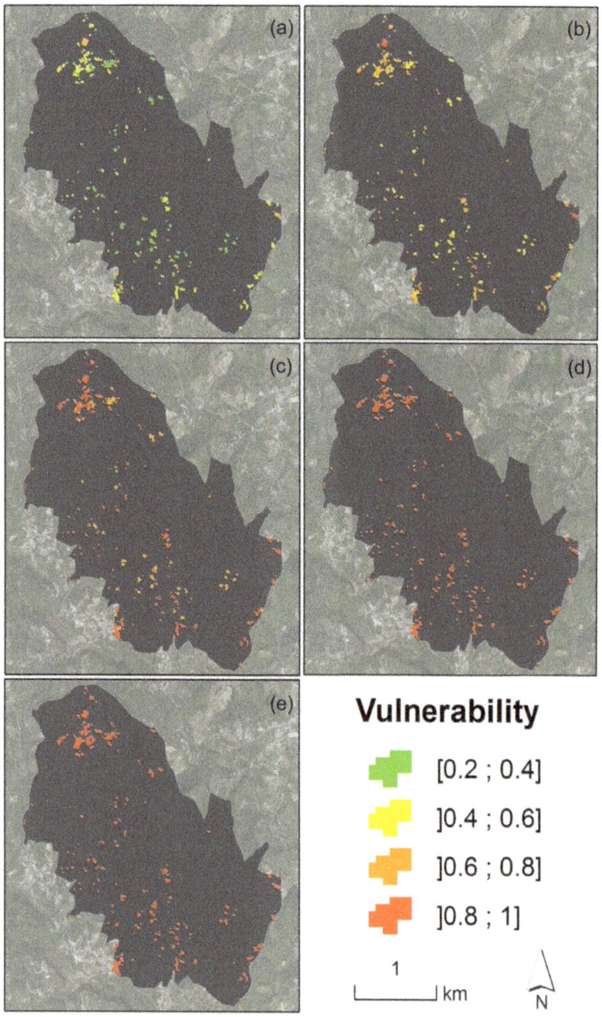

Figure 9. Vulnerability of buildings inventoried in the fieldwork area, located on a landslide body with a slip surface depth of **(a)** 1 m, **(b)** 3 m, **(c)** 5 m, **(d)** 10 m, and **(e)** 20 m.

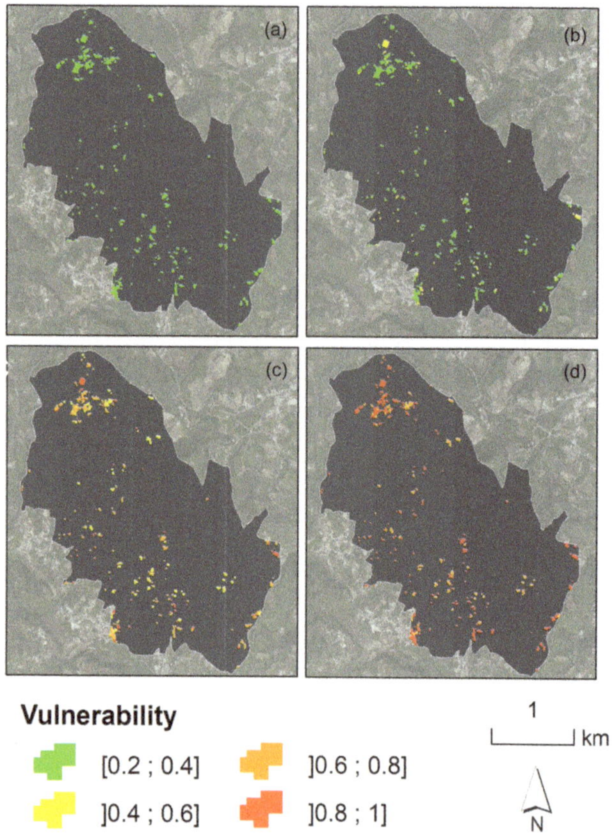

Figure 10. Vulnerability of buildings inventoried in the fieldwork area, located on a landslide foot with an affected material height of **(a)** 0.5 m, **(b)** 1 m, **(c)** 3 m, and **(d)** 5 m.

Sacavém (Figs. 3 and 12). The civil parishes of Santo António dos Cavaleiros, Loures, Santa Iria de Azóia, São João da Talha, and Bobadela also have high economic value buildings. Most of them are recent residential and industrial buildings located near social facilities.

4.4 Landslide risk

Figures 13 and 14 illustrate the risk for buildings according to the spatiotemporal landslide probability, the landslide magnitude, and the building vulnerability and value. The buildings have been transformed into a raster in order to multiply the potential losses associated with the buildings by the hazard values. The value of risk is the value per pixel, and each pixel has an area of 25 m^2. The total area of the buildings in the vector is 9.25 km^2, and the total area of the buildings in the raster is 9.00 km^2. The 0.25 km^2 which was lost during the transformation from the vector to the raster only represent 2.7 % of the total area of the buildings; thus, even if the transformation changes the shape of the buildings slightly, their surface is almost the same, which has little influence on the risk estimates. Figures 13 and 14 show that the risk values are closely related to the landslide susceptibility values. As the buildings have similar economic values, the ones that were constructed in high or very high susceptibility zones have a higher risk in comparison to the ones constructed in the "low" or "very low" susceptibility zones.

The box plots of the risk values were plotted for each scenario in order to compare them (Fig. 15). Outliers have been considered, but their values are too high to be shown on this figure (the maximum value is EUR 25.68 per pixel, for a 3 m deep slide). Figure 15 and Table 9 show that the maximum values of risk correspond to landslides that are 3 m deep, for which 741 pixels buildings (that is 0.2 % of the buildings of the Loures municipality) have a risk above EUR 5 per pixel, and for which there is an annual risk of EUR 96 693 for the Loures municipality, that is EUR 109 per hectare of buildings (Table 9). Indeed, these landslides are the ones which combine a relatively high frequency in the Loures municipality (magnitude probability = 0.34, cf. Table 3) with a substantial

Table 9. Landslide risk per civil parish. Vulnerability data obtained with a sub-pool of landslide experts who know the study area.

				Body depth: 1 m		Body depth: 3 m		Body depth: 5 m		Body depth: 10 m		Body depth: 20 m	
Civil parish ID (cf. Fig. 3a)	Civil parish name	Civil parish area (ha)	Area of buildings (ha)	Total risk (EUR)	EUR ha^{-1} of buildings	Total risk (EUR)	EUR ha^{-1} of buildings	Total risk (EUR)	EUR ha^{-1} of buildings	Total risk (EUR)	EUR ha^{-1} of buildings	Total risk (EUR)	EUR ha^{-1} of buildings
1	Bucelas	3397	66	772	12	6671	101	4856	73	2004	30	467	7
2	Lousa	1653	46	1150	25	4349	96	3236	71	1364	30	347	8
3	Fanhões	1162	25	832	34	3112	126	2354	95	1024	41	286	12
4	Loures	3300	141	4778	34	16 310	115	12 404	88	5456	39	1501	11
5	Sto Antão do T.	1513	45	866	19	2702	60	2053	46	890	20	235	5
6	São Julião do T.	1328	56	695	12	2411	43	1817	32	772	14	164	3
7	Sto António dos C.	363	23	2265	97	5730	246	4521	194	2109	91	627	27
8	Frielas	556	25	299	12	2802	110	2225	87	1020	40	286	11
9	Apelação	140	12	382	32	1494	124	1163	97	523	44	146	12
10	Unhos	451	39	705	18	4698	121	3653	94	1652	42	457	12
11	S João da T.	652	83	1413	17	6546	79	5033	61	2249	27	566	7
12	Sta Iria de A.	756	87	1502	17	7077	82	5538	64	2494	29	653	8
13	Camarate	566	83	2025	24	9653	117	7341	89	3237	39	860	10
14	Bobadela	382	25	592	24	2534	101	1960	78	875	35	230	9
15	Prior Velho	131	35	662	19	2769	80	2134	61	966	28	258	7
16	Sacavém	379	51	1950	39	8465	167	6681	132	3075	61	889	18
17	Portela	102	21	1126	53	4594	217	3617	171	1684	80	488	23
18	Moscavide	102	22	1144	51	4775	213	3708	165	1685	75	489	22
–	Loures municipality	16 934	886	23 158	26	96 693	109	74 293	84	33 078	37	8946	10
–	Loures municipality with vulnerability data from the pool of the 52 landslide European experts			26 871	30	92 581	105	67 389	76	28 642	32	7551	9
–	Difference (%)			−16.0		4.3		9.3		13.4		15.6	

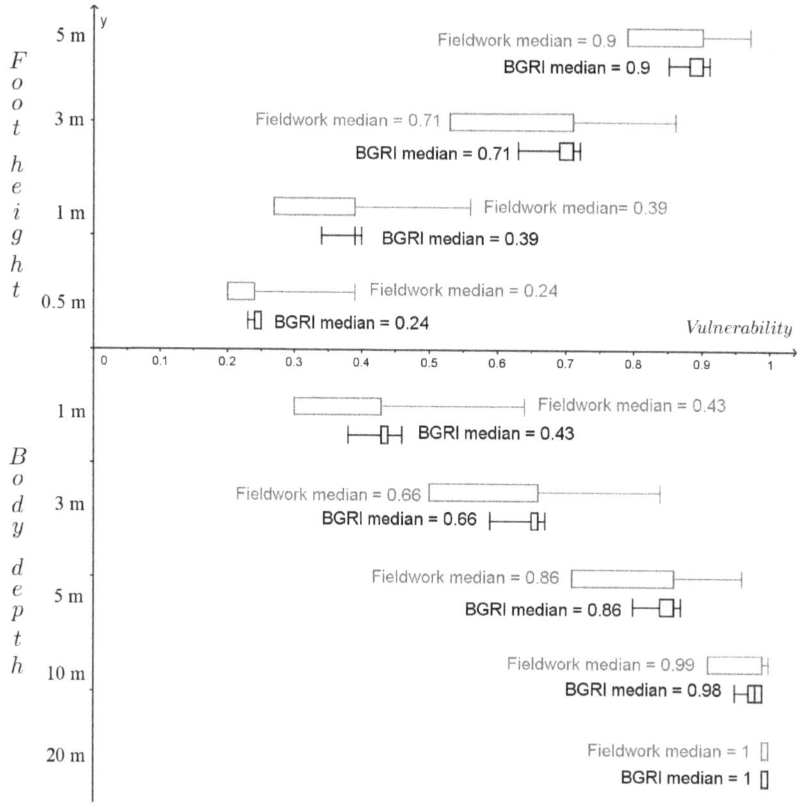

Figure 11. Box plots of the vulnerability of the test site buildings for each scenario, for the buildings inventoried by fieldwork (in grey) and for the buildings of the BGRI subsections (in black).

potential damage (the median vulnerability value associated with them is 0.66; cf. Fig. 11). More frequent landslides have a lower magnitude and are less destructive, whereas the ones which have a higher magnitude have a very low frequency; for example, the annual probability of a landslide with a depth of 20 m or more in the Loures municipality is 0.02 (cf. Fig. 4 and Table 3). Therefore, despite the high median vulnerability associated with these landslides (1; cf. Fig. 11), the risk associated with them is quite low (the median value is 0.04; cf. Fig. 15). The risk was calculated for each civil parish for the five scenarios considering the different landslide body depths (1, 3, 5, 10, and 20 m). The risk in euros per hectare of buildings was also calculated for each civil parish (Table 9). The maximum annual risk value was computed for the Loures civil parish (EUR 16 310), and the maximum value of risk per area of buildings was obtained for the Santo António dos Cavaleiros civil parish (246 EUR ha^{-1}). The Loures civil parish has the highest number of buildings within the municipality and it also has the highest risk values for the five scenarios summarized in Table 9. The Sacavém and Camarate civil parishes also have a high risk, which can be explained by the high economic value of their built environment.

The last two lines of Table 9 show the annual risk values for the municipality obtained using the average vulnerability given by the pool of European landslide experts and the differences for risk values obtained with the average vulnerability given by the sub-pool of study area experts. For low-magnitude landslides (landslides that are 1 m deep), the study area experts gave lower vulnerabilities for the SBT3 and SBT4 buildings than the European experts (Table 7); these buildings represent 97.5 % of all the buildings of the Loures municipality (Table 1) and their low vulnerability implies a lower risk at the municipality scale. For high-magnitude landslides, the study area experts gave higher vulnerabilities for any structural building types than the European experts, which implies a generalized higher risk for the municipality.

Finally, the risk was calculated considering different time periods. Figure 16 shows the risk to landslides that are 10 m deep in a part of the Loures municipality, for 1, 10, 25, and 50 years. In this zone which was zoomed in on, the annual risk is between EUR 1 and 5 per pixel in the very high susceptibility zones, and below EUR 1 per pixel in the rest of the zoomed area. However, the risk increases when we consider longer periods of time; for instance, for a 50-year period, risk values are above EUR 20 per pixel for high and very high susceptibility zones and between EUR 5 and 20 per pixel for low susceptibility zones.

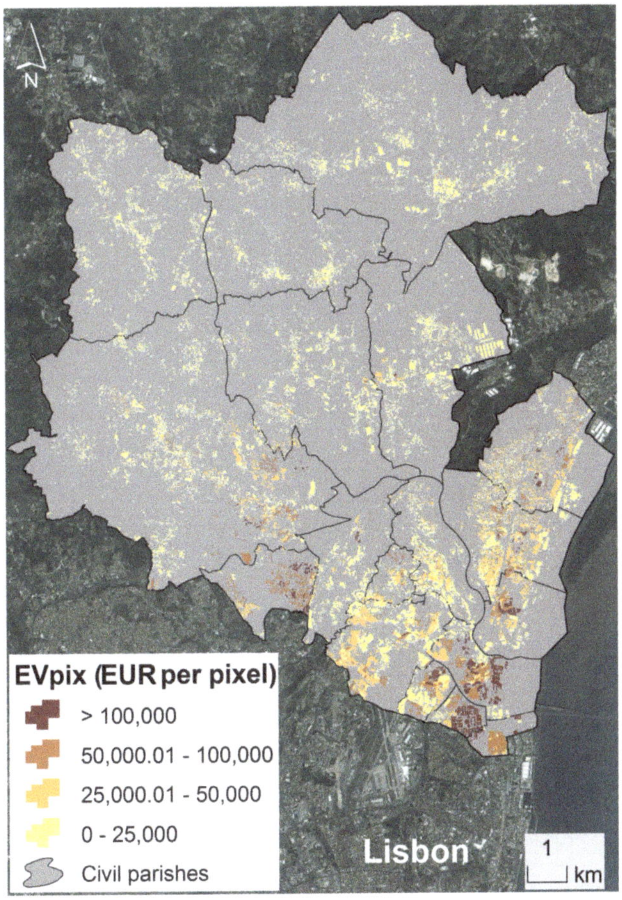

Figure 12. Economic value of buildings per 5 m pixel in the Loures municipality.

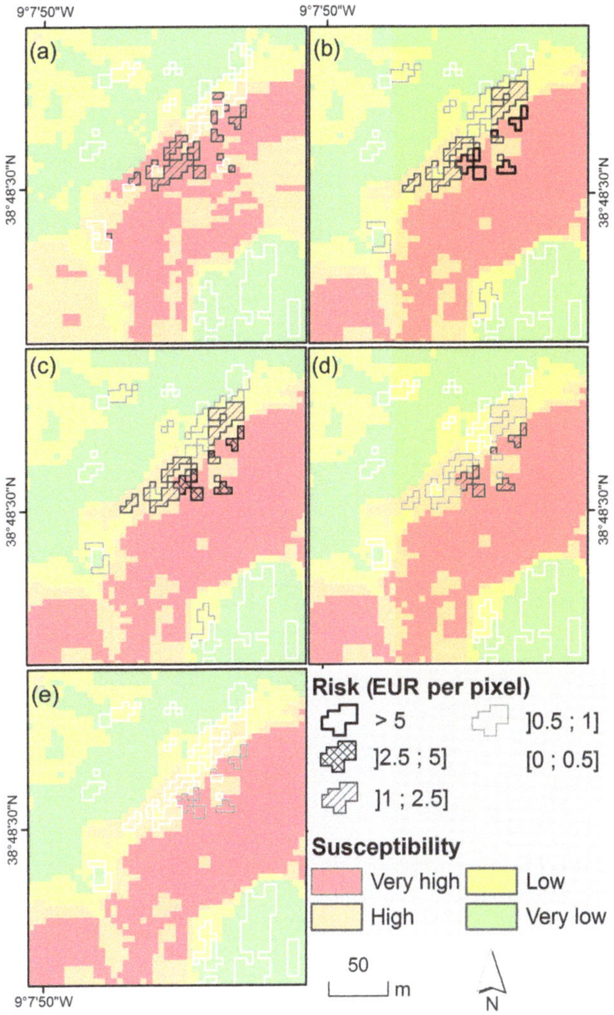

Figure 13. Detail of annual risk for buildings of the Loures municipality located on a landslide body, for a slip surface depth of **(a)** 1 m, **(b)** 3 m, **(c)** 5 m, **(d)** 10 m, and **(e)** 20 m. Pixel size: 5 m. For location, see Fig. 6.

5 Discussion

The vulnerability values obtained in this study are in agreement with the ones found in the literature. Indeed, we found that in general, landslides smaller than $\sim 1500\,m^2$ resulted in negligible to significant damage to buildings, corresponding to a vulnerability of 0.6 or less, whereas landslides larger than $\sim 7000\,m^2$ produced significant to very severe damage, corresponding to a vulnerability of 0.6 or higher, which is in agreement with the results found by Galli and Guzzetti (2007). Moreover, in terms of accumulated material height, the landslides that have a 5 m depth of accumulated material, produce an average damage for the four structural building types corresponding to a vulnerability of 0.91. For comparison, the vulnerability curves computed by Papathoma-Köhle et al. (2012b) using a Weibull distribution show that debris flows produce a total destruction (vulnerability = 1) when the accumulated material reaches 3.5 m high. Considering that the debris flows' intensity is increased by their velocity, it is understandable that their potential for damage is higher than the potential for damage of the slow landslides considered in the present study.

The answers obtained by the sub-pool of experts with a deep knowledge of the landslides and built environment of the study area have low standard deviation; they are more consistent than the answers obtained by the whole European experts pool, given that they know the typical landslide characteristics in the study area (e.g. landslide velocity, affected material, height of landslide scarps) as well as the characteristics of the built environment that may influence the physical vulnerability (e.g. age, state of conservation, construction materials) better, and are better able to assess the degree of loss produced by the impact of landslides. This shows that the vulnerability is in part a site-specificity parameter, and it has to be taken into account during vulnerability assessment by a questionnaire.

The standard deviation tends to be higher for lower magnitude landslides, for which the potential damage is more

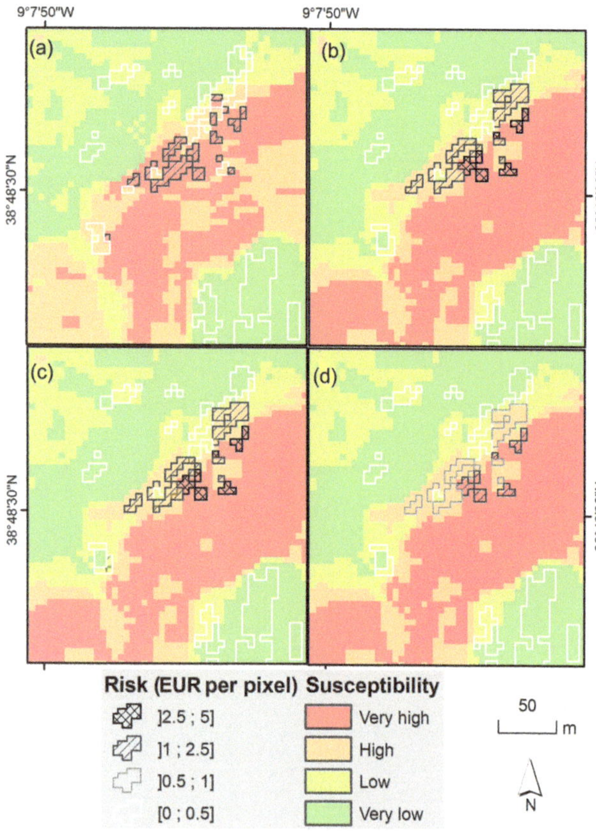

Figure 14. Detail of annual risk for buildings of the Loures municipality located on a landslide foot, for affected material that is **(a)** 0.5 m, **(b)** 1 m, **(c)** 3 m, and **(d)** 5 m high. Pixel size: 5 m. For location, see Fig. 6.

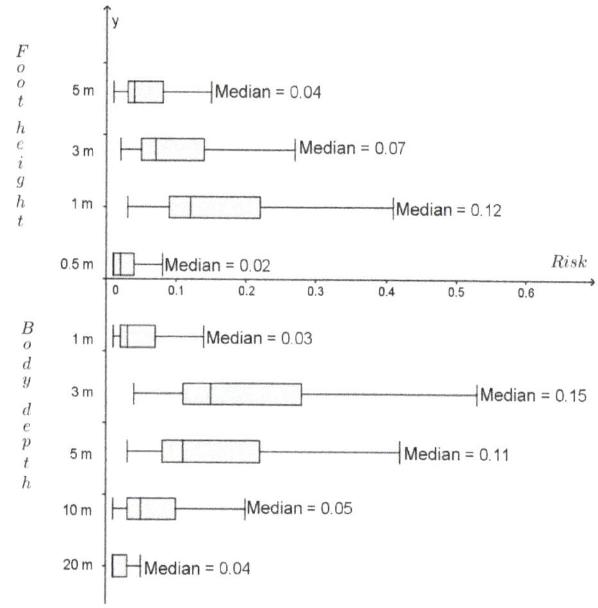

Figure 15. Box plots of the risk for the buildings per 5 m pixel, for each scenario. Outliers are not shown. The maximum outlier values are 8.35 (foot height: 5 m), 12.81 (foot height: 3 m), 19.58 (foot height: 1 m), 5.46 (foot height: 0.5 m), 8.2 (body depth: 1 m), 25.68 (body depth: 3 m), 20.38 (body depth: 5 m), 9.62 (body depth: 10 m), and 2.99 (body depth: 20 m).

difficult to assess than for the higher magnitude landslides, which are considered as highly destructive by the large majority of experts within the sub-pool of experts. Implications of high standard deviation for final risk calculation may be relevant. For example, assessing the risk for a SBT1 building, with a value of EUR 100,000, affected by a landslide with 0.5 m high accumulated material located in the highest landslide susceptibility class, the annual risk is EUR 33.6 considering the average vulnerability. However, the risk may range between EUR 18 and 49 considering the standard deviation value, which means a difference of 46 % to the average value.

If we consider that the sub-pool experts have a more accurate opinion of the building vulnerability to landslides in the Loures municipality, we can state that the pool of European landslide experts overestimated the low-magnitude landslides and underestimated the high-magnitude landslides. Regarding the vulnerability assessment by the European landslide experts, most of them merely completed the questionnaire, but some of them expressed doubts that arose while filling in the questionnaire or made some comments. Whenever necessary, emails were exchanged before the experts completed the questionnaire. Most of the experts who had doubts expressed that it was difficult to assess the potential damage caused by a landslide to a building based only on the depth of the landslide slip surface or the height of accumulated material. Additionally, the structure of the building and its position on the landslide body or foot were referred as major concerns. However, it was not useful to give them more detailed information about the building position or about the characteristics of the landslides (e.g. the velocity of the landslide, the type of affected material, the height of the scarp) as they requested, because such information was not available for the complete landslide inventory and the aim of this study is to assess the vulnerability of the buildings of a whole municipality in a systematic fashion. One adopted solution was to consider the worst case scenario for the potential damage assessment; i.e. the height of the scarp is slightly smaller than the depth of the slip surface; the building is partly within the body and partly outside (on the scarp); the foot is perpendicular to length of the building; and the building is well within the foot, not simply touched by it. This model is quite conservative in that in more favourable situations, damage would logically be lower. But as some of the experts expressed the potential damage as maximum, and the others as medium, the average values provide a model that is not too conservative, but not too low either in terms of expected potential damage, and this is what the authors were seeking.

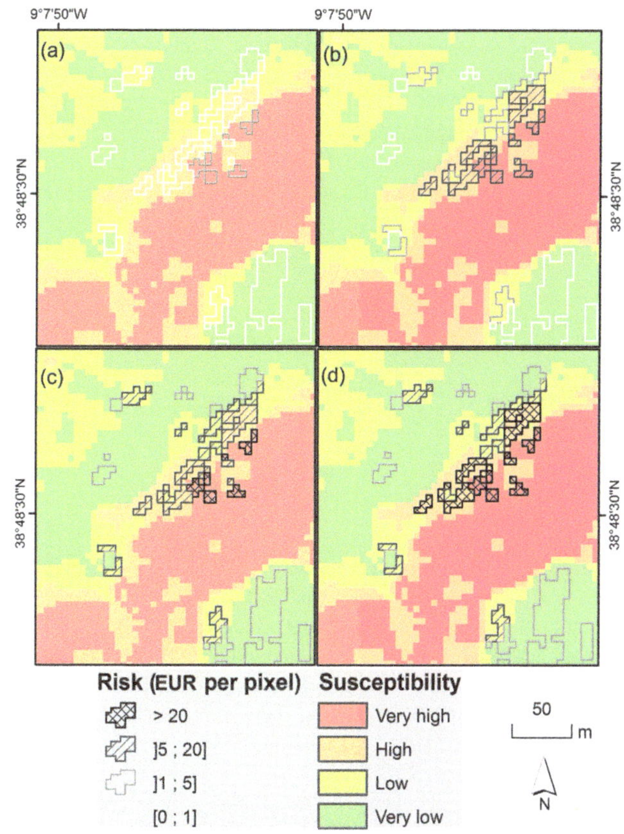

Figure 16. Detail of multiannual risk for buildings of the Loures municipality located on a landslide body with a 10 m deep slip surface, for a hazard of **(a)** 1 year, **(b)** 10 years, **(c)** 25 years, and **(d)** 50 years. Pixel size: 5 m. For location, see Fig. 6.

Regarding the representation of the buildings' vulnerability at the municipal scale, the vulnerability approach based on statistical mapping units is satisfactory. This approach is time-saving and provides correct results when the structural building types within the BGRI subsections are homogeneous. In the BGRI subsections where the structural building types are very heterogeneous, it is useful to take time to identify the structural building type of each building, by fieldwork.

The vulnerability assessment developed in this study has three main advantages: first, the method can be applied to the buildings of the whole Loures municipality despite its huge number (more than 30 000) and the few data available for these buildings; second, the variability of results can be assessed by calculating the standard deviation of the attributed vulnerabilities; third, the vulnerability assessment method developed in this study was applied to the Loures municipality, but it can be reproduced in another municipality or a region with similar landslide types and built environment in a reasonable time.

However, the risk analysis presented here has some limitations and drawbacks involving both the hazard assessment and the potential damage assessment. In relation to the hazard assessment, the spatiotemporal probabilities were overestimated as they were calculated on the landslide areas as a whole. Therefore, the risk calculated for a building constructed on a landslide body on the one hand, or on a landslide foot on the other hand was also amplified because the potential damage was assessed separately for the body and the foot. In addition, the spatiotemporal probabilities were calculated on the basis of the total areas of the inventoried landslides, considering that the 686 landslides of the Loures municipality were the only ones that occurred from 1967 (first landslides inventoried and dated) until 2004 (date of the orthophoto maps used to complete the inventory); in reality, it is obvious that the real total affected area is larger because we could not have inventoried all the landslides that occurred in the Loures municipality during this period. An annual inventory of the whole municipality and extensive fieldwork from 1967 to 2004 could be the solution to obtain a complete landslide inventory. From this point of view, the hazard was underestimated. In addition, changes in the frequency of occurrence of landslides associated with climate change increase the uncertainty of probabilities computed for 10, 25, and 50 years.

In relation to the potential damage assessment, the element at risk values were underestimated. Indeed, the value of the contents inside the buildings was not considered as they were not known. Moreover, indirect costs linked to the function of the building are difficult to quantify and were not considered in this study, although they play an important role in a complete risk analysis. Some examples of these indirect costs would be the costs linked to the temporary or definitive resettling of families whose house has been destroyed by a landslide, as well as the eventual additional costs of transportation if their resettled home is farther from their work place. Another example of indirect costs is the capital lost by the cessation of activity in case an industry or an office were destroyed or damaged by a landslide. Last but not least, it would be even worse if the destroyed building was a strategic building such as a hospital or a school; the vital and sensitive role of these kinds of buildings was not considered in this study, which is another limitation.

The risk analysis is based on the assumption that future landslides will have similar characteristics to the past ones; however, if the landslide preparatory and triggering conditions change (e.g. due to climate change or direct human interference on slopes), the number of landslides and their magnitude would increase, as would the associated damage, and that would have to be considered.

Finally, the risk is underestimated for the scenarios of 10, 25, or 50 years, because it was calculated for the buildings that exist presently, without taking the urban expansion into account, which is a factor of element at risk exposure, and is thus responsible for an increasing risk.

6 Concluding remarks

An assessment of buildings' vulnerability to landslides, based on an inquiry of nine magnitude scenarios by a pool and a sub-pool of landslide experts, was developed and applied to Loures, a municipality within the greater Lisbon area. The obtained vulnerabilities vary from 0.2 to 1 as a function of the structural building types and increase with the landslide magnitude, being maximal for a 10 m or a 20 m deep landslide. The annual and multiannual landslide risk has also been computed for the nine magnitude scenarios; the maximum annual risk occurs for the landslides that are 3 m deep, with a maximum value of EUR 25.68 per 5 m pixel.

For the other magnitude scenarios, risk values are low, but they should not be confused with the potential loss values. Indeed, the risk values of the landslides that are 5, 10, or 20 m deep are low because the magnitude probabilities of these landslides are low; nevertheless, when these landslides occur, they produce severe or very severe damages to the buildings.

The analysis of the landslide risk for the buildings of the Loures municipality enables the stakeholders to focus on the buildings for which the landslide vulnerability and the landslide risk are high. All the magnitude scenarios must be taken into account for accurate planning. The landslides that have a low-magnitude being more frequent, the risk they imply has to be considered for short-term planning, whereas the risk implied by high-magnitude landslides has to be considered for long-term planning.

Landslide risk analysis performed in this work may be very useful for insurance companies, which are interested in risk values for buildings, but it may not be so useful for end users dealing with spatial planning and civil protection. Indeed, for spatial planning stakeholders, it is crucial to know where future landslides will occur in order to select the safest zones for development purposes. Therefore, a validated landslide susceptibility assessment, as the one which was presented by Guillard and Zêzere (2012), is a very useful tool for spatial planning, which can be improved with additional data on landslide magnitude and landslide frequency. On the other hand, the civil protection stakeholders need to know the landslide risk for buildings that have a vital or strategic role (e.g. hospitals, schools), but also the location of the population that need to be protected, including the most vulnerable groups of people. Therefore, the landslide hazard assessment and mapping is not enough for civil protection and should be complemented by the assessment of the specific risk (hazard × vulnerability), namely for critical structures and infrastructures, which might be more useful and less time-consuming than the complete risk analysis for the complete built environment.

Acknowledgements. Clémence Guillard-Gonçalves and Susana Pereira were supported by the Portuguese Foundation for Science and Technology (FCT) through grants SFRH/BD/64973/2009 and SFRH/BPD/69002/2010, respectively. This work is part of the project FORLAND – Hydro-geomorphologic Risk in Portugal: Driving Forces and Application for Land Use Planning (PTDC/ATPGEO/1660/2014), funded by FCT. We are grateful to two anonymous reviewers whose comments and suggestions greatly helped to improve the quality of this paper.

References

AGS – Australian Geomechanics Society Sub-Committee on Landslide Risk Management: Landslide risk management concepts and guidelines, Aust. Geomech. J., 35, 49–92, 2000.

Alexander, D.: Landslide damage to buildings, Environ. Geol. Water Sci., 8, 147–151, doi:10.1007/BF02509902, 1986.

Alexander, D.: Vulnerability to landslides, in: Landslide Hazard and Risk, edited by: Glade, T., Anderson, M., and Crozier, M. J., John Wiley & Sons Ltd., Chichester, UK, 175–198, 2005.

Cardinali, M., Reichenbach, P., Guzzetti, F., Ardizzone, F., Antonini, G., Galli, M., Cacciano, M., Castellani, M., and Salvati, P.: A geomorphological approach to the estimation of landslide hazards and risks in Umbria, Central Italy, Nat. Hazards Earth Syst. Sci., 2, 57–72, doi:doi:10.5194/nhess-2-57-2002, 2002.

Ciurean, R. L., Schröter, D., and Glade, T.: Conceptual Frameworks of Vulnerability Assessments for Natural Disasters Reduction, in: Approaches to Disaster Management – Examining the Implications of Hazards, Emergencies and Disasters, edited by: Tiefenbacher, J., InTech, Rijeka, 3–32, 2013.

Corominas, J., van Westen, C., Frattini, P., Cascini, L., Malet, J.-P., Fotopoulou, S., Catani, F., Van Den Eeckhaut, M., Mavrouli, O., Agliardi, F., Pitilakis, K., Winter, M. G., Pastor, M., Ferlisi, S., Tofani, V., Hervás, J., and Smith, J. T.: Recommendations for the quantitative analysis of landslide risk, Bull. Eng. Geol. Environ., 209–263, doi:10.1007/s10064-013-0538-8, 2014.

Corominas, J., Einstein, H., Davis, T., Strom, A., Zuccaro, G., Nadim, F., and Verdel, T.: Glossary of Terms on Landslide Hazard and Risk, in: Engineering Geology for Society and Territory – Volume 2, Landslide Processes, edited by: Lollino, G., Giordan, D., Crosta, G. B., Corominas, J., Azzam, R., Wasowski, J., and Sciarra, N., Springer, Cham, 1775–1779, doi:10.1007/978-3-319-09057-3_314, 2015.

Cruden, D. and Varnes, D.: Landslides types and processes, in: Landslides, Investigation and Mitigation, edited by: Turner, A. K. and Schuster, R. L., Transportation Research Board, Special Report 247, National Academy Press, Washington, D.C., 36–75, 1996.

Cutter, S. L. and Finch, C.: Temporal and spatial changes in social vulnerability to natural hazards, P. Natl. Acad. Sci. USA, 105, 2301–2306, doi:10.1073/pnas.0710375105, 2008.

Dai, F. C., Lee, C. F., and Ngai, Y. Y.: Landslide risk assessment and management?: an overview, Eng. Geol., 64, 65–87, 2002.

Douglas, J.: Physical vulnerability modelling in natural hazard risk assessment, Nat. Hazards Earth Syst. Sci., 7, 283–288, doi:10.5194/nhess-7-283-2007, 2007.

Du, J., Yin, K., Nadim, F., and Lacasse, S.: Quantitative vulnerability estimation for individual landslides, in: 18th International Conference on Soil Mechanics and Geotechnical Engineering, Paris, 2181–2184, 2013.

Fell, R., Corominas, J., Bonnard, C., Cascini, L., Leroi, E., and Savage, W. Z.: Guidelines for landslide susceptibility, hazard and risk zoning for land use planning, Eng. Geol., 102, 85–98, doi:10.1016/j.enggeo.2008.03.022, 2008.

Galli, M. and Guzzetti, F.: Landslide vulnerability criteria: a case study from Umbria, central Italy, Environ. Manage., 40, 649–664, doi:10.1007/s00267-006-0325-4, 2007.

Garcia, R. A. C.: Metodologias de avaliação de perigosidade e risco associado a movimentos de vertente. Aplicação na bacia da ribeira de Alenquer, PhD thesis, Institute of Geography and Spatial Planning (IGOT), Universidade de Lisboa, Lisbon, Portugal, 469 pp., 2012.

Glade, T., Anderson, M. and Crozier, M. J. (Eds.): Landslide Hazard and Risk, John Wiley & Sons Ltd., Chichester, UK, 2005.

Godfrey, A., Ciurean, R. L., van Westen, C. J., Kingma, N. C., and Glade, T.: Assessing vulnerability of buildings to hydro-meteorological hazards using an expert based approach – An application in Nehoiu Valley, Romania, Int. J. Disaster Risk Reduct., 13, 229–241, doi:10.1016/j.ijdrr.2015.06.001, 2015.

Guillard, C. and Zêzere, J.: Landslide susceptibility assessment and validation in the framework of municipal planning in Portugal: the case of Loures Municipality, Environ. Manage., 50, 721–735, doi:10.1007/s00267-012-9921-7, 2012.

Guillard-Gonçalves, C., Cutter, S. L., Emrich, C. T., and Zêzere, J. L.: Application of Social Vulnerability Index (SoVI) and delineation of natural risk zones in Greater Lisbon, Portugal, J. Risk Res., 18, 651–674, doi:10.1080/13669877.2014.910689, 2015.

Guzzetti, F., Carrara, A., Cardinali, M., and Reichenbach, P.: Landslide hazard evaluation: a review of current techniques and their application in a multi-scale study , Central Italy, Geomorphology, 31, 181–216, 1999.

Highland, L. M. and Bobrowsky, P.: The Landslide Handbook – A Guide to Understanding Landslides, Reston, Virginia, 2008.

Hungr, O.: Some methods of landslide intensity mapping, in: Landslide risk assessment, Proceedings of the International Workshop on Landslide Risk Assessment, 19–21 February 1997, Honolulu, edited by: Cruden, D. M. and Fell, R., Balkema, Rotterdam, 215–226, 1997.

INE – Instituto Nacional de Estatísticas: Censos 2001 – resultados definitivos – Lisboa (Census 2001 – Definitive Results – Lisbon), http://www.ine.pt/xportal/xmain?xpid=INE&xpgid=ine_publicacoes&PUBLICACOESpub_boui=377750&PUBLICACOEStema=00&PUBLICACOESmodo=2 (last access: December 2013), 2002.

INE – Instituto Nacional de Estatística: Portuguese population grids for the years 2001, 2006 and 2011, available at: http://www.efgs.info/data/portugal/Metodology_Portuguese_GridData2001_2006_2011.docx/view (last access: 13 July 2015), 2011.

Jaiswal, P., van Westen, C. J., and Jetten, V.: Quantitative assessment of direct and indirect landslide risk along transportation lines in southern India, Nat. Hazards Earth Syst. Sci., 10, 1253–1267, doi:10.5194/nhess-10-1253-2010, 2010.

Jaiswal, P., van Westen, C. J., and Jetten, V.: Quantitative assessment of landslide hazard along transportation lines using historical records, Landslides, 8, 279–291, doi:10.1007/s10346-011-0252-1, 2011a.

Jaiswal, P., van Westen, C. J., and Jetten, V.: Quantitative estimation of landslide risk from rapid debris slides on natural slopes in the Nilgiri hills, India, Nat. Hazards Earth Syst. Sci., 11, 1723–1743, doi:10.5194/nhess-11-1723-2011, 2011b.

Lateltin, O., Haemmig, C., Raetzo, H., and Bonnard, C.: Landslide risk management in Switzerland, Landslides, 2, 313–320, doi:10.1007/s10346-005-0018-8, 2005.

Lee, E. M. and Jones, D. K. C.: Landslide risk assessment, Tilford, London, 2004.

Léone, F.: Caractérisation des vulnérabilités aux catastrophes "naturelles": contribution à une évaluation géographique multirisque (mouvements de terrain, séismes, tsunamis, éruptions volcaniques, cyclones) – Mémoire d'Habilitation à Diriger des Recherches (HDR), Section 23 (géographie), Université Paul Valéry – Montpellier III, Laboratoire GESTER, Montpellier, 2007.

Léone, F., Aste, J. P., and Leroi, E.: Vulnerability assessment of elements exposed to mass-movement: working toward a better risk perception, in: Landslides, edited by: Senneset, K., Balkema, Rotterdam, 263–270, 1996.

Li, Z., Nadim, F., Huang, H., Uzielli, M., and Lacasse, S.: Quantitative vulnerability estimation for scenario-based landslide hazards, Landslides, 7, 125–134, doi:10.1007/s10346-009-0190-3, 2010.

Macquarie, O., Thiery, Y., Malet, J. P., Weber, C., Puissant, A., and Wania, A.: Current practices and assessment tools of landslide vulnerability in mountainous basins-identification of exposed elements with a semi- automatic procedure, in: Landslides: evaluation and stabilization, edited by: Lacerda, W. A., Ehrlich, M., Fontoura, S. A. B., and Sayao, A. S. F., Taylor and Francis Group, London, 171–176, 2004.

Malamud, B. D., Turcotte, D. L., Guzzetti, F., and Reichenbach, P.: Landslide inventories and their statistical properties, Earth Surf. Proc. Land., 29, 687–711, doi:10.1002/esp.1064, 2004.

Michael-Leiba, M., Baynes, F., and Scott, G.: Quantitative landslide risk assessment of Cairns, AGSO Records 1999/36, Australian Geological Survey Organisation, Department of Industry, Science & Resources, Canberra, 40 pp., 1999.

Papathoma-Köhle, M., Kappes, M., Keiler, M., and Glade, T.: Physical vulnerability assessment for alpine hazards: state of the art and future needs, Nat. Hazards, 58, 645–680, doi:10.1007/s11069-010-9632-4, 2011.

Papathoma-Köhle, M., Totschnig, R., Keiler, M., and Glade, T.: A new vulnerability function for debris flow – The importance of physical vulnerability assessment in Alpine areas, in: 12th Congress Interpraevent, Grenoble, France, 2012a.

Papathoma-Köhle, M., Keiler, M., Totschnig, R., and Glade, T.: Improvement of vulnerability curves using data from extreme events: debris flow event in South Tyrol, Nat. Hazards, 64, 2083–2105, doi:10.1007/s11069-012-0105-9, 2012b.

Peng, L., Xu, S., Hou, J., and Peng, J.: Quantitative risk analysis for landslides: the case of the Three Gorges area, China, Landslides, 12, 943–960, doi:10.1007/s10346-014-0518-5, 2015.

Remondo, J., Bonachea, J., and Cendrero, A.: A statistical approach to landslide risk modelling at basin scale: from landslide susceptibility to quantitative risk assessment, Landslides, 2, 321–328, doi:10.1007/s10346-005-0016-x, 2005.

Remondo, J., Bonachea, J., and Cendrero, A.: Quantitative landslide risk assessment and mapping on the basis of recent occurrences, Geomorphology, 94, 496–507, doi:10.1016/j.geomorph.2006.10.041, 2008.

Santos, J. G.: Landslide susceptibility and risk maps of Regua (Douro basin, NE Portugal), in: Proceeding of the IAG and IGU-C12 Regional Conference "Geomorphic hazards; towards the prevention of disasters", Mexico City, Mexico, 2003.

Silva, M. and Pereira, S.: Assessment of physical vulnerability and potential losses of buildings due to shallow slides, Nat. Hazards, 72, 1029–1050, doi:10.1007/s11069-014-1052-4, 2014.

Sterlacchini, S., Frigerio, S., Giacomelli, P., and Brambilla, M.: Landslide risk analysis: a multi-disciplinary methodological approach, Nat. Hazards Earth Syst. Sci., 7, 657–675, doi:10.5194/nhess-7-657-2007, 2007.

Tinti, S., Tonini, R., Bressan, L., Armigliato, A., Gardi, A., Guillande, R., Valencia, N., and Scheer, S.: Handbook of Tsunami Hazard and Damage Scenarios, JRC Scientific and Technical Reports, European Commission, Luxembourg, 41 pp., doi:10.2788/21259, 2011.

Uzielli, M., Nadim, F., Lacasse, S., and Kaynia, A. M.: A conceptual framework for quantitative estimation of physical vulnerability to landslides, Eng. Geol., 102, 251–256, doi:10.1016/j.enggeo.2008.03.011, 2008.

Uzielli, M., Catani, F., Tofani, V., and Casagli, N.: Risk analysis for the Ancona landslide – II: estimation of risk to buildings, Landslides, 12, 83–100, doi:10.1007/s10346-014-0477-x, 2015.

van Westen, C. J., Asch, T. W. J., and Soeters, R.: Landslide hazard and risk zonation – why is it still so difficult?, Bull. Eng. Geol. Environ., 65, 167–184, doi:10.1007/s10064-005-0023-0, 2005.

van Westen, C. J., Castellanos, E., and Kuriakose, S. L.: Spatial data for landslide susceptibility, hazard, and vulnerability assessment: An overview, Eng. Geol., 102, 112–131, doi:10.1016/j.enggeo.2008.03.010, 2008.

Varnes, D. and the International Association of Engineering Geology Commission on Landslides and other Mass-Movements: Landslide hazard zonation: a review of principles and practice, UNESCO Pre., Paris, 1984.

Winter, M. G., Smith, J. T., Fotopoulou, S., Pitilakis, K., Mavrouli, O., Corominas, J., and Argyroudis, S.: An expert judgment approach to determining the physical vulnerability of roads to debris flow, Bull. Eng. Geol. Environ., 73, 291–305, doi:10.1007/s10064-014-0570-3, 2014.

Yin, K. and Yan, T.: Statistical prediction models for slope instability of metamorphosed rocks, edited by: Bonnard, C., Proceedings of 5th ISL, Rotterdam, 1269–1272, 1988.

Záruba, Q. and Mencl, V.: Landslides and their control, Elsevier, Amsterdam, 1982.

Zêzere, J. L.: Landslide susceptibility assessment considering landslide typology. A case study in the area north of Lisbon (Portugal), Nat. Hazards Earth Syst. Sci., 2, 73–82, doi:10.5194/nhess-2-73-2002, 2002.

Zêzere, J. L., De Brum Ferreira, A., and Rodrigues, M.: The role of conditioning and triggering factors in the occurrence of landslides: a case study in the area north of Lisbon (Portugal), Geomorphology, 30, 133–146, doi:10.1016/S0169-555X(99)00050-1, 1999.

Zêzere, J. L., Reis, E., Garcia, R., Oliveira, S., Rodrigues, M. L., Vieira, G., and Ferreira, A. B.: Integration of spatial and temporal data for the definition of different landslide hazard scenarios in the area north of Lisbon (Portugal), Nat. Hazards Earth Syst. Sci., 4, 133–146, doi:10.5194/nhess-4-133-2004, 2004.

Zêzere, J. L., Oliveira, S., Garcia, R., and Reis, E.: Landslide risk analysis in the area North of Lisbon (Portugal): evaluation of direct and indirect costs resulting from a motorway disruption by slope movements, Landslides, 4, 123–136, doi:10.1007/s10346-006-0070-z, 2007.

Zêzere, J. L., Garcia, R., Oliveira, S., and Reis, E.: Probabilistic landslide risk analysis considering direct costs in the area north of Lisbon (Portugal), Geomorphology, 94, 467–495, 2008.

14

Discussing the role of tropical and subtropical moisture sources in cold season extreme precipitation events in the Mediterranean region from a climate change perspective

S. O. Krichak[1], S. B. Feldstein[2], P. Alpert[1], S. Gualdi[3], E. Scoccimarro[3], and J.-I. Yano[4]

[1]Department of Geosciences, Raymond and Beverly Sackler Faculty of Exact Sciences, Tel Aviv University, Tel Aviv, Israel
[2]Department of Meteorology, The Pennsylvania State University, University Park, USA
[3]Istituto Nazionale di Geofisica e Vulcanologia, INGV, Centro Euro-Mediterraneo sui Cambiamenti Climatici, CMCC, Bologna, Italy
[4]CNRM/GAME UMR 3589, Météo-France and CNRS, 31057 Toulouse CEDEX, France

Correspondence to: S. O. Krichak (shimonk@post.tau.ac.il)

Abstract. This paper presents a review of a large number of research studies performed during the last few decades that focused on the investigation of cold season extreme precipitation events (EPEs) in the Mediterranean region (MR). The publications demonstrate the important role of anomalously intense transports of moist air from the tropical and subtropical Atlantic in the occurrence of EPEs in the MR. EPEs in the MR are directly or indirectly connected to narrow bands with a high concentration of moisture in the lower troposphere, i.e., atmospheric rivers, along which a large amount of moisture is transported from the tropics to midlatitudes. Whereas in a significant fraction of the EPEs in the western MR moisture is transported to the MR from the tropical Atlantic, EPEs in the central, and especially the eastern, MR are more often associated with intense tropical moisture transports over North Africa and the Red Sea. The moist air for the EPEs in the latter part of the MR also mainly originates from the tropical Atlantic and Indian oceans, and in many cases it serves as a temporary moisture reservoir for future development. The paper is supplemented by the results of a test for a possible connection between declining Arctic sea ice and the climatology of intense precipitation in the eastern MR. Based on the results of the evaluation supporting those from the earlier climate change analyses and modeling studies, it is concluded that a further anthropogenic global warming may lead a greater risk of higher rainfall totals and therefore larger winter floods in western and central parts of the MR as a consequence of stronger and more numerous Atlantic atmospheric rivers, possibly accompanied by a decline in the number of EPEs in the eastern part of the MR.

1 Introduction

The cold season, which for the Mediterranean region (MR) may be defined as September–May, is characterized by a substantial number of exceptionally intense extreme precipitation events (EPEs) with precipitation rates being comparable to those of hurricanes (Clarke and Rendell, 2006; Lionello et al., 2006b; Pfahl and Wernli, 2012). These EPEs have been selected as the main focus of the current review. The cold season MR EPEs are almost exclusively associated with cyclones (e.g., Reale and Lionello, 2013). Heavy rains during the MR EPEs are often associated with suddenly occurring flash floods that represent significant threats to society, human lives and infrastructure (e.g., Lastoria et al., 2006; Lionello, et al., 2006a; Jansa et al., 2014; Trigo et al., 2016).

The formation of a cold season rainy event in the MR typically results from the interplay between two acting factors – high water vapor content that induces mesoscale moist convection and larger-scale baroclinic dynamics. Obviously, moisture does not automatically induce convection, as the air must become positively buoyant. In the MR, moist convection is often associated with the effects of topography (Reale and Lionello, 2013).

Among the famous MR EPEs are such cases as the "century" floods in Florence, Italy, on 4 November 1966, with up to 750 mm of rain in 24 h (e.g., De Zolt et al., 2006); floods in Gandía (eastern Spain, province of Valencia) on 3–4 November 1987, with 817 mm of rain in 24 h (Romero et al., 2000); the event on 15 November 1985 with more than 200 mm of rain in 2 h in the Balearic Islands (Romero et al., 1998); the rainy day of 9 October 1994 with 450 mm in 24 h in the province of Valencia, Spain (Ramis et al., 1998); heavy rains in Egypt, Israel and northern Italy during 1–6 November 1994 with more than 300 mm of rain in 36 h (Buzzi et al., 1998; Krichak and Alpert, 1998; Jansa et al., 2000); a case with catastrophic floods (11 fatalities) in central Greece and in Athens with more than 140 mm of rain in 24 h on 21 October 1994 (Lagouvardos et al., 1996); a case with devastating floods in central and southern Greece with more than 300 mm of rain in 24 h during 11–12 January 1997 (Kotroni et al., 1999); a torrential rain event in the Valencia region in October 2007 that exceeded 400 mm of rain in 24 h (Pastor et al., 2010); a heavy precipitation event in Israel with over 260 mm in 24 h during 4 December 2001 (e.g., Krichak et al., 2004, 2007); a flood event in Antalya, Turkey, on 4–6 December 2002 with more than 230 mm in 24 h (Kotroni et al., 2006); and many others (e.g., Ramis and Llasat 1994; Doswell et al., 1998; Romero et al., 1999, 2000; Kahana et al., 2004; Homar et al., 2007; Ramis et al., 2009; de Vries et al., 2013).

A number of international research projects have been devoted to attaining a better understanding of the physical mechanisms responsible for the formation of EPEs. The list of research efforts includes ALPEX (Davies and Pichler, 1990; McGinley and Zhupanski, 1990; Alpert et al., 1996; Buzzi et al., 1998), the Mediterranean Experiment (MEDEX, Jansa et al., 2014), Climate Change and Impact Research: the Mediterranean Environment (CIRCE FP6-EU, Navarra and Tubiana, 2013) project, the Mediterranean CLImate VARiability and Predictability (MedCLIVAR) network (e.g., Lionello et al., 2012, 2014; Garcia-Herrera et al., 2014) and the HYdrological cycle in the Mediterranean EXperiment (HyMeX; Drobinski et al., 2014).

In this paper, we present a discussion of the investigations on very intense cold season high-impact EPEs in the MR performed mainly during the last 30 years. The review was prepared in accordance with the call for a special NHESS issue on "Climate change, extreme events and hazards in the Mediterranean region" (edited by P. Lionello, V. Artale, D. Gomis and H. Saaroni).

The paper is organized as follows: an extended summary of the research that focused on the physical mechanisms responsible for the formation of the MR EPEs is presented in Sect. 2. The results of investigations on the role of tropical moisture as well as the mechanisms for its transport to the MR are discussed in Sect. 3. A discussion of the research that addressed the role of global warming for recent EPEs and the expected future MR EPEs trends is presented in Sect. 4. An extended discussion of the major trends of MR EPEs including an analysis of the possible role of the melting of Arctic ice, and the summary and conclusions are given in Sect. 5.

2 Identification of acting mechanisms

The cold season cyclones associated with EPEs in the extratropics are often characterized by an extreme pressure fall ("explosive" or "bomb" cyclones) as well as strong surface winds (e.g., Roebber, 1984; Wernli et al., 2002; Fink et al., 2009; Liberato et al., 2011). Such events can be especially damaging. The earliest analyses of these cyclones focused on individual cases (e.g., Sanders and Gyakum, 1980; Zhu and Newell, 1994; Alpert et al., 1996; Schmith et al., 1998; Massacand et al., 1998; Tsidulko and Alpert, 2001). These studies demonstrated the importance of mountains, upper-level jets, low-level fronts as well as air–sea interaction. Later analyses (e.g., Krichak and Alpert, 1998; Ulbrich and Christoph, 1999; Sickmöller et al., 2000; Ferraris et al., 2001; Ulbrich et al., 2001) identified a notable role for moisture transported from extra-Mediterranean regions.

The arrival of moist air from the tropical Atlantic into the Mediterranean basin is typically followed by an intense low-level moisture flux convergence over the area of the EPE formation (Ferraris et al., 2001). This series of processes appears to characterize the majority of EPEs in the MR. In contrast to rainy events of ordinary intensity, the MR EPEs are more likely to originate from a specific atmospheric process (or a group of processes) associated with the formation of hurricanes or intense cyclones over the Atlantic Ocean. Such synoptic developments are characterized by intense convergence of moist air from the tropics (serving as an unlimited moisture reservoir; Ralph and Dettinger, 2011) which is followed by a fast intrusion of moist air into the midlatitudes and eventually to the MR without significant mixing with the surrounding air.

The possibility of identifying moisture sources in the subtropics or tropics has stimulated the application of a Lagrangian approach for trajectory calculations. Reale et al. (2001) applied this method to investigate a series of 1998 flood events in the Mediterranean. Their analysis revealed a notable role for the advection of moisture from the North Atlantic in the formation of the EPEs. Further evaluations (Berto et al., 2004; Turato et al., 2004; Rudari et al., 2005; Ziv et al., 2005; De Zolt et al., 2006; Malguzzi et al., 2006) have supported this conclusion. In particular, by studying a precipitation event in the Trentino region (northern Italy), Berto et al. (2004) found contributions of moisture not only from the Mediterranean but also from North Africa and the eastern and central Atlantic. Turato et al. (2004) studied a heavy precipitation event in Piedmont in 2000. They identified a variety of moisture sources that contribute to the EPE, including those in the North Atlantic. Specifically, more than

50 % of the evaporation contributing to the precipitation during the event comes from the Atlantic Ocean. Rudari et al. (2005), De Zolt et al. (2006) and Malguzzi, et al. (2006) also arrived at the conclusion on the importance of strong advection of moist air into the central Mediterranean Sea prior to the formation of MR EPEs.

A number of intercomparison studies based on different reanalysis data sets (e.g., Trigo, 2006; Wang et al., 2006; Raible et al., 2008; Ulbrich et al., 2009; Hodges et al., 2011) have further addressed this issue. In particular, Pinto et al. (2009) have significantly improved our understanding of the role of external sources of moisture in the formation of several intense cyclones in the Euro-Mediterranean region. According to this study, moisture originating from the storm track region over the North Atlantic plays a notable role in the intensification of explosive cyclones.

A strong relationship between EPEs in the MR and the large-scale atmospheric circulation at the upper, middle and lower troposphere has been also demonstrated by Toreti et al. (2010). Here, a two-step classification procedure has been applied for the identification of anomaly patterns over the western–central and eastern Mediterranean Basin. Over the western Mediterranean, an anomalous surface to mid-tropospheric southwesterly flow induces enhanced moisture transport from the Atlantic. For the eastern Mediterranean EPEs, the anomaly patterns suggest warm and moist air advection into the region induced by anomalously intense moist air ascent.

The issue of the origin of the moisture affecting the Mediterranean during and prior to the formation of EPEs has been addressed in a series of studies (Gimeno et al., 2010a, b, 2012; Nieto et al., 2010). North Atlantic moisture sources were found to be dominant during winter. While the northern slope of the Alps is dominated by moisture from the North Atlantic and Central Europe, both with a clear seasonality and limited variability, the EPEs in the southern Alps have a stronger influence from the highly variable Mediterranean and extra-Mediterranean moisture sources. These evaluations were based on the methodology of Stohl and James (2004) who used a Lagrangian particle dispersion model. The atmosphere was divided homogeneously into a large number of particles, which have a constant mass. These particles were advected numerically. The evaluations using the Lagrangian tracking confirmed the earlier detected remote moisture sources for the EPEs. An area extending from the Gulf of Mexico to Africa has been found to be especially important for the EPEs in the western and central MR (Stohl and James, 2004; Stohl et al., 2008; Lionello et al., 2006a). Analyses by Duffourg and Ducrocq (2011) and Liberato et al. (2011) have additionally identified a notable contribution from moisture sources in the subtropical Atlantic for EPEs over southeastern France and Portugal, respectively.

A comprehensive analysis of tropical moisture exports based on the calculation of 7-day forward trajectories starting daily from the tropical lower troposphere (Knippertz and Wernli, 2010; also Knippertz et al., 2013) has allowed the identification of the existence of four distinct activity maxima with different seasonal behavior over the Northern Hemisphere. One of the maxima is located over the North Atlantic (mainly western, but in some cases reaching the MR) showing a small amplitude annual cycle with a maximum in winter and autumn. The result explains a link between tropical moisture sources and the climatology of precipitation and occasionally explosive cyclogenesis over the MR (e.g., Sodemann and Zubler, 2010; Pfahl and Wernli, 2012)

Application of a water vapor tagging approach into a limited-area numerical weather prediction model (Sodemann et al., 2009; Winschall et al., 2012) has complemented these studies. The method consists of labeling water vapor originating from different regions along its atmospheric pathway, including transport, phase changes and eventual precipitation. By keeping tag of water vapor origin, it is thus possible to gain quantitative information on the contribution of specific predefined water sources to a precipitation event. Using this method over the Atlantic Ocean, Winschall et al. (2012) simulated moisture transport and moisture source contributions to precipitation at high resolution, including all relevant parameterized processes, such as cloud microphysics. This diagnosis has demonstrated the importance of the coupling between surface evaporation and moisture advection for the EPEs. Pinto et al. (2013) confirmed this conclusion by performing an objective identification and cluster ranking of extraordinary rainfall events over northwestern Italy using time series of annual precipitation maxima for 1938–2002 at over 200 MR stations. The two top clusters are characterized by strong and persistent upper air troughs that induce not only moisture advection from the North Atlantic into the western Mediterranean but also a strong northward flow towards the southern Alpine ranges. Moisture transport from the North Atlantic was found to be less important for the weaker clusters.

By considering the 50 strongest precipitation events in the Alpine area during 1989–2009, Winschall et al. (2014) revealed the importance of evaporative hot spots over the eastern North Atlantic. Their results indicate that for MR EPEs in autumn and winter the Mediterranean Sea surface is only one of several source regions and is typically matched by moisture from the eastern North Atlantic.

The duration for the maximum moisture uptake varies between a few hours to more than 1 week before a precipitation event takes place. Whereas moisture uptakes that occur only a few hours or days prior to the EPE can be regarded as causally related to the dynamics of the EPE itself, the much earlier uptakes over the subtropical Atlantic were found not to be linked dynamically to the triggering of the MR EPEs (Winschall et al., 2014).

Additional analyses (e.g., Fiori et al., 2014; Dacre et al., 2015) have revealed a high sensitivity of the modeling results to the accuracy in the representation of the timing of the acting mesoscale processes. In particular, Dacre et al. (2015)

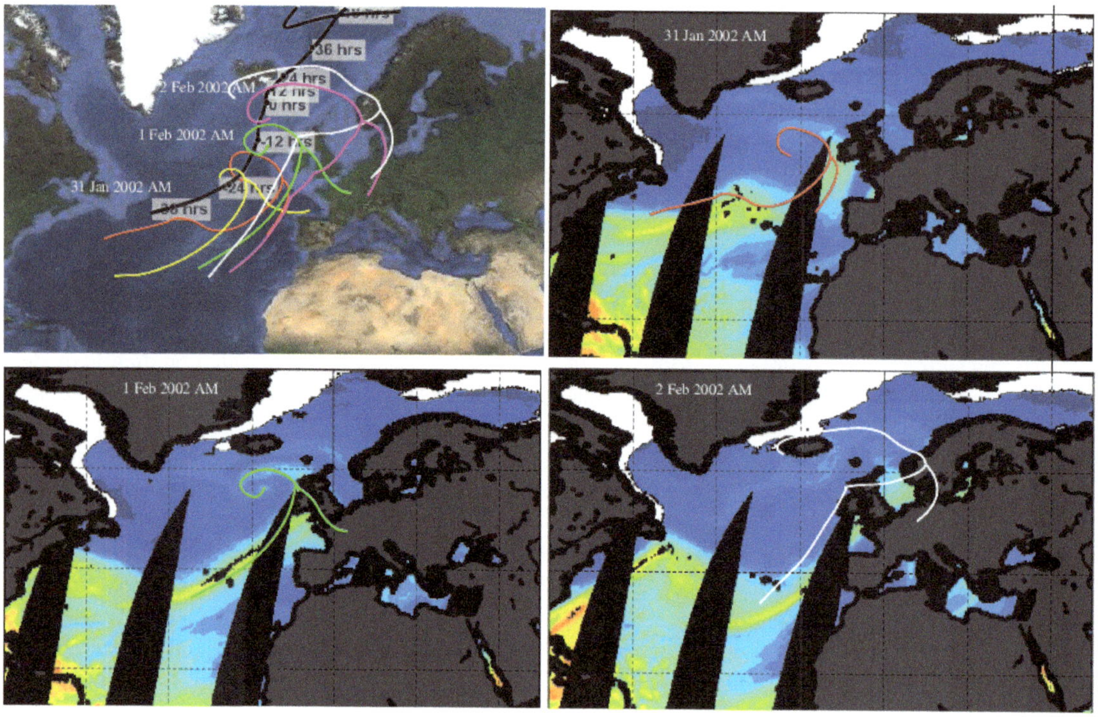

Figure 1. **(a)** Schematic showing positions of surface fronts at successive 12-hourly intervals, starting at 06:00 UTC on 31 January 2002, and Special Sensor Microwave Imager (SSM/I F13) integrated water vapor on **(b)** 31 January 2002 overlaid with 06:00 UTC on 31 January 2002 frontal positions, **(c)** on 1 February 2002 overlaid with 06:00 UTC on 1 February 2002 frontal positions and **(d)** on 2 February 2002 overlaid with 06:00 UTC on 2 February 2002 frontal positions (Fig. 8 in Dacre et al., 2015; used with permission ©American Meteorological Society).

demonstrated the importance of tropical moisture transport in the climatology of extratropical cyclones by investigating the spatial distribution of the 200 most intense extratropical cyclones during 1979–2009 (identified and tracked in the ERA-Interim reanalysis using 850 hPa relative vorticity). The individual terms in the water vapor budget equation were calculated for each of the 200 cyclones. Their results showed that evaporation from the sea surface, occurring mostly behind the cold front, contributes significantly to the total cyclone water vapor throughout the entire cyclone life cycle. The total cyclone integrated water vapor decreases throughout its movement as the water vapor lost from the atmosphere by precipitation exceeds that gained via evaporation or water vapor convergence. In this study, water vapor convergence into and out of the system was found to be negligible and even negative during the most rapidly intensifying stage of the cyclone evolution, showing that water vapor is actually exported from the system, leaving a water vapor footprint behind the cyclone as it travels polewards. It was also concluded that, as the cold front catches up with the warm front causing the warm sector to narrow, local convergence of water vapor occurs along the cold front and is thus responsible for creating the band of high total column water vapor (TCWV). Figure 1 shows the relative positions of the surface fronts and regions of high TCWV in the cyclone case study of Dacre et al. (2015, their Fig. 8). The band of high integrated water vapor narrows as the cold front catches up with the warm front, sweeping up water vapor as it travels. The location of the band of high water vapor travels farther from the cyclone center as the cyclone evolves due to frontal fracture. By the decaying stage of the cyclone evolution, the band is found within 1000 km distance from the cyclone center. The filaments of high water vapor content seen in the special sensor microwave imager (SSM/I) satellite imagery represent the footprints left behind as the cyclone channels atmospheric moisture into a narrow band when it travels poleward from its origin in the subtropics.

In summary, a large number of scientific publications suggest a crucial role of anomalously intense transport of moist air from the tropical and subtropical Atlantic in the occurrence of MR EPEs.

3 The role of atmospheric rivers (ARs)

The physical mechanism responsible for fast transport of tropical air into midlatitudes has also been actively addressed in the research performed. It has been discovered (Krichak et al., 2015) that a significant number of the MR EPEs appear to take place during (or immediately after) Atlantic hurricanes or storms (Krichak et al., 2015). The following is noteworthy: the 1966 "century" flood in Florence coincided with Hurricane Lois (4–11 November); the flood that occurred in Valencia on 3–4 November 1987 took place during a time period characterized by Tropical Depression 14 (duration of 31 October–4 November; peak intensity of $55\,\mathrm{km\,h^{-1}}$) in the Atlantic; the flood events in Egypt, Israel and Italy occurred during the same period as Hurricane Florence (2–8 November 1994); the heavy rainy event of 10 November 2001 in Algeria occurred immediately after Hurricane Noel (4–6 November 2001); the 3–4 December 2001 flood event in Israel occurred nearly concomitantly with Hurricane Olga (24 November–4 December 2001); and the flood event in Antalya, Turkey, on 5 December 2002 took place during the development of a powerful North Atlantic storm south of the Canadian Maritime Provinces on 5–9 December. An important contribution of moist tropical and/or subtropical air mass transport by ARs during the formation of the synoptic developments of the type has been demonstrated (Gimeno, et al., 2014; Krichak et al., 2015; Ramos et al., 2015).

ARs are defined as narrow bands in the warm sector of extratropical cyclones, characterized by strong (greater than $12.5\,\mathrm{m\,s^{-1}}$) wind speeds, high values (more than 20 mm) of vertically integrated water vapor (IWV) and values of vertically integrated vapor transport (IVT) in the layer from 300 to 1000 hPa of about (or greater than) $250\,\mathrm{kg\,m^{-1}\,s^{-1}}$ (Zhu and Newell, 1998; Ralph et al., 2004; Neiman et al., 2011; Ralph and Dettinger, 2011; Gimeno, et al., 2014). These transient filamentary regions often occur within the warm conveyor belt (WCB; Browning, 2004; Sodemann and Stohl, 2013) of extratropical cyclones. At any one time, there are approximately four or five narrow, elongated ARs across the midlatitudes (Fig. 2; Fig. 1 in Gimeno et al., 2014), which account for approximately 90 % of the total global poleward atmospheric water vapor transport (Zhu and Newell, 1998; Lavers et al., 2011).

A large amount of water vapor transported by ARs often leads to heavy precipitation and floods when an AR makes landfall, especially when the moisture-laden air is forced to rise over mountains (Dettinger, 2011; Lavers et al., 2011; Ralph and Dettinger, 2011; Gimeno et al., 2014). A strong link between ARs and extreme precipitation over a large part of western Europe has been established (e.g., Stohl et al., 2008; Knippertz and Wernli, 2010; Lavers et al., 2011; Liberato et al., 2012; Lu et al., 2013; Liberato, 2014; Dacre et al., 2015). It has been demonstrated that ARs are responsible for a majority of extreme precipitation (8 out of the 10 largest daily rainfall events) particularly in autumn and winter in Britain, France and Norway (e.g., Lavers and Vallarini, 2013). The regions experiencing major effects of ARs are characterized by hills and mountainous relief, which provide the necessary uplift leading to a significant rainfall.

A study undertaken by Sodemann and Stohl (2013) considered a monthly time period characterized by EPEs in Norway (December 2006). Here, the authors applied a water vapor tagging approach to gain new insight into the processes

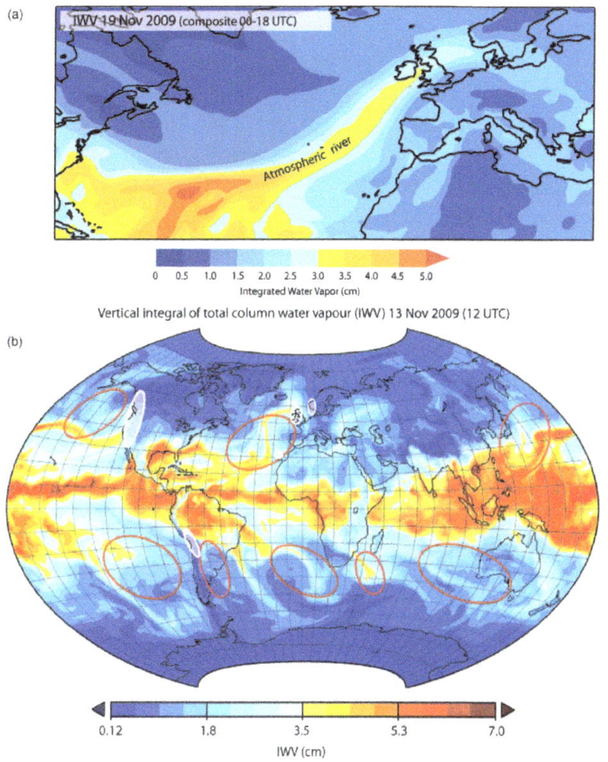

Figure 2. **(a)** Composite integrated total column of water vapor (IWV) between 00:00 and 18:00 UTC on 19 November 2009 showing an atmospheric river (AR) associated with extreme precipitation events that affected the United Kingdom. Data: ERA-Interim. **(b)** A general distribution of areas of occurrence of ARs (red contours) based on Zhu and Newell (1998). White contours showed the continental areas where there are reported cases of ARs linked with extreme precipitation and floods (Fig. 1 in Gimeno et al., 2014; used with permission from Frontiers in Earth Science Editorial Office).

leading to heavy precipitation in western Scandinavia and, more generally, of water vapor transport due to extratropical cyclones during winter. New process understanding has been obtained, which warrants corroboration on longer timescales and by a larger number of cases to attain a more general validity.

The results of the analysis are summarized schematically by Fig. 3 (Fig. 10 in Sodemann and Stohl, 2013). Here, the Thorncroft et al. (1993) classification of cyclone life cycles (LC1 and LC2) is adopted. The two life cycles, corresponding to "basic" and "anomalous" cases, illustrate two extreme and contrasting types of upper-air behavior ("anticyclonic" and "cyclonic"). Anticyclonic behavior dominates late stages of LC1 and is characterized by backward-tilted, thinning troughs being advected anticyclonically and equatorward. Cyclonic behavior dominates LC2 and is characterized by forward-tilted, broadening troughs wrapping themselves up cyclonically and poleward, producing major cut-off cyclones in high latitudes. When a pronounced wave pattern is present at upper levels, resembling the anticyclonic (LC1)

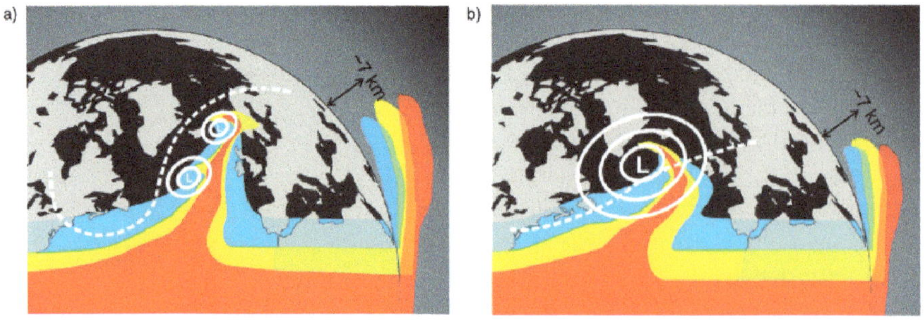

Figure 3. Schematic view of the two moisture transport configurations. **(a)** Anticyclonic (LC1-like) wave breaking with a meridional upper-level jet and **(b)** cyclonic (LC2-like) wave breaking with a zonal upper-level jet. Dashed white line shows the orientation of the upper-level jet, solid white lines show SLP, and shaded colors indicate oceanic moisture of different latitudes. On the right-hand side, quasi-vertical projections of the moisture tracers are shown. (Fig. 10 in Sodemann and Stohl, 2013; used with permission ©American Meteorological Society.)

life cycle (Fig. 3a, white dashed line), the meridional jet orientation enhances the poleward moisture advection throughout the troposphere, leading to the formation of sometimes irregular-shaped ARs (Fig. 3a, colored areas). Due to advection of the air masses along moist adiabats, a vertical stack of water vapor is thereby formed with water vapor originating from lower latitudes later residing at higher altitudes. At the western edge of the AR, small cyclones develop from frontal waves. Their cold fronts add moisture to the AR by zonal advection. At the same time, the cyclones tap the moisture available from the AR region for their WCB airstream, thus contributing to the AR at one end and feeding their spin-up and development at the other. Subsequent cyclones thus profit from the moisture transported poleward by previous cyclones, leading to a "handover" of moisture between subsequent short-lived cyclones.

A weak wave pattern at the tropopause, resembling the LC2 life cycle, leads to predominantly zonal flow throughout the troposphere (Fig. 3b, white dashed line). In this case a single, large, slow-moving cyclone develops, with more limited poleward advection of moisture to lower altitudes, as the AR is slowly advected eastward. As follows from Fig. 3, the LC1 regime is characterized by more intense intrusions of moist tropical and/or subtropical air in western Europe and, in some cases, the MR.

A role for ARs was also identified in the occurrence of many MR EPEs. The first to suggest a possible link between these events were Berto et al. (2005) and Malguzzi et al. (2006) in their analysis of the 3–5 November 1966 floods. Figure 4 confirms their findings: the AR is represented here by an elongated area with exceptionally high IWV extending from North Africa to Italy, high values of IVT and strong 850 hPa winds at 18:00 UTC, 4 November 1966. An IWV tongue connecting the AR with the air moisture reservoir in equatorial Africa may be also noted in the figure (see Krichak et al., 2015, for additional references).

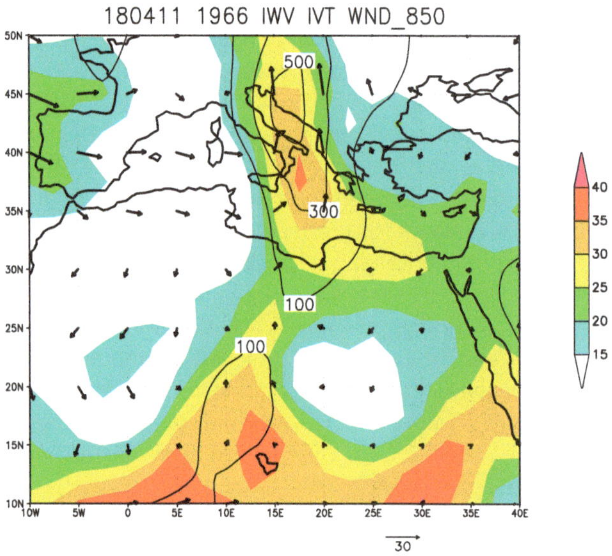

Figure 4. IWV (mm) starting from 15 mm shaded IVT ($kg\,m^{-1}\,s^{-1}$) and 850 hPa winds ($m\,s^{-1}$) at 18:00 UTC, 4 November 1966; based on the data from the National Centers for Environmental Prediction (NCEP) – National Center for Atmospheric Research (NCAR) reanalysis project (NNRP) (Kalnay et al., 1996).

Gimeno et al. (2010a, b), Krichak et al. (2012), Liberato et al. (2012), Winschall et al. (2012), Buzzi et al. (2014), Lu et al. (2013) and Liberato (2014) provided additional evidence of the important role for ARs in the formation of MR EPEs. The link between the export of moist tropical air and the multiyear trend in EPEs in this region has also been well documented.

Krichak et al. (2015) further addressed the role of ARs for MR EPEs by analyzing the data for 50 cold season MR EPEs. Their analysis identified a role for ARs during all the events considered. It was found that the location of ARs in the Atlantic controls that of specific EPEs in the MR. It was also found that the export of moist air of subtropical origin from the North Atlantic in the direction of Europe is more typical for the occurrence of EPEs in the western MR, whereas EPEs over the eastern MR are more strongly affected by the export of humid air over northeastern Africa.

An automated AR detection algorithm was adopted by Ramos et al. (2015) for the North Atlantic Ocean for the identification and a comprehensive characterization of the major AR events that affected the Iberian Peninsula over the 1948–2012 period. The extreme precipitation days in the Iberian Peninsula and their association with the occurrence of ARs was analyzed. The results of their evaluation demonstrated a noteworthy association between ARs and extreme precipitation days in the western part of the Peninsula, while for the eastern and southern parts of the Iberian Peninsula the impact of ARs was found to be reduced.

In summary, it may be concluded that, as was found earlier for western Europe, a majority of the most damaging cold season flood events in the MR are connected to ARs in the Atlantic along which the large flux of moisture is transported from the tropics to the midlatitudes.

4 Understanding climate trends

The recent past trends in climate extremes over different parts of the world, including Europe, have also been extensively investigated (e.g., Groisman et al., 2005; Zolina et al., 2010; Giorgi et al., 2011; van den Besselaar et al., 2013; Kunkel et al., 2013). The results have consistently demonstrated a rise in annual totals and in the frequency of cold season EPEs in western Europe and some parts of the MR. A change in western European precipitation, with longer wet periods and more abundant precipitation during the last decades, has also been highlighted (Zolina et al., 2010).

The situation appears to be more complex in the MR. Many climate change studies for the MR have been based on the results from reanalysis data sets and climate modeling (e.g., Xoplaki et al., 2012; Ulbrich et al., 2012; Gualdi et al., 2013; Kharin et al., 2013; Toreti et al., 2013; Mariotti et al., 2008, 2015; Saaroni et al., 2015). It should be noted, however, that the observation and climate modeling data, as well as methodologies used in these evaluations, were not necessarily optimal for a satisfactory assessment of EPEs. Discrepancies between observed and simulated trends in the MR and neighboring areas found in these studies are especially notable (Barkhordarian et al., 2013). They tend to vary on a regional basis and tend to be larger for seasonal means (Barkhordarian et al., 2013).

The result appears to be a consequence of the fact that general circulation models still cannot adequately capture the frequency, intensity, tendency and spatial distribution of observed precipitation extremes over large regions of the world (e.g., Toreti et al., 2012). Global warming implies an increase of atmospheric water vapor content at a rate of about $7\,\%\,K^{-1}$, through the Clausius–Clapeyron equation. Hence, a comparable increase in extreme precipitation would be expected over the next several decades (e.g., Kharin et al., 2013). Application of more sophisticated object-oriented methodologies focusing on extremes or frequencies of extreme events appears to be a helpful strategy for obtaining reliable estimates of the expected future changes in MR EPEs.

Progress in understanding the role of the export of moist air from the tropics and/or subtropics and that of ARs for cold season EPEs has motivated a number of additional climatological analyses addressing the issue of extreme precipitation in the MR. The recent past and expected future trends in synoptic developments have been investigated.

4.1 Evaluations based on the observation data

A number of studies evaluated the available observations on extreme precipitation in the MR during recent years (e.g., Vicente-Serrano et al., 2009; Gallego et al., 2011). In particular, Gallego et al. (2011) examined the trends in three indices designed to highlight changes in the frequency of precipitation at 27 stations covering the Iberian Peninsula for the period 1903–2003 (PT) and the two subperiods 1903–1953 (P1) and 1954–2003 (P2). Their analysis demonstrated a notable decrease in precipitation during the P2 subperiod. This trend is at least partially linked to a significant increase of the Arctic Oscillation (AO) and North Atlantic Oscillation (NAO) indices throughout the second half of the 20th century (Gallego et al., 2011).

Another recent study (Casanueva et al., 2014) focused on the variability of precipitation indices and their link to atmospheric processes over the western MR and Europe. This work, in particular, focused on the evaluation of the trend of three extreme precipitation indicators: the number of consecutive dry and wet days (CDD and CWD) and contribution of the very wet days to total precipitation R95pTOT. The authors noted that CWD and CDD are more related to the large-scale atmospheric circulation, while R95pTOT has a convective origin and depends more on local processes and moisture fluxes. The authors considered the roles of different acting factors in order to understand the extremes' variability by linking large-scale processes and precipitation extremes. Opposite associations with the NAO in winter and summer and the relationships with the Scandinavian Oscillation (SCAND) and East Atlantic patterns, as well as El Niño Southern Oscillation (ENSO) events in spring and autumn, have been found. The study also showed a significant correlation between the Atlantic Multidecadal Oscillation and R95pTOT during the entire year, apparently demonstrating the role of tropical moisture exports in the climatology of EPEs in the region.

It may be noted here that the results for the western MR are in agreement with other recent analyses that focus on the eastern MR (e.g., Givati and Rosenfeld, 2013; Krichak et al., 2014). Namely, in the eastern MR, four teleconnection regimes – AO, SCAND, East Atlantic/Western Russia

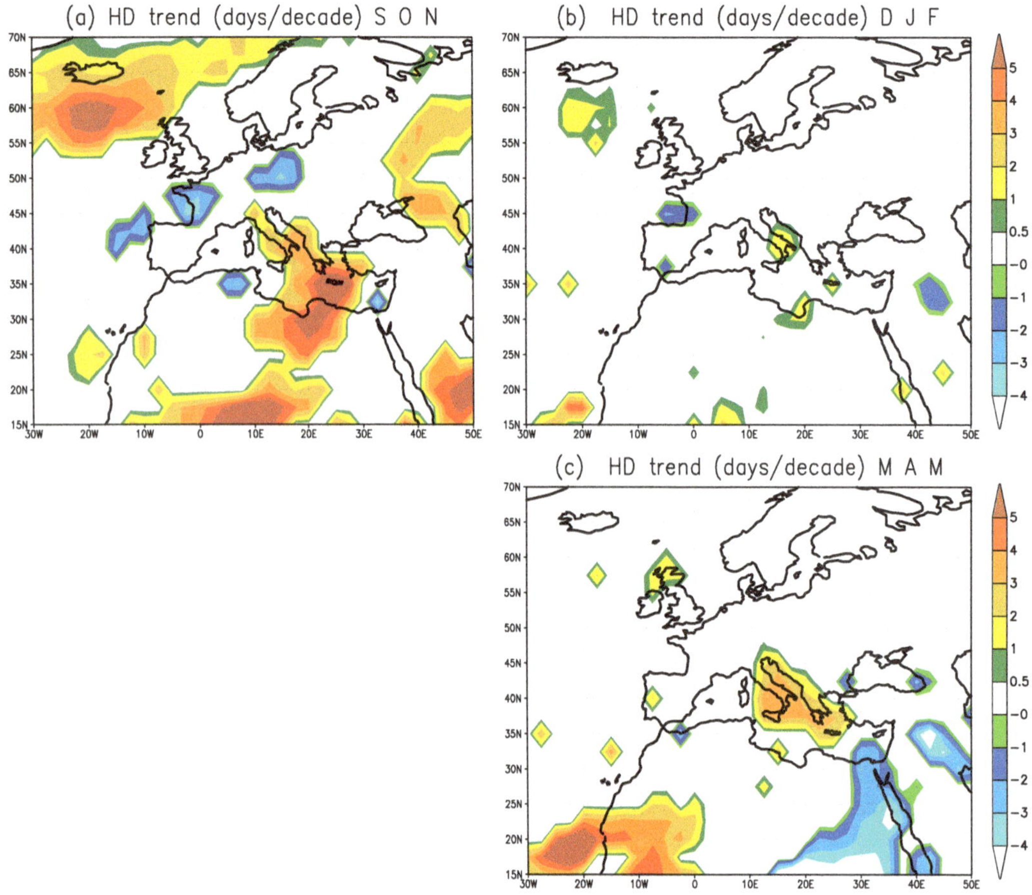

Figure 5. Linear trend in the frequency of occurrence of humid days (in days decade^{-1}) during 1979–2013 according to the NNRP data: **(a)** September–November; **(b)** December–February; **(c)** March–May (Fig. 13 in Krichak et al., 2015; used with permission from Springer).

Oscillation and ENSO – jointly influence the extreme precipitation.

These findings are also in agreement with that from Krichak et al. (2015), who detected a positive trend in the frequency of days exceeding 20 mm IWV values (frequency of humid days, HDs) over a large area in the Atlantic Ocean and western Europe during 1979–2013. The patterns showing the linear HD trends (Krichak et al., 2015) during 1979–2013 (for September–November, SON; December–February, DJF; and March–May, MAM) are presented in Fig. 5 (shaded values indicate statistical significance at $p < 0.05$; Fig. 13 in Krichak et al., 2015). The positive (negative) values of the trend indicate a rise (decline) in the frequency of days with HD conditions during the period. For SON (Fig. 5a), a region with a positive HD trend (up to $\sim$4 days decade^{-1}) for the North Atlantic Ocean west of the British Isles can be seen. Another two zones with positive trend values ($\sim$2–4 days decade^{-1}) are found from central Africa to the central MR and the southern Arabian Peninsula.

The positive trend zones are separated by areas with a negative HD trend over western Europe (−1–1.5 days decade^{-1}) and over the southeastern Mediterranean (−1.0 day decade^{-1}). For DJF (Fig. 5b), there are almost no statistically significant HD frequency trends for central and eastern Europe. A zone with positive trend values can be seen extending from central Africa to the central Mediterranean. Another region with positive statistically significant trend values (1 day decade^{-1}) is found for the North Atlantic.

During the MAM season (Fig. 5c), the tropical eastern Atlantic and west Africa are characterized by strong (up to 3–4 days decade^{-1}) HD frequency trends. Another zone with positive values of the trend (3.5 days decade^{-1}) occurs over the central MR. A zone with small positive trend values over the North Atlantic is also present. A large area from tropical east Africa to northeastern Africa and the Red Sea basin to the Middle East (Middle East–North Africa, MENA) is characterized by negative values of the HD trend (−1.0 to −4.0 days decade^{-1}). The existence of a dipole with mainly positive HD trend values over North Atlantic Ocean west of the British Isles (most notable during SON) and negative trend values over the MENA region (most notable during MAM) (Fig. 5a–c) appears to be of a special importance for EPE climatology over the MR. This is addressed in the following section.

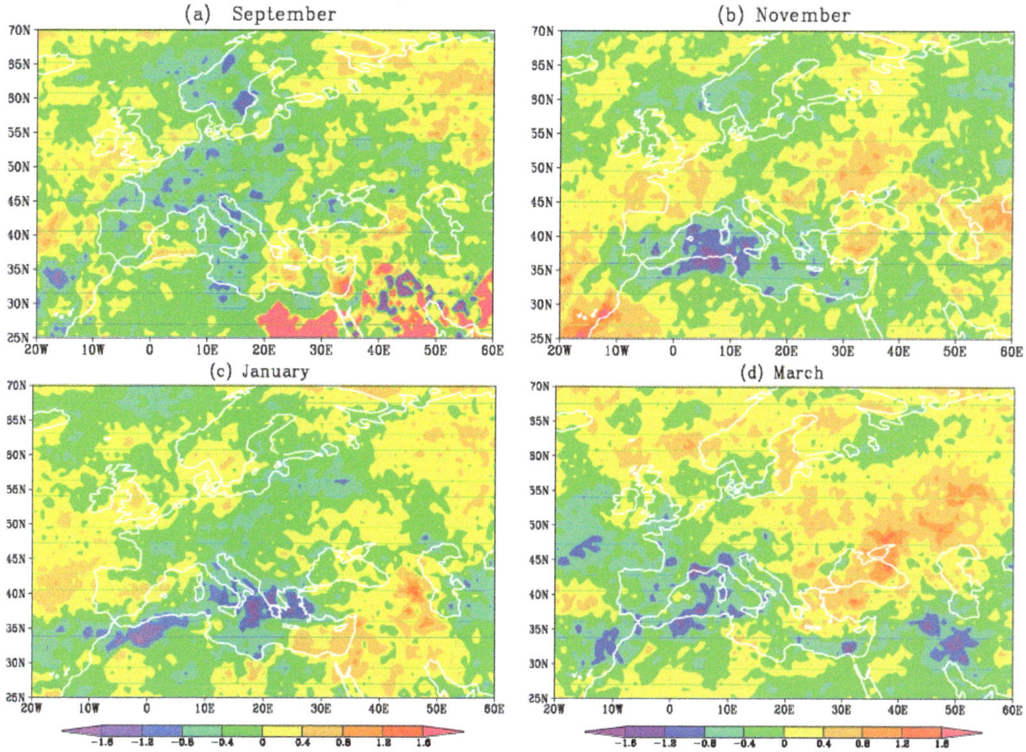

Figure 6. Climate change trend (2021–2050 minus 1961–1990) in frequency of days with extreme precipitation in **(a)** September, **(b)** November, **(c)** January and **(d)** March. The color bar presents values from −1.6 to + 1.6 days month $^{-1}$ with 0.4 interval (Fig. 3.16 in Gualdi et al., 2012; used with permission from Springer).

4.2 Projections of future changes

Time periods characterized by frequent days with above-normal precipitation over some areas are quite typical for Europe. The intensity of such climate events is significantly affected by atmospheric dynamics. Climate change trends in the frequency of days with extreme precipitation in the MR have been investigated with gridded data from a climate simulation experiment performed for the EU CIRCE Project with an INGV climate model (Gualdi et al., 2012) according to a two time slice approaches for 1961–1990 (20C) and 2021–2050 (A1B) and A1B greenhouse gas (GHG) emission scenarios. Time series with the frequencies (number of days within each month) of extreme values of precipitation (ExtPrecF), dynamic tropopause pressure and IWV were constructed according to a recently suggested approach (Carril et al., 2008; Krichak et al., 2014). Climate change trends, calculated as the difference between 30-year mean ExtPrecF values (A1B − 20C), are given in Fig. 6 (Fig. 3.16 in Gualdi et al., 2012) for September, November, January and March.

The results demonstrate a notable tendency toward an increase (up to 2 days month^{-1}) in the frequency of days with extreme precipitation over the MENA region during September (Fig. 6a). Taking into account the aridity of the area during September the modeling result indicates a tendency to a rise in the frequency of humid days. A rise in the frequency of days with extreme precipitation (0.8 days month^{-1}) is projected for the eastern Atlantic, northwestern Iberian Peninsula and southwestern Europe for November (Fig. 6b). A decrease in the frequency of extreme precipitation days (down to −1 day month^{-1}) is projected for the MENA region for March (Fig. 6d).

In another research effort (Hertig et al., 2013, 2014), model outputs from six "state of science" climate change simulation experiments for 1950–2100 (with A1B and B1 GHG emission scenarios for 2001–2000) were adopted for statistical downscaling of the hydrodynamic climate modeling results. The evaluation focused on above-normal precipitation intensity – exceeding the 95th percentile of daily precipitation from the reference period 1961–1990 (R95N). Hertig and colleagues also analyzed the changes in the total amount of precipitation from these events (R95AM) projected at the end of the 21st century for the considered scenarios. The analysis showed a decrease in the amount of extreme precipitation in autumn over many parts of the Iberian Peninsula with the strongest reductions over the eastern Mediterranean coast of Spain and parts of Mediterranean Algeria. Decreases are also projected for Tunisia, parts of Italy, the eastern coast of the Adriatic Sea and some eastern Mediterranean regions. In contrast, increases of R95AM have been found over the northern coast of the Ligurian Sea, Greece, and the southern coast of Turkey. In winter, increases of R95AM were projected in the study up until the end of the 21st century. Widespread decreases of R95AM have been projected in the MR for the spring season. The strongest declines were projected for the northwestern Iberian Peninsula, northwestern Africa, parts of Greece and Albania as well as over southeastern Turkey. At the same time, noticeable increases of R95AM in spring have been found for parts of northern and western Italy, southern France and southern Greece (Hertig et al., 2014). The conclusions are consistent in major details with the results by Gualdi et al. (2013).

It has to be noted, however, that the above presented projections of the climate change estimates have been only indirectly focusing on the effects of ARs. Only a limited number of studies have addressed the role of climate change in future trends of tropical and subtropical moisture sources and ARs in the Atlantic. Although the analyses have focused on western Europe, due to the geographic position of the region as well as predominant cyclone tracks, their results allow an interpretation relevant for the MR.

Lavers et al. (2013) examined future changes of ARs and their implications for winter floods over all of western Europe including the MR. A new set of CMIP model simulations (Taylor et al., 2012) was performed using state-of-the-art climate models and new GHG concentration scenarios (Representative Concentration Pathways or RCPs) under the CMIP Phase 5 (CMIP5) modeling framework (e.g., Taylor et al., 2012). Lavers et al. (2013) evaluated the results of the CMIP5 simulations (according to the RCP4.5 and RCP8.5 emission scenarios). Gridded reanalysis data were adopted for identification of the ARs. For each of the ARs identified in the historical (1980–2005) and the RCP4.5 and RCP8.5 projections (2074–2099), the maximum values of IVT between 1000 and 300 hPa (averaged over the lifetime of the AR) were calculated.

A sensitivity analysis was performed to determine the thermodynamic contribution to future AR changes. With this purpose for the historical and scenario (RCP) runs of each model, the average winter surface temperature (October to March over 1980–2005 and 2074–2099, respectively) over the North Atlantic Ocean (20–60° N; 60° W–0°) was calculated. The change in the regional temperature from the historical to the RCPs was then used to artificially scale up the specific humidity accordingly in the historical runs. This experiment allows testing how many additional ARs would be detected over the model dependent threshold if the historical circulation pattern were combined with the increased water vapor expected from the increased temperature in the RCP simulations.

The analysis leads to the conclusion that under the climate change scenarios used (RCP4.5, RCP8.5), the strongest ARs directed to western Europe and also the MR are projected to become more intense and – for any given intensity threshold – more frequent, indicating an increase of EPEs. A large part of these changes is thermodynamic in origin, suggesting that they are a relatively robust response to anthropogenic climate forcing. Consequently, the peak multi-day precipitation totals associated with extratropical cyclones are likely to be intensified over western Europe and the western MR, with more frequent and larger winter flood episodes under climate change.

The changes in the frequency and intensity of ARs and their potential contribution to an enhanced flood risk from ARs were also investigated by Baatsen et al. (2015). This analysis showed that baroclinic instability did not decrease significantly in future storms. The inflow of moisture increased greatly, however, resulting in more intense storms that carry more moisture. There are also additional changes in the future climate. Tropical cyclones become stronger under warmer climate conditions. The general tendency for extratropical cyclones to form in the western North Atlantic favors the development of LC1 type storms (Thorncroft et al., 1993) typically associated with a stronger influx of tropical air into Europe (see Sect. 3 for reference) has also been noted. This implies that not only the frequency of AR associated storms over the eastern Atlantic but also their intensity and the affected area may increase. In fact, these storms are more likely to reach Europe and the MR, because the transit region between the tropics and the baroclinic zone becomes smaller. Especially in the latter part of the 21st century, the expansion and eastward shift of the intense storm genesis region would result in more storms curving towards Europe and the MR and impacting a larger region (including the MR).

Baatsen et al. (2015) have also concluded that, in general, tropical air would have a greater impact on future European weather through more severe autumn storms. The projection for an increased availability of water in a warmer climate has also been supported by a larger vertically integrated water vapor content in climate change projections (Scoccimarro et al., 2015).

The results presented indicate a rise in the potential contribution of ARs to an enhanced flood risk in some parts of the mainly western MR associated with global warming.

5 Discussion and conclusion

The above review shows that research studies performed during the last several decades have revealed the notable role of tropical and subtropical moist air transport for MR EPEs. As in the case of the EPEs in western Europe, a significant part of extreme cyclonic events with heavy floods in the MR is associated with the effects of ARs. Atlantic ARs play a major role in the advection of tropical moisture which fuels the EPEs in the western MR. The situation is more complex for EPEs in the central and eastern MR. Moisture for these events also originates from the Atlantic or Indian oceans but ARs transport the moisture not to the MR directly but to other neighboring areas (mainly in North Africa), sometimes becoming a temporary moisture reservoir for future EPEs.

The identification (as demonstrated in Fig. 5) of a dipole with positive (over the northwest Atlantic Ocean, being most intense during SON) and negative (over the MENA region, most intense during MAM) trends in the frequency of HDs appears to be important for understanding the recent changes in the climatology of above-normal precipitation in the MR. A contribution arising from the reduction in the area of Arctic sea ice (Barnes, 2013) during the time period (1979–2013) could be among the factors influencing these trends.

To test for the possible connection between declining Arctic sea ice and the climatology of extreme precipitation in the MR, we have investigated the relationship between the frequency of EPEs in the MENA region and changes in Arctic sea ice. We focus on the MENA area as it is this part of the eastern Mediterranean that has been observed to undergo the strongest negative trend in the number of humid days during MAM (Krichak et al., 2015).

In this evaluation, monthly averaged data for Arctic sea ice area and extent[1] from the National Snow and Ice Data Center (http://nsidc.org) have been correlated with the monthly mean frequency (Krichak et al., 2014) of EPEs (referred to as the intense precipitation frequency) in the MENA region. In accordance with Carril et al., 2008) the data on the frequencies of days with above-normal precipitation amount have been determined (Krichak et al., 2014). The analysis was focused on daily precipitation amounts that exceed a particular threshold value (a 90 % threshold was used). With this purpose, at all grid points in the NNRP reanalysis data within the Euro-Mediterranean region, the total number of days within each month that the daily precipitation was found to exceed a particular threshold value (90 %) has been counted and the time series of the frequency of days per month with precipitation events of above-normal intensity in the region have been calculated.

The analysis of the relationship between Arctic ice extent and the frequency of days with above-normal precipitation has been performed for a target subregion represented by six locations (grid points) with longitudes 33.75° E and 35.615° E and latitudes 28.42° N, 30.33° N and 32.24° N.

The results of the evaluation are presented in Fig. 7. Here, the data for frequencies of days (per month) with above-normal precipitation are those that were previously calculated (Krichak et al., 2014). As can be seen from the figure, statistically significant ($p < 0.05$) positive lagged correlations are found between the intense precipitation frequency

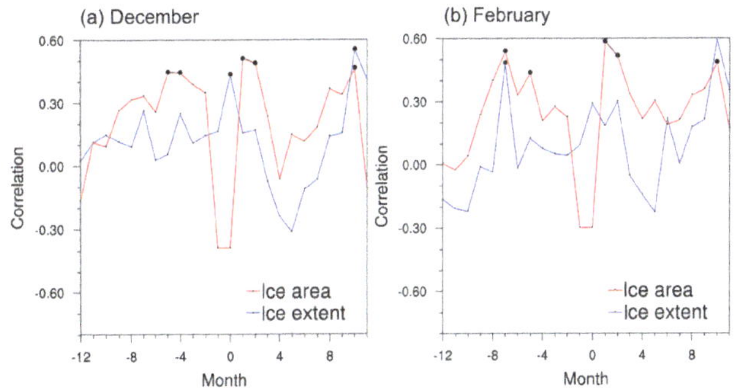

Figure 7. Correlation between monthly averaged Arctic sea ice area and extent with the monthly mean frequency of extreme precipitation events in the MENA region for **(a)** December (x axis time lag (month), y axis correlation) and **(b)** February.

in the region of analysis in December (Fig. 7a) and February (Fig. 7b) (although not significant in November and January) and Arctic sea ice area at lags −4 and −5 months for December and lags −7 and −5 months for February. This result supports the suggestion that Arctic sea ice may be seen as an important factor causing changes in the frequency of intense precipitation events in the MENA with a lead time of up to 6 months.

Indeed, this linkage between Arctic sea ice and intense precipitation in the MENA may be manifested through the excitation of the NAM/AO (and the very similar NAO). Many studies have shown that an increase (decrease) in Arctic sea ice area during the late summer and early autumn is followed 2 to 3 months later by the establishment of the positive (negative) phase of the NAM/AO (and the very similar NAO) (Deser et al., 2007; Francis et al., 2009; Francis and Vavrus, 2012; Honda et al., 2009; Smith et al., 2011; Liu et al., 2012). Furthermore, Krichak et al. (2014) found statistically significant positive correlations between the NAM/AO and precipitation frequency in the MENA area.

On the same line, three recent studies (Peings and Magnusdottir, 2014; Feldstein and Lee, 2014; Kim et al., 2014), using both observational and model data, show that a reduction in Arctic sea ice is followed by an increase in the vertical propagation of planetary-scale wave activity into the extratropical stratosphere and a deceleration of the stratospheric polar vortex. This enhancement in the vertical wave activity propagation was shown to arise from constructive interference (Garfinkel et al., 2010) between transient planetary waves excited by the loss of sea ice and the climatological planetary-scale stationary wave field (see also Smith et al., 2011; Cohen et al., 2014).

The weakening of the strength of the stratospheric polar vortex coincides with the excitation of the negative phase of the stratospheric NAM (Baldwin and Dunkerton, 1999). This is followed 2 months later by the excitation of the negative NAM/AO throughout the troposphere, likely through downward control and a positive eddy feedback (e.g., Polvani and

[1]Sea ice area is defined as the area of the ocean surface that is covered by sea ice. The contribution to the sea ice area from each grid cell comes from the portion of the grid cell that is covered in sea ice. For sea ice extent, the grid cell is defined to be either ice covered or ice free, depending upon whether a threshold of 15 % ice concentration is exceeded (for more information, see https://support.nsidc.org/forums).

Kushner, 2002; Kushner and Polvani, 2004; Song and Robinson 2004; Simpson et al., 2009). Feldstein and Lee (2014) have shown that the opposite sequence of processes links an increase in Arctic sea ice to the positive NAM/AO in the troposphere.

It must be noted here that the above analysis represents one of the first attempts to understand the processes responsible for the trends in extremes in the MR in the context of the global dynamics. The interplay between the Arctic ice and NAM/NAO/AO appears to be an important object for further systematic studies. It also appears possible that the extension of the Arctic ice cover may be just one of the factors responsible for the trends in the EPEs.

There are also a number of physical problems that remain to be resolved. As pointed out by Baatsen et al. (2015), not only the moisture export by ARs but also baroclinic instability is important for the (re-)intensification of the MR EPE storms. The latter is most important for extratropical development while the former provides heat and moisture that enhances the latent heat release in a newly forming warm core as well as in the warm conveyor belt, thus speeding up the intensification. Under appropriate conditions, many Mediterranean storms are likely to develop at the synoptic scale baroclinically.

Another issue for further analyses is to determine what part of the water that rains during an EPE originates from the tropics. Analysis by Dacre et al. (2015) of North Atlantic extratropical cyclones shows that ARs are formed by the cold front which sweeps up water vapor in the warm sector as it catches up with the warm front. This causes a narrow band of high water vapor content to form ahead of the cold front at the base of the warm conveyor belt. Thus, water vapor in the cyclone's warm sector, and not long-distance transport of water vapor from the subtropics, is responsible for the generation of filaments of high water vapor content in ARs reaching northwestern Europe. The ARs reaching the MR, however, may be characterized by different peculiarities.

Finally, we conclude that the last decade has been characterized by a growing understanding of the critical role played by the export of moisture from the tropics and subtropics in the formation of cold season MR EPEs. This progress is in line with that in the investigation of disastrous flooding events in western Europe and other regions of the world. The moisture exports are typically taking place in narrow elongated zones characterized by high water vapor content and strong winds in the lower troposphere referred to as "atmospheric rivers".

In the current review we addressed this issue by discussing the major trends found in the investigations performed during the last 3 decades. The studies show that in spite of significant differences between the different parts of the MR, the export of large amounts of subtropical and/or tropical moisture entering coastal areas during the landfall of ARs may be seen as a critical factor responsible for the exceptionally intense precipitation during EPEs over the MR. Progress in understanding the role of ARs in the MR EPEs also allows a new perspective for projecting future climate changes in the MR, which appears to be especially important since contemporary global climate models still cannot adequately capture the frequency, intensity, tendency and spatial distribution of observed precipitation extremes over large regions in the world. Only a limited number of such evaluations have so far been performed. Current climate change analyses as well as the modeling studies show that anthropogenic global warming may lead to stronger and more numerous Atlantic ARs in the next few decades. This implies a greater risk of higher rainfall totals and therefore larger winter floods in some parts of the MR accompanied by a greater risk of droughts in the other parts of the MR.

The authors believe that the above discussions and recommendations also have policy relevance. Additionally, it is suggested that monitoring and reporting changes in tropical and subtropical moisture exports and AR activity may allow for reducing the risk of extreme floods in the MR under current climate conditions.

Acknowledgements. This manuscript has been greatly improved thanks to the careful reviews and insightful comments by the anonymous reviewers that the authors gratefully acknowledge. The financial support for this work has been partly provided by the joint project The Virtual Institute DEad SEa Research Venue (DESERVE) of the Helmholtz Centers KIT, GFZ and UFZ and their partners from Germany, Israel, Jordan and Palestine, the United States–Israel Binational Science Foundation (BSF) under research grant no. 2008436, the United States National Science Foundation (NSF) under grants AGS-1036858 and AGS-1401220, research grant 4500568707 for 2011–2013 by the Israeli Water Authority of the Ministry of Infrastructures, the European Union Sixth Framework programme project Climate Change and Impact Research: Mediterranean Environment (CIRCE, http://www.cmcc.it/projects/circe-climate-change-and-impact-research-the-mediterranean-environment) under contract GOCE-036961 and the Cyprus Institute and Tel Aviv University cooperative initiative. This work is also a contribution to the Hydrological Cycle in Mediterranean Experiment (HyMeX) and the Northern Eurasian Earth Science Partnership Initiative (NEESPI). The authors S. Gualdi and E. Scoccimarro thank the projects GEMINA and NextData funded by the Italian Ministries of University and Research (MIUR) and of Environment (MATT).

Discussions supported through the European Cooperation in Science and Technology (COST) Earth System Science and Environmental Management (ESSEM) Action ES0905 "Basic concepts for convection parameterization in weather forecast and climate models" (J.-I. Yano and S. O. Krichak) and the WMO MEDEX and European Science Foundation MedCLIVAR programs are also acknowledged. Finally, we acknowledge the use of the NCEP Reanalysis data provided by the

NOAA/OAR/ESRL PSD, Boulder, Colorado, USA, from their web site at http://www.esrl.noaa.gov/psd/.

References

Alpert, P., Krichak, S. O., Krishnamurti, T. N., Stein, U., and Tsidulko, M.: The relative role of lateral boundaries, initial conditions and topography in mesoscale simulations of lee cyclogenesis, J. Appl. Meteorol., 35, 1091–1099, 1996.

Baatsen, M., Haarsma, R. J., Van Delden, A. J., and de Vries, H.: Severe autumn storms in future western Europe with a warmer Atlantic ocean, Clim. Dynam. 45, 949–964, doi:10.1007/s00382-014-2329-8, 2015.

Baldwin, M. P. and Dunkerton, T. J.: Propagation of the Arctic Oscillation from the stratosphere to the troposphere, J. Geophys. Res., 104, 30937–30946, 1999.

Barkhordarian, A., von Storch, H., and Bhend J.: The expectation of future precipitation change over the Mediterranean region is different from what we observe, Clim. Dynam. 40, 225–244, doi:10.1007/s00382-012-1497-7, 2013.

Barnes, E. A.: Revisiting the evidence linking Arctic amplification to extreme weather in mid-latitudes, Geophys. Res. Lett., 40, 4734–4739. doi:10.1002/grl.50880, 2013.

Berto, A., Buzzi, A., and Zardi, D.: Back-tracking water vapor contributing to a precipitation event over Trentino: a case study, Meteorol. Z., 13, 189–200, 2004.

Berto, A., Buzzi, A., and Zardi, D.: A warm conveyor belt mechanism accompanying extreme precipitation events over northeastern Italy, Proceedings of the 28th ICAM, The annual Scientific MAP Meeting, Zadar, Croatia, Extended Abstracts, Hrv. Meteorol. Casopis (Croat. Meteorol. J.), 40, 338–341, 2005.

Browning, K. A.: The sting at the end of the tail: Damaging winds associated with extratropical cyclones, Q. J. Roy. Meteor. Soc., 130, 375–399, doi:10.1256/qj.02.143, 2004.

Buzzi, A., Tartaglione, N., and Malguzzi, P.: Numerical simulations of the1994 Piedmont flood: role of orography and moist processes, Mon. Weather Rev., 126, 2369–2383, 1998.

Buzzi, A., Davolio, S., Malguzzi, P., Drofa, O., and Mastrangelo, D.: Heavy rainfall episodes over Liguria in autumn 2011: numerical forecasting experiments, Nat. Hazards Earth Syst. Sci., 14, 1325–1340, doi:10.5194/nhess-14-1325-2014, 2014.

Carril, A. F., Gualdi, S., Cherchi, A., and Navarra, A.: Heat waves in Europe: areas of homogeneous variability and links with the regional to large-scale atmospheric and SSTs anomalies, Clim. Dynam., 30, 77–98, 2008.

Casanueva, A., Rodríguez-Puebla, C., Frías, M. D., and González-Reviriego, N.: Variability of extreme precipitation over Europe and its relationships with teleconnection patterns, Hydrol. Earth Syst. Sci., 18, 709–725, doi:10.5194/hess-18-709-2014, 2014.

Clarke, M. L. and Rendell, H. M.: Hindcasting extreme events: the occurrence and expression of damaging floods and landslides in Southern Italy, Land Degrad. Dev., 17, 365–380, doi:10.1002/ldr.743, 2006.

Cohen, J., Screen, J. A., Furtado, J. C., Barlow, M., Whittleston, D., Coumou, D., Francis, J., Dethloff, K., Entekhabi, D., Overland, J., and Jones, J.: Recent Arctic amplification and extreme mid-latitude weather, Nat. Geosci., 7, 627–637, doi:10.1038/ngeo2234, 2014.

Dacre, H. F., Clark, P. A., Martinez-Alvarado, O., Stringer, M. A., and Lavers, D. A.: How do atmospheric rivers form?, B. Am. Meteorol. Soc., doi:10.1175/BAMS-D-14-00031.1, 2015.

Davies, H. C. and Pichler, H.: Mountain meteorology and ALPEX: an introduction, Meteorol. Atmos. Phys., 43, 3–4, 1990.

Deser, C., Tomas, R. A., and Peng, S.: The transient atmospheric circulation response to North Atlantic SST and sea ice anomalies, J. Climate, 20, 4751–4767, 2007.

Dettinger, M.: Climate change, atmospheric rivers, and floods in California – a multimodel analysis of storm frequency and magnitude changes, J. Am. Water Resour. As., 47, 514–523, 2011.

de Vries, A. J., Tyrlis, E., Edry, D., Krichak, S. O., Steil, B., and Lelieveld, J.: Extreme precipitation events in the Middle East: dynamics of the Active Red Sea Trough, J. Geophys. Res.-Atmos., 118, 7087–7108, doi:10.1002/jgrd.50569, 2013.

De Zolt, S., Lionello, P., Nuhu, A., and Tomasin, A.: The disastrous storm of 4 November 1966 on Italy, Nat. Hazards Earth Syst. Sci., 6, 861–879, doi:10.5194/nhess-6-861-2006, 2006.

Doswell, C. A., Ramis, C., Romero, R., and Alonso, S.: A diagnostic study of three heavy precipitation episodes in the western Mediterranean region, Weather Forecast, 13, 102–124, 1998.

Drobinski, P., Ducrocq, V., Alpert, P., Drobinski, P., Ducrocq, V., Alpert, P., Anagnostou, E., Béranger, K., Borga, M., Braud, I., Chanzy, A., Davolio, S., Delrieu, G., Estournel, C., Filali Boubrahmi, N., Font, J., Grubisiè, V., Gualdi, S., Homar, V., Ivanèan-Picek, B., Kottmeier, C., Kotroni, V., Lagouvardos, K., Lionello, P., Llasat, M. C., Ludwig, W., Luto, C., Mariotti, A., Richard, E., Romero, R., Rotunno, R., Roussot, O., Ruin, I., Somot, S., Taupier-Letage, I. Tintore, J., Uijlenhoet, R., and Wernli, H.: HYMEX A 10-Year Multidisciplinary Program on the Mediterranean Water Cycle, B. Am. Meteorol. Soc., 7, 1063–1081, 2014.

Duffourg, F. and Ducrocq, V.: Origin of the moisture feeding the Heavy Precipitating Systems over Southeastern France, Nat. Hazards Earth Syst. Sci., 11, 1163–1178, doi:10.5194/nhess-11-1163-2011, 2011.

Ferraris, L., Reale, O., and Turato, B.: Synoptic and hydrological analysis of a flood event, Phys. Chem. Earth. Pt. B, 26, 655–661, 2001.

Fink, A. H., Brücher, T., Ermert, V., Krüger, A., and Pinto, J. G.: The European storm Kyrill in January 2007: synoptic evolution, meteorological impacts and some considerations with respect to climate change, Nat. Hazards Earth Syst. Sci., 9, 405–423, doi:10.5194/nhess-9-405-2009, 2009.

Fiori, E., Comellas, A., Molini, L., Rebora, N., Siccardi, F., Gochis, D. J., Tanelli, S., and Parodi, A.: Analysis and hindcast simulations of an extreme rainfall event in the Mediterranean area: the Genoa 2011 case, Atmos. Res. 138, 13–29, 2014.

Feldstein, S. and Lee, S.: Intraseasonal and interdecadal jet shifts in the Northern Hemisphere: The role of warm pool tropical convection and sea ice, J. Climate, 27, 6497–6518, doi:10.1175/JCLI-D-14-00057.1, 2014.

Francis, J. A. and Vavrus, S. J.: Evidence linking Arctic amplification to extreme weather in mid-latitudes, Geophys. Res. Lett., 39, L06801, doi:10.1029/2012GL051000, 2012.

Francis, J. A., Chan, W., Leathers, D. J., and Miller, J. R.: Winter northern hemisphere weather patterns remember summer arctic sea-ice extent, Geophys. Res. Lett., 36, L07503, doi:10.1029/2009GL037274, 2009.

Gallego, M. C., Trigo, R. M., Vaquero, J. M., Brunet, M., García, J. A., Sigró, J., and Valente, M. A.: Trends in frequency indices of daily precipitation over the Iberian Peninsula during the last century, J. Geophys. Res., 116, D02109, doi:10.1029/2010JD014255, 2011.

Garfinkel, C. I., Hartmann, D. L., and Sassi, F.: Tropical precursors of anomalous Northern Hemisphere stratospheric polar vortices, J. Climate, 23, 3282–3299, 2010.

Garcia-Herrera, R. F., Lionello, P., and Ulbrich U.: Preface: Understanding dynamics and current developments of climate extremes in the Mediterranean region, Nat. Hazards Earth Syst. Sci., 14, 309–316, doi:10.5194/nhess-14-309-2014, 2014.

Gimeno, L., Drumond, A., Nieto, R., Trigo, R. M., and Stohl, A.: On the origin of continental precipitation, Geophys. Res. Lett, 37, L13804, doi:10.1029/2010GL043712, 2010a.

Gimeno, L., Nieto, R., Trigo, R., Vicente-Serrano, S., and López-Moreno, J. I.: Where does the Iberian Peninsula moisture come from? An answer based on a Lagrangian approach, J. Hydrometeorol., 11, 421–436, doi:10.1175/2009JHM1182.1, 2010b.

Gimeno, L., Stohl, A., Trigo, R. M., Dominguez, F., Yoshimura, K., Yu, L., Drumond, A., Durán-Quesada, A. M., and Nieto, R.: Oceanic and terrestrial sources of continental precipitation, Rev. Geophys., 50, RG4003, doi:10.1029/2012RG000389, 2012.

Gimeno, L., Nieto, R., Vázquez, M., and Lavers, D. A.: Atmospheric rivers: a mini-review Front, Earth Sci., 5, 2:2, doi:10.3389/feart.2014.00002, 2014

Giorgi, F., Im, E.-S., Coppola, E., Diffenbaugh, N. S., Gao, X. J., Mariotti, L., and Shi, Y.: Higher hydroclimatic intensity with global warming, J. Climate, 24, 5309–5324, 2011.

Givati, A. and Rosenfeld, D.: The Arctic Oscillation, climate change and the effects on precipitation in Israel, Atmos. Res., 132–133, 114–124, 2013.

Groisman, P., Knight, R., Easterling, D., Karl, T., Hegerl, G., and Razuvaev, V.: Trends in intense precipitation in the climate record, J. Climate, 18, 1326–1350, 2005.

Gualdi, S., Somot, S., May, W., Castellari, S., Déqué, M., Adani, M., Artale, V., Bellucci, A., Breitgand, J. S., Carillo, A., Cornes, R., Dell'Aquila, A., Dubois, C., Efthymiadis, D., Elizalde, A., Gimeno, L., Goodess, C. M., Harzallah, A., Krichak, S. O., Kuglitsch, F. G., Leckebusch, G. C., L'Heveder, B., Li, L., Lionello, P., Luterbacher, J., Mariotti, A., Nieto, R., Nissen, K. M., Oddo, P., Ruti, P., Sanna, A., Sannino, G., Scoccimarro, E., Sevault, F., Struglia, M. V., Toreti, A., Ulbrich, U., and Xoplaki, E.: Future Climate Projections, in: Regional Assessment of Climate Change in the Mediterranean, edited by: Navarra, A. and Tubiana, L., Springer, Dordrecht, the Netherlands, 870 pp., 2012.

Gualdi, S., Somot, S., Li, L., Artale, V., Adani, M., Bellucci, A., Braun, A., Calmanti, S., Carillo, A., Dell'Aquila, A., Déqué, M., Dubois, C., Elizalde, A., Harzallah, A., Jacob, D., L'Hévéder, B., May, W., Oddo, P., Ruti, P., Sanna, A., Sannino, G., Scoccimarro, E., Sevault, F., and Navarra, A.: The CIRCE Simulations: Regional Climate Change Projections with Realistic Representation of the Mediterranean Sea, B. Am. Meteorol. Soc., 94, 65–81, 2013.

Hertig, E., Seubert, S., Paxian, A., Vogt, G., Paeth, H., and Jacobeit, J.: Changes of total versus extreme precipitation and dry periods until the end of the 21st century: statistical assessments for the Mediterranean area, Theor. Appl. Climatol., 111, 1–20, 2013.

Hertig, E., Seubert, S., Paxian, A., Vogt, G., Paeth, H., and Jacobeit, J.: Statistical modeling of extreme precipitation for the Mediterranean area under future climate change, Int. J. Climatol., 34, 1132–1156, doi:10.1002/joc.3751, 2014.

Hodges, K. I., Lee, R. W., and Bengtsson L.: A Comparison of extratropical cyclones in recent reanalyses ERA-Interim, NASAMERRA, NCEP-CFSR, and JRA-25, J. Climate, 24, 4888–4906, doi:10.1175/2011JCLI4097.1, 2011.

Homar, V., Jansà, A., Campins, J., Genovés, A., and Ramis, C.: Towards a systematic climatology of sensitivities of Mediterranean high impact weather: a contribution based on intense cyclones, Nat. Hazards Earth Syst. Sci., 7, 445–454, doi:10.5194/nhess-7-445-2007, 2007.

Honda, M., Inoue, J., and Yamane, S.: Influence of low Arctic sea-ice minima on anomalously cold Eurasian winters, Geophys. Res. Lett., 36, L08707, doi:10.1029/2008GL037079, 2009.

Jansa, A., Genoves, A., and Garsia-Moya, J. A.: Western Mediterranean cyclones and heavy rain. Part 1: numerical experiment concerning the Piedmont flood case, Meteorol. Appl., 7, 323–333, 2000.

Jansa, A., Alpert, P., Arbogast, P., Buzzi, A., Ivancan-Picek, B., Kotroni, V., Llasat, M. C., Ramis, C., Richard, E., Romero, R., and Speranza, A.: MEDEX: a general overview, Nat. Hazards Earth Syst. Sci., 14, 1965–1984, doi:10.5194/nhess-14-1965-2014, 2014.

Kahana, R., Ziv, B., Dayan, U., and Enzel, Y.: Atmospheric predictors for major floods in the Negev desert, Israel, Int. J. Climatol., 24, 1137–1147, 2004.

Kalnay, E., Kanamitsu, M., Kistler, R., Collins, W., Deaven, D., Gandin, L., Iredell, M., Saha, S., White, G., Woollen, J., Zhu, Y., Chelliah, M., Ebisuzaki, W., Higgins, W., Janowiak, J., Mo, K. C., Ropelewski, C., Wang, J., Leetmaa, A., Reynolds, R., Jenne, R., and Joseph, D.: The NCEP/NCAR 40-years Reanalysis Project, B. Am. Meteorol. Soc, 77, 437–471, 1996.

Kharin, V. V., Zwiers, F. W., Zhang, X., and Wehner, M.: Changes in temperature and precipitation extremes in the CMIP5 ensemble, Climatic Change, 119, 345–357, doi:10.1007/s10584-013-0705-8, 2013.

Kim, B.-M., Son, S.-W., Min, S.-K, Jeong, J.-H., Kim, S.-J., Zhang, X., Shim, T., and Yoon, J.-H.: Weakening of the stratospheric polar vortex by Arctic sea-ice loss, Nat. Commun., 5, 4646, doi:10.1038/ncomms5646, 2014.

Knippertz, P. and Wernli, H.: A Lagrangian climatology of tropical moisture exports to the northern hemispheric extratropics, J. Climate, 23, 987–1003, 2010.

Knippertz, P., Wernli, H., and Gläser, G.: A Global Climatology of Tropical Moisture Exports, J. Climate, 26, 3031–3045, doi:10.1175/JCLI-D-12-00401.1, 2013.

Kotroni, V., Lagouvardos, K., Kallos, G., and Ziakopoulos, D.: Severe flooding over central and southern Greece associated with pre-cold frontal orographic lifting, Q. J. Roy. Meteor. Soc., 125, 967–991, 1999.

Kotroni, V., Lagouvardos, K., Defer, E., Dietrich, S., Porcu, F., Medaglia, C. M., and Demitras, M.: The Antalya 5 December

2002 storm: observation and model analysis, J. Appl. Meteorol. Clim., 45, 576–590, 2006.

Krichak, S. O. and Alpert, P.: Role of large scale moist dynamics in November 1–5, 1994, hazardous Mediterranean weather, J. Geophys. Res., 103, 19453–19468, 1998.

Krichak, S. O., Alpert, P., and Dayan, M.: The role of atmospheric processes associated with hurricane Olga in the December 2001 floods in Israel, J. Hydrometeorol., 5, 1259–1270, 2004.

Krichak, S. O., Alpert, P., and Dayan, M.: A southeastern Mediterranean PV streamer and its role in December 2001 case with torrential rains in Israel, Nat. Hazards Earth Syst. Sci., 7, 21–32, doi:10.5194/nhess-7-21-2007, 2007.

Krichak, S. O., Breitgand, J. S., and Feldstein, S. B.: A conceptual model for the identification of active Red Sea Trough synoptic events over the southeastern Mediterranean, J. Appl. Meteorol. Clim., 5, 962–971, doi:10.1175/JAMC-D-11-0223.1, 2012.

Krichak, S. O., Breitgand, J. S., Gualdi, S., and Feldstein, S. B.: Teleconnection-extreme precipitation relationships over the Mediterranean region, Theor. Appl. Climatol., 117, 679–692, doi:10.1007/s00704-013-1036-4, 2014.

Krichak, S. O., Barkan, J., Breitgand, J. S., Gualdi, S., and Feldstein, S. B.: The role of the export of tropical moisture into mid-latitudes for extreme precipitation events in the Mediterranean region, Theor. Appl. Climatol., 121, 499–515, doi:10.1007/s00704-014-1244-6, 2015.

Kunkel, K. E., Karl T. R., Easterling, D. R., Redmond, K., Young, J., Yin X., and Hennon, P.: Probable maximum precipitation and climate change, Geophys Res. Lett., 40, 1402–1408, doi:10.1002/grl.50334, 2013.

Kushner, P. J. and Polvani, L. M.: Stratosphere-troposphere coupling in a relatively simple AGCM. The role of eddies, J. Climate, 17, 629–639, 2004.

Lagouvardos, K., Kotroni, V., Dobricic, S., Nickovic, S., and Kallos, G.: On the storm of 21–22 October 1994 over Greece: Observations and model results, J. Geophys. Res., 101, 26217–26226, 1996.

Lastoria, B., Simonetti, M. R., Casaioli, M., Mariani, S., and Monacelli, G.: Socio-economic impacts of major floods in Italy from 1951 to 2003, Adv. Geosci., 7, 223–229, doi:10.5194/adgeo-7-223-2006, 2006.

Lavers, D. A. and Villarini, G: The nexus between atmospheric rivers and extreme precipitation across Europe, Geophys. Res. Lett., 40, 3259–3264, doi:10.1002/grl.50636, 2013.

Lavers, D. A., Allan, R. P., Wood, E. F., Villarini, G., Lloyd-Hughes, B., Brayshaw, D. J., and Wade A. J.: Winter floods in Britain are connected to atmospheric rivers, Geophys. Res. Lett., 38, L23803, doi:10.1029/2011GL049783, 2011.

Lavers, D. A., Allan, R. P., Villarini, G., Lloyd-Hughes, B., Brayshaw, D. J., and Wade, A. J.: Future changes in atmospheric rivers and their implications for winter flooding in Britain, Environ. Res. Lett., 8, 034010, doi:10.3389/feart.2014.00002, 2013.

Liberato, M. L. R.: The 19 January 2013 windstorm over the North Atlantic: large-scale dynamics and impacts on Iberia, Weather and Climate Extremes, 5–6, 16–28, doi:10.1016/j.wace.2014.06.002, 2014.

Liberato, M. R. L., Pinto, J. G., Trigo, I. F., and Trigo, R. M.: Klaus – an exceptional winter storm over Northern Iberia and Southern France, Weather, 66, 330–334, doi:10.1002/wea.755, 2011.

Liberato, M. L. R., Ramos, A. M., Trigo, R. M., Trigo, I. F., Durán-Quesada, A. M., Nieto, R., and Gimeno, L.; Moisture sources and large-scale dynamics associated with a flash flood event, in: Lagrangian Modeling of the Atmosphere, edited by: Lin, J., Brunner, D., Gerbig, C., Stohl, A., Luhar, A., Webley, P., American Geophysical Union, Washington, DC, USA, available at: http://dx.doi.org/10.1029/2012GM001244, 2012.

Liberato, M. L. R., Pinto, J. G., Trigo, R. M., Ludwig, P., Ordóñez, P., Yuen, D., and Trigo, I. F.: Explosive development of winter storm Xynthia over the subtropical North Atlantic Ocean, Nat. Hazards Earth Syst. Sci., 13, 2239–2251, doi:10.5194/nhess-13-2239-2013, 2013.

Lionello, P., Malanotte-Rizzoli, P., Boscolo, R., Alpert, P., Artale, V., Li, L., Luterbacher, J., May, W., Trigo, R., Tsimplis, M., Ulbrich, U., and Xoplaki, E.: The Mediterranean Climate: An Overview of the Main Characteristics and Issues, in: Mediterranean climate variability, edited by: Lionello, P., Malanotte-Rizzoli, P., and Boscolo, R., Elsevier, Amsterdam, the Netherlands, 1–26, 2006a.

Lionello, P., Bhend, J., Buzzi, A., Della-Marta, P. M., Krichak, S. O., Jansa, A., Maheras, P., Sanna, A., Trigo, I. F., and Trigo, R.: Cyclones in the Mediterranean region: climatology and effects on the environment, in: Mediterranean climate variability, edited by: Lionello, P., Malanotte-Rizzoli, P., and Boscolo, R., Elsevier, Amsterdam, the Netherlands, 325–372, 2006b.

Lionello, P., Gacic, M., Gomis, D., Garcia-Herrera, R., Giorgi, F., Planton, S., Trigo, R., Theocharis, A., Tsimplis, M. N., Ulbrich, U., and Xoplaki, E.: Program focuses on climate of the Mediterranean region, EOS, 93, 105–106, doi:10.1029/2012EO100001, 2012.

Lionello, P., Abrantes, F., Gacic, M., Planton, S., Trigo, R., and Ulbrich, U.: The climate of the Mediterranean region: research progress and climate change impacts, Reg. Environ. Change, 14, 1679–1684, doi:10.1007/s10113-014-0666-0, 2014.

Liu, J., Curry, J. A., Wang, H., Song, M., and Horton, R. M.: Impact of declining Arctic sea ice on winter snowfall, P. Natl. Acad. Sci. USA, 109, 4074–4079, doi:10.1073/pnas.1114910109, 2012.

Lu, M., Lall, U., Schwartz, A., and Kwon H.: Precipitation predictability associated with tropical moisture exports and circulation patterns for a major flood in France in 1995, Water Resour. Res., 49, 6381–6392, doi:10.1002/wrcr.20512, 2013.

Malguzzi, P., Grossi, G., Buzzi, A., Ranzi, R., and Buizza, R.: The 1966 "century" flood in Italy: A meteorological and hydrological revisitation, J. Geophys. Rev., 111, D24106, doi:10.1029/2006JD007111, 2006.

Mariotti, A., Zeng, N., Yoon, J.-H., Artale, V., Navarra, A., Alpert, P., and Li, L. Z. X.: Mediterranean water cycle changes: transition to drier 21st century conditions in observations and CMIP3, Environ. Res. Lett., 3, 044001, doi:10.1088/1748-9326/3/4/044001, 2008.

Mariotti, A., Pan, Y., Zeng, N., and Alessandri, A.: Long-term climate change in the Mediterranean region in the midst of decadal variability, Clim. Dynam., 44, 1437–1456, doi:10.1007/s00382-015-2487-3, 2015.

Massacand A. C., Wernli, H., and Davies H. C.: Heavy precipitation on the Alpine southside: an upper-level precursor, Geophys. Res. Lett, 25, 1435–1438, 1998.

McGinley, J. A. and Zupanski, M.: Numerical Analysis of the Influence of Jets, Fronts, and Mountains, Meteorol. Atmos. Phys. 43, 7–20, 1990.

Navarra, A. and Tubiana, L.:, Regional Assessment of Climate Change 3 in the Mediterranean, Adv. Glob. Change Res., 50, doi:10.1007/978-94-007-5781-3_1, Springer Science+Business Media, Dordrecht, the Netherlands, 2013.

Neiman, P. J., Schick, L. J. Ralph, F. M., Hughes, M., and Wick, G. A.: Flooding in western Washington: The connection to atmospheric rivers, J. Hydrometeorol., 12, 1337–1358, doi:10.1175/2011JHM1358.1, 2011

Nieto, R., Gimeno, L., Drumond, A., and Hernandez, E.: A Lagrangian identification of the main moisture sources and sinks affecting the Mediterranean area, WSEAS Transactions on Environment and Development, 5, 365–374, 2010.

Pastor, F., Gómez, I., and Estrela, M. J.: Numerical study of the October 2007 flash flood in the Valencia region (Eastern Spain): the role of orography, Nat. Hazards Earth Syst. Sci., 10, 1331–1345, doi:10.5194/nhess-10-1331-2010, 2010.

Peings, Y. and Magnusdottir, G.: Response of the wintertime Northern Hemisphere atmospheric circulation to current and projected Arctic sea ice decline: A numerical study with CAM5, J. Climate, 27, 244–264, 2014.

Pfahl, S. and Wernli, H.: Quantifying the Relevance of Cyclones for Precipitation Extremes, J. Climate, 25, 6770–6780, 2012.

Pinto, J. G., Zacharias, S., Fink, A. H., Leckebusch, G. C., and Ulbrich, U.: Factors contributing to the development of extreme North Atlantic cyclones and their relationship with the NAO, Clim. Dynam., 32, 711–737, doi:10.1007/s00382-008-0396-4, 2009.

Pinto, J. G., Ulbrich, S., Parodi, A., Rudari, R., Boni, G., and Ulbrich, U.: Identification and ranking of extraordinary rainfall events over northwest Italy: the role of Atlantic moisture, J. Geophys. Res.-Atmos., 118, 2085–2097, doi:10.1002/jgrd.50179, 2013.

Polvani, L. M. and Kushner, P. J.: Tropospheric response to stratospheric perturbations in a relatively simple general circulation model, Geophys. Res. Lett., 111, 18.1–18.4, doi:10.1029/2001GL014284, 2002.

Raible, C. C., Della-Marta, P. M., Schwierz, C., Wernli, H., and Blender, R.: Northern hemisphere extratropical cyclones: a comparison of detection and tracking methods and different reanalyses, Mon. Weather. Rev., 136, 880–897, 2008.

Ralph, F. M. and Dettinger, M. D.: Storms, floods, and the science of atmospheric rivers, EOS T. Am. Geophys. Un., 92, 265–272, 2011.

Ralph, F. M., Neiman, P. J., and Wick, G. A.: Satellite and CALJET aircraft observations of atmospheric rivers over the eastern North Pacific Ocean during the winter of 1997/98, Mon. Weather Rev. 132, 1721–1745, doi:10.1175/1520-0493(2004)132<1721:SACAOO>2.0.CO;2, 2004.

Ramis, C. and Llasat, M. C.: The October-1987 floods in Catalonia: synoptic and mesoscale mechanisms, Meteorol. Appl., 1, 337–350, 1994.

Ramis, C., Romero, R., Homar, V., Alonso, S., and Alarcón, M.: Diagnosis and numerical simulation of a torrential precipitation event in Catalonia (Spain), Meteorol. Atmos. Phys., 69, 1–21, 1998.

Ramis, C., Romero, R., and Homar, V.: The severe thunderstorm of 4 October 2007 in Mallorca: an observational study, Nat. Hazards Earth Syst. Sci., 9, 1237–1245, doi:10.5194/nhess-9-1237-2009, 2009.

Ramos, A. M., Trigo, R. M., Liberato, M. L. R., and Tome, R.: Daily Precipitation Extreme Events in the Iberian Peninsula and Its Association with Atmospheric Rivers, J. Hydrometeorol, 16, 579–597, doi:10.1175/JHM-D-14-0103.1, 2015.

Reale, M. and Lionello, P.: Synoptic climatology of winter intense precipitation events along the Mediterranean coasts, Nat. Hazards Earth Syst. Sci., 13, 1707–1722, doi:10.5194/nhess-13-1707-2013, 2013.

Reale, O., Feudale, L., and Turato, B.: Evaporative moisture sources during a sequence of floods in the Mediterranean region, Geophys. Res. Lett., 28, 2085–2088, 2001.

Roebber, P. J.: Statistical analysis and updated climatology of explosive cyclones, Mon. Weather Rev, 112, 1577–1589, 1984.

Romero, R., Ramis, C., Alonso, S., Doswell, C. A., and Stensrud, D. J.: Mesoscale model simulations of three heavy precipitation events in the western Mediterranean region, Mon. Weather Rev., 126, 1859–1881, 1998.

Romero R., Sumner G., Ramis C., and Genove A.: A classification of the atmospheric circulation patterns producing significant daily rainfall in the Spanish Mediterranean area, Int. J. Climatol., 19, 765–785, 1999.

Romero, R., Doswell, C. A., and Ramis, C.: Mesoscale numerical study of two cases of long-lived quasi-stationary convective systems over eastern Spain, Mon. Weather Rev., 128, 3731–3751, 2000.

Rudari, R., Entekhabi, D., and Roth, G.: Large-scale atmospheric patterns associated with mesoscalefeatures leading to extreme precipitation events in Northwestern Italy, Adv. Water Res. 28, 601–614, 2005.

Saaroni, H., Ziv, B., Lempert, J., Gazit, Y., and Morin, E.: Prolonged dry spells in the Levant region: climatologic-synoptic analysis, Int. J. Climatol., 35, 2223–2236, doi:10.1002/joc.4143, 2015.

Sanders, F. and Gyakum, J. R.: Synoptic-dynamic climatology of the bomb, Mon. Weather Rev., 108, 1589–1606, 1980.

Scoccimarro E., Gualdi, S., Bellucci, A., Zampieri, M., and Navarra, A.: Heavy precipitation events in a warmer climate: results from CMIP5 models, J. Climate, 26, 7902–7911, doi:10.1175/JCLI-D-12-00850.1, 2013.

Scoccimarro, E., Gualdi, S., Villarini, G., Vecci, G. A., Zhao, M., Walsh, K., and Navarra, A.: Intense precipitation events associated with land falling tropical cyclones in response to a warmer climate and increased CO_2, J. Climate, 27, 4642–4654, doi:10.1175/JCLI-D-14-00065.1, 2015.

Schmith, T., Kaas, E., and Li, T.-S.: Northeast Atlantic winter storminess 1875–1995 reanaysed, Clim. Dynam., 14, 529–536, 1998.

Smith, K. L., Kushner, P. J., and Cohen, J.: The role of linear interference in northern annular mode variability associated with Eurasian snow cover extent, J. Climate, 24, 6185–6202, 2011.

Sodemann, H. and Stohl, A.: Moisture origin and meridional transport in atmospheric rivers and their association with multiple cyclones, Mon. Weather Rev., 141, 2850–2868, 2013.

Sodemann, H. and Zubler, E.: Seasonal and inter-annual variability of the moisture sources for Alpine precipitation during 1995–2002, Int. J. Climatol., 30, 947–961, 2010.

Sodemann, H., Wernli, H., and Schwierz, C.: Sources of water vapor contributing to the Elbe flood in August 2002 – A tagging study in a mesoscale model, Q. J. Roy. Meteor. Soc., 135, 205–223, 2009.

Song, Y. and Robinson, W. A.: Dynamical mechanisms for stratospheric influences on the troposphere, J. Atmos. Sci., 61, 1711–1725, 2004.

Stohl, A. and James, P.: A Lagrangian analysis of the atmospheric branch of the global water cycle. Part 1: Method description, validation, and demonstration for the August 2002 flooding in central Europe, J. Hydrometeorol., 5, 656–678, 2004.

Stohl, A., Forster, C., and Sodemann, H.: Remote sources of water vapor forming precipitation on the Norwegian west coast at 60N – a tale of hurricanes and an atmospheric river, J. Geophys. Res., 113, D05102, doi:10.1029/2007JD009006, 2008.

Sickmöller, M., Blender, R., and Fraedrich, K.: Observed winter cyclone tracks in the northern hemisphere in re-analysed ECMWF data, Q. J. Roy. Meteor. Soc., 126, 591–620, 2000.

Simpson I. R., Blackburn M., and Haigh J. D.: The role of eddies in driving the tropospheric response to stratospheric heating perturbations, J. Atmos. Sci., 66, 1347–1365, 2009.

Taylor, K. E., Stouffer, R. J., and Meehl, G. A.: An Overview of CMIP5 and the experiment design, B. Am. Meteorol. Soc., 93, 485–498, doi:10.1175/BAMS-D-11-00094.1, 2012.

Thorncroft, C. D., Hoskins, B. J., and McIntyre, M. E.: Two paradigms of baroclinic-wave life-cycle behavior, Q. J. Roy. Meteor. Soc., 119, 17–55, 1993.

Toreti, A., Xoplaki, E., Maraun, D., Kuglitsch, F. G., Wanner, H., and Luterbacher, J.: Characterisation of extreme winter precipitation in Mediterranean coastal sites and associated anomalous atmospheric circulation patterns, Nat. Hazards Earth Syst. Sci., 10, 1037–1050, doi:10.5194/nhess-10-1037-2010, 2010.

Toreti, A., Naveau, P., Zampieri, M., Schindler, A., Scoccimarro, E., Xoplaki, E., Dijkstra, H. A., Gualdi, S., and Luterbacher, J.: Projections of global changes in precipitation extremes from Coupled Model Intercomparison Project Phase 5 models, Geophys. Res. Lett., 40, 4887–4892, doi:10.1002/grl.50940, 2013.

Trigo, I. F.: Climatology and interannual variability of storm-tracks in the Euro–Atlantic sector: a comparison between ERA-40 and NCEP/NCAR reanalyses, Clim. Dynam., 26, 127–143, 2006.

Trigo, R. M., Ramos, C., Pereira, S. S., Ramos, A. M., Zezere, J., and Liberato, M. L. R.: The deadliest storm of the 20th century striking Portugal: Flood impacts and atmospheric circulation, J. Hydrol., doi:10.1016/j.jhydrol.2015.10.036, in press, 2016.

Tsidulko, M. and Alpert, P.: Synergism of upper-level potential vorticity and mountains in Genoa lee cyclogenesis – A numerical study, Meteorol. Atmos. Phys., 78, 261–285, 2001.

Turato, B., Reale, O., and Siccardi, F.: Water vapor sources of the October 2000 Piedmont flood, J. Hydrometeorol., 5, 693–712, 2004.

Ulbrich, U. and Christoph, M.: A shift of the NAO and increasing storm track activity over Europe due to anthropogenic greenhouse gas forcing, Clim. Dynam., 15, 551–559, 1999.

Ulbrich, U., Fink, A. H., Klawa, M., and Pinto J. G.: Three extreme storms over Europe in December 1999, Weather, 56, 70–80, 2001.

Ulbrich, U., Leckebusch, G. C., and Pinto, J. G.: Extra-tropical cyclones in the present and future climate: a review, Theor. Appl. Climatol. 96, 117–131, 2009.

Ulbrich, U., Lionello, P., Belušić, D., Jacobeit, J., Knippertz, P., Kuglitsch, F. G., Leckebusch, G. C., Luterbacher, J., Maugeri, M., Maheras, P., Nissen, K. M., Pavan, V., Pinto, J. G., Saaroni, H., Seubert, S., Toreti, A., Xoplaki, E., and Ziv, B.: Climate of the Mediterranean: synoptic patterns, temperature, precipitation, winds, and their extremes. in: The climate of the Mediterranean region: from the past to the future, edited by: Lionello, P., Elsevier, Amsterdam, the Netherlands, 301–346, doi:10.1016/B978-0-12-416042-2.00005-7, 2012.

van den Besselaar, E. J. M., Klein Tank, A. M. J., and Buishand, T. A.: Trends in European precipitation extremes over 1951–2010, Int. J. Climatol. 33, 2682–2689, 2013.

Vicente-Serrano, S. M., Begueria S., Lopez-Moreno, J. I., El Kenawy, A. M., and Angulo-Martinez, M.: Daily atmospheric circulation events and extreme precipitation risk in northeast Spain: Role of the North Atlantic Oscillation, the Western Mediterranean Oscillation, and the Mediterranean Oscillation, J. Geophys. Res-Atmos., 114, D08106, doi:10.1029/2008JD011492, 2009.

Wang, X. L. L., Swail, V. R., and Zwiers, F. W.: Climatology and changes of extra- tropical cyclone activity: comparison of ERA-40 with NCEP-NCAR reanalysis for 1958–2001, J. Climate, 19, 3145–3166, 2006.

Wernli, H., Dirren, S., Liniger, M., and Zillig, M.: Dynamical aspects of the life cycle of the winter storm "Lothar" (24–26 December 1999), Q. J. Roy. Meteor. Soc., 128, 405–429, 2002.

Winschall, A., Pfahl, S., Sodemann, H., and Wernli, H.: Impact of North Atlantic evaporation hot spots on southern Alpine heavy precipitation events, Q. J. Roy. Meteor. Soc., 138, 1245–1258, 2012.

Winschall, A., Sodemann, H., Pfahl, S., and Wernli, H.: How important is intensified evaporation for Mediterranean precipitation extremes?, J. Geophys. Res.-Atmos., 119, 5240–5256, doi:10.1002/2013JD021175, 2014.

Xoplaki, E., Trigo R. M., García-Herrera, R. F., Barriopedro, D., D'Andrea, F., Fischer, E. M., Gimeno, L., Gouveia, C., Hernández, E., Kuglitsch, F. G., Mariotti, A., Nieto, R., Pinto, J. G., Pozo-Vázquez, D., Saaroni, H., Toreti, A., Trigo, I. F., Vicente-Serrano, S. M., Yiou, P., and Ziv, B.: Large-scale atmospheric circulation driving extreme climate events in the Mediterranean and its related impacts, in: The climate of the Mediterranean region: from the past to the future, edited by: Lionello, P., Elsevier, Amsterdam, the Netherlands, 347–417, doi:10.1016/B978-0-12-416042-2.00006-9, 2012.

Zolina, O., Simmer, C., Gulev, S. K., and Kollet, S.: Changing structure of European precipitation: longer wet periods leading to more abundant rainfalls, Geophys. Res. Lett, 37, L06704, doi:10.1029/2010GL042468, 2010.

Zhu, Y. and Newell, R. E.: Atmospheric rivers and bombs, Geophys. Res. Lett., 21, 1999–2002, 1994.

Zhu, Y. and Newell, R. E.: A proposed algorithm for moisture fluxes from atmospheric rivers, Mon. Weather Rev., 126, 725–735, doi:10.1175/1520-0493(1998)126<0725:APAFMF>2.0.CO;2, 1998.

Ziv, B., Dayan, U., and Sharon, D.: A mid-winter, tropical extreme flood-producing storm in southern Israel: synoptic scale analysis, Meteorol. Atmos. Phys., 88, 53–63, doi:10.1007/s00703-003-0054-7, 2005.

15

Storms or cold fronts: what is really responsible for the extreme waves regime in the Colombian Caribbean coastal region?

L. J. Otero, J. C. Ortiz-Royero, J. K. Ruiz-Merchan, A. E. Higgins, and S. A. Henriquez

Applied Physics Group – Ocean and Atmosphere Area – Physics Department, Universidad del Norte, Barranquilla, Colombia

Correspondence to: L. J. Otero (ljotero@uninorte.edu.co)

Abstract. The aim of this study is to determine the contribution and importance of cold fronts and storms to extreme waves in different areas of the Colombian Caribbean in an attempt to determine the extent of the threat posed by the flood processes to which these coastal populations are exposed. Furthermore, the study wishes to establish the actions to which coastal engineering constructions should be subject. In the calculation of maritime constructions, the most important parameter is the height of the wave. For this reason, it is necessary to establish the design wave height to which a coastal engineering structure should be resistant. This wave height varies according to the return period considered. The significant height values for the areas focused on in the study were calculated in accordance with Gumbel's extreme value methodology. The methodology was evaluated using data from the reanalysis of the spectral National Oceanic and Atmospheric Administration (NOAA) WAVEWATCH III® (WW3) model for 15 points along the 1600 km of the Colombian Caribbean coastline (continental and insular) between the years 1979 and 2009. The results demonstrated that the extreme waves caused by tropical cyclones and those caused by cold fronts have different effects along the Colombian Caribbean coast. Storms and hurricanes are of greater importance in the Guajira Peninsula (Alta Guajira). In the central area (consisting of Baja Guajira, and the cities of Santa Marta, Barranquilla, and Cartagena), the strong impact of cold fronts on extreme waves is evident. However, in the southern region of the Colombian Caribbean coast (ranging from the Gulf of Morrosquillo to the Gulf of Urabá), the extreme values of wave heights are lower than in the previously mentioned regions, despite being dominated mainly by the passage of cold fronts. Extreme waves in the San Andrés and Providencia insular region present a different dynamic from that in the continental area due to their geographic location. The wave heights in the extreme regime are similar in magnitude to those found in Alta Guajira, but the extreme waves associated with the passage of cold fronts in this region have lower return periods than those associated with the hurricane season.

1 Introduction

The principal source of coastal energy and cause of sediment dynamics are waves. The waves generated by "sea"-type local winds and those generated offshore by atmospheric disturbances (swells) determine the average and extreme maritime climate of a region. The activities which develop in coastal or maritime areas, such as fishing, transport, oil exploitation, structure design, and even recreational activities (surfing etc.), require knowledge of waves. The maritime climate (wind in particular) is a dynamic phenomenon which presents short-period (daily, weekly etc.), seasonal, and long-term variations. Considering the fact that wind is the main agent of wave generation, wave variation is also based on this pattern. Given the impact of the wave climate on the operation and construction costs of maritime infrastructure, it is imperative to design such structures with knowledge of the temporal variation of this phenomenon at an oceanic and local scale in mind.

Cold fronts occur when two air masses at different temperatures and densities come into proximity. The cold air mass with higher density pushes the hot air, making it ascend. The rising hot air cools down, producing clouds, storms, strong winds, and the pressure gradient in the area. Furthermore, it increases the wave height or swell through its passage along the Colombian Caribbean coast (Ortiz et al., 2013).

The incidence of cold fronts in the Colombian Caribbean coast has not been widely studied. Cold fronts are noted in the marine meteorological monthly reports of the Center of Oceanographic and Hydrographic Research (Centro de Investigaciones Oceanográficas e Hidrográficas, CIOH) of the General Maritime Direction (Dirección General Marítima, DIMAR). Their potential impact on the instances of extreme waves in the region was described by Ortiz et al. (2014), where the most important marine meteorological event of the last 10 years in the Colombian Caribbean was reconstructed. This event was the passage of a cold front in March 2009, when the associated waves caused 200 m of the pier at Puerto Colombia to collapse.

Ortiz et al. (2013) presented a characterization of cold fronts in the central region of the Colombian Caribbean and their relationship with extreme waves. The findings concluded that cold fronts have been the cause of the major waves in this region over the last 15 years.

Ortiz (2012) and Ortiz et al. (2015) also described the threat of hurricanes in the Colombian Caribbean. Hurricane Joan in 1988 and Hurricane Lenny in 1999 were two of the most important events which affected the continental Caribbean coast. San Andrés and the insular region have, for their part, been consistently impacted over the past 50 years (Hurricane Hattie in 1961, Hurricane Alma in 1970, Hurricane Joan in 1988, Hurricane Cesar in 1996, Hurricane Katrina in 1999, and Hurricane Beta in 2005). This demonstrates that the San Andrés y Providencia Archipelago is the zone in the Colombian Caribbean which is most vulnerable to hurricane threats.

Within this context, the traditional method of constructing the extreme wave regime has consisted of predictions of waves based on historical information for a period of no less than 10 years (Martínez and Coria, 1993; Naveau et al., 2005). Using this information, the most unfavorable extreme values of each year are selected, and the results are then extrapolated to different return periods.

Although there is no physical, theoretical, or empirical evidence of what the selection or adjustment of a probability distribution function should be for the calculation of the design wave height, the Gumbel or Fisher Tippett I and Weibull distribution functions are widely used for such purposes, as reported by Martínez and Coria (1993), Katz et al. (2002), García et al. (2004), and Naveau et al. (2005).

In response to the lack of information on the variability of extreme waves along the coast of the Colombian Caribbean, the main objective of this study is to determine the importance and contribution of cold fronts and tropical storms to extreme waves in order to ascertain the actions to which coastal and port structures in different regions of the Colombian Caribbean should be subjected and furthermore to assess the degree of threat to which coastal populations are exposed due to flood processes caused by wave run-up.

2 Description of the study area

The coastline of the Colombian Caribbean has a wide territory which extends from 8 to 13° N and from the country's border with Panamá in the southwest (SW) at longitude 79° W to the Guajira Peninsula in the northeast (NE) at longitude 71° W. The Colombian coastal area is approximately 1600 km in length and includes important cities from both an economic and touristic perspective: Riohacha, Santa Marta, Barranquilla, and Cartagena.

The insular Colombian Caribbean is formed by the Archipelago of San Andrés, Providencia, Santa Catalina, the Roncador, Quitasueño, Serranía, and other cays, adjacent islets, and deep-water coral reefs.

San Andrés Island is located in the Caribbean Sea, 180 km to the east of Nicaragua and northeast of Costa Rica. The island is 480 km northwest of the Colombian coast (Parsons, 1954). The island of San Andrés is the largest of the islands that form part of the Archipelago of San Andrés, Providencia, and Santa Catalina, with a total extension of 26 km^2. It is surrounded on its northwestern side by a small coral reef and several cays that are home to marine flora and fauna (Geister, 1973).

It is interesting to review and analyze the phenomenon responsible for the extreme wave regime in this area due to the fact that the area of the Colombian Caribbean in its entirety is not located at the same latitude.

The Colombian climate is characterized by two seasons in the year that are marked according to precipitation: a dry season and a wet season. The main dry season in the region occurs from December to May. In this season, swells produced by cold fronts may occur. The wet season occurs in the rest of the year, interrupted by a relative minimum in June and July known as an Indian summer (Andrade and Barton, 2001). The dry season and the Indian summer are associated with the jet stream winds of San Andrés and the northeastern (NE) trade winds. When the jet stream of San Andrés is stronger, the dry season occurs on the entire Caribbean coast, coinciding with very intense NE trade winds. The same occurs with the Indian summer (Andrade and Barton, 2001). By contrast, the wet season, which ranges from August to November, coincides with the time of the most intense jet stream winds from Chocó and the less intense jet stream winds from San Andrés (Poveda, 2004).

The annual wave cycle presents a bimodal form of behavior influenced by the NE trade winds, consisting of two wind periods, intense waves, and low precipitation in the dry season, and two wind periods, weak waves, and high precipitation during the wet season (Mesa, 2010).

3 Methodology

In the area under study, there are no extensive historical records of waves of an experimental nature (Meza, 2010).

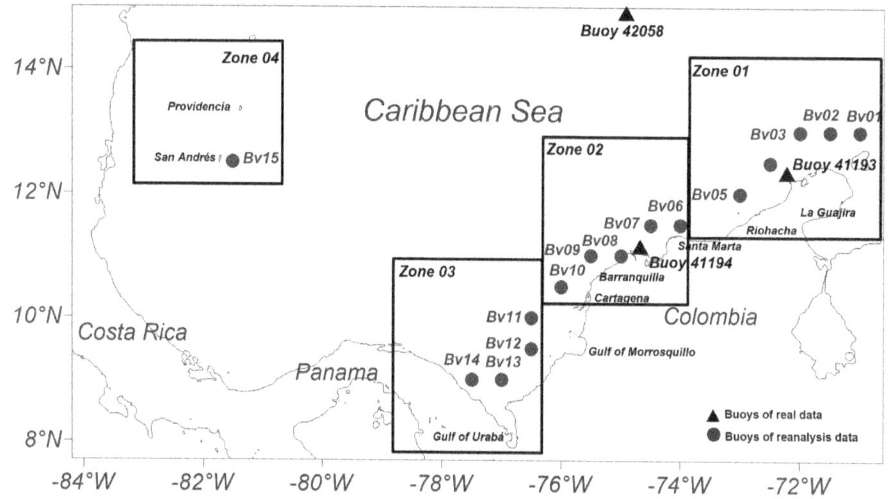

Figure 1. Location of the DIMAR and NOAA buoys on the Colombian Caribbean coast, and location of the 15 virtual buoys generated using WW3, including the zoning proposed by Ortiz (2012) based on the threat degree of hurricanes.

Buoys of the National Oceanic and Atmospheric Administration (NOAA) are extremely far from the coastal zone of the Colombian Caribbean. The DIMAR controls several directional buoys near the Colombian coasts in the Caribbean and the Pacific. For the Caribbean, the DIMAR has two buoys, 41193 and 41194, which are located near the coasts of Barranquilla and Riohacha, respectively. In the central Caribbean, the NOAA controls scalar buoy 42058, as shown in Fig. 1.

The National Data Buoy Center (NDBC) database has records for 2008, 2009, and 2010. Directional buoy 41194, which began operation in 2007, is located near Bocas de Ceniza, which is the mouth of the Magdalena River in the Caribbean Sea. It is located near the city of Barranquilla. Directional buoy 41193 is near Puerto Bolívar (Guajira Department) and has records since 2010, while buoy 42058 has been collecting data since 2005. Therefore, in order to characterize the wave climate, it is necessary to implement data obtained from models, satellites, or visual waves (Agudelo et al., 2005). According to Martínez and Coria (1993), for an analysis of extreme waves it is necessary to have at least 10 years of wave data in the area of interest. For this purpose, a historic database of 30 years (1979–2009) from the reanalysis of 15 virtual buoys distributed along the Caribbean coastline from the NCEP Climate Forecast System Reanalysis (CFSR) winds, was initially analyzed (Chawla et al., 2011). These data were obtained online (http://polar.ncep.noaa.gov/waves/CFSR_hindcast.shtml). The spatial resolution of wave data for the Caribbean Sea is 1/6° × 1/6°. The wave model used at NCEP is a third-generation wind wave model, WAVEWATCH III® (Tolman, 2009). The wind fields with which the model was forced originate from a new reanalysis of the atmospheric, oceanic, sea ice, and land data between 1979 and 2010, as well as a reforecast run with this reanalysis (Saha et al., 2010). A detailed description of the data is carried out in Chawla et al. (2013).

The locations of these virtual buoys (BV) are shown in Fig. 1. This image also demonstrates the segmentation of the continental and insular Colombian Caribbean coast into 4 zones, as proposed by Ortiz (2012), in accordance with their level of vulnerability to the threat of hurricanes.

Given the limited information of oceanographic buoys in the area of study, detailed parametric calibration and validation were applied to the reanalyzed wave series for data coinciding with the significant wave heights (H_s) in undefined depths for the zone in the Caribbean where data from oceanographic buoys existed. In this study, the data from buoy 42058 were used for the calibration of the database, and the data from buoy 41194 were used for validation. Buoy 42058 was selected to perform the calibration process for two main reasons: the higher extension of continuous data available and its location in undefined depths.

From the data on significant wave heights (H_s) from the WW3 model for the point 14.9° N, 75° W and the data collected by oceanographic buoy 42058, a detailed parametric calibration method was applied for coinciding data using power and linear regression models (Tomas, 2009). With the coefficients obtained in each regression model, the data from the series corresponding to buoy BV05 were corrected. These data were validated using the values of significant wave heights (H_s), as measured by oceanographic buoy 41194 of Barranquilla, Colombia.

With the aim of verifying the degree of adjustment of the measured and calculated time series, the D adjustment index and the average deviation P, described by Willmott (1981), were applied. In order to determine the adjustment degree of the regression models, the r^2 correlation coefficient was used.

In accordance with Wornon and Welsh (2002a, b), the D index is very sensitive to errors in the mean square root of the difference between predictive and observed values. $D = 0$ shows total dissociation, while $D = 1$ shows a perfect association between the measured and calculated data. Willmott's D index is defined as

$$D = 1 - \frac{\sum_{n=1}^{N} (\text{Pn} - \text{On})^2}{\sum_{n=1}^{N} (|\text{Pn} - O| + |\text{On} - O|)^2}. \tag{1}$$

Here, Pn is the prediction performed, On represents the measurements, and O is the mean of the measurements performed.

The average deviation P is defined as

$$\text{bias}(P) = \frac{\sum_{n=1}^{N} (\text{Pn} - \text{On})^2}{\sum_{n=1}^{N} \text{On}}. \tag{2}$$

For example, a deviation of -0.05 demonstrates mean underestimations of the order of 5 %.

To differentiate the influence of cold fronts on hurricanes, calculations of extreme values were performed in two seasons of the year: the hurricane season of the Caribbean from June to November, and the cold front season from December to May.

For the calculation of extreme values, Gumbel's maximum distribution or Fisher-Tippet's type I distribution were selected. This distribution is suitable for random, identically distributed, and independent variables and is widely used to characterize extreme regimes of oceanographic geophysical variables. Gumbel's maximum distribution is a bi-parametric function with a location parameter λ and a scale parameter δ (García et al., 2004):

$$F(x) = e^{-e^{-b}}, \tag{3}$$

where

$$b = \frac{x - u}{\alpha} \tag{4}$$

$$\alpha = \frac{S_x}{\lambda} \tag{5}$$

$$u = \bar{x} - \delta\alpha. \tag{6}$$

Here, x is the data series, S_x is the standard deviation, and $\bar{x}$ the arithmetic mean of the data series. The values of λ and δ are selected according to the number of data. In this case, $\lambda = 0.53622$ and $\delta = 1.11237$.

From the series of annual maximums, the characteristic waves for return periods of 5, 10, 25, 50, and 100 years were calculated using Gumbel's maximum distribution.

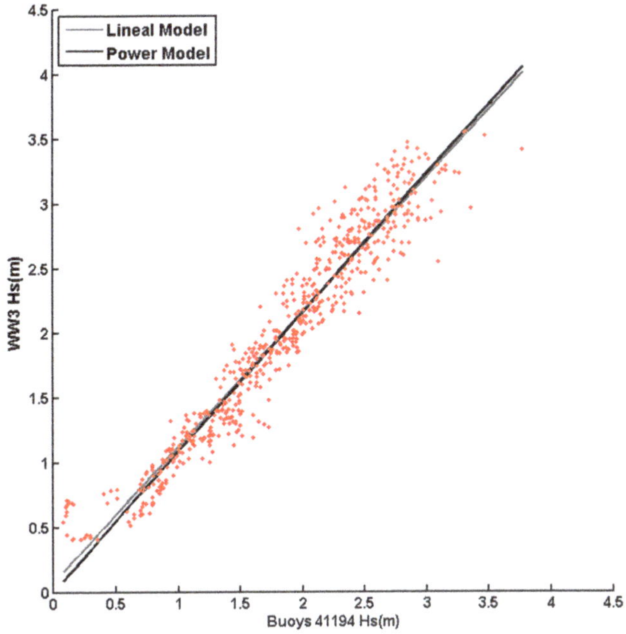

Figure 2. Scalar calibration using power and linear regression models of H_s WW3 with coinciding data from the central Caribbean buoy ($n = 594$).

4 Analysis and results

The results of the calibration of the H_s WW3 reanalysis data and the data from the central Caribbean oceanographic buoy (42058) can be seen in Fig. 2. The lines represent the linear regression Eq. (7) and power regression Eq. (8) models. The dispersion diagram shows that both models present a good correlation with the data, with r^2 values of 0.92 and 0.93 for the power and linear regression models, respectively. The calibration equations obtained were the following:

$$H_{\text{scal}} = \alpha H_{\text{sbuoys}} + \beta (\alpha = 1.047 \pm 0.024; \\ \beta = 0.07619 \pm 0.047) \tag{7}$$

$$H_{\text{scal}} = a H_{\text{sbuoys}}^{b} (a = 1.096 \pm 0.0265; \\ b = 0.9847 \pm 0.0284). \tag{8}$$

Based on the previous calibration relations, the H_s data from the WW3 model in the coordinate point 11.25° N–74.75° W (coinciding with the location of buoy 41194) were corrected.

The uncorrected H_s values of the WW3 model and the data from buoy 41194 are compared in Fig. 3a, with $D = 0.97$ and a bias$(P) = -0.06$. Figure 3b and c compare the H_s series from buoy 41194 with the H_s series corrected using a linear and power regression model, respectively. Based on the adjustment index D and Willmott's average deviation P, the improvement in the adjustment degree in the H_s time series is shown in Table 1. As demonstrated in the table, the indices show improvements of equal proportion for both models. In this sense, the linear regression model was selected for the

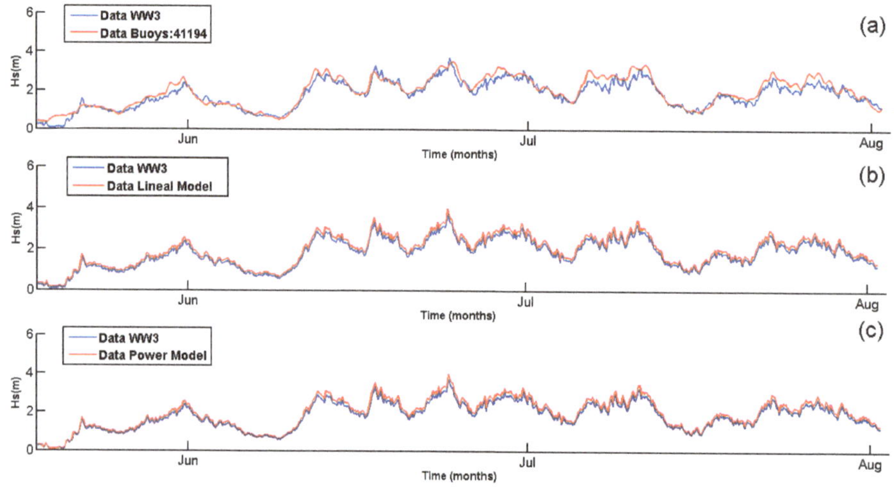

Figure 3. Validation of the H_s time series (m) corresponding to the year 2009 ($n = 767$). **(a)** Buoy 41194 and WW3 data without calibration. **(b)** Buoy 41194 and data corrected with the linear model. **(c)** Buoy 41194 and data corrected with the power model.

correction of the WW3 H_s series from the 15 points selected for the analysis.

Once the information of the data for each of the 15 buoys was calibrated and validated, we proceeded to perform a detailed analysis from which a total of eight buoys were selected as representative samples of each of the zones of interest based on the H_s value and the location of the cities and ports of interest.

As noted above, tropical cyclones occur in the season from June to November, and cold fronts occur from December to May. Therefore, the H_s time series of the eight buoys were divided for these two seasons with the objective of achieving independently performed extreme wave analysis. That is, the extreme regime was calculated considering only the H_s information for the months between June and November, and another extreme regime was calculated, only taking into account the months from December to May. Additionally, the extreme regime was calculated considering the complete series in an effort to identify which seasons of the year contributed the most to this regime. This method of presenting the results allows a comparative analysis to identify in which seasons of the year the most extreme waves for the area occur and also to establish which atmospheric phenomenon is responsible for generating these extreme waves.

The results of the application of Gumbel's theory for the significant height H_s on each of the eight buoys are shown in Figs. 4, 7, 8, and 9 in the same order in which the previous results were presented.

Virtual buoys BV01–BV05 are located in the arid zone of the Alta Guajira, in the northeast of the Colombian Caribbean (zone 1 in Fig. 1). The bar diagrams in Fig. 4 explicitly show that the height of extreme waves is clearly influenced by tropical cyclones. During this period, the extreme waves associated with return periods of 5 and 10 years oscillate between 3.2 and 4.5 m. For return periods of 25, 50, and 100 years,

Table 1. Adjustment index D and Willmott's average deviation P, data from buoy 41194 and WW3 model.

Significant height $H_S(m)$	D	Bias (P)
WW3 vs. buoy (41194)	0.97	−0.06
WW3 corrected with linear Regression vs. buoy (41194)	0.99	0.02
WW3 corrected with power Regression vs. buoy (41194)	0.99	0.02

there is a probability of exceedance of 4, 2, and 1 % respectively, varying between 5.5 and 6.6 m in height. The higher waves for this region occur because Alta Guajira is the closest zone to the trajectory of tropical cyclones passing through the Colombian Caribbean (Ortiz, 2012).

Conversely, the dry season of the year (from December to May) is influenced by the increase of trade winds from the northeast (Bernal et al., 2006; Paramo et al., 2011) and the passage of cold fronts coming from the Northern Hemisphere, an event which produces strong tides (Ortiz et al., 2013). Extreme wave heights covering return periods of 5 and 10 years show significant heights which oscillate between 3 and 4.2 m. Meanwhile, the extreme waves for return periods of 25, 50, and 100 years take significant height values which vary between 3.6 and 4.8 m.

Regarding the direction of the waves, it is considered that this is likely from E and ENE with heights generally between 1 and 3 m.

When comparing the extreme regime for virtual buoys BV01 and BV03 located in the Alta Guajira area, it can be observed that the significant heights (for return periods of 5 and 10 years which take the complete series of annual maximums) show values greater than the significant heights ob-

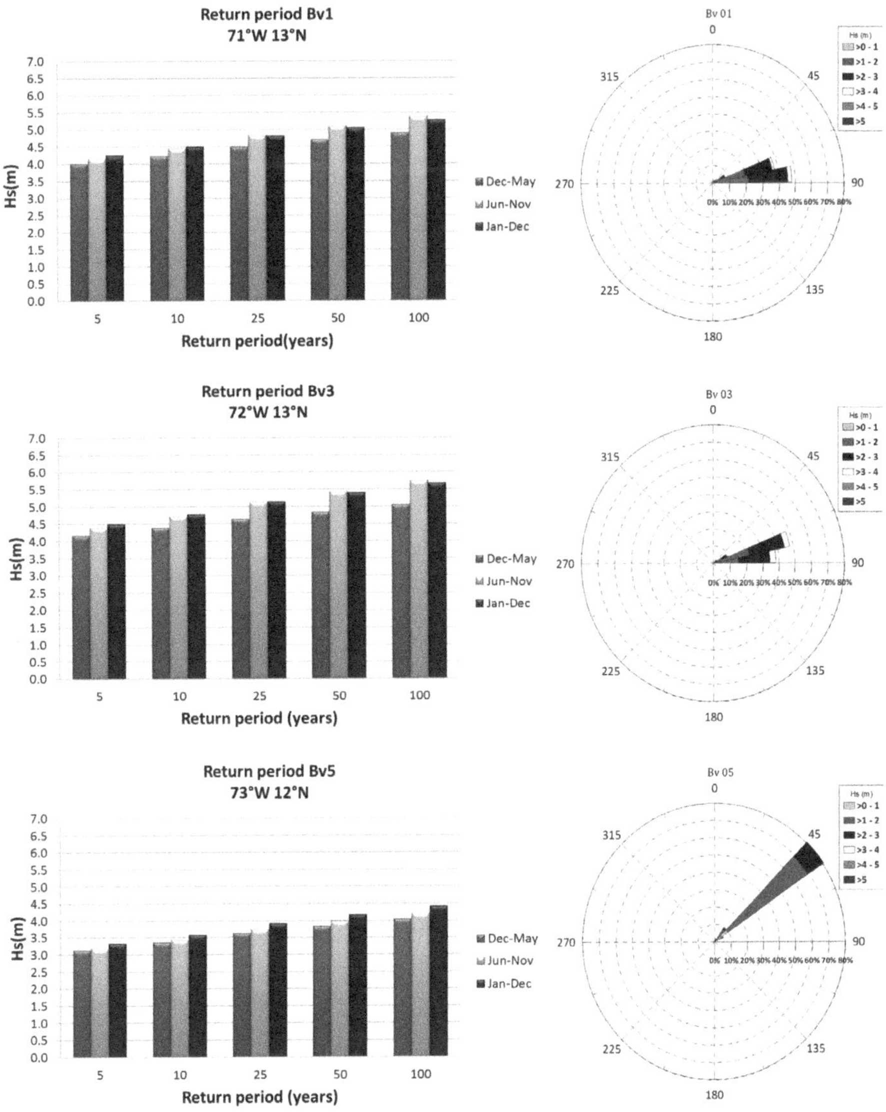

Figure 4. Return periods for H_S of virtual buoys BV01–BV05 (zone 1).

tained when dividing the series by season. However, for a return period of more than 50 years, the significant height extrapolated from the cyclone season data surpasses the significant height obtained within the annual maximums. This occurs because, over the 30 years of the registry, there were 16 years in which the maximum significant height occurred in hurricane season. These values were the highest of the record (between 4.9 and 5.2 m). In the remaining 14 years, the maximum significant height in cyclone season is lower in comparison to the wave height registered in the cold front season. This variation in the significant height for each period generates a greater dispersion in the maximum point of the season from June to November. The difference in the standard deviation of data causes an increase in the slope of the line H_S against the return period in the semi-logarithmic space (Fig. 5) of the model for the cyclone season. In the remaining virtual buoys, there is no variation in the extreme regime when increasing the return period because all the extreme wave events are found to be related to one of the two seasons as shown in Fig. 6.

The diagrams of Fig. 7 correspond to the central zone of the Colombian Caribbean formed by Baja Guajira, Santa Marta, Barranquilla, and Cartagena (zone 2 in Fig. 1). It can be seen that in BV06, for both the cyclone season (June–November) and the cold front season (December–May), there are similar magnitudes in the extreme waves. However they present slightly higher values in the dry season, especially in the return periods of 5 and 25 years. From BV09, it can be observed that the most energetic waves in the coastal zone of this region are clearly influenced by cold fronts, whereas the less energetic waves are dominated by tropical cyclones. Therefore, the extreme waves for return

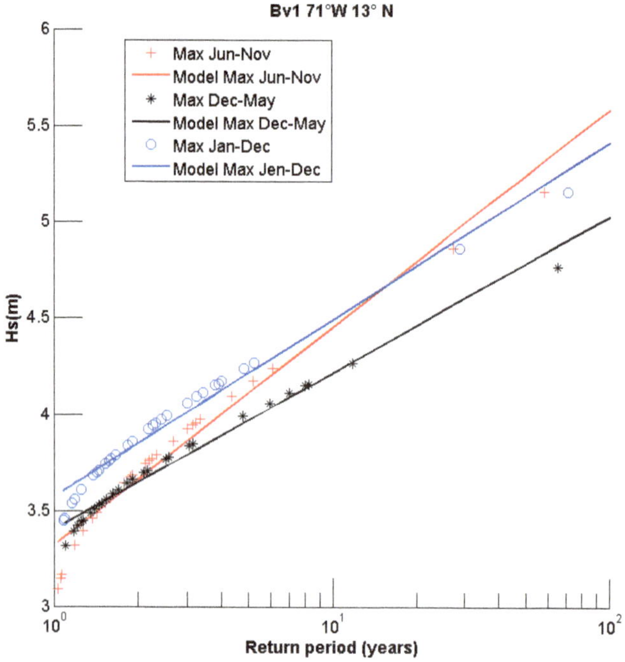

Figure 5. Comparison of the extreme wave regime in virtual buoy 01, considering the maximums of the complete series, the maximums of the cold front season, and the maximums of the hurricane season.

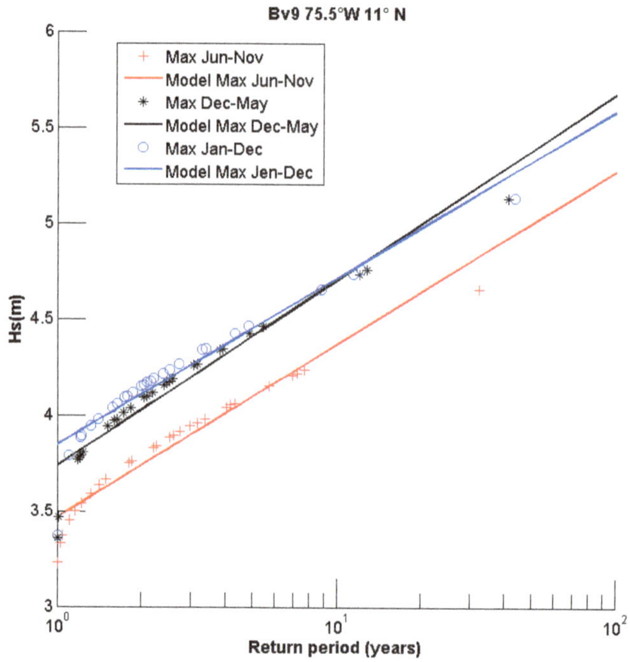

Figure 6. Comparison of the extreme wave regime in virtual buoy 09, considering the maximums of the complete series, the maximums of the cold front season, and the maximums of the hurricane season.

periods of 5 and 10 years register significant heights between 3.5 and 4.5 m, and the extreme waves for return periods of 25, 50, and 100 years present values that vary between 4 and 5.5 m in height.

In this transition region, the behavior of extreme waves is modified because a clearer and stronger preponderance of cold fronts begins, decreasing the influence of tropical storms on this regime. This modification can clearly be seen in Fig. 7, where the extreme regime calculated considering the complete series is influenced by the maximum wave heights which occur in the cold front season.

It is important to highlight that the extreme wave height regime in one area of the central zone of the Caribbean is higher than in Alta Guajira (i.e., buoy 09). This zone of the Colombian coast ceases to be directly influenced by tropical cyclones and the extreme regime prevails due to the passage of cold fronts.

In relation to the direction of the waves, it is reported that in this sector of the central Colombia Caribbean, the NE direction is predominant with heights mainly between 1 and 3 m. With respect to zone 1, the energy in the average regime is similar, and the direction of the wave trains is affected by the orientation of the coastline, which stretches from NE to SW.

The data shown in Fig. 8 correspond to the zone known as the western Caribbean, which consists of the area between the Morrosquillo Gulf and the Urabá Gulf (zone 3 in Fig. 1). In this zone, regardless of the season, the wave energy is lower in comparison to the other zones of the Colombian Caribbean. This is because the Morrosquillo Gulf is shielded from winds by Boquerón Island and the other islands of San Bernardo. For its part, the Urabá Gulf is protected from waves in the eastern zone, and Colombia bay is protected by its closed "U" shape.

The bar diagrams in Fig. 8 show that the least energetic waves of the entire Colombian coast, which are mainly influenced by the passage of cold fronts, can be found in this area. The significant height for return periods of 5 and 10 years show values oscillating between 4 and 4.5 m, and the extreme waves for return periods of 25, 50, and 100 years assume values which vary between 4.5 and 5 m in height. On the other hand, the least energetic extreme waves are dominated by tropical cyclones. In these graphics, it is evident that the extreme wave height is clearly influenced by cold fronts, a fact which is confirmed when the values of the extreme wave heights are calculated taking into account the complete series of maximum heights. Regarding the direction of waves in the southwestern sector of the Colombian Caribbean, NE and NNE are predominant, with heights generally ranging between 1 and 2 m. Regarding zones 01 and 02, the energy in the average regime is lower.

BV15 is located in the insular region of San Andrés and Providencia (zone 4 in Fig. 1). In Fig. 9, it can be seen that extreme waves for return periods of 5 and 10 years have sig-

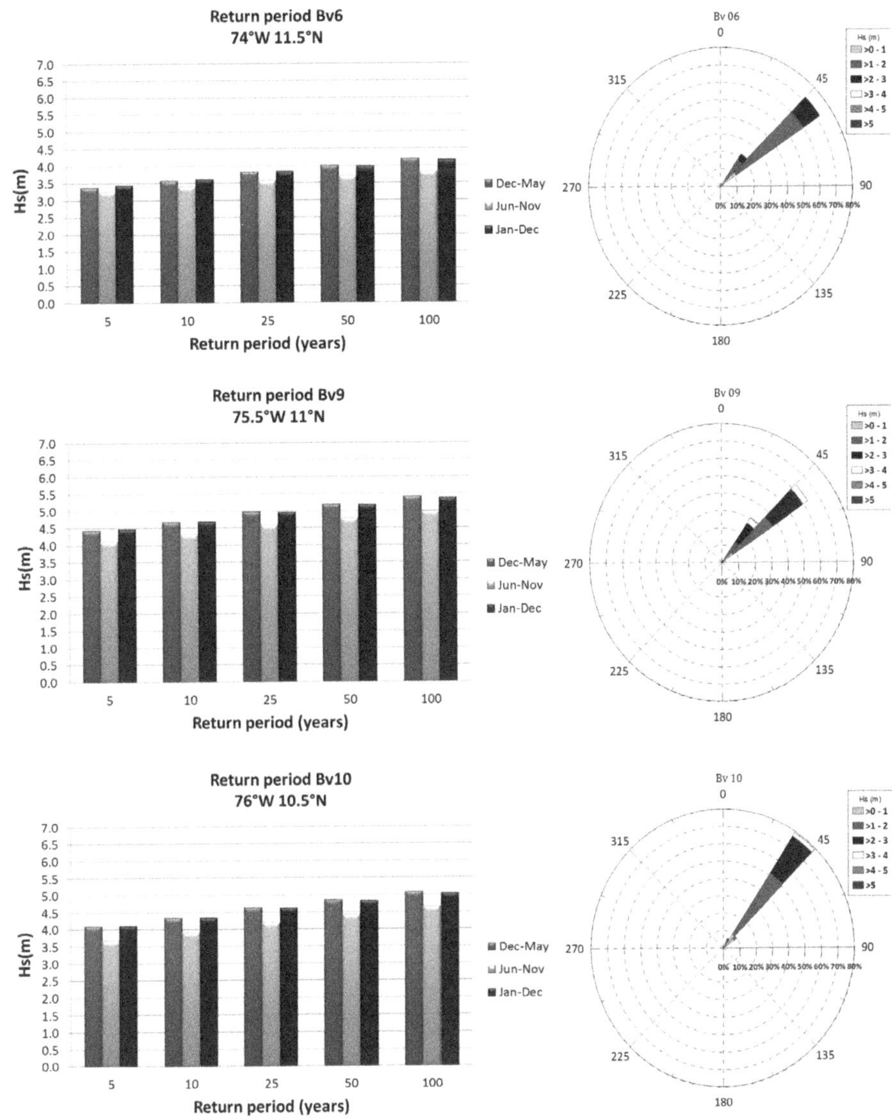

Figure 7. Return periods for H_S of virtual buoys BV06–BV10 (zone 2).

nificant height values between 5 and 5.5 m for the wet season (on account of hurricanes and tropical storms), while for the dry season (with the influence of cold fronts), the values oscillate between 4.5 and 4.7 m. The extreme waves for return periods of 25, 50, and 100 years assume values which vary between 5.1 and 5.7 m in height for the dry season (as influenced by cold fronts), while for the wet season, the values for extreme waves oscillate between 6.2 and 7 m. Due to their geographic location in the Caribbean Sea, the islands of San Andrés and Providencia exhibit high vulnerability to tropical storms and the passage of cold fronts (Ortiz et al., 2014b). Regardless of the analyzed series, it is clearly evident that the waves generated by tropical storms are the waves which govern the extreme regime. This is due to the location of San Andrés and Providencia islands in an area of high hurricane frequency.

Previous studies have revealed that, according to historical records for the years 1900 to 2013, a total of 17 hurricanes have affected San Andrés and Providencia. Furthermore, in seven of these phenomena, the eye of the storm was located less than 150 km from the islands. Of these seven hurricanes, Hurricane Joan in 1988 generated the highest wave heights, with significant height values of 5 m SE of the island of San Andrés in the 50 m isobaths (Ortiz et al., 2015). It is important to highlight that, over the last 50 years, of all the hurricanes which have affected the island of San Andrés, only Hurricane Joan has generated wave heights of 7.8 m (Ortiz et al., 2015). Nevertheless, values higher than 5 m have been registered during that season due to the passage of hurricanes. In the analysis of 30 years of the dry season for BV15, heights superior to 5 m were not encountered. This finding suggests that wave heights in the extreme regime associated

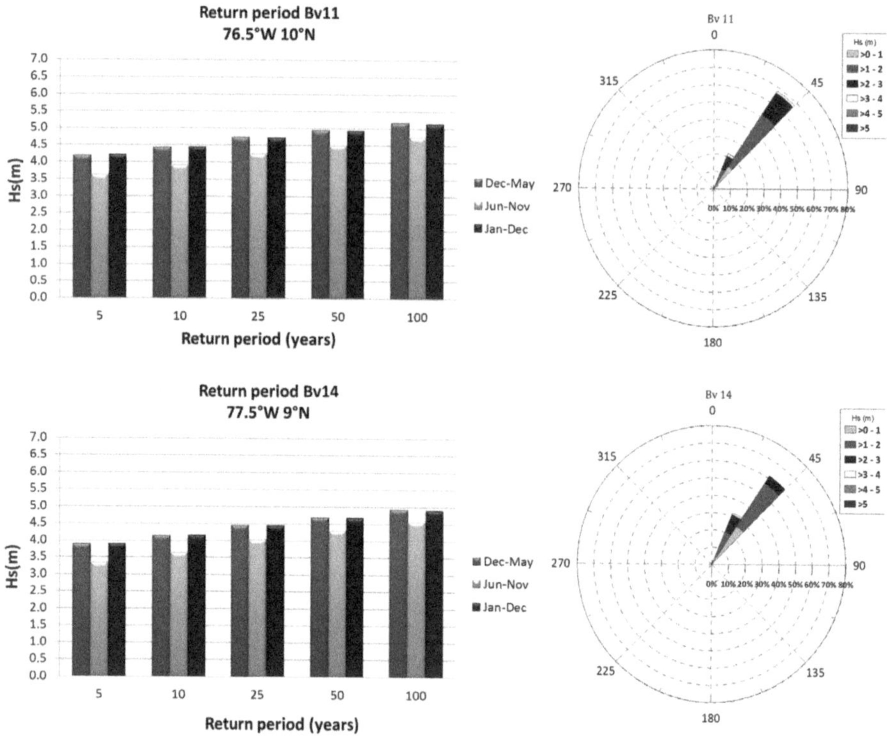

Figure 8. Return period for H_S of virtual buoys BV11–BV14 (zone 3).

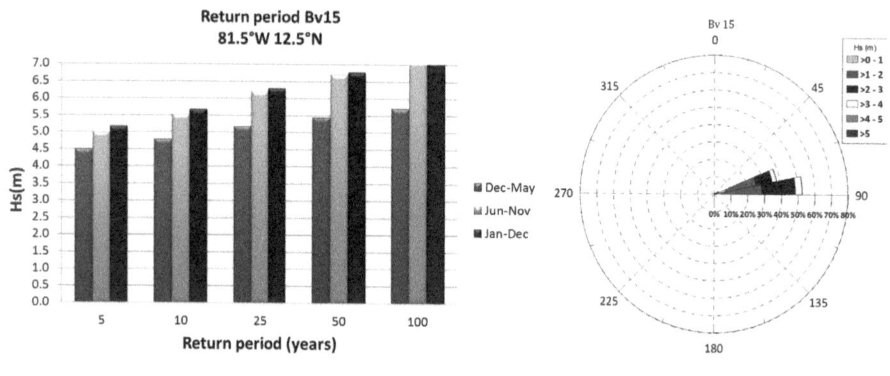

Figure 9. Return periods for H_S of virtual buoys BV15 (zone 4).

with the passage of tropical storms in zone 4 have lower return periods than the wave heights associated with the cold fronts season.

5 Conclusions

The present study has led to the conclusion that extreme waves along the Colombian Caribbean coasts (continental and insular) are not simply generated by the same forces. The geomorphology of the 1600 km of coastline of the Colombian Caribbean and its geographical position ensure that the waves generated by extreme events do not have the same magnitude of influence in all areas. There is a clear difference between the north zone (Alta Guajira), central zone (Baja Guajira, Santa Marta, Barranquilla, and Cartagena), west zone (Morrosquillo Gulf and Urabá Gulf), and the Archipelago of San Andrés because the extreme regime of the waves is influenced by the spatial and temporal changes of the atmospheric phenomena which develop over a particular zone on the Colombian Caribbean coast, depending on the dry and wet climatic seasons.

Based on the analysis of the variability of extreme waves in the continental coast of the Colombian Caribbean, it can be concluded that in the higher part of the Guajira (zone 1), there are higher extreme wave heights caused by tropical cyclones. This fact is consistent with the findings reported in Ortiz (2012). This area is followed by the central zone (zone

2), where the influence of extreme waves is associated with the passage of cold fronts. Less energetic extreme waves occur in the western part of the Caribbean (zone 3), where they are mainly forced by the passage of cold fronts.

The extreme waves in the insular region of San Andrés and Providencia (zone 4) show a dynamic different from the continental region due to its geographic location. Although the heights of the waves in the extreme regime are similar in magnitude to those found for zone 1 (Alta Guajira), the extreme waves associated with the passage of tropical storms in this zone have lower return periods than the extreme waves associated with the cold fronts season.

The aforementioned results make a valuable contribution towards determining the extreme regime and the design waves for the construction of coastal infrastructure. Furthermore, the results provide an important component for the planning of alert and mitigation strategies against marine meteorological threats. In addition, because different zones of the Colombian Caribbean are affected by coastal erosion, this study constitutes a source of data for the further analysis of how, and to what extent, these events are related to erosion.

Acknowledgements. The authors wish to express their gratitude to the Directorate of Research and Projects (Dirección de Investigaciones y Proyectos (DIDI)) of the Universidad del Norte and to the Departamento Administrativo de Ciencia y Tecnología e Innovación de Colombia (COLCIENCIAS) for their support in conducting this research.

References

Agudelo, P., Restrepo, A., Molares. A., Tejada. C., Torres. R., and Osorio, A.: Determinación del clima medio y extremal en el Caribe Colombiano [Determination of the average and extreme climate in the Colombian Caribbean]. Boletín Científico CIOH, 23, 33-45, 2005.

Andrade, C. and Barton, E.: Eddy development and motion in the Caribbean Sea, J. Geophys. Res., 105, 26191–26201, 2001.

Bernal, G., Poveda, G., Roldán, P., and Andrade, C.: Patrones de variabilidad de las temperaturas superficiales del mar en la Costa Caribe Colombiana [Variability patterns of the sea surface temperatures in the Colombian Caribbean coast], Rev. Acad. Colomb. Cienc., 30, 195–208, 2006.

Chawla, A., Spindler, D., and Tolman, H. L.: A thirty year wave hindcast using the latest NCEP Climate Forecast System Reanalysis winds, Proc. 12th Int. Workshop on Wave Hindcasting and Forecasting, Waikoloa, HI, Environment Canada, I1, 2011.

Chawla, A., Spindler, D., and Tolman, H. L.: Validation of a thirty year wave hindcast using the Climate Forecast System Reanalysis winds, Ocean Modell., 70, 189–206, 2013.

García, V., Garcia-Bartual, R., Cabrera, E., Arregui, F., and García, J.: Stochastic Model to Evaluate Residential Water Demands, J. Water Plan. Manage., 130, 386–394, 2004.

Geister, J.: Los arrecifes de la isla de San Andres (Mar Caribe, Colombia) [The coral reefs in the island of San Andrés (Caribbean Sea, Colombia)], Mitt Inst. Colombo Alemán Invest. Cient., 7, 211–228, 1973.

Katz, R. W., Parlange, M. B., and Naveau, P.: Statistics of extremes in hydrology, Adv. Water Resour., 25, 1287–1304, 2002.

Martínez, A. and Coria, P.: Distribución de probabilidad de la altura del oleaje dentro de la bahía de Todos los Santos, B.C. [Probability distribution of waves' height inside the Todos los Santos Bay, B.C.], México, Ciencias Marinas, 19, 203–218, 1993.

Mesa, J. C.: Metodología para el reanálisis de series de oleaje para el caribe colombiano [Methodology for the re-analysis of wave series for the Colombian Caribbean]. Tesis Doctoral, Universidad Nacional de Colombia, Colombia, Medellin, 5–12, 2010.

Naveau, P., Nogaj, M., Ammann, C., Yiou, P., Cooley, D., and Jomelli, V.: Statistical methods for the analysis of climate extremes, C. R. Geosci., 337, 1013–1022, 2005.

Ortiz, J. C.: Exposure of the Colombian Caribbean Coast, including San Andrés Island, to Tropical Storms and Hurricanes, 1900–2010, Nat. Hazard J., 61, 815–827, 2012.

Ortiz-Royero, J. C., Otero, L. J., Restrepo, J. C., Ruiz, J., and Cadena, M.: Cold fronts in the Colombian Caribbean Sea and their relationship to extreme wave events, Nat. Hazards Earth Syst. Sci., 13, 2797–2804, doi:10.5194/nhess-13-2797-2013, 2013.

Ortiz, J. C., Salcedo B., and Otero, L.: Investigating the collapse of the Puerto Colombia Pier (Colombian Caribbean Coast) in March of 2009: methodology for the reconstruction of extreme events and their impact on the coastal infrastructure, J. Coast. Res., 30, 291–300, 2014.

Ortiz, J. C., Plazas, J. M., and Lizano, O.: Evaluation of Extreme Waves Associated with Cyclonic Activity on San Andrés Island in the Caribbean Sea since 1900, J. Coast. Res., J. Coast. Res., 31, 557–568, 2015.

Paramo, J., Correa, M., and Núñez, S.: Evidencias de desacople físico-biológico en el sistema de surgencia en La Guajira, Caribe colombiano [Evidence of the physico-biological decoupling of the upwelling system in La Guajira, Colombian Caribbean]. Revista de biología marina y oceanografía, 46, 421–430, 2011.

Parsons, J. J.: English Speaking Settlement of the Western Caribbean, Yearbook of the Association of Pacific Coast Geographers, 16, 3–16, 1954.

Poveda, G.: La hidroclimatología de Colombia: una síntesis desde la escala inter-decadal hasta la escala diurna [The hydroclimatology of Colombia: a synthesis from the inter-decade scale to the diurnal scale], Rev. Acad. Colomb. Cienc., 28, 201–222, 2004.

Saha, S., Moorthi, S., Pan, H. L., Wu, X., Wang, J., Nadiga, S., Tripp, P., Kistler, R., Woollen, J., Behringer, D., Liu, H., Stokes, D., Grumbine, R., Gayno, G., Wang, J., Hou, Y. T., Chuang, H. Y., Juang, H. M. H., Sela, J., Iredell, M., Treadon, R., Kleist, D., Van Delst, P., Keyser, D., Derber, J., Ek, M., Meng, J., Wei, H., Yang, R., Lord, S., Van Den Dool, H., Kumar, A., Wang, W., Long, C., Chelliah, M., Xue, Y., Huang, B., Schemm, J.-K., Ebisuzaki, W., Lin, R., Xie, P., Chen, M., Zhou, S., Higgins, W., Zou, C. Z., Liu, Q., Chen, Y., Han, Y., Cucurull, L., Reynolds, R. W., Rutledge, G., and Goldberg, M.: The NCEP climate forecast system reanalysis, B. Am. Meteor. Soc., 91, 1015–1057, 2010.

Tolman, H. L.: User manual and system documentation of WAVEWATCH III version3.14, Tech. Note 276, NOAA/NWS/NCEP/MMAB, 220 pp., 2009.

Tomas, A.: Metodologías de calibración de bases de datos de reanálisis de clima marítimo [Methodologies of the calibration of marine climate reanalysis database]. Tesis Doctoral, Universidad de Cantabria., Spain, 69–89, 2009 .

Willmott, C.: On the validation of models Phys. Geog., 2, 183-194, 1981.

Wornon, S. and Welsh, D.: An MPI quasi time-accurate approach for nearshore wave prediction using SWAN code – Part I, Coast. Eng. J., 44, 247–256, 2002a.

Wornon, S. and Welsh, D.: An MPI quasi time-accurate approach for nearshore wave prediction using SWAN code – Part II, Coast. Eng. J., 44, 257–280, 2002b.

PERMISSIONS

All chapters in this book were first published in NHESS, by Copernicus Publications; hereby published with permission under the Creative Commons Attribution License or equivalent. Every chapter published in this book has been scrutinized by our experts. Their significance has been extensively debated. The topics covered herein carry significant findings which will fuel the growth of the discipline. They may even be implemented as practical applications or may be referred to as a beginning point for another development.

The contributors of this book come from diverse backgrounds, making this book a truly international effort. This book will bring forth new frontiers with its revolutionizing research information and detailed analysis of the nascent developments around the world.

We would like to thank all the contributing authors for lending their expertise to make the book truly unique. They have played a crucial role in the development of this book. Without their invaluable contributions this book wouldn't have been possible. They have made vital efforts to compile up to date information on the varied aspects of this subject to make this book a valuable addition to the collection of many professionals and students.

This book was conceptualized with the vision of imparting up-to-date information and advanced data in this field. To ensure the same, a matchless editorial board was set up. Every individual on the board went through rigorous rounds of assessment to prove their worth. After which they invested a large part of their time researching and compiling the most relevant data for our readers.

The editorial board has been involved in producing this book since its inception. They have spent rigorous hours researching and exploring the diverse topics which have resulted in the successful publishing of this book. They have passed on their knowledge of decades through this book. To expedite this challenging task, the publisher supported the team at every step. A small team of assistant editors was also appointed to further simplify the editing procedure and attain best results for the readers.

Apart from the editorial board, the designing team has also invested a significant amount of their time in understanding the subject and creating the most relevant covers. They scrutinized every image to scout for the most suitable representation of the subject and create an appropriate cover for the book.

The publishing team has been an ardent support to the editorial, designing and production team. Their endless efforts to recruit the best for this project, has resulted in the accomplishment of this book. They are a veteran in the field of academics and their pool of knowledge is as vast as their experience in printing. Their expertise and guidance has proved useful at every step. Their uncompromising quality standards have made this book an exceptional effort. Their encouragement from time to time has been an inspiration for everyone.

The publisher and the editorial board hope that this book will prove to be a valuable piece of knowledge for researchers, students, practitioners and scholars across the globe.

LIST OF CONTRIBUTORS

D. Andronico
Istituto Nazionale di Geofisica e Vulcanologia, Osservatorio Etneo, Piazza Roma 2, 95125 Catania, Italy

P. Del Carlo
Istituto Nazionale di Geofisica e Vulcanologia, Sezione di Pisa, Via Della Faggiola 32, 56126 Pisa, Italy

S. Gentile
Department of Physical and Chemical Sciences/CETEMPS, University of L'Aquila, Via Vetoio, 67010 Coppito, L'Aquila, Italy

R. Ferretti
Department of Physical and Chemical Sciences/CETEMPS, University of L'Aquila, Via Vetoio, 67010 Coppito, L'Aquila, Italy
Danish Meteorological Institute, Lyngbyvej 100, 2100 Copenhagen, Denmark

W. Kim, N. Kerle and M. Gerke
Faculty of Geo-Information Science and Earth Observation, University of Twente, Enschede, the Netherlands

C. Violante and E. Esposito
Istituto per l'Ambiente Marino Costiero (IAMC), Consiglio Nazionale delle Ricerche (CNR), Naples, Italy

G. Braca and G. Tranfaglia
Istituto Superiore per la Protezione e la Ricerca Ambientale (ISPRA), Rome, Italy

S. Pereira and J. L. Zêzere
Centre for Geographical Studies, Institute of Geography and Spatial Planning, Universidade de Lisboa, Lisboa, Portugal

R. M. Trigo and A. M. Ramos
Instituto Dom Luiz (IDL), Faculdade de Ciências, Universidade de Lisboa, Lisboa, Portugal

J. M. Vaquero
Departamento de Física, Universidad de Extremadura, Mérida, Spain

U. Ulbrich
Institute of Meteorology, Freie Universität Berlin, Berlin, Germany

R. Osinski
CNRM/GAME, Météo-France and CNRS, Toulouse, France
Institute of Meteorology, Freie Universität Berlin, Berlin, Germany

P. Lorenz
Meteo Service Weather Research GmbH, Berlin, Germany
Institute of Meteorology, Freie Universität Berlin, Berlin, Germany

T. Kruschke
GEOMAR Helmholtz Centre for Ocean Research Kiel, Kiel, Germany
Institute of Meteorology, Freie Universität Berlin, Berlin, Germany

M. Voigt
GFZ German Research Centre for Geosciences, Potsdam, Germany
Institute of Meteorology, Freie Universität Berlin, Berlin, Germany

G. C. Leckebusch
School of Geography, Earth and Environmental Sciences, University of Birmingham, Birmingham, UK
Institute of Meteorology, Freie Universität Berlin, Berlin, Germany

T. Hofherr and E. Faust
Munich Re, Munich, Germany

D. Majewski
German Weather Service, Offenbach, Germany

A. Blais-Stevens and P. Behnia
Geological Survey of Canada, 601 Booth Street, Ottawa, K1A0E8, Canada

U. C. Nkwunonwo
Department of Geoinformatics and Surveying, University of Nigeria, Enugu Campus, Nigeria
School of Earth and Environmental Sciences, University of Portsmouth, Portsmouth, UK

M. Whitworth
School of Earth and Environmental Sciences, University of Portsmouth, Portsmouth, UK

B. Baily
Department of Geography, University of Portsmouth, Portsmouth, UK

J. M. Brown
The National Oceanography Centre, Liverpool, UK

P. Ciavola
University of Ferrara, Ferrara, Italy

G. Masselink
Plymouth University, Plymouth, UK

R. McCall
Plymouth University, Plymouth, UK
Deltares, Delft, the Netherlands

A. J. Plater
University of Liverpool, Liverpool, UK

J. Zhang
School of Energy and Power Engineering, Key Laboratory of Fluid and Power Machinery, Ministry of Education, Xihua University, Chengdu, 610039, China
State Key Laboratory of Hydraulics and Mountain River Engineering, Sichuan University, Chengdu, 610065, China

Z. X. Guo
State Key Laboratory of Hydraulics and Mountain River Engineering, Sichuan University, Chengdu, 610065, China

D. Wang
State Key Laboratory of Hydraulics and Mountain River Engineering, Sichuan University, Chengdu, 610065, China
Highway Planning Survey&Design Institute, Sichuan Communications Department, Chengdu, 610041, China

H. Qian
Chengdu Municipal Engineering Design and Research Institute, Chengdu, 610015, China

I. Lehtonen, A. Venäläinen, M. Kämäräinen and H. Gregow
Finnish Meteorological Institute, Helsinki, Finland

H. Peltola
University of Eastern Finland, School of Forest Sciences, Joensuu, Finland

X. Mo, J. J. Wu, Q. Wang and H. Zhou
Academy of Disaster Reduction and Emergency Management, MOCA/MOE, Beijing Normal University, Beijing 100875, China

C. Guillard-Gonçalves, J. L. Zêzere, S. Pereira and R. A. C. Garcia
RISKam, CEG, Institute of Geography and Spatial Planning, Universidade de Lisboa, Lisbon, Portugal

S. O. Krichak and P. Alpert
Department of Geosciences, Raymond and Beverly Sackler Faculty of Exact Sciences, Tel Aviv University, Tel Aviv, Israel

S. B. Feldstein
Department of Meteorology, The Pennsylvania State University, University Park, USA

S. Gualdi and E. Scoccimarro
Istituto Nazionale di Geofisica e Vulcanologia, INGV, Centro Euro-Mediterraneo sui
Cambiamenti Climatici, CMCC, Bologna, Italy

J.-I. Yano
CNRM/GAME UMR 3589, Météo-France and CNRS, 31057 Toulouse CEDEX, France

L. J. Otero, J. C. Ortiz-Royero, J. K. Ruiz-Merchan, A. E. Higgins and S. A. Henriquez
Applied Physics Group - Ocean and Atmosphere Area - Physics Department, Universidad del Norte, Barranquilla, Colombia

Index